Metal Ions in Life Sciences

Volume 5

Metallothioneins and Related Chelators

Metal Ions in Life Sciences

edited by

Astrid Sigel,[1] Helmut Sigel,[1] and Roland K. O. Sigel[2]

[1] *Department of Chemistry*
Inorganic Chemistry
University of Basel
Spitalstrasse 51
CH-4056 Basel, Switzerland

[2] *Institute of Inorganic Chemistry*
University of Zürich
Winterthurerstrasse 190
CH-8057 Zürich, Switzerland

Volume 5

Metallothioneins and Related Chelators

RSC Publishing

The figure on the dustcover is Figure 9c of Chapter 5 by Eva Freisinger.

ISBN: 978-1-84755-899-2
ISSN: 1559-0836
DOI: 10.1039/9781847558992

A catalogue record for this book is available from the British Library

Published by The Royal Society of Chemistry,
Thomas Graham House, Science Park, Milton Road,
Cambridge CB4 0WF, UK

Registered Charity Number 207890

For further information see our web site at www.rsc.org

Historical Development and Perspectives of the Series *Metal Ions in Life Sciences**

It is an old wisdom that metals are indispensable for life. Indeed, several of them, like sodium, potassium, and calcium, are easily discovered in living matter. However, the role of metals and their impact on life remained largely hidden until inorganic chemistry and coordination chemistry experienced a pronounced revival in the 1950s. The experimental and theoretical tools created in this period and their application to biochemical problems led to the development of the field or discipline now known as *Bioinorganic Chemistry*, *Inorganic Biochemistry*, or more recently also often addressed as *Biological Inorganic Chemistry*.

By 1970 *Bioinorganic Chemistry* was established and further promoted by the book series *Metal Ions in Biological Systems* founded in 1973 (edited by H.S., who was soon joined by A.S.) and published by Marcel Dekker, Inc., New York, for more than 30 years. After this company ceased to be a family endeavor and its acquisition by another company, we decided, after having edited 44 volumes of the *MIBS* series (the last two together with R.K.O.S.) to launch a new and broader minded series to cover today's needs in the *Life Sciences*. Therefore, the Sigels new series is entitled

Metal Ions in Life Sciences.

After publication of the first four volumes (2006–2008) with John Wiley & Sons, Ltd., Chichester, UK, we are happy to join forces now in this still new endeavor with the Royal Society of Chemistry, Cambridge, UK; a most experienced Publisher in the *Sciences*.

---

* Reproduced with some alterations by permission of John Wiley & Sons, Ltd., Chichester, UK (copyright 2006) from pages v and vi of Volume 1 of the series *Metal Ions in Life Sciences* (*MILS*-1).

The development of *Biological Inorganic Chemistry* during the past 40 years was and still is driven by several factors; among these are (i) the attempts to reveal the interplay between metal ions and peptides, nucleotides, hormones or vitamins, etc., (ii) the efforts regarding the understanding of accumulation, transport, metabolism and toxicity of metal ions, (iii) the development and application of metal-based drugs, (iv) biomimetic syntheses with the aim to understand biological processes as well as to create efficient catalysts, (v) the determination of high-resolution structures of proteins, nucleic acids, and other biomolecules, (vi) the utilization of powerful spectroscopic tools allowing studies of structures and dynamics, and (vii), more recently, the widespread use of macromolecular engineering to create new biologically relevant structures at will. All this and more is and will be reflected in the volumes of the series *Metal Ions in Life Sciences*.

The importance of metal ions to the vital functions of living organisms, hence, to their health and well-being, is nowadays well accepted. However, in spite of all the progress made, we are still only at the brink of understanding these processes. Therefore, the series *Metal Ions in Life Sciences* will endeavor to link coordination chemistry and biochemistry in their widest sense. Despite the evident expectation that a great deal of future outstanding discoveries will be made in the interdisciplinary areas of science, there are still "language" barriers between the historically separate spheres of chemistry, biology, medicine, and physics. Thus, it is one of the aims of this series to catalyze mutual "understanding".

It is our hope that *Metal Ions in Life Sciences* proves a stimulus for new activities in the fascinating "field" of *Biological Inorganic Chemistry*. If so, it will well serve its purpose and be a rewarding result for the efforts spent by the authors.

Astrid Sigel, Helmut Sigel
Department of Chemistry
Inorganic Chemistry
University of Basel
CH-4056 Basel
Switzerland

Roland K. O. Sigel
Institute of Inorganic Chemistry
University of Zürich
CH-8057 Zürich
Switzerland

October 2005

and October 2008

Preface to Volume 5
Metallothioneins and Related Chelators

Metallothioneins are cysteine-rich low molecular mass (5 to 10 kDa) proteins, occurring from bacteria to humans, having, in the case of vertebrates, commonly 7 metal ions incorporated. This wide distribution already emphasizes the importance of these proteins, which are in the focus of this book. Relevant research is going on now for more than 50 years and its historical development, concentrating on mammalian metallothioneins (MTs) and their role in cadmium toxicology, is summarized in Chapter 1. Chapter 2 provides an overview of our current knowledge on the expression and regulation of MT genes: The intracellular concentration of MTs is adjusted to cellular demand. Upon heavy metal load, metallothionein gene transcription is often strongly induced. Indeed, all organisms use elaborate systems to regulate the levels of bioavailable zinc, copper, and other essential metal ions. Thus, MTs play pivotal roles in metal homeostasis as well as in detoxification reactions. Their high cysteine content enables MTs to avidly bind toxic metal ions and also to influence the cellular redox balance and radical scavenging. These points are further highlighted throughout the volume.

Chapters 3 to 5 give an account of bacterial MTs, MTs in yeast and fungi, and MTs in plants. Most astonishingly, the MTs of bacteria and plants contain next to cysteine also histidine residues and thus, metal ions are not only sulfur- but also imidazole-coordinated which gives rise to zinc finger-like structures. Remarkably, most yeast and fungal MTs are Cu(I) rather than Zn(II) or Cd(II) binding proteins. Next, Chapters 6 through 9 discuss the MTs of dipteran insects, including the model organism *Drosophila*

Metal Ions in Life Sciences, Volume 5 Edited by Astrid Sigel, Helmut Sigel, and Roland K. O. Sigel

Published by the Royal Society of Chemistry, www.rsc.org DOI: 10.1039/9781847558992-FP007 *melanogaster*, earthworms and nematodes, as well as echinoderms, crustaceans, molluscs, and fish. Actually, aquatic animals, both vertebrates and invertebrates, have the potential to be used for monitoring metal contamination in aquatic ecosystems. Interestingly, in Chapter 9 the remarkable speculation is presented that under chronic natural exposure conditions the animals establish a trade-off between the "cost" of detoxifying non-essential metal ions and the "cost" of allowing some of these metal ions to spill over onto metal-sensitive sites. This contrasts with laboratory experiments involving aquatic animals, where the toxicity of non-essential metal ions normally exhibits a threshold response: at low exposure concentrations the organisms can detoxify the incoming metal ion and thus, tolerate the exposure, whereas at concentrations above the threshold, the detoxification mechanism is no longer able to protect the organism completely.

The structure and function of vertebrate MTs is detailed in Chapter 10, centering on MT-1 and MT-2. MT-3, discovered in 1991 and also known as the neuronal growth inhibitory factor, is dealt with in Chapter 11: it plays a vital role in zinc and copper homeostasis in the brain. Furthermore, MT-3 is involved in the protection against copper-mediated toxicity in Alzheimer's disease and the control of abnormal metal-protein interactions in other neurodegenerative disorders. The next two chapters address the role of MTs in protecting cells from injury due to toxic metal ions, oxidants, and electrophiles. In fact, a poor ability to produce MT in response to metal ion exposure may predispose certain individuals to carcinogenesis by some, though not all, inorganic carcinogens.

The final two chapters deal with "relatives" of metallothioneins. Chapter 14 is devoted to thioredoxins and glutaredoxins, which represent the major cellular systems for the reduction of protein disulfides and protein deglutathionylation, respectively. They take part in many aspects of human health, e.g., by controlling and maintaining the cellular redox state, and accumulating evidence suggests a close relationship between the redoxins and the cellular iron pool. Phytochelatins, which are dealt with in the terminating Chapter 15, are produced by plants, fungi, and algae (as well as nematodes) to maintain the homeostasis of essential metal ions in different cellular compartments and to regulate metal tolerance and detoxification mechanisms.

Astrid Sigel
Helmut Sigel
Roland K. O. Sigel

Contents

Metal Ions in Life Sciences, Volume 5 Edited by Astrid Sigel, Helmut Sigel, and Roland K. O. Sigel

Published by the Royal Society of Chemistry, www.rsc.org DOI: 10.1039/9781847558992-FP009

Contributors to Volume 5

Numbers in parentheses indicate the pages on which the authors' contributions begin.

Eric Achterberg School of Ocean and Earth Sciences, National Oceanography Centre, University of Southampton, European Way, Southampton SO14 3ZH, UK, Fax: +44-23-8059-3059
<eric@noc.soton.ac.uk> (441)

Sílvia Atrian Department of Genetics, Faculty of Biology, Universitat de Barcelona, Av. Diagonal 645, E-08028 Barcelona, Spain, Fax: +34-93-403-4420 <satrian@ub.edu> (155)

Kuppusamy Balamurugan Institute of Molecular Biology, University of Zürich, Winterthurerstrasse 190, CH-8057 Zürich, Switzerland
<bala_105@hotmail.com> (31)

Carsten Berndt The Medical Nobel Institute for Biochemistry, Department of Medical Biochemistry and Biophysics, Karolinska Institutet, S-17177 Stockholm, Sweden <carsten.berndt@ki.se> (413)

Claudia A. Blindauer Department of Chemistry, University of Warwick, Gibbet Hill Road, Coventry, CV4 7AL, UK, Fax: +44-24-765-28112 <c.blindauer@warwick.ac.uk> (51)

Peter G. C. Campbell Université du Québec, Institut National de la Recherche Scientifique, INRS-ÉTÉ, 490, de la Couronne, Québec, QC, G1K 9A9, Canada, Fax: +1-418-654-2600 <peter.campbell@ete.inrs.ca> (239)

Roger Chung NeuroRepair Group, Menzies Research Institute, University of Tasmania, Tasmania, Australia <rschung@utas.edu.au> (279)

Aurélie Devez School of Ocean and Earth Sciences, National Oceanography Centre, University of Southampton, European Way, Southampton SO14 3ZH, UK, Fax: +44-23-8059-3059 <aude@noc.soton.ac.uk> (441)

Benedikt Dolderer Anorganische-Biochemie, Interfakultäres Institut für Biochemie, Universität Tübingen, Hoppe-Seylerstrasse 4, D-72076 Tübingen, Germany (83)

Eva Freisinger Institute of Inorganic Chemistry, University of Zürich, Winterthurerstrasse 190, CH-8057 Zürich, Switzerland, Fax: +41-44-635-6802 <freisinger@aci.uzh.ch> (107)

Martha Gledhill School of Ocean and Earth Sciences, National Oceanography Centre, University of Southampton, European Way, Southampton SO14 3ZH, UK <mzg@noc.soton.ac.uk> (441)

Landis Hare Université du Québec, Institut National de la Recherche Scientifique, INRS-ÉTÉ, 490, de la Couronne, Québec, QC, G1K 9A9, Canada <landis.hare@ete.inrs.ca> (239)

Hans-Jürgen Hartmann Anorganische-Biochemie, Interfakultäres Institut für Biochemie, Universität Tübingen, Hoppe-Seylerstrasse 4, D-72076 Tübingen, Germany (83)

Juan Hidalgo Unidad de Fisiología Animal, Facultad de Ciencias, Universidad Autonoma de Barcelona, Bellaterra, E-08193 Barcelona, Spain, Fax: +34-93-581-2390 <juan.hidalgo@uab.es> (279)

Susan Krezoski Department of Chemistry and Biochemistry, University of Wisconsin-Milwaukee, 3210 N. Cramer Street, Milwaukee, WI 53201, USA <suek@uwm.edu> (353)

Christopher Horst Lillig Department of Clinical Cytobiology and Cytopathology, Phillips University, D-35037 Marburg, Germany and The Medical Nobel Institute for Biochemistry, Department of Medical Biochemistry and Biophysics, Karolinska Institutet, S-17177 Stockholm, Sweden <horst@lillig.de> (413)

Jie Liu Inorganic Carcinogenesis Section, Laboratory of Comparative Carcinogenesis, National Cancer Institute at NIEHS, 111 Alexander Drive, Mail Drop F0-09, Research Triangle Park, NC 27709, USA <liu6@niehs.nih.gov> (399)

Gabriele Meloni Department of Biochemistry, University of Zürich, Winterthurerstrasse 190, CH-8057 Zürich, Switzerland (319)

Gunnar F. Nordberg Environmental Medicine, Department of Public Health and Clinical Medicine, Umeå University, SE-901 87 Umeå, Sweden <gunnar.nordberg@envmed.umu.se> (1)

Monica Nordberg Institute of Environmental Medicine, Karolinska Institutet, SE-171 77 Stockholm, Sweden, Fax: +46-8-314-124 <monica.nordberg@ki.se> (1)

Milena Penkowa Panum Institute, Faculty of Health Sciences, University of Copenhagen, Blegdamsvej 3, 18.1.44, DK-2200 Copenhagen, Denmark <m.penkowa@mai.ku.dk> (279)

David H. Petering Department of Chemistry and Biochemistry, University of Wisconsin-Milwaukee, 3210 N. Cramer Street, Milwaukee, WI 53201, USA <petering@uwm.edu> (353)

Walter Schaffner Institute of Molecular Biology, University of Zürich, Winterthurerstrasse 190, CH-8057 Zürich, Switzerland, Fax: +41-44-635-6811 <walter.schaffner@molbio.uzh.ch> (31)

Stephen R. Stürzenbaum School of Biomedical and Health Sciences, Department of Biochemistry, Pharmaceutical Sciences Division, King's College London, 150 Stamford Street, London, SE1 9NH, UK <stephen.sturzenbaum@kcl.ac.uk> (183)

Niloofar M. Tabatabai Division of Endocrinology, Metabolism and Clinical Nutrition and Kidney Disease Center, Medical College of Wisconsin, Milwaukee, WI 53226, USA <ntabatab@mcw.edu> (353)

Milan Vašák Department of Biochemistry, University of Zürich, Winterthurerstrasse 190, CH-8057 Zürich, Switzerland, Fax: +41-44-635-6805 <mvasak@bioc.uzh.ch> (279), (319)

Laura Vergani Department of Biology, University of Genova, Corso Europa 26, I-16132 Genova, Italy, Fax: +39-010-353-7584 <laura.vergani@unige.it> (199)

Michael P. Waalkes Inorganic Carcinogenesis Section, Laboratory of Comparative Carcinogenesis, National Cancer Institute at NIEHS, 111

Alexander Drive, Mail Drop F0-09, Research Triangle Park, NC 27709, USA <waalkes@niehs.nih.gov> (399)

Ulrich Weser Anorganische Biochemie, Interfakultäres Institut für Biochemie, Universität Tübingen, Hoppe-Seylerstrasse 4, D-72076 Tübingen, Germany, Fax +49-7071-295564 <ulrich.weser@uni-tuebingen.de> and Centro di Risonanze Magnetiche, University of Florence, Via Luigi Sacconi 6, I-50019 Sesto Fiorentino (Firenze), Italy (83)

Titles of Volumes 1–44 in the *Metal Ions in Biological Systems* Series

edited by the SIGELs
and published by Dekker/Taylor & Francis (1973–2005)

Contents of Volumes in the *Metal Ions in Life Sciences* Series

edited by the SIGELs

Volumes 1–4

published by John Wiley & Sons, Ltd., Chichester, UK (2006–2008)

<www.Wiley.com/go/mils>

and from Volume 5 on

by the Royal Society of Chemistry, Cambridge, UK (since 2009)

<www.rsc.org/book>

Volume 2: Nickel and Its Surprising Impact in Nature

Volume 3: The Ubiquitous Roles of Cytochrome P450 Proteins

Volume 4: Biomineralization. From Nature to Application

Volume 6: Metal-Carbon Bonds in Enzymes and Cofactors
(in press)

Volume 7: Organometallics in Environment and Toxicology
(tentative contents)

Comments and suggestions with regard to contents, topics, and the like for future volumes of the series are welcome.

Met. Ions Life Sci. **2009**, *5*, 1–29

1

Metallothioneins: Historical Development and Overview

Monica Nordberg [1] ***and Gunnar F. Nordberg*** [2]

[1] Institute of Environmental Medicine, Karolinska Institutet, SE-171 77 Stockholm, Sweden
<monica.nordberg@ki.se>

[2] Environmental Medicine, Department of Public Health and Clinical Medicine, Umeå University, SE-901 87 Umeå, Sweden
<gunnar.nordberg@envmed.umu.se>

Metal Ions in Life Sciences, Volume 5 Edited by Astrid Sigel, Helmut Sigel, and Roland K. O. Sigel

Published by the Royal Society of Chemistry, www.rsc.org DOI: 10.1039/9781847558992-00001

ABSTRACT: The history on research of metallothionein is reviewed. Various methods for isolation, characterization, and quantification are evaluated. The role of metallothionein in metal metabolism and toxicity is explained. Gender differences and polymorphism as well as possible relationships with diseases are discussed. The review is based on data from the literature and on own original experimental and epidemiological data. Aspects on future research within the metallothionein field are indicated.

KEYWORDS: disease · metabolism · metallothionein · metals · toxicity · transport

1. INTRODUCTION

Metallothioneins (MT) are sulfur-rich low molecular mass (6–7 kDa) proteins with 7 metal ions that constitute a natural part in forming the three-dimensional structure. Research on MT has been going on for more than 50 years and this chapter displays the progress made; it reviews the biochemical and experimental methods used in this research since the beginning.

Many metal binding proteins are known with a specific function of the metal ion(s), e.g., hemoglobin with Fe as a central metal ion is in charge of supplying oxygen to the cell. Metallothionein is a protein as well and importantly, it also serves functions in the cell; sometimes it is mentioned in relation to chelators because of its capacity to bind metal ions. Already in the late 1970's it has been proposed to administer to humans the amino acid sequence, i.e., the thionein part of the protein, or even the MT rich in zinc, to treat renal effects caused by Cd in the kidney; however, this is likely to be harmful because an exchange of Zn^{2+} by Cd^{2+} may take place. When cadmium-metallothionein (CdMT) is given to laboratory animals renal damage is seen already at a Cd level of 10 μg/g wet tissue because MT goes straight to the kidney which has a much lower capacity for the synthesis of MT than liver. So far no positive role of MT has been shown in the treatment of cadmium poisoning that would be similar to that seen with EDTA in lead poisoning, or BAL and DMSA in mercury poisoning.

Below we give an overview on the historical development of MT research. We concentrate mainly on mammalian metallothioneins and indicate also several of the aspects discussed in other chapters. Since the early days, both isolation and characterization as well as the role of MT in Cd toxicology [1,2] have contributed to the understanding of MT and its role as a general sequestering protein for toxic metals also reducing cellular occurrence of reactive oxygen species. Future developments in the application of molecular "omics" technologies (genomics, proteomics, and metabolomics) will undoubtedly lead to a further understanding of the role of MT in biology and help in biomonitoring of environmental exposures.

2. HISTORY OF METALLOTHIONEINS

The first publication in 1957 on a cadmium-binding protein in equine tissue [3] was initiated by a report in the form of an abstract [4] dealing with cadmium in human organs. Small amounts of cadmium had been shown to be present in tissues and body fluids in several animal species. Various hypotheses were postulated to explain this unexpected finding: Either would cadmium be coordinated to a macromolecule and then have a natural function in biological systems or else cadmium could just be a contaminant. In 1960 the first detailed report on metallothionein was published [5,6]. The cadmium-containing protein, isolated from equine renal tissue, was described and named "metallothionein" because of its extremely high sulfur content of 4.1%/g dry weight and 2.9% of Cd and 0.6% of Zn. Isolation was performed from five frozen horse kidneys with for that time conventional methods.

Later studies reported data on physical properties and the molecular weight was estimated to be 10000 ± 260. The specific absorption at 250 nm was explained by cadmium mercaptide charge transfer bonds. Metallothionein was assumed to lack aromatic amino acids as indicated by the absence of an absorption at 280 nm.This was later verified by amino acid analyses [7,8] which also showed that the high sulfur content was due to cysteine. At that time reactive mercapto groups in proteins were determined by titration with silver ions, CMB, and N-ethylmaleimide. Amino acids were identified by two-dimensional paper chromatography and ion exchange chromatography. Cysteine residues were quantified as cysteic acid after oxidation of metallothionein with performic acid and as derivates of N-ethylmaleimide. The sedimentation constant was determined via a Schlieren diagram by sedimentation in an ultracentrifuge at 1.75 s (s020w). The diffusion constant, i.e., the partial specific volume, and the friction ratio were also reported. The estimated molecular weight of the protein was still varying from 9790–10500. This was in part explained by the formation of various artefacts during preparation. Metal analyses gave 5.2 g atoms/mol or 5.9% of MT weight for Cd and 3.3 g atoms/mol or 2.2% by weight for zinc. Some exchange between zinc and cadmium was obviously taking place. It was suggested that bonding with three deprotonated SH-groups and one atom of either cadmium or zinc occurred.

As part of the research of the Swedish group on the health effects caused by cadmium, a study on rabbits [9] showed that cadmium-metallothionein could be induced by repeated injections of small doses of cadmium. A single high dose of cadmium [1] was found to be more toxic to the organism giving rise to liver damage and lethality, while the same dose administered as several small doses during a prolonged exposure time gave no such effects. In fact, animals with induced metallothionein synthesis by pretreatment with smaller doses of cadmium developed resistance to acute toxicity to the liver [1] and the testes

[10]. Isolation of the cadmium-binding protein from livers of cadmium exposed rabbits showed an increase of metallothionein in relation to the administered dose or amount of cadmium present [7]. In animals protected by pretreatment, cadmium in the target tissues, liver and testis, was bound to a low molecular weight protein corresponding to metallothionein.

Techniques newly developed in the 1960's were used for isolation of the protein. After homogenization of the tissues rivanol was applied to precipitate high molecular weight proteins and cell fragments. Several steps of precipitation, dialyses, and various gel chromatography steps were carried out, as Sephadex gel had previously been introduced into protein chemistry. The initial assumptions by Piscator in 1964 [9] were later confirmed in these animal experiments which demonstrated indeed that exposure to cadmium increased the concentration of metallothionein in the liver. These findings gave further support to the original ideas of metallothionein induction as a mechanism of making tissue less sensitive to cadmium toxicity. In this group [11–13], working with the toxicity and kinetics of cadmium, it was known that cadmium gave rise to adverse health effects upon increasing exposure, particularly to renal damage. Metallothionein research now continued or developed into two tracks – one in protein chemistry and another one focusing on kinetics and toxicity of cadmium and other metal ions.

However, all studies demanded pure and well characterized metallothionein and this was prepared with techniques modern at that time. Tissue was homogenized in a buffer system, mostly of Tris-HCl in sodium chloride with a pH of 8.1. This step was followed by ultracentrifugation at 105000gav and the supernatant was taken for gel chromatography (Sephadex gel G-75). If the absorption ratio at 250 and 280 nm was low, improvement could be achieved in one step by Sephadex G-50 used for preparative purposes. When the fractions eluted as MT on G-75 Sephadex were separated on G-50, a protein was isolated with a high absorption at 280 nm and no metal content [11]. Further separation by isoelectric focusing or ion exchange chromatography after concentrating and desalting by ultrafiltration on UM-2 filters with a cut off level for a molecular weight of 1000 [7] showed different fractions containing MT. Further separation by isoelectric focusing of rabbit liver revealed at least three major forms of MT with pI 3.9, 4.5, and 6.0. Two of these were characterized by amino acid analyses [7] and identified as form I and II of MT. To be successful with the preparation of metallothionein from tissue it became obvious quite early that avoidance of oxidation of the protein by rapid preparation and working at a cool temperature was crucial. Mercaptoethanol could, however, restore oxidized metallothionein [12,13] as shown by gel chromatograpy on Sephadex G-75 where metallothionein showed up at the ordinary position after treatment with mercaptoethanol. An important contribution to the tertiary structure was made [14] when two metal clusters were described, i.e., an α- and a β-domain with four and three metals, respectively, as part of the structure. The α-domain constitutes the C-terminal and the β-domain the N-terminal end of the protein.

The other track of research already indicated above expanded to the importance of MT for different metal ions, in particular for copper [15] and mercury [16–18]. Pioneering work [10] showed that MT could protect against testicular damage caused by cadmium. Knowledge on MT and its involvement in the transport of cadmium and that cadmium is partly present in blood [2] bound to MT led to a metabolic model for cadmium toxicity. Further work focused on the speciation of cadmium and it was found that CdMT is taken up in the renal tubules and causes renal damage at cadmium concentrations as low as 10 μg/g renal tissue [19].

The identification of a cadmium-binding protein in mammals, which was first believed to have a high molecular mass, turned out to be a low molecular mass protein (see also Chapter 10). As part of the research on cadmium and adverse health effects in Sweden, a project on MT was developed and it was shown that MT is a most important protein in the metabolism and kinetics of cadmium in animals and humans. Methods for isolation and characterization of MT were developed. To study the history of MT also means to consider available analytical techniques and methods. In the late 1960's and early 1970's only three full length articles were available, two in English and one in Swedish. The combination of available knowledge about protein separation and radioactive techniques made it possible to isolate, characterize, and study the role of MT [1].

In the 1970's and early 1980's only a limited number of groups performed research related to metallothionein. However, an increasing number of publications in which a different nomenclature for MT was used made it clear that an evaluation of the knowledge available at that time would be of importance. Hence, a workshop with approximately 25 invited participants, who had submitted background manuscripts, was arranged and a tentative report [20] was prepared and distributed in advance to each participant. A consensus report was agreed during the meeting held in Zürich in 1978 [21].

Agreement on the nomenclature of MT in the mentioned first workshop [20,21] stimulated interest in MT research. For a long time after the first workshop Roman numbers, like MT-I and MT-II, were used to identify different MTs. This is no longer consistent with the nowadays accepted terminology, developed by Kägi and coworkers [21], which uses Arabic numbers. The official designation in the SwissProt (proteins) data base and the MCBI data base (Medical Center for Biotechnology Information), which deals with the genome and also gives proteins, uses Arabic numbers, e.g., MT-1 and MT-2. However, in all these data bases the older designations like MT-I or MT-II as well as names given during the discovery of a protein are given as synonyms (SwissProt) or aliases in MCBI. Perhaps more importantly, the

Human Genome Organization (data base) also approved the symbols MT-1, MT-1A, MT-2 and the like. Further issues on nomenclature can be found on the website: http://www.expasy.org/cgi-bin/lists?metallo.txt.

During the mentioned first international meeting on metallothioneins held in Zürich consensus was reached not only about the nomenclature, but also about other issues, like methods for preparing the proteins. This first meeting was followed by another one in Zürich in 1985, which was more open but still of a workshop type. In-between a meeting had been arranged in Aberdeen in 1981. Other meetings have focused on various areas of interests, e.g., cadmium-binding proteins in non-mammalian species was brought to attention in an international meeting in 1984 [22] and pharmaceutical interests led to the third international meeting held in Japan in 1992 [23]. A variety of meetings with different themes and approaches followed (Table 1) [24–27].

Table 1. History of metallothionein and important workshops.

1941, 1957	Metallothionein, discovery	[3,4]
1960, 1961	Metallothionein, details about	[5,6]
1964	MT induction by cadmium	[9]
1971	Modification of Cd toxicity	[1]
1972	Amino acid composition	[7]
1976	Sequence	[24]
1978	1st International Metallothionein Meeting, Zürich, Switzerland; consensus on nomenclature	[21]
1979	Radioimmunoassay	[25]
1981	2nd International Metallothionein Meeting, Aberdeen, Scotland	
1983	1st International Meeting on Metallothionein and Cadmium Nephrotoxicity, National Institute of Environmental Health Sciences, Research Triangle Park, North Carolina, USA	[26]
1984	High affinity metal-binding proteins in non-mammalian species	[22]
1985	International Workshop on metallothionein, Zürich, Switzerland	[27]
1992	3rd International Metallothionein Meeting with pharmaceutical implications, Tsukuba, Japan	[23]
1996	1st International Workshop on Metallothionein JRC/IRMM EU, Geel, Belgium	
1997	4th International Metallothionein Meeting, Kansas City, USA	
1999	2nd International Workshop on Metallothioneins (Euroconference) JRC/IRMM EU, Geel, Belgium	
2005	5th International Metallothionein Conference (MT-2005) (Metals and Metallothionein in Biology and Medicine), Beijing, China	

Research on metallothionein has now been going on for more than 50 years. Initially it followed two tracks, i.e., strict protein chemistry and toxicokinetics of the metals that constituted structural parts of the protein and those that turned on the synthesis. Among the metal ions especially cadmium and zinc were in the focus, but to some extent also copper. Of course, matters developed further and in 1986 a high intake of Cd-containing seafood and shellfish by human consumers in New Zealand was reported [28] and it was also shown that the chemical species containing Cd was different in two species of oysters. Cd-binding proteins identified in foodstuff have been reviewed by Petering and Fowler [29]. The chemical species containing Cd, particularly its binding to metallothionein-like proteins, is of importance for the uptake, distribution, and toxicity of Cd. These insights are of relevance with regard to the outcome of human exposure to metals in the form of MT.

The number of publications per year has increased over time and a recent search on Medline for the years 1950 to late 2007 gives nearly 7000 publications and the corresponding search in Pubmed provides almost 8200 hits. New techniques developed from molecular biology have confirmed earlier findings, opened new aspects, and made this rapid progress possible.

3. PROTEIN CHEMISTRY AND METAL BINDING

Metallothionein is a low molecular mass protein, characteristically with 6 to 7 KDa. Major hallmarks of MT are the amino acids that vary between 61–68. The typical MT consists of 20 cysteines (30%), methionine (N-terminal), alanine (C-terminal), no aromatics, no histidine and it has a unique amino acid sequence with a tertiary structure forming two domains of metal clusters, i.e., the α- and β- clusters. The metal content of Zn, Cd, Hg, and Cu can vary and may constitute 11% of its weight, the metal ions being bound by several sulfhydryl groups [30,31]. Specific absorption occurs at 225 (Zn), 250 (Cd), 300 (Hg), and 275 nm (Cu). Synthesis of MT-1 and -2 is induced by Cd^{2+} and Zn^{2+}. There are no disulfide bonds and MT is regarded as heat-stable. It is mainly localized in the cytoplasm. Metallothioneins exist in four major forms, MT-1 to MT-4. MT-3, present in brain and renal tissue (see Chapters 10 and 11), is not inducible by Cd as are MT-1 and 2. MT-1 also occurs in several isoforms and MT-4 is expressed in keratinocytes. In humans the gene is localized on chromosome 16 and in the mouse on chromosome 8 (see also Chapter 2). Metallothionein has been isolated from the liver and several other tissues of animals. Its synthesis is induced by Cd^{2+}, Zn^{2+}, and other metal ions or stress [20,32,33] (see Chapter 12).

The structure of MT-1 and -2 has two domains consisting of one cluster with 3 and one with 4 metal atoms and was first described by Winge and Miklossy [14]. The metallothionein gene is located on chromosomes varying with species. For humans and other primates it is on chromosome 16 [34]. The protein consists of a number of isoforms coded by various alleles. The ratio of mRNA for MT-1 and MT-2 genes remains constant during induction by metals, e.g., Cd, Cu, and Zn. It was found that 1.4 times more MT-1 RNA than MT-2 RNA exists, indicating that the transcription rate is slightly higher for the MT-1 gene compared to the MT-2 gene [34]. MT-1A and MT-2A genes seem to be differentially regulated by metals. Lack of MT gene expression makes the organism sensitive to toxic effects. The MT gene becomes transcriptionally inactive as a consequence of DNA methylation.

Cells with extra copies of MT genes can be selected by exposure to a toxic concentration of cadmium. In metal-exposed mammals zinc is the dominating metal ion in MT and at least one zinc seems always to be present in MT (see Chapter 10). Various biological factors influence metal ion composition such as tissue of origin, age, and stage of development, This means that renal MT is higher in Cd and Cu in exposed animals than liver MT from the same organism. In an evaluation of several gels Sephadex G-75 and G-50 have been proven to be still the most efficient technique for purification of MT. Sometimes it is necessary to add mercaptoethanol to the samples in order to reduce MT back to non-polymerized MT [12,13].

Transgenic mice [35] have been introduced in order to gain new knowledge on MT and metal toxicity. Experimental laboratory animals (mainly mice) have shown that species differences with regard to sensitivity and resistance to metal toxicity exist [36]. By introduction of transgenic animals it is expected that the mechanism behind these differences will be elucidated.

4. METHODS FOR QUANTIFICATION OF METALLOTHIONEIN

A number of methods for quantification of MTs was reviewed in 2002 [37]. Several methods for measuring and quantifying metallothioneins are available [36,38] as displayed in Table 2 [39,40]. A review of various methods has also been published [41]. In addition to the information given in Table 2 it should be mentioned that metal-binding assays involving the binding of metals such as cadmium [39,42], mercury [17], and silver, together with pulse polarography [43] and immunoassays such as radioimmunoassay (RIA) and enzyme-linked immunosorbent assay (ELISA), based on the use of antibodies [40] are employed. Each of these methods has advantages and disadvantages. The radioimmunoassay needs specific antibodies that do not cross-react and this also applies to the immunoassay ELISA.

Most cadmium in urine of humans occurs bound to metallothionein. Urinary cadmium and metallothionein concentrations correlate well as shown in elderly women exposed to cadmium in the general environment

Table 2. Methods for the quantification of MT and MT mRNA.

After isolation:	
Freeze-drying	[7,30]
Calculation of amino acid analyses	[7,8]
Biuret method	[7]
In crude tissue fractions:	
In vitro binding of Hg	[17]
In vitro binding of Cd	[39]
Radioimmunoassay	[27]
ELISA	[40]

[44,45] and in occupationally exposed male cadmium workers [46,47], for whom values were reported of 2–155 ng MT/g urine and for plasma 2–11 ng MT/g. The level of MT by RIA in healthy humans is reported to be 1–16 μg/L in plasma or serum and 5–400 μg/g creatinine in urine [44,45]. In cadmium-exposed persons metallothionein in urine is a good indicator of urine concentration of cadmium and can also be assumed to be an index of the burden of cadmium.

Immunohistochemical staining of metallothionein in placental tissue indicated that metallothionein reflects the concentration of copper in this tissue. The method can be used for pre- and postnatal diagnosis of Menkes disease [48].

A gender difference is observed: Women have a higher metallothionein concentration in urine compared to men even at similar cadmium levels [49] as shown with RIA [49,50]. As presented in Table 3 [40,51–55,57] the range of normal concentrations of MT is defined by concentrations found among humans without proteinuria.

The normal concentrations of metallothionein in rat tissues as determined by ELISA are 18 μg/g in liver [53], 30 μg/g in kidney [53], and 35 μg/g in kidney cortex [56]. The detection limit for ELISA has been reported to be 100 pg MT [53].

The determination of metallothionein concentration in urine and blood has been found to be related with problems not observed in tissue analyses. This is likely due to the techniques used for sampling and storage of samples, procedures which are most crucial for the results of analyses. Usually urine samples are treated with bactericides in order to prevent bacterial contamination with a reduced pH as result. At low pH metallothionein loses the metals and free thionein is obtained; a change in configuration and instability of the protein follows. This is likely to influence the results. In experimental studies [57] excretion of MT, cadmium, calcium, and various

Table 3. Levels of metallothionein in humans.[a]

Method	Media	MT Concentration	Status	Reference
RIA	Sera	0.01–1 ng/g	Normal[b]	[46]
RIA	Sera (human)	>2 ng/g	Abnormal[b]	[46]
RIA	Urine (human)	1–10 ng/g	Normal[b]	[46]
RIA	Urine (human)	>10 ng/g	Abnormal[b]	[46]
RIA	Urine (human)	1880 μg/g CR	Itai-itai patients	[44,45]
		880 μg/g CR	Cd-polluted area	
		394 μg/g CR	Non-polluted areas	
ELISA	Urine (human)	120 and 210 μg/g CR	Normal[b]	[55]
		320 and 1050 μg/g CR	Abnormal[b]	

[a] Collected from data published in [51,52].

[b] Occupational exposure, abnormal denotes presence of low molecular weight proteinuria.

enzyme markers in urine were followed in male rats exposed to cadmium. MT was determined by ELISA. Interference with luminal and basolateral membranes and handling of calcium was demonstrated [58,59]. This is in accordance with the suggested model that Cd is released from CdMT after it is catabolized in lysosomes and appears in the cytoplasm of renal tubule cells, where it may change the electrochemical gradient across the luminal membrane giving rise to decreased calcium absorption.

Thus, it is necessary to evaluate and standardize the methods for urine and blood. As the body fluids are used for biological monitoring of many metals, it would be most valuable to measure metallothionein as well and to relate the concentration of metals to the concentration of metallothionein.

One problem with the estimation of metallothionein in tissues and body fluids is how to manage to express the true concentration and relate it to biological events. For a long time this has been done in the form of μg MT/g wet weight tissue. Since the concentration of MT varies with many factors this has to be further expressed in relation to something that is stable in the cell. The cellular concentration of MT is age-dependent and also dependent on exposure to numerous agents [29]. Several methods and pitfalls with various methods for MT determination have been summarized by contributions from many scientists in MT research [41]. It is urgent to develop a method for MT quantification with high precision, accuracy, and specificity. To test the specificity, a known amount of MT may be added to the samples. In spiking the samples, several questions are raised: How should the various forms of MT be quantitated? They might reflect various biological functions that are related to both age and exposure as suggested previously [12,13]. Isoelectric focusing [60] is a rapid, quick, and good method for the preparation

Table 4. Media and methods for the detection of MT and MT mRNA upon exposure to Cd.

Metallothionein and MT mRNA as Biomarker in Environmental and Occupational Cadmium Exposure	
Peripheral lymphocytes	MT mRNA RT-PCR
Plasma or Serum	MT by ELISA or RIA
Urine	MT by ELISA or RIA

of the various isoforms of metallothionein. Commercially available metallothionein has to be checked with regard to purity even if a well recognized method has been used for preparation. It has been noticed that the metal concentration in some shippings from commercial purchases has been extremely low, indicating a low purity of the protein.

Commercial antibodies are available for MT-1, -2 and -3. Monoclonal antibodies are believed not to be specific enough to be taken for ELISA assays.

Methods from molecular biology offer possibilities to quantitate MT in eukaryotes and prokaryotes. MT gene transcription and RT-PCR to measure MT mRNA can be used. It may be mentioned that it is also possible to use a DNA probe for RNA synthesis and translated DNA [37]. MT-2 expression in lymphocytes can be performed by PCR techniques [37].

A question to be solved is how the determination and estimation of MT in body fluids and tissues should be performed and how to relate the concentration of MT to effects? A RT-PCR method for MT mRNA primers and oligo probes are commercially available. Basal and *in vitro* induced MT mRNA is significantly higher in Cd-exposed groups than in controls. Quantification of MT by ELISA demands a good antibody. Detection of expression of MT by Western blot demands a specific antibody. Detection of mRNA expression by RT-PCR demands specific primers (see Table 4).

5. ROLE OF METALLOTHIONEIN IN METAL METABOLISM AND TOXICOLOGY

Metallothionein in the physiological system has not only one but several roles, especially in the metabolism and kinetics of metals (see also Chapter 10). These are

- transport of metal ions
- detoxification of metal ions
- protection from metal toxicity

– free radical scavenging
– storage of metal ions
– metabolism of essential metal ions
– immune response
– genotoxicity and carcinogenicity

The mechanisms by which MT protects cells from toxicity include binding of metals to proteins and localization in the cell where MT is mainly present in the cytoplasm, but also in the nucleus and in the lysosomes of the kidney. MT stands for detoxification! The protection against adverse effects caused by cadmium exposure follows from the ratio of non-MT-bound to MT-bound cadmium [1]. Such binding may also occur with other metals, e.g., Hg [16], and is thus a mechanism for detoxification.

Another function of metallothionein, in addition to its involvement to transport metals, is as a free radical scavenger and it also stores metals like Zn, Cd, Cu, and Hg. In the immune response MT acts as a Zn donator. Most likely MT is also involved in genotoxicity and carcinogenicity processes (see Chapter 13).

After absorption from the lungs or the gut, cadmium is transported via blood to other parts of the body. In blood cadmium is mainly found in the blood cells [36], where a high molecular weight and a low molecular weight fraction occurs [2]. Further studies [12,13,61] have shown that the latter fraction is similar to metallothionein, which also binds cadmium in plasma [2] and which has an important role in the transport of cadmium in the body of animals and humans [62,63]. The low molecular weight of metallothionein enables this protein to be filtered through the kidney glomerular membrane. Like other proteins in the primary urine, metallothionein is reabsorbed into proximal tubular cells. The transport of cadmium bound to metallothionein from blood to renal tubular cells is rapid and almost complete [19,64]. Cadmium not bound to metallothionein does not enter the kidneys to the same extent. A similar difference was seen in animals fed cadmium-metallothionein and cadmium chloride [65]. The former gave rise to a much higher accumulation of cadmium in the renal cortex than the latter, most probably because Cd^{2+} from chloride binds to albumin in blood plasma [12,13].

Cadmium exposure induces the synthesis of metallothionein in a number of tissues [31]. During the first 12 hours after a high acute exposure to cadmium (not bound to metallothionein), there will be an increase over time of cadmium bound to metallothionein due to the increased production of the protein [1,66]. As the transport of cadmium to the kidney is dependent on its metallothionein binding in plasma, the distribution of cadmium within the body found after an acute exposure will be different from that found after repeated exposures. Figure 1 summarizes the transport of Cd in blood and its uptake in kidney tubules.

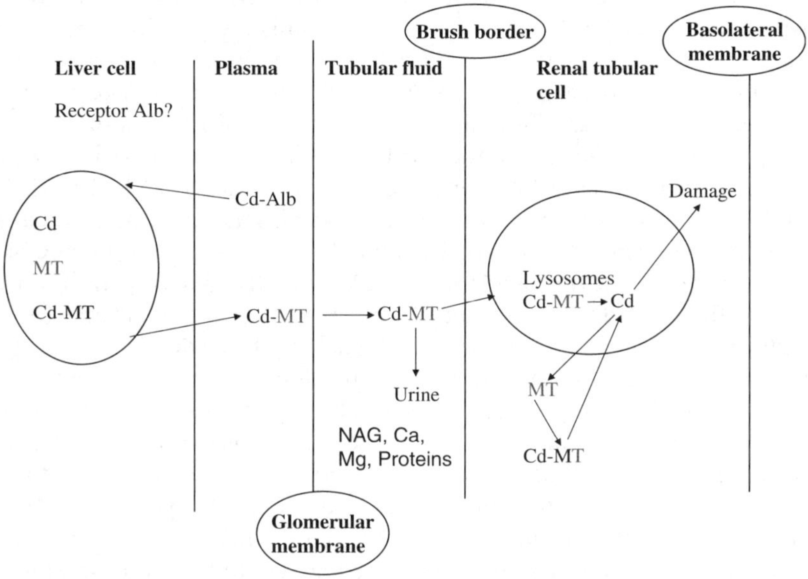

Figure 1. Transport of Cd in blood and uptake in kidney tubules.

Metallothionein-bound cadmium in plasma is filtered through the renal glomeruli and reabsorbed in the tubuli, where cadmium is released. If CdMT is injected in animals the released cadmium causes damage to the kidney tubule because there is insufficient tissue MT available to protect the kidney [67]. Upon a long-term exposure, the unbound cadmium stimulates new metallothionein production which binds cadmium and protects the renal tubular cells. When this process is insufficient, toxic effects occur, possibly because of cadmium interference with zinc-dependent enzymes and membrane functions.

The scheme in Figure 1 was suggested in early studies by Nordberg et al. in 1971 [1,2] and continued to be used for cadmium [68,69] as well as later on for copper as summarized by Bremner in 1987 [70].

A long time after a single exposure, or in long-term exposure, a considerable proportion of plasma Cd is bound to metallothionein [12,13]. Uptake of CdMT may become more efficient in cells pre-exposed to Cd compared to non pre-Cd-exposed cells [71]. In long-term exposure there is a slow release of CdMT from the liver to the blood. This transport phenomenon has gained support from studies where Cd-containing livers were transplanted to non-Cd-exposed animals, which showed a gradual uptake of Cd in the kidney [53] and from studies demonstrating a lower Cd accumulation in kidneys of MT-null mice [72]. Inorganic cadmium compounds are known to cause toxicity to the kidney after long-term exposure. In animal experiments it has been shown that Cd administered as MT may cause similar renal damage at a tissue concentration of Cd at 10 μg/g wet weight [19] compared to 70–200 μg/g wet weight when administered as Cd ions.

In addition to the early findings regarding the modifications of cadmium toxicity by metallothionein induction [1,9], data have also been provided concerning the binding of Cd to metallothionein in blood [12,13,19,27]. The identification of bound forms of Cd in blood plasma and studies by autoradiography showing that Cd is distributed selectively to the kidney after administration of CdMT, while it is predominantly taken up by the liver after injection as Cd^{2+} or as Cd-albumin, provided a background for the mechanistic model of cadmium kinetics [19,68,69]. Immediately after uptake in blood, cadmium is bound to albumin in blood plasma, distributed and taken up in the liver (Figure 1). It is speculated that an albumin receptor is present on the surface of liver cells. Once in the liver, cadmium binds to already present metallothionein by exchanging zinc. Then, cadmium induces the synthesis of metallothionein and the newly synthesized MT is sequestering cadmium from other binding sites, thus protecting liver cells from toxicity. Cadmium-metallothionein is released to the blood stream and transported to the kidney where it is filtered through the glomerulus and taken up by adsorptive endocytosis [73]. Metallothionein is catabolized in the lysosomes of the tubules [74] and the free cadmium ions induce then the new synthesis of metallothionein in the cell and, of course, cadmium may also react with other sensitive sites.

Cadmium has a biological half-time in humans of 10–15 years which is regarded as very long. This observation may be explained by the property of Cd^{2+} to induce synthesis of metallothionein. The described model has been further developed [75,76]: CdMT-induced kidney damage was shown to decrease uptake and binding of calcium in membrane vesicles isolated from animals injected with CdMT [75]. Rats given a combined exposure by injection of CdMT (0.25 mg/kg), Zn (12.5 mg/kg), and Cu (6.25 mg/kg) had considerably increased levels of MT in the renal cortex, i.e., up to 746 μg/g wet weight. These high concentrations of MT were considered to be of major importance explaining the protection against renal damage from Cd in these animals [56].

Repeated injections of CdMT given with a short time interval gave rise to a considerably prolonged and possibly irreversible calciuria in rats [76]. Increased excretion of magnesium has been found in rats with CdMT-induced nephrotoxicity [77]. A possible contribution of endogenous intestinal metallothionein to renal accumulation of cadmium was studied in rats fed with cadmium [78]. To distinguish between exogenous and endogenous metallothionein isoforms from rat and pig, differences in chromatographic behavior were used [79]. Cadmium may also possibly be bound in small

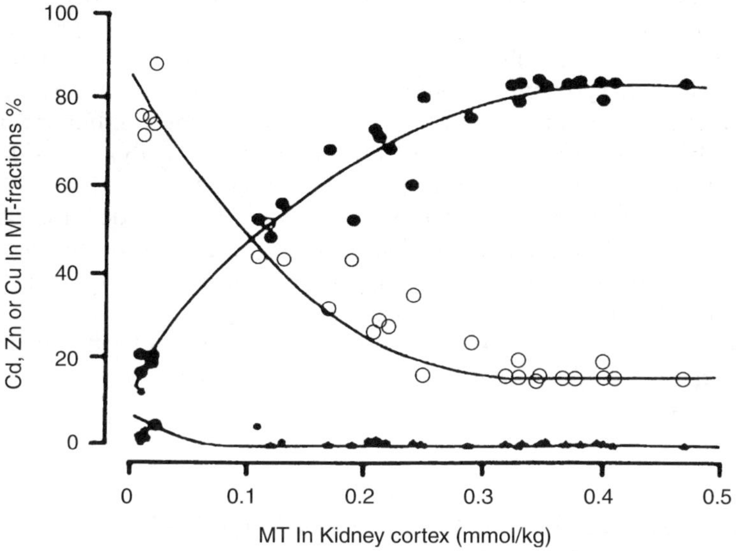

Figure 2. Relative concentrations (%) of cadmium, zinc, and copper in MT fractions in relation to the total MT concentration. MT isolated from kidney cortex of rabbits with varying exposure to Cd. (●) Cd; (○) Zn; (∗) Cu. Reproduced from [81] by permission of Elsevier; copyright (1987).

amounts to low molecular weight SH-rich compounds such as glutathione and cysteine [80] (see also Chapter 14) although evidence for such binding in plasma of mammals exposed to cadmium salts is limited and the main transporting protein for cadmium to the kidney most probably is metallothionein [63]. As mentioned above, a similar pathway was shown for copper [70].

In normal human beings, the increase in cadmium in the renal cortex with age is accompanied by an equimolar increase in zinc. This is thought to be due to the metallothionein stored in the kidney, which contains equimolar amounts of the two metals. In Figure 2 it is seen that the intersection of Cd and Zn in MT in kidney cortex gives a MT concentration in kidney which is equal to the critical concentration of Cd [81].

Another important function for metallothionein is the cellular defense mechanism against free radicals where methionine might serve as free radical scavenger as discussed in many publications [68,70]. Furthermore, metallothionein may protect DNA by sequestering copper and preventing its participation in redox reactions and thus inhibit the formation of free radicals as pointed out by Cai, Koropatnick, and Cherian [82].

Metallothionein regulates the toxicity of various metals and trace elements, and as we have seen, copper and zinc are examples of this. With regard to high-tech metals, e.g., gallium, germanium, indium, antimony, tellurium, yttrium, niobium, thallium, and bismuth and some more recently introduced high-tech compounds used in superconductors such as yttrium-barium-copper-oxide (YBCO) and $Bi_2Sr_2Ca_2Cu_3O_{10}$ (BSCCO), it may well be, if taken up, that their content of Bi and Cu may induce MT and that BiMT and CuMT are formed. It cannot be excluded that among exposed people MT levels might increase and an evaluation in relation to MT is warranted in the electronics industry when handling some of the semiconductor and superconductor materials [83].

Another aspect of toxicity is exposure via food. MT and MT-like proteins have been described to occur in various foodstuff. Questions related to such problems have received some attention [50,84].

6. METALLOTHIONEIN AND DNA, GENETIC POLYMORPHISM, GENDER PERSPECTIVES

Does genetic polymorphism, i.e., several genes for MT on the same chromosome, code for specific MT functions? Literature data on metallothionein, genetic polymorphism, and gender perspectives are scarce. Metallothionein is a ligand for zinc, cadmium, and various other metals. In humans fourteen different genes are located in a gene cluster of about 82 kb [85] on chromosome 16. Six of these have been identified to be functional and two are not [86]. The genes have been identified on the basis of nucleotide sequencing. As several genes coding for metallothionein are present on the same chromosome, this might indicate that the various codes are for a specific purpose, i.e., a specific biological function as was suggested already during the first international meeting on metallothionein [21] (see also Chapter 2).

An age-dependent change of metal composition in metallothionein also indicates specific functions. In the fetus no cadmium is found, but the concentration of metallothionein is high. During gestation and in the newborn a high concentration of intracellular metallothionein rich in copper and zinc is mainly present in the liver [50,87]. This copper probably has an important function in providing Cu during the first period of life when the tissue copper concentration declines to the concentration characteristic of adult life if no metabolic disorders are present. A similar situation occurs for zinc-metallothionein. Immunohistochemical localization of MT shows MT in the nucleus of hepatocytes in neonates for several days and later on MT is present as a cytoplasmic protein during postnatal development as studied in rats [88].

That DNA synthesis and cell growth is stimulated by very low concentrations of cadmium [89] was shown in cultured mammalian cells. Human brain tissue has been found to be rich in MT-3 [90]. A growth inhibitory factor (GIF) from this tissue was identified to be a metallothionein. MT-3 shows tissue specificity and a structure that differs by having six glutamic acids inserted near the C-terminal and one additional insert in the N-terminal. Expression of MT-3 is not regulated by metals. MT-3 is downregulated in Alzheimer's disease. So far this is the only metallothionein that has a function in relation to growth. The GIF/MT-3 gene like other MT genes is located on chromosome 16 in humans.

The occurrence of metallothionein in growing tissue in tumors was reported by Cherian [91] (see also Chapter 13). Metallothionein concentrations measured by the Cd saturation method displays a clear age dependence. A decline of metallothionein in kidneys of humans after age 60 is in accordance with findings for cadmium kinetics. It has been postulated [33] that the capacity of renal tissue to produce metallothionein is age-dependent and that protein synthesis is possibly less efficient at older biological age. Epidemiological and experimental studies in laboratory animals show that females are more vulnerable to cadmium toxicity than men [49,53] (see also Chapter 10).

Women have a higher cadmium concentration in blood [92] compared to men even if differences for cadmium levels in blood is overwhelmed by smoking habits [92] and they also have a higher concentration in the liver. The iron status is of importance for cadmium and metallothionein concentration: A low iron status increases the absorption of cadmium [36]. Iron deficiency also increases the concentration of MT-1 in bone marrow in rats exposed to cadmium revealing an effect on the bone marrow, as was also suggested by Piscator [9] for rabbits with hemolytic anemia. The concentration of MT in liver is unchanged but the renal concentration is reduced in animals with iron deficiency [93]. However, that the cadmium concentration seems to be higher in aged women compared to men is contradictory to the observation that men have higher MT levels than women. A gender perspective is present in MT.

The involvement in signal transduction has not yet been described in the literature. MT-3 or -4 might, however, be involved in this. MT-4 is expressed in stratified squamous epithelia differentiating cells. The metallothionein part from the transport of metals such as cadmium and copper in the cell also functions as a free radical scavenger.

Regulation of MT gene expression in mammalian species involves the metal-responsive transcription factor (MTF1), a nuclear receptor [94]. MT gene expression was studied in humans either exposed to cadmium in the working environment or in the general environment. In these studies MT mRNA levels were measured using RT-PCR [95–98]. As will be described in more detail in

Section 7, the results indicate that metallothionein gene expression in peripheral blood lymphocytes (PBLs) may be used as a biomarker for cadmium exposure and the related susceptibility to renal dysfunction.

7. METALLOTHIONEINS AND DISEASE

7.1. General Aspects and Disease Etiology

The role of MT in the etiology of kidney disease has been described in Section 5. Binding of Cd to MT implies efficient transport to the kidney where, however, de novo synthesized MT counteracts the toxic effect of Cd and a high tissue level of Cd can be tolerated, particularly in persons that are able to synthesize MT efficiently. Those who have a poor ability to synthesize MT will suffer kidney damage at a lower tissue concentration of Cd.

There are diseases that are linked to metabolic disorders of handling metals genetically transferred from generation to generation. Examples of such diseases are Menkes' and Wilson's disease (Table 5). Menkes disease, an inherited X-linked recessive disturbed copper metabolism with mutation in ATP7A results in an accumulated copper concentration in many tissues; however, in spite of the accumulation of CuMT, the lack of ATP7A makes copper unavailable for many copper-dependent enzymes. Another inherited disorder with regard to copper accumulation is Wilson's disease with a

Table 5. Metallothionein and disease.[a,b]

Agent/ Metal	Organ	Illness
Cadmium	Kidney	Proteinuria, calciuria
		Itai-itai disease
	Kidney, pancreatic islets	Diabetes type-2/increased kidney disease
Copper	Liver	Wilson's disease
		Indian liver cirrhosis
	Intestine	MT-induction in treatment of Wilson's disease
	Placenta	Menkes disease
GIF[c]	CNS	Neurodegenerative diseases
	Brain	Alzheimer's disease

[a] Modified and compiled from [100].
[b] Cancer of several organs: MT as marker of malignancy and tumor sensitivity to chemotherapy (see also Chapter 13).
[c] Growth inhibitor factor, MT-3.

mutation in ATP7B. Patients suffering from Wilson's disease have a failure in excretion of excess copper in bile from liver leading to excessive copper accumulation in the liver, CNS, and other organs and toxicities result, e.g., in neurological disorders (see Chapter 11). The excessive Cu is stored in tissues partly bound to MT [99,100].

Metabolic disorders related to copper have been studied in animal models with similar genetic deficiencies as in human Wilson's disease. It is proposed that non-MT bound copper is transferred to ceruloplasmin. The toxicity from tissue accumulation of copper is explained by participation of MT and active oxygen species that are produced upon reactions involving copper [101].

Livers of patients with diabetes mellitus have a high concentration of metallothionein. In humans approximately 600 μg MT/g and 60 μg Zn/g have been reported in the liver. The susceptibility of spontaneously diabetic mice to CdMT nephrotoxicity was studied [102]: In these mice CdMT injections gave rise to renal damage at considerable lower doses than in normal mice [102]. Iron status in subjects exposed to various metals like Cd is linked to metal toxicity with an increased uptake of the metal and thus an increased concentration of CdMT will be seen. Iron deficiency increases MT-1 in bone marrow, with unchanged MT in liver, and decreased MT in kidney which indicates a mechanism that needs further explanation [93].

Ushida et al. [90] found that the concentration of growth inhibiting factors (GIF), shown to be identical to MT-3, is decreased in brain tissue from Alzheimer patients. Neurotropic activity of neonatal rat brain tissue was inhibited by GIF/MT-3 [103].

Presently much attention is paid to an interesting research direction on metallothionein and its potential role in neurodegenerative disorders. Molecular pathways of neuroprotection and regeneration are metallothionein-mediated. MTs expressed in astrocytes after CNS injury are reported to serve both neuroprotective and neuroregenerative roles critical for the outcome of recovery [104]. MTs lacking signal peptides, scavenge free radicals and bind toxic metals and because of this they have a neuroprotective function intracellularly. Neuroprotective functions of MTs may also involve an extracellular component. A possible significant therapeutic potential of MT in the context of current understanding of the role of MT in astrocyte-neuron interactions in the injured brain has been brought forward [105]. MT-3 is predominantly expressed in Zn-containing neurons of the hippocampus. Disturbed MT homeostasis can lead to changes in brain concentrations of Zn (see Chapters 10 and 11). Since the intracellular concentration of Zn is mediated by complexing with apo-thionein to form MT, there has been great interest in ascertaining whether disordered MT regulation plays a role in the etiology of neurodegenerative disorders. Abnormalities in MT and/or Zn homeostasis have been reported in multiple neurological disorders, e.g., Alzheimer's disease [106]. A possible role of MT and metals in other neurodegenerative diseases, e.g., in amyotrophic lateral sclerosis is suggested by the finding of increased levels of metals in cerebrospinal fluid (CSF) among such cases [107]. MT-binding was studied by chromatographic separation and metal analyses [108]. Silencing of genes changes the expression of the protein that is encoded by the gene. If the protein is not present in the tissue, a different physiological/toxicological pattern is seen even if the gene is present. Thus it can be questioned what happens when MT genes are silenced. MT-3 is seen in the brain but mRNA for MT-3 synthesis has been identified in the kidney though no MT-3 has been found. This raises the question if MT-3 can be involved in neurodegenerative disorders by disturbed expression or silencing of the gene.

7.2. Metallothionein-Related Biomonitoring in Diseases

Metallothionein in urine correlates well with Cd in urine. Both measurements can be used as a biomarker of cumulative cadmium doses in long-term exposures [52]. Basal and induced MT gene expression levels in PBLs are closely related with Cd exposure. MT gene expression in PBLs thus may be used as biomarker of Cd exposure, but more direct methods for this purpose are determinations of Cd in blood or urine [52].

7.2.1. Biomarker of Susceptibility

Induced MT mRNA levels in PBLs seem to reflect the renal ability for MT induction, thus providing a possible index of susceptibility to the adverse effect of Cd on the kidney [96–98]. Because it is not possible in routine biomonitoring to measure the renal MT gene expression *in vivo*, *in vitro* induced MT gene expression levels in PBLs would serve as a potential suitable index for this. A dose-effect relationship between the internal dose of cadmium and the MT mRNA level confirmed the validity of MT gene expression in PBLs as a biomarker of cadmium exposure. Both studies on Cd workers and environmentally Cd-exposed persons have measured the *in vitro* induced MT mRNA level in PBLs sampled from exposed persons as an indicator reflecting the ability of the body to synthesize MT upon cadmium exposure. A negative correlation between urinary N-acetyl-β-D-glucosaminidase (UNAG), a sensitive indicator for renal effects of cadmium exposure and the *in vitro* induced MT mRNA level in the subjects with high UCd level (over 10 μg/g creatinine) was shown. The lower proportion of individuals expected to have exceeded their individual critical concentration in the renal cortex at UCd levels below 10 μg/g creatinine explains that no statistically significant correlation was observed between the *in vitro* induced MT mRNA level and UNAG in 2–10 μg/g creatinine UCd groups.

A reverse relationship between *in vitro* induced MT mRNA level in PBLs and UNAG indicates that MT gene expression in PBLs can be used as a biomarker inversely related to the susceptibility to renal toxicity of cadmium. It is suggested to apply MT gene expression in PBLs for the risk assessment of cadmium exposure [96–98]. Studies of MT mRNA in human lymphocytes provide evidence of a lower prevalence of tubular proteinuria among Cd-exposed persons with high ability to express tissue metallothionein compared to persons with such a low ability. These studies provide support in humans for the important role of MT in Cd toxicology [96–98]. Metallothionein in urine can be used as a sensitive biomarker for metal induced nephrotoxicity [109].

Metallothionein gene expression in peripheral blood lymphocytes as a biomarker of susceptibility to renal Cd toxicity in humans is an example of how advances in molecular epidemiology may increase delineation of human health risks from exposure to this element.

The presence of MT-ab in blood plasma is a significant indicator for the occurrence of tubular dysfunction among diabetic subjects. Among Cd-exposed workers and among persons suffering type-2 diabetes, elevated levels of MT-ab was associated with a higher prevalence of tubular dysfunction compared to those with lower MT-ab levels [109–111].

It was shown that the cellular localization of MT is different in some cancer cells compared to normal cells (see Chapter 13). This observation forms the basis for the use of MT as biomarker of cancer malignancy [91]. There is an extensive recent literature on the use of MT immunostaining in tumor diagnosis, e.g., in adenoid cystic carcinomas MT immunolocalization may be important [112].

7.3. Metallothionein in the Treatment of Diseases

Different MT synthesis in tumor tissue compared to normal tissue has been suggested as a basis for treatment. The possible use of cadmium as a cancer treatment agent in some forms of liver tumors was discussed [113] based on observations in mice. Animals that were treated with combinations of Cd and NDEA (N-nitrosodiethylamine) did not develop liver tumors in contrast to animals treated only with the tumorigenic agent NDEA. MT levels were markedly reduced in the livers of tumor bearing animals compared to normal animals. The effect of cadmium treatment on these liver tumors, thus can be explained by their higher sensitivity to Cd due to the lack of MT expression.

It has been shown in animal experiments that bismuth treatment will induce MT in the kidney and reduce toxicity of *Cisplatin* to the kidney

[114,115]. This observation may serve as a basis for administering higher doses of this therapeutic agent with retained patient safety when the combination with bismuth is used in cancer therapy.

Oral treatment with zinc has been suggested [116] and is now recommended as a standard treatment in Wilson's disease. The conditions of patients improved by a mechanism that zinc induces synthesis of intestinal metallothionein that sequesters copper for the structure and thus blocks the intestinal uptake of Cu.

8. FUTURE ASPECTS ON METALLOTHIONEINS

Already at the first meeting on MT in Zürich in 1978 it was discussed to set up a bank or producer of MT in order to have pure and standardized MT probes available. Discussions dealt with the question if interlaboratory exchange and a quality control program would solve some difficulties that might occur due to reports on the varying purity of MT. In order to be able to compare results in MT research and reported quantifications from different laboratories it is necessary to include some quality control. This should include standard material or reference material of MT. Due to the lack of the latter mostly commercially available MT is employed as a substitute of a standard in MT quantification. However, the quality of MT employed as such can be questioned. Several reports about the quality have ended with results that are difficult to interpret. The large number of isoforms and subisoforms within the MT family makes application of immunological techniques to biological samples difficult. In order to increase the selectivity and sensitivity of these methods, the specificity of the antibodies is important. However, the complete family of MT lacks common antibodies for the different isoforms and subisoforms, and this could lead to underestimations when total MT is measured by, e.g., ELISA [37]. Purification of MT by the individual research groups seems to yield high quality MT.

In order to gain more information on MT it is necessary to harmonize methods for its quantification. Several methods for estimating and measuring the concentration of metallothionein in tissues and body fluids have been developed and are at present more or less successfully in use. It is necessary to have in mind factors that influence the concentration of metallothionein in tissues and fluids. Metallothionein is increasingly used, but reported levels in tissues and fluids vary among research groups. Like other biomarkers, MTs may have to be related to some other factors. Exposure to many agents induce MT synthesis and metals are inducers with high potency; cadmium being the strongest one.

Little is known about the biological aging in synthesizing metallothionein. Studies on cadmium concentration in renal tissue have shown a decrease after the age of 50–60 years in human beings. A method to measure the concentration of metallothionein should be protein-specific and manage to measure changes such as increase or decrease of the normal or, more precisely, basic concentration. Cadmium concentration increases upon exposure and particularly with age as the new born has a low concentration of cadmium since this metal does not pass the placental barrier. The concentration of metallothionein in the cell is influenced by so many parameters and factors that a method to quantitate it has to be standardized with respect to many factors. Just to measure the concentration in relation to exposure to some inducing agent, many of the previously mentioned methods are quite acceptable. However, in order to use MT values in routine biomonitoring, "normal" values in matrices such as blood, urine, and CSF need to be established. Measurement of MT mRNA in peripheral blood lymphocytes has been suggested as a biomarker of susceptibility in metal exposures. In order for this method to be used more widely, it has to be standardized and adapted to field conditions. Metallothionein autoantibodies in blood plasma appear to be a strong indicator of susceptibility to kidney effects of cadmium and may serve a similar role in other metal-induced diseases. Future studies concerning such uses would be of great interest.

Metallothionein is still a protein that demands further research and attracts scientists from many fields. A number of aspects have been brought to attention. The state of present knowledge and seen from the historical point of view indicates that results of future research on metallothionein will contribute to explain many rather different biological effects.

9. CONCLUSIONS

Summarizing half a century of research on MT shows that metallothionein remains a fascinating protein with several physiological functions and that MT can be described as a Camelot protein. MTs are a family of ubiquitous low molecular weight proteins with a high thiolate sulfur and metal content (Zn(II), Cu(I)), on the order of 10% (w/w). In conclusion, the experimental difficulties with the quantification of MT from the perspectives of three different fields, that is electrochemistry, chromatography, and immunochemistry, shows that even if advances in the development of the equipment occurred during the last five years, problems still exist in the determination of MT and particularly in the quantification in various complex media such as tissues and biological fluids. A number of newly developed techniques and equipments with increasing sensitivity and specificity has recently been applied to MT determinations in various media and greatly improved the possibilities for accurate measurements. Although MT is stable under specific circumstances it can be degraded or complexed with other proteins under other commonly occurring circumstances.

Knowledge and training in handling the protein is most crucial in order to avoid confusing results that are tricky to interpret. MT can be used as an indicator in both environmental and biological monitoring reflecting exposure to metals, and as a good biomarker of renal dysfunction. When values can be set for the normal concentration of MT, the protein could also serve in relation to medical aspects and assist in the calculation of an allowable intake or exposure limit for Cd. MT is an established biomarker in biomontoring of human Cd exposure and may also be useful in the risk assessment of other metal exposures. MT mRNA in lymphocytes in humans has been suggested as an indicator of susceptible groups in relation to metal exposure; the development of practical procedures to measure MT in biological samples, e.g., blood, urine, biopsies of tumors, etc., is highly desirable. MT-ab appears to be an important biomarker for Cd-related tubular dysfunction.

ABBREVIATIONS

Alb	albumin
BAL	British anti-lewisite, 2,3-dimercaptopropanol (dimercaprol)
CdMT	metallothionein containing Cd^{2+}
CMB	p-chloromercuribenzoate
CNS	central nervous system
CR	creatinine
CSF	cerebrospinal fluid
DMSA	meso-2,3-dimercaptosuccinic acid (succimer)
EDTA	ethylenediamine-N,N,N′,N′-tetraacetic acid
ELISA	enzyme-linked immunosorbent assay
GIF	growth inhibitor factor
MCBI	Medical Center for Biotechnology Information
MT	metallothionein
MT-ab	metallothionein autoantibody
MTF	metal-responsive transcription factor
NDEA	N-nitrosodiethylamine
PBLs	peripheral blood lymphocytes
PCR	polymerase chain reaction
RIA	radio-immunoassay
RT-PCR	real-time polymerase chain reaction
Tris	tris-(hydroxymethyl)-aminomethane
UCd	urinary cadmium
UNAG	urinary N-acetyl-β-D-glucosaminidase
YBCO	yttrium-barium-copper-oxide

REFERENCES

1. G. F. Nordberg, M. Piscator and B. Lind, *Acta Pharmacol. Toxicol.*, 1971, **29**, 456–470.
2. G. F. Nordberg, M. Piscator and M. Nordberg, *Acta Pharmacol. Toxicol.*, 1971, **30**, 289–295.
3. M. Margoshes and B. L. Vallee, *J. Am. Chem. Soc.*, 1957, **79**, 4813–4814.
4. D. P. Maliuga, *Doklady Akad. Nauk. USSR*, 1941, **31**, No 2 145 and "Cadmium and Organismus", *Chemical Abstracts* 1943, 37.1. 894, Translation RJ–296.
5. J. H. R. Kägi and B. L. Vallee, *Biochemistry*, 1960, **235**, 3460–3465.
6. J. H. R. Kägi and B. L. Vallee, *Biochemistry*, 1961, **236**, 2435–2442.
7. G. F. Nordberg, M. Nordberg, M. Piscator and O. Vesterberg, *Biochemical J.*, 1972, **126**, 491–498.
8. J. H. R. Kägi, *8th Int. Congr. Biochem. Abstract*, 1970, 130–131.
9. M. Piscator, *Nordisk hygienisk tidskrift*, 1964 **XLV** 76–82 (english summary).
10. G. F. Nordberg, *Environ. Physiol.*, 1971, **1**, 171–187.
11. M. Nordberg, G. F. Nordberg and M. Piscator, *Environ. Physiol. Biochem.*, 1975, **5**, 396–403.
12. M. Nordberg, *Studies on Metallothionein and Cadmium*, Doctoral thesis, Dept. of Environmental Hygiene, Karolinska Institutet, Stockholm, Sweden, 1977.
13. M. Nordberg, *Environ. Res.*, 1978, **15**, 381–404.
14. D. R. Winge and K.-A. Miklossy, *J. Biol. Chem.*, 1982, **257**, 3471–3476.
15. I. Bremner, in *Metallothionein*, Ed. J. H. R. Kägi and M. Nordberg, Birkhäuser Verlag, Basel/Boston/Stuttgart, 1979, pp. 273–280.
16. M. Nordberg, B. Trojanowska and G. F. Nordberg, *Environ. Physiol. Biochem.*, 1974, **4**, 149–158.
17. J. K. Piotrowski, W. Bolanowska and A. Sapota, *Acta Biochim. Pol.*, 1973, **20**, 207–215.
18. L. Björkman, B. Palm, M. Nylander and M. Nordberg, *Biol. Trace Element Res.*, 1994, **40**, 255–265.
19. M. Nordberg and G. F. Nordberg, *Environ. Health Perspect.*, 1975, **12** 103–108.
20. M. Nordberg and Y. Kojima, in *Metallothionein*, Ed. J. H. R. Kägi and M. Nordberg, Birkhäuser Verlag, Basel/Boston/Stuttgart, 1979, pp. 41–116.
21. *Metallothionein*, Ed. J. H. R. Kägi and M. Nordberg, Birkhäuser Verlag, Basel, Boston, Stuttgart, 1979, pp. 1–378. As far as the nowadays accepted nomenclature is concerned, see P.-A. Binz and J. H. R. Kägi, *Metallothionein: Molecular Evolution and Classification, in Metallothionein* IV, Ed. C. D. Klaassen, Birkhäuser Verlag, Basel, 1999, pp. 7–13, and see also the website given in the text: http://www.expasy.org/cgi-bin/lists?metallo.txt.
22. *Proceedings of the High Affinity Metal-Binding Proteins in Non-Mammalian Species Conference 1984*, National Institute of Environmental Health Sciences, Research Triangle Park, NC, USA, Ed. B. A. Fowler, *Environ. Health Perspect.*, 1986, **65**, 3–224.

23. K. T. Suzuki, M. Imura and M. Kimura, *Metallothionein III. Biological Roles and Medical Implications, ALS Advances in Life Sciences*, Birkhäuser Verlag, Basel, Boston, Berlin, 1993, pp. 1–479.
24. Y. Kojima, C. Berger, B. L. Vallee and J. H. R. Kägi, *Proc. Natl. Acad. Sci. USA*, 1976, **73**, 3413–3417.
25. R. J. Vandermallie and J. S. Garvey, *J. Biol. Chem.*, 1979, 8416–8421.
26. *Proceedings of the Metallothionein and Cadmium Nephrotoxicity Conference 1983*, National Institute of Environmental Health Sciences, Research Triangle Park, NC, USA, Ed. R. A. Goyer, B. A. Fowler, G. F. Nordberg, G. Shepard and L. Moustafa, *Environ. Health Perspect.*, 1984, **54**, 1–295.
27. J. H. R. Kägi and Y. Kojima, *Metallothionein II*, Birkhäuser Verlag, Basel, Boston, Stuttgart, 1987, pp. 1–755.
28. M. Nordberg, I. Nuottaniemi, M. G. Cherian, G. F. Nordberg, T. Kjellström and J. S. Garvey, *Environ. Health Perspect.*, 1986, **65**, 57–62.
29. D. H. Petering and B. A. Fowler, *Environ. Health Perspect.*, 1986, **65**, 217–224.
30. J. H. R. Kägi, S. R. Himmelhoch, P. D. Whanger, J. L. Bethune and B. L. Vallee, *J. Biol. Chem.*, 1974, **249**, 3537–3542.
31. C.-G. Elinder and M. Nordberg, in *Cadmium and Health*, Vol. **I**, Ed. L. Friberg, C.-G. Elinder, T. Kjellström and G. F. Nordberg, CRC Press, Boca Raton, Fla,1985, pp. 65–79.
32. M. Nordberg, in *Ethnic Differences in Reactions to Drugs and Xenobiotics*, Ed. W. Kalow and H. E. Goedde, Alan R. Liss Inc., New York, 1986, pp. 401–410.
33. M. Nordberg, *Talanta*, 1998, **46**, 243–254.
34. R. Palmiter, in *Metallohionein II*, Ed. J. H. R. Kägi and Y. Kojima, Birkhäuser Verlag, Basel, Boston, Stuttgart, 1987, pp. 63–80.
35. Y. Liu, J. Liu, R. D. Palmiter and C. D. Klaassen, *Toxicol. Appl. Pharmacol.*, 1996, **137**, 307–315.
36. G. F. Nordberg, T. Kjellström and M. Nordberg, in *Cadmium and Health*, Vol. **I**, Ed. L. Friberg, C.-G. Elinder, T. Kjellström and G. F. Nordberg, CRC Press, Boca Raton, Fla, 1985, pp. 103–197.
37. M. Dabrio, A. R. Rodriguez, G. Bordin, M. J. Bebianno, M. De Ley, I. Sestakova, M. Vašák and M. Nordberg, *J. Inorg. Biochem.*, 2002, **88**, 123–134.
38. M. Nordberg, *Environ. Health Perspect.*, 1984, **54**, 13–20.
39. S. Onosaka and M. G. Cherian, *Toxicology*, 1981, **22**, 91–101.
40. J. S. Garvey, *Environ. Health Perspect.*, 1984, **54**, 117–127.
41. J. F. Riordan and B. L. Vallee, *Methods in Enzymology*, Academic Press, New York, 1991, **205**, 3–681.
42. D. L. Eaton and B. F. Toal, *Toxicol. Appl. Pharmacol.*, 1982, **66**, 134–142.
43. R. W. Olafson and R. G. Sim, *Anal. Biochem.*, 1979, **100**, 343–351.
44. C. Tohyama, Z. A. Shaikh, A. Nogawa, E. Kobayashi and R. Honda, *Toxicology*, 1981, **20**, 289–297.
45. C. Tohyama, Z. A. Shaikh, K. J. Ellis and S. H. Cohn, *Toxicology*, 1981, **22**, 181–191.
46. G. F. Nordberg, J. S. Garvey and C. C. Chang, *Env. Res.*, 1982, **28**, 179–182.
47. H. A. Roels, R. R. Lauwreys, J. P. Buchet, A. Bernard, J. S. Garvey and H. J. Linton, *Int. Arch. Occup. Environ. Health*, 1983, **52**, 159–166.

48. T. Haerslev, G. K. Jacobsen, N. Horn and E. Damsgaard, *Acta Pathol. Microbiol. Immunol. Scand.*, 1995, **103**, 568–573.
49. WHO/IPCS Environmental Health Criteria Document 134 Cadmium, WHO, Geneva 1992.
50. M. G. Cherian and M. Nordberg, *Toxicology*, 1983, **28**, 1–15.
51. M. Nordberg and G. F. Nordberg, in *Handbook of Heavy Metals in the Environment*, Ed. B. Sarkar, Marcel Dekker, Inc, New York, 2002, pp. 231–270.
52. *Handbook on the Toxicology of Metals*, Ed. G. F. Nordberg, B. A. Fowler, M. Nordberg and L.T. Friberg, Elsevier 2007, pp. 1–975, see pp. 445–486.
53. H. M. Chan, L.-F. Zhu, R. Zhong, D. Grant, R. A. Goyer and M. G. Cherian, *Toxicol. Appl. Pharmacol.*, 1993, **123**, 89–96.
54. X. Liu, T. Jin, G. F. Nordberg, M. Sjöström and Y. Zhou, *Toxicol. Appl. Pharmacol.*, 1994, **126**, 84–90.
55. Z. A. Shaikh, C. Tohyama and C. V. Nolan, *Arch Toxicol.*, 1987, **59**, 360–364.
56. X. Liu, T. Jin, G. F. Nordberg, M. Sjöström and Y. Zhou, *Toxicol. Appl. Pharmacol.*, 1994, **126**, 84–90.
57. T. Jin, M. Nordberg and G. F. Nordberg, *Toxicology*, 1992, **75**, 29–37.
58. G. F. Nordberg, T. Jin and M. Nordberg, *Environ. Health Perspect.*, 1994, **102**(suppl. 3), 191–194.
59. T. Jin, K.-F. Norrback, M. Nordberg, X. Liu and G. F. Nordberg, *The Role of Cadmium and Zinc Pretreatment on the Distribution of Cadmium Bound to Proteins in Renal Membrane and Microsome*, Abstract, Commission on Toxicology, Vancouver 1995.
60. M. Nordberg, in *Methods in Enzymology*, Ed. J. F. Riordan and B. L. Vallee, Academic Press, New York, 1991, **205**, 247–252.
61. M. Garty, K.-L. Wong and C. D. Klaassen, *Toxicol. Appl. Pharmacol.*, 1981, **59**, 548–554.
62. G. F. Nordberg, *Environ. Phys. Biochem.*, 1972, **2**, 7–36.
63. M. Nordberg and G. F. Nordberg, *Cell. Mol. Biol.*, 2000, **46**, 451–463.
64. D. R. Johnson and E. G. Foulkes, *Environ. Res.*, 1980, **21**, 360–365.
65. M. G. Cherian, R. A. Goyer and L. S. Valberg, *Toxicol. Environ. Health*, 1978, **4**, 861–868.
66. A. P. Leber and T. S. Miya, *Toxicol. Appl. Pharmacol.*, 1976, **37**, 403–414.
67. G. F. Nordberg, R. A. Goyer and M. Nordberg, *Arch. Pathol.*, 1975, **99**, 192–197.
68. M. Nordberg and G. F. Nordberg, in *Mechanisms of Cell Injury Implications for Human Health*, Ed. B. F. Fowler, Dahlem Konferenzen, John Wiley & Sons Ltd, Chichester, 1987, pp. 53–65.
69. M. Nordberg, T. Jin and G. F. Nordberg, in *Cadmium in the Human Environment: Toxicity and Carcinogenicity*, Ed. G. F. Nordberg, L. Alessio and R. F. M. Herber, International Agency for Research on Cancer, Lyon, France (IARC), Scientific Publications No. 118, 1992, pp. 293–297.
70. I. Bremner, in *Metallothionein II*, Ed. J. H. R. Kägi and Y. Kojima, Birkhäuser Verlag, Basel, Boston, Stuttgart, 1987, 81–107.
71. T. Jin, G. F. Nordberg and M. Nordberg, *J. Appl. Toxicol.*, 1986, **6**, 397–400.
72. J. Liu and C. D. Klaassen, *Fundam. Appl. Toxicol.*, 1996, **29**, 294–300.

73. A. M. Bernard, A. A. Ouled and R. R. Lauwerys, *Toxicol. Appl. Pharmacol.*, 1987, **87**, 440–445.
74. B. A. Fowler and G. F. Nordberg, *Toxicol. Appl. Pharmacol.*, 1978, **46**, 609–623.
75. T. Y. Jin, P. Leffler and G. F. Nordberg, *Toxicology*, 1987, **45**, 307–317.
76. P. Leffler, T. Jin and G. F. Nordberg, *Toxicology*, 1996, **16**, 112, 151–156.
77. P. Leffler, T. Jin and G. F. Nordberg, *Toxicol. Appl. Pharmacol.*, 1990, **103**, 180–184.
78. B. Elsenhans, K. Kolb, K. Schumann and W. Forth, in *Cadmium in the Human Environment: Toxicity and Carcinogenicity*, Ed. G. F. Nordberg, L. Alessio and R. F. M. Herber, International Agency for Research on Cancer, Lyon, France (IARC), Scientific Publications No. 118, 1992, pp. 225–230.
79. J. P. Groten, E. Hissink and P. J. van Bladeren, in *Cadmium in the Human Environment: Toxicity and Carcinogenicity*, Ed. G. F. Nordberg, L. Alessio and R. F. M. Herber, International Agency for Research on Cancer, Lyon, France (IARC), Scientific Publications No. 118, 1992, pp. 219–224.
80. R. K. Zalups and S. Ahmad, *Toxicol. Appl. Pharmacol.*, 2003, **186**, 163–188.
81. C. G. Elinder, M. Nordberg, B. Palm, L. Björck and L. Jönsson, *Environ. Res.*, 1987, **42**, 553–562.
82. L. Cai, J. Koropatnick and M. G. Cherian, *Chem.-Biol. Interactions*, 1995, **96**, 143–155.
83. M. Nordberg and G. F. Nordberg, *Environ. Sci.*, 1996, **4**, 187–197.
84. M. Nordberg and G. F. Nordberg, in *Natural Antioxidants and Food Quality in Atherosclerosis and Cancer Prevention*, Ed. J. T. Kumpulainen and J. T. Salonen, The Royal Society of Chemistry, Cambridge, UK, 1996, 412–416.
85. A. K. West, R. Stallings, C. E. Hildebrand, R. Chia, M. Karin and R. J. Richard, *Genomics*, 1990, **8**, 513–518.
86. B. L. Vallee and W. Maret, in *Metallothionein III. Biological Roles and Medical Implications*, Ed. K. T. Suzuki, M. Imura and M. Kimura, *ALS Advances in Life Sciences*, Birkhäuser Verlag, Basel, Boston, Berlin, 1993, pp. 1–27.
87. B. R. Sonawane, M. Nordberg, G. F. Nordberg and G. W. Lucier, *Environ. Health Perpect.*, 1975, **12**, 97–102.
88. M. G. Cherian, D. M. Templeton, K. R. Gallant and D. Banerjee, *Experientia, Suppl.*, 1987, **52**, 499–505.
89. T. von Zglinicki, C. Edvall, E. Östlund, B. Lind, M. Nordberg, N. R. Ringertz and J. Wroblewski, *J. Cell Science*, 1992, **103**, 1073–1081.
90. Y. Uchida, in *Metallothionein III. Biological Roles and Medical Implication*, Ed. K. T. Suzuki, M. Imura and M. Kimura, *ALS Advances in Life Sciences*, Birkhäuser Verlag, Basel, Boston, Berlin, 1993, pp. 315–328.
91. M. G. Cherian, *Environ. Health. Perspect.*, 1994, **102** (suppl. 3), 131–135.
92. C.-G. Elinder, *Normal Values for Cadmium in Human Tissues, Blood and Urine in Different Countries*, in *Cadmium and Health*, Vol. I, Ed. L. Friberg, C.-G. Elinder, T. Kjellström and G. F. Nordberg, CRC Press, Boca Raton, Fla., 1985, p. 82.
93. A. Robertson, J. N. Morrison, A. M. Wood and I. Bremner, *J. Nutr.*, 1989, **119**, 439–445.

94. A. Selvaraj, K. Balamurugan, H. Yepiskoposyan, H. Zhou, D. Egli, O. Georgiev, D. J. Thiele and W. Schaffner, *Genes Dev.*, 2005, **19**, 891–896.
95. S. Ganguly, E. Taioli, B. Baranski, B. Cohen, P. Toniolo and S. J. Garte, *Cancer Epidemiol. Biomarkers Prev.*, 1996, **5**, 297–301.
96. J. Lu, T. Jin, G. Nordberg and M. Nordberg, *Cell Stress and Chaperones*, 2001, **6**, 97–104.
97. J. Lu, T. Jin, G. F. Nordberg and M. Nordberg, *BioMetals*, 2004, **17**, 569–570.
98. J. Lu, T. Jin, G. F. Nordberg and M. Nordberg, *Toxicol. Appl. Pharmacol.*, 2005, **206**, 150–156.
99. D. G. Ellingsen, N. Horn and J. Aaseth, in *Handbook on the Toxicology of Metals*, Ed. G. F. Nordberg, B. A. Fowler, M. Nordberg and L. T. Friberg, Elsevier, 2007, pp. 529–546.
100. G. F. Nordberg, T. Jin, P. Leffler, M. Svensson, T. Zhou and M. Nordberg, *Analusis*, 2000, **28**, 396–400.
101. K. T. Suzuki, *Res. Comm. Mol. Pathol. Pharmacol.*, 1995, **89**, 221–240.
102. T. Jin, G. F. Nordberg, O. Sehlin, P. Leffler and J. Wu, *Toxicology*, 1994, **89**, 81–90.
103. J. G. Erickson, G. Hollopeter, S. A. Thomas, G. J. Froelick and R. D. Palmiter, *J. Neurosci.*, 1997, **17**, 1271–1281.
104. R. S. Chung, J. Hidalgo and A. K. West, *J. Neurochem.*, 2008, **104**, 14–20.
105. M. Ambjørn, J. W. Asmussen, M. Lindstam, K. Gotfryd, C. Jacobsen, V. V. Kiselyov, S. K. Moestrup, M. Penkowa, E. Bock and V. Berezin, *J. Neurochem.*, 2008, **104**, 21–37.
106. M. Aschner and A. K. West. J. *Alzheimers Dis.*, 2005, **8**, 139–145; discussion pp. 209–215.
107. P. M. Roos, O. Vesterberg and M. Nordberg, *J. Exper. Biol. Med.*, 2006, **231**, 1481–1487.
108. K. Gellein, P. M. Roos, L. Evje, O. Vesterberg, T. P. Flaten, M. Nordberg and T. Syversen, *Brain Res.*, 2007, **1174**, 136–142.
109. L. Chen, T. Jin, B. Huang, G. F. Nordberg and M. Nordberg, *Toxicol. Appl. Pharmacol.*, 2006, **215**, 93–99.
110. L. Chen, L. Lei, T. Jin, M. Nordberg and G. F. Nordberg, *Diabetes Care*, 2006, **29**, 2682–2687.
111. L. Chen, T. Jin, B. Huang, X. Chang, L. Lei, G. F. Nordberg and M. Nordberg, *Toxicol. Sci.*, 2006, **91**, 104–112.
112. S. M. Alves, S. V. Cardoso, V. de Fátima Bernardes, V. C. Machado, R. A. Mesquita, M. A. Vieira do Como and M. C. Ferreira Aguiar, *Oral.Oncol.*, 2007, **43**, 252–256.
113. M. P. Waalkes, B. A. Diwan, C. M. Weghorst, R. M. Bare, J. M. Ward and J. M. Rice, *Toxicol. Appl. Pharmacol.*, 1991, **110**, 327–335.
114. Y. Kondo, M. Satoh, N. Imura and M. Akimoto, *Anticancer Res.*, 1992, **12**, 2303–2308.
115. Y. Kondo, S. Himeno and M. Satoh, *Cancer Chemother. Pharmacol.*, 2004, **53**, 33–38.
116. G. J. Brewer, *Drugs*, 1995, **50**, 240–249.

Met. Ions Life Sci. **2009**, *5*, 31–49

2

Regulation of Metallothionein Gene Expression

Kuppusamy Balamurugan and Walter Schaffner
Institute of Molecular Biology, University of Zürich, Winterthurerstrasse, 190, CH-8057 Zürich, Switzerland
<bala_105@hotmail.com>
<walter.schaffner@molbio.uzh.ch>

Metal Ions in Life Sciences, Volume 5 Edited by Astrid Sigel, Helmut Sigel, and Roland K. O. Sigel

Published by the Royal Society of Chemistry, www.rsc.org DOI: 10.1039/9781847558992-00031

ABSTRACT: Organisms from bacteria to humans use elaborate systems to regulate levels of bioavailable zinc, copper, and other essential metals. An excess of them, or even traces of non-essential metals such as cadmium and mercury, can be highly toxic. Metallothioneins (MTs), short, cysteine-rich proteins, play pivotal roles in metal homeostasis and detoxification. With their sulfhydryl groups they avidly bind toxic metals and also play a role in cellular redox balance and radical scavenging. The intracellular concentration of MTs is adjusted to cellular demand primarily via regulated transcription. Especially upon heavy metal load, metallothionein gene transcription is strongly induced. From insects to mammals, the major regulator of MT transcription is MTF-1 (metal-responsive transcription factor 1), a zinc finger protein that binds to specific DNA sequence motifs (MREs) in the promoters of *MT* genes and other metal-regulated genes. This chapter provides an overview of our current knowledge on the expression and regulation of *MT* genes in higher eukaryotes, with some reference also to fungi which apparently have independently evolved their own regulatory systems.

KEYWORDS: copper transporter · heavy metal toxicity · metal homeostasis · metallothioneins · metal-regulated gene transcription · MTF-1

1. INTRODUCTION

All organisms need to cope with adverse environmental conditions such as heavy metal load, oxidative stress, and UV irradiation. Heavy metal toxicity may be caused by a wide range of interactions at the molecular level [1,2]. On the one hand, binding of reactive metals to histidines or sulfhydryl groups in proteins can lead to a reduction/ablation of their activity; on the other hand the metals may stimulate the formation of free radicals and reactive oxygen species (ROS), causing extensive cellular damage [1–3]. At the same time, metals such as zinc and copper are pivotal for normal life. Zinc is essential for proper functioning of a variety of proteins including zinc-dependent enzymes and many transcriptional regulators [4,5]. A distorted zinc metabolism can culminate in clinical conditions including *acrodermatitis enteropathica*, a genetic disorder caused by malabsorption of zinc. Genetic impairment of copper transport leads to life-threatening diseases such as Menkes and Wilson's diseases [6–8].

Organisms from bacteria to humans use elaborate systems to regulate levels of bioavailable zinc, copper, and other essential metals [9–11]. Metallothionein proteins (MTs) play pivotal roles in metal homeostasis and detoxification [12,13]. Their high cysteine content enables MTs to avidly bind toxic metals and also to play a role in cellular redox balance and radical scavenging. The intracellular concentration of MTs is adjusted to cellular demand. Especially upon heavy metal load, metallothionein gene transcription is strongly induced [12–15]. This chapter provides an overview of our current knowledge on the expression and regulation of MT genes.

2. METALLOTHIONEINS ARE ENCODED BY A FAMILY OF SHORT GENES

Downregulation of metal importers may not be fast enough to cope with a sudden increase in extracellular metal concentration. Intracellular sequestration of toxic heavy metals, an important aspect of heavy metal homeostasis, is mainly achieved by metallothioneins [9–11]. Metallothionein proteins, discovered more than 50 years ago by Margoshes and Vallee [18], are found in all phyla. They are particularly widespread in animals, fungi, and plants. In higher eukaryotes, MTs consist of some 60 amino acids with up to 30% cysteines and usually no aromatic amino acids [12–15]. Metallothioneins are able to bind as many as 18 different metals, but with variable affinity, whereby copper, cadmium, silver, mercury, bismuth, and lead can displace zinc [13,15,16]. (For convenience, the designations copper or Cu, cadmium or Cd, silver or Ag, mercury or Hg, bismuth or Bi, lead or Pb and zinc or Zn are also used here to denote Cu^{2+} and Cd^{2+}, Ag^{+}, Hg^{2+} and Bi^{3+}, Pb^{2+} and Zn^{2+}, respectively).

Several studies have shown that MTs can bind seven zinc or cadmium or up to 12 copper ions [13–17,35]. In mammals under normal conditions, MTs predominantly exist as a zinc-complex, while in *Drosophila* and fungi they are typically found complexed with copper [19,20]. In contrast to vertebrate MTs, the size of metallothioneins varies considerably among other species. For example, MT of the cyanobacterium *Synechococcus* contains 56 amino acids, *Drosophila MtnA* (also termed *Mtn*) and *MtnB* (=*Mto*) genes encode proteins of 43 and 40 amino acids, respectively, and the filamentous fungus *Neurospora* harbors a metallothionein of only 26 amino acids [19,21,22]. Of note, metallothioneins from higher plants contain 60–85 amino acids [23].

Elevated concentrations of both essential and non-essential heavy metals in the soil can inhibit the growth of most plants. To cope with this, plants have a range of potential mechanisms at the cellular level including high-affinity ligands such as amino acids and organic acids, and two classes of peptides, the phytochelatins and the MTs [24,25]. The phytochelatins (PCs) or metal-complexing peptides have a general structure (γ-Glu Cys)$_n$-Gly where n = 2–11 (see also Chapter 15). Phytochelatin synthesis is rapidly induced in plants by exposure to heavy metals, especially cadmium, which is detoxified mostly by PCs [25]. Whereas phytochelatins are synthesized enzymatically by the enzyme phytochelatin synthase, MTs are encoded by genes [13,24,26]. The budding yeast *Saccharomyces cerevisiae* contains two MTs, Cup1, and CRS5 [27]. Inactivation of these two MTs in yeast leads to copper sensitivity; accordingly, MT overexpression confers resistance to copper toxicity [27,28]. Cup1 is a clear-cut copper-thionein, but CRS5, which also contributes to neutralizing excess intracellular copper, has recently been demonstrated to be significantly closer to Zn-thioneins than to Cu-thioneins with respect to its binding properties. In line with this finding, CRS5 can protect yeast from zinc toxicity [29].

Drosophila contains four metallothioneins, namely, MtnA, MtnB, MtnC, and MtnD [30–32]. MtnA is found in late stage embryos, larvae, and in adult flies within the gut, Malpighian tubules, fat body, and hemocytes while MtnB is primarily expressed during embryogenesis. The MtnC and MtnD proteins each are closely related to MtnB (>67% amino acid identity) [30–33]. Even though the three genes have obviously arisen by duplication events, MtnC and MtnD play only minor roles in the protection against metal toxicity. A "family knockout" of all four MT genes in *Drosophila* led to increased sensitivity of flies to copper and cadmium load [34]. Extensive analysis of individual *Drosophila* metallothioneins showed that MtnA and MtnB are of major importance in the defense against heavy metal exposure, whereby MtnA preferentially protects against copper load and MtnB primarily against cadmium load [35]. Strikingly, *Drosophila* metallothionein expression coincides with the sites of copper accumulation in the cytoplasm of so-called copper cells near the midgut constriction and also in the posterior midgut. Upon copper load, these copper cells, or cuprophilic cells, display an orange luminescence resulting from a copper-metallothionein complex. Copper cells are the sites of copper sequestration and they seem to provide copper to the growing organism. In line with such a scenario, copper disappears rapidly from copper cells if *Drosophila* larvae are transferred to copper-depleted food [34].

In vertebrates, there are four MT types, each represented by at least one member, designated MT-1 to MT-4 (or MT-I to MT-IV in older nomenclature) [13–15]. Human MT-1 is expanded into a large sub-family with at least seven members (MT-1A, MT-1B, MT-1E, MT-1F, MT-1G, MT-1H and MT-1X). Under physiological conditions about 5–20% total cellular zinc is complexed to MTs [15,36]. MTs are primarily localized to the cytoplasm, but under stress conditions such as UV irradiation they can translocate to the nucleus where they help to activate zinc finger-containing transcription factors by promoting zinc exchange [37]. Except for some impaired immune responses, *MT* knockout animals are viable and fertile [38,39]. MT-1 and MT-2 are expressed at all stages of development in most cell types and function as a reservoir of zinc. MT-2 has been shown to be a neuroprotective gene in cerebral ischemia [40,41]. MT-3 is constitutively expressed, predominantly in neurons but also in glia and male reproductive organs. Elimination of MT-3 in the mouse increases susceptibility to seizures [42]. MT-4 expression is restricted to squamous differentiated epithelia such as skin, tongue and esophagus [43]. Recently, several reports have indicated important roles of MTs in carcinogenic and apoptotic processes of some tumors [44,45].

3. REGULATION OF METALLOTHIONEIN EXPRESSION IS MOSTLY TRANSCRIPTIONAL

In organisms ranging from yeast to humans, metallothionein genes are expressed at a basal level but their transcription is strongly induced upon heavy metal load [46–48]. This induction is mediated by specific metal-responsive transcription factors that bind to metallothionein gene promoters and thereby boost their transcription. As an exception to this, transcription of the cyanobacterial *Synechococcus* metallothionein smtA is negatively regulated by a repressor protein called smtB [49].

In the budding yeast *S. cerevisiae*, Ace1 is the transcription factor responsible for MT induction. The amino-terminal half of Ace1 is rich in cysteines and basic amino acids and harbors a copper-dependent DNA binding domain, while the carboxyl-terminal half mediates contacts to the transcription apparatus [50]. Ace1 mediates transcriptional activation in response to copper load by binding to *cis*-regulatory DNA elements UAS_{Cu} of the core consensus sequence 5′-GCTG-3′ in the yeast metallothionein gene promoters [50,51]. Upon copper deprivation, another transcription factor called Mac1 is activated in budding yeast [52]. Interestingly, in the fission yeast *Schizosaccharomyces pombe*, the transcription factor Cuf1 is also activated by copper starvation, even though Cuf1 is more closely related to *S. cerevisiae* Ace1 than to Mac1 [53]. In the nematode *C. elegans*, cadmium, but not copper or zinc, was shown to be a potent inducer of mRNAs for intestinal metallothioneins (*mtl-1 and mtl-2*) which are regulated by hitherto unkown transcription factor(s) [54] (see also Section 4). In mammals, expression of *MT-1* and *MT-2* genes is also induced by other stress conditions including glucocorticoids, interleukin, interferon, and UV irradiation [13,14,48]. From insects to mammals, the major transcription regulator for handling heavy metal load and, at least in insects, also copper starvation, is MTF-1 (metal response element binding transcription factor 1, or metal-responsive transcription factor 1; see on page 37) [11,55,56].

4. METAL RESPONSE ELEMENTS IN THE UPSTREAM PROMOTER–ENHANCER REGION CONFER METAL INDUCIBILITY

The promoter regions of all MT genes in mammals and *Drosophila* contain multiple copies of metal response elements (MREs), short DNA sequence motifs of a core consensus sequence TGCRCNC or its complement GNGYGCA (where R = A or G, Y = C or T and N = any base). MREs are necessary and sufficient for the transcriptional regulation of MT genes [32,57]. Recent studies in a plant, the bean *Phaseolus vulgaris*, have shown that the promoter of a stress-related gene 2 (PvSR2) contains MRE-like sequences (TGCAGGC) which are required for the response to heavy metal load, even though plants do not contain an obvious homolog of MTF-1 [58]. The general consensus of MREs in vertebrates and MREs in *Drosophila* MTs and the copper importer *Ctr1B* [59,60] are depicted in Figure 1 and Table 1. Metallothionein genes of other species, such as the nematode *C. elegans* (*mtl-1 and mtl-2*) and the fungus *Neurospora crassa* (CuMT), lack MREs in their promoter-enhancer regions [54,61].

4.1. Metal-Induced Transcription is Mediated via Metal Response Elements

Metal response elements were identified by comparing the promoters of a number of metallothionien genes [32,57,58]. When multiple candidate MREs were fused to a minimal heterologous promoter, transcription of a reporter gene was strongly inducible upon heavy metal load [32,57,58,62]. These findings demonstrate that MREs are the *cis*-acting DNA sequences for metallothionein expression in response to heavy metal stress [32,57,58,62]. An

B)

	MRE1
D.melanogaster	TTTGCGCACGTC
D.yakuba	TTTGCGCACGTC
D.pseudobscura	TTTGCGCACGGC
D.virilis	TTTGCGCACGCC
consensus	- -TGCRCNCG C

Figure 1. Metal response elements. (A) Consensus DNA sequence of the MRE. (B) The MRE1-3 cluster upstream of the Ctr1B copper transcription unit is highly conserved among all *Drosophila* species. MRE1 conservation in four of the *Drosophila* species is presented here (see also [60]).

Table 1. Metal responsive elements in *Drosophila* metallothioneins.

MREs	Gene				T	G	C	R	C	N	C				Location
1	*MtnA*	t	t	t	T	G	C	A	C	A	C	g	c	c	–88
2		t	t	c	T	G	C	A	C	A	C	g	t	c	–125
1	*MtnB*	t	t	t	T	G	C	A	C	T	C	g	t	t	–82
2		g	c	t	T	G	C	A	C	A	C	g	a	c	–141
3		t	t	t	T	G	C	G	C	A	C	g	c	g	–202
4		c	g	g	T	G	C	A	C	A	C	g	a	a	–229
5		t	t	t	T	G	C	G	C	A	C	a	a	t	downstream
1	*MtnC*	t	t	t	T	G	C	A	C	A	C	c	a	t	–45
2		t	t	t	T	G	C	A	C	A	C	g	t	t	–76
3		t	g	t	T	G	C	A	C	C	C	a	a	c	–103
4		a	g	g	T	G	C	A	C	T	C	a	g	a	–191
5		t	t	a	T	G	C	G	C	A	C	t	t	g	–542
6		t	c	t	T	G	C	A	C	T	C	g	c	a	downstream
7		a	a	t	T	G	C	A	C	T	C	c	t	a	downstream
1	*MtnD*	c	t	t	T	G	C	G	C	A	C	c	a	a	–130
2		a	t	t	T	G	C	A	C	A	C	g	c	c	–165
3		t	t	t	T	G	C	G	C	A	C	g	c	c	–207
4		a	c	t	T	G	C	A	C	A	C	g	g	c	–225
5		a	a	t	T	G	C	A	C	A	C	c	c	a	downstream
6		t	a	t	T	G	C	A	C	A	C	a	c	a	downstream

exception to this was the finding that a transcription enhancer associated with the gene for copper importer Ctr1B harbors MREs, yet is activated by copper starvation [60]. In other words, both *Ctr1B* and *MT* genes in *Drosophila* are transcriptionally regulated by MTF-1 but under opposite conditions. However, the arrangement of MREs in *Ctr1B* was found to be critical for this unusual response to copper deprivation. A synthetic *Ctr1B*-like "mini-promoter" composed of the four MREs but closely spaced (without the intervening and flanking sequences), unexpectedly was no longer induced at low copper but highly responded to copper load, i.e., behaved simply like a classical MT promoter at high copper conditions [60] (see also Section 5.7). These findings suggest that the unique constellation of MREs in the enhancer/promoter region of *Ctr1B* gene contributes to its regulatory characteristics.

5. METAL RESPONSE ELEMENT BINDING TRANSCRIPTION FACTOR (MTF-1)

In the first three decades following the discovery of metallothioneins, little was known about the precise mechanisms of transcriptional induction of

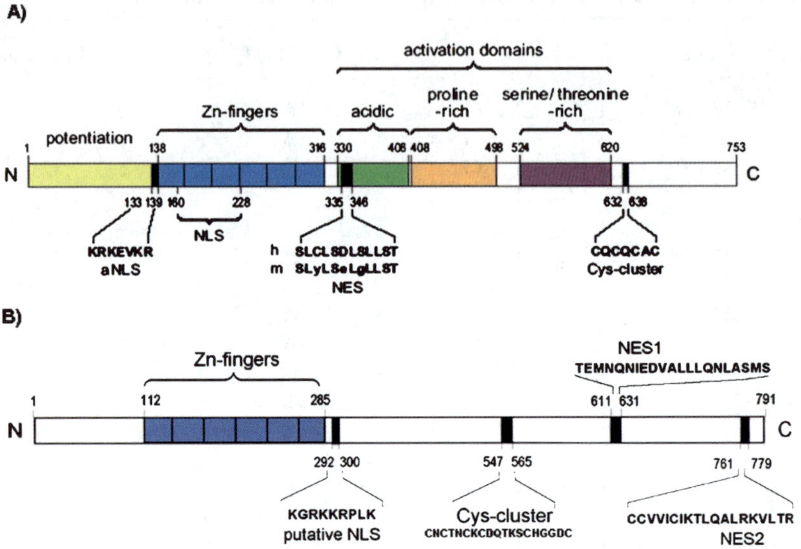

Figure 2. Schematic representation of (A) human and (B) *Drosophila* MTF-1. Note that the zinc finger region is highly conserved both in human and *Drosophila* MTF-1 and both contain a metal-sensing cysteine cluster near the C-terminus. The remainder of the protein sequences are highly divergent.

MTs under heavy metal load. In 1988, it was discovered that a specific nuclear protein binds to MREs derived from metallothionein promoters and that binding parallels transcription activity [63]. This protein preferentially binds to MREs at elevated zinc concentrations and was dubbed MRE binding transcription factor (MTF-1) [63]. Later studies in our and other labs showed that MTF-1 is the major transcription factor handling heavy metal excess [32,55,56,64]. In humans it is encoded by a gene on the short arm of chromosome 1 (1p33) and the protein contains 753 amino acids [65]. *Drosophila* has a homolog of mammalian MTF-1, termed dMTF-1, which is a 791 amino acid protein (Figure 2) [32].

Although dMTF-1 shares 39% amino acid identity with full length hMTF-1, it differs in two crucial aspects. Firstly, Cd(II) and Zn(II) are the most potent inducers of mammalian metallothioneins, while *Drosophila* metallothioneins are primarily induced by Cu(I) and Cd(II) and to a lesser extent by Zn(II) [32,62]. Secondly, knockout of *MTF-1* in mammals leads to embryonic lethality, while *dMTF-1* mutant flies are viable but are highly sensitive to heavy metal load [66,67].

5.1. MTF-1 Binds to DNA in a Zinc-Dependent Way

MTF-1, which is highly conserved from insects to humans, requires elevated zinc concentrations for optimal DNA binding. Six C_2H_2-type zinc fingers constitute the DNA-binding domain and contribute to zinc sensing [63,68]. It has been shown that zinc fingers 1–4 are most important for binding to the MRE core sequence [56,68,69]. More recently, zinc fingers 1, 5, and 6 were found to be required for efficient *in vivo* recruitment of MTF-1 to metallothionein promoters. Thereby, a stable MTF-1-chromatin complex is formed that serves as a rate-limiting step in metal-induced activation of gene expression by MTF-1 [70]. The same group also found a zinc sensing function associated with the linker peptide sequence (RGEYT) between zinc fingers 1 and 2 which affected DNA binding by MTF-1 [71,72]. This sequence motif is present in all vertebrate MTF-1 orthologs.

Both human and *Drosophila* MTF-1 are well-conserved in the DNA-binding zinc finger region but vary considerably outside of it [32,62]. In cell-free transcription experiments it was demonstrated that zinc activates human MTF-1 directly, whereas copper, cadmium, and hydrogen peroxide activate transcription indirectly by displacing zinc from zinc-saturated metallothioneins and probably also from other zinc-binding cellular proteins [73]. Nevertheless, this *in vitro* system does not reflect the complexity of the *in vivo* situation where MTF-1 is also subject to phosphorylation/dephosphorylation and nucleo-cytoplasmic shuttling [74,75].

5.2. MTF-1 Regulates Metallothionein Gene Induction

To ascertain the role of MTF-1 in regulation of its target genes, we measured the mRNA levels of *MT-1* and *MT-2* genes in wild-type and *MTF-1* knockout mice. Both basal and metal-induced *MT-1* and *MT-2* mRNA levels were essentially abrogated in mouse embryonic fibroblast (Dko7) cells devoid of functional *MTF-1*, or in the *MTF-1* knockout mice [66,76]. The same loss of metallothionein transcription was found in *Drosophila* lacking *MTF-1* [62,67]. Notably, transcription of *MT* genes was boosted by transient overexpression of MTF-1 in mammalian cells or in transgenic flies harboring an extra copy of dMTF-1 [62,67]. Taken together, these findings show that *MT* genes are prominent targets of MTF-1.

5.3. Domains of MTF-1 and Their Role in General Activity and Metal Induction

MTF-1 from all species analyzed so far contains six zinc fingers of the C_2H_2 type. Human MTF-1 also contains a "potentiation domain" close to its N-terminus. Deletion of this segment does not appreciably change the properties of MTF-1 except resulting in an overall reduction of activity [69,77]. The C-terminal half of MTF-1 contains three distinct transactivation domains, namely an acidic, a serine/threonine-rich and a proline-rich region. These three have different activation properties [69]. Besides these three regions, both human and *Drosophila* MTF-1 contain a cysteine cluster near their C-terminus. Recent studies have shown that these cysteines are metal sensing elements and are important for heavy metal-induced expression of metallothioneins [78].

5.4. Upon Stress, MTF-1 Translocates from Cytoplasm to Nucleus

In resting or non-stressed cells, MTF-1 predominantly localizes to the cytoplasm. Upon heavy metal load, it translocates to the nucleus, binds to MREs in target gene promoters and activates transcription [74]. MTF-1 contains an extended nuclear localization signal (NLS) sequence that overlaps with the zinc finger domain. Furthermore, just at the N-terminal end of zinc finger 1 of MTF-1 there is a cluster of basic residues of the sequence KRKEVKR. Mutation of this latter motif delays the nuclear import of MTF-1 but does not prevent it. Thus we refer to it as an auxiliary nuclear localization signal, or aNLS. This aNLS is also not sufficient to confer import when fused to a "neutral" indicator protein [74,79].

MTF-1 also contains a conserved nuclear export signal (NES) sequence (LCLSDLSLL) overlapping with the major, acidic activation domain [74]. This NES is essential for nucleo-cytoplasmic transport of MTF-1 and may contribute to metal inducibility. Mutations in the NES impair the relocation of MTF-1 to the cytoplasm. We also found that NES function was sensitive to the anti-fungal drug leptomycin B (LMB), implying that nuclear export is mediated via the export protein CRM1 [74]. Nuclear accumulation of MTF-1 is not *per se* sufficient for transcriptional activation upon metal load, suggesting that nuclear accumulation and transcriptional activation functions can be separated [74]. *Drosophila* MTF-1 also possesses both NLS and NES, but the nucleo-cytoplasmic shuttling is poorly understood [80].

5.5. Developmental Role of MTF-1 in the Mouse

To gain more insights into the *in vivo* function of MTF-1, our group has generated mice devoid of functional MTF-1 by targeted disruption of the *MTF-1* genomic locus [66]. In *MTF-1* mutant embryos at day E14.5, we observed severe liver degeneration and loss of liver-specific cytokeratin expression. There is extensive swelling and necrotic death of hepatocytes [66]. In the absence of other apparent developmental defects, liver degeneration is the major, if not the only, cause of embryonic death. Although the molecular mechanism underlying the lethal phenotype of *MTF-1* knockout mice remains to be determined, the observed reduction of transcripts of liver-enriched factors such as C/EBPalpha and alpha-fetoprotein might make a contribution [66].

We have also quantified the transcript levels of MTF-1 targets in total RNA isolated from these animals and found that both the basal and heavy metal-induced expression of all of MTF-1 targets, including MT-1, MT-2, and possibly γ-GCS, is reduced [66,81]. In agreement with these findings, we also observed that *MTF-1* knockout cells are more susceptible to cadmium and hydrogen peroxide compared to wild-type cells [81]. Moreover, the phenotype of *MTF-1* mutant animals shows similarity to the lethal knockout phenotype of c-Jun and p65/RelA, underlining an essential role of MTF-1 in embryogenesis [80,82,83].

5.6. MTF-1 in the Fly *Drosophila* Confers Resistance to Heavy Metal Load, Mostly via Metallothioneins

Unlike the situation in mammals, *Drosophila MTF-1* knockout flies were found to be viable but highly sensitive to copper, cadmium, and zinc load, and unexpectedly also to copper-depleted food [67]. Recently we have also demonstrated that a "family" knockout of all four metallothionein genes in flies leads to elevated sensitivity to copper or cadmium load [34]. Interestingly, ectopic overexpression of the copper importer Ctr1B in the *Drosophila* eye led to a rough-eye phenotype which could be rescued by co-overexpression of either MTF-1, MtnA, or MtnB [35]. These results indicate that the resistance of flies to excess copper is mainly conferred by metallothioneins.

5.7. Copper Starvation in *Drosophila*: MTF-1 is Required, but not Metallothioneins

As mentioned, flies lacking MTF-1 are not only sensitive to excess copper but also to copper starvation. These findings led to the identification of the major larval copper importer Ctr1B as an MTF-1 target gene. Altogether *Drosophila* has three copper importers termed Ctr1A, Ctr1B, and Ctr1C, which are encoded by separate genes, each with a distinct expression pattern [59]. Whereas Ctr1A is expressed constitutively at an intermediate level at all stages [59,84], Ctr1B is strongly expressed at late embryonic and early larval stages [59]. The expression pattern and function of Ctr1C remain to be elucidated. In the upstream enhancer region of the *Ctr1B* gene, there are three uniquely spaced, strong MREs which, somewhat counter-intuitively, mediate transcriptional induction upon copper starvation [59,60]. This MRE cluster is highly conserved among several *Drosophila* species [60]. Moreover, the low-copper phenotype in flies lacking Ctr1B is akin to that of *MTF-1* mutants, in that *Ctr1B* mutants are highly sensitive to

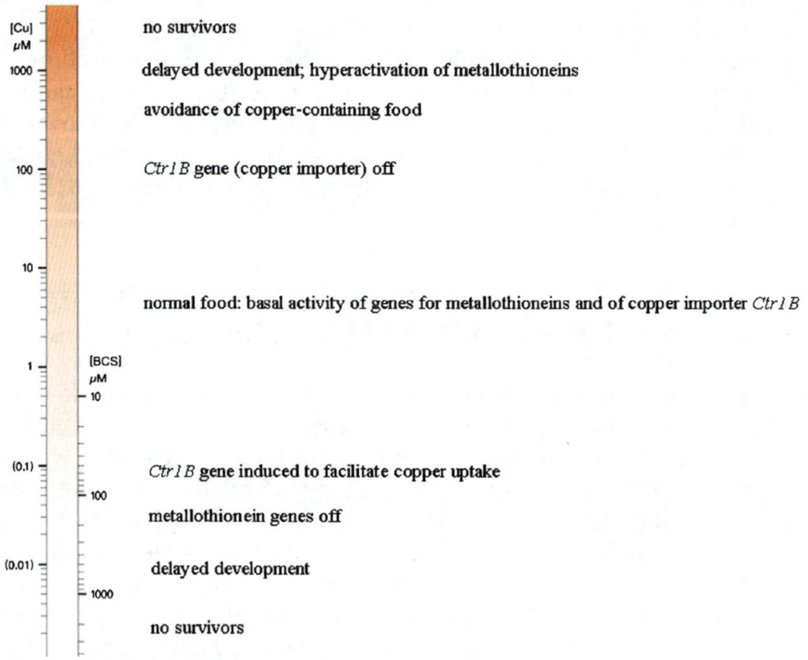

Figure 3. Scheme of cellular components, including metallothioneins, that maintain copper homeostasis. Normal food contains copper in the low-micromolar range (center). Under these conditions, the copper importer (Ctr1B) and the metal scavengers (metallothioneins) are expressed at basal level, and the copper transporter ATP7 exports copper into the trans-Golgi network. Increasing copper concentrations (upper half) result in the induction of metallothionein gene expression via the transcription factor MTF-1, activation of copper export via translocation of ATP7 to the cell membrane, and downregulation of the copper importer Ctr1B. At 0.5 mM copper and more, larvae avoid copper-containing food. *Drosophila* can develop in food containing up to maximally 3 mM copper salt, but already at lower concentrations, development is delayed in relation to copper concentration. Upon copper starvation (lower half), transcription of the Ctr1B copper importer is boosted via MTF-1, while metallothionein expression is abolished. Towards lower copper concentrations, development again becomes progressively delayed until no larvae are able to survive [11,86]. Note that the concentration of available copper in copper-depleted food (supplemented with a high-affinity chelator) is an estimate.

copper depletion [59,60]. The MREs in Ctr1B are bound by MTF-1 both at low and high copper, and due to their specific arrangement, possibly in conjunction with auxiliary factors, confer activity upon copper starvation (Figure 3).

Unlike the situation in metallothionein genes, where a few tandemly arranged MREs are sufficient to confer metal-inducibility, the MREs associated with *Ctr1B* are necessary but not sufficient to confer low-copper induction, as mentioned also in Section 4.1 [60]. The latter function requires additional flanking sequences of around 300 bp from its enhancer-promoter region [85]. Of note, the two copper importers Ctr1 and Ctr2 of mammals are not regulated by MTF-1. Also by sequence anlysis we did not find any MTF-1 consensus binding sites in the promoters of these mammalian copper importers. Unlike *Drosophila* Ctr1B, yeast and human Ctrs are regulated posttranscriptionally in that excess copper stimulates rapid protein degradation and/or endocytosis [59,60,86–88].

6. TRANSCRIPTION FACTORS IN OTHER SPECIES IMPLICATED IN HEAVY METAL HANDLING

In lower eukaryotes, other transcription factors mediate metal-responsive gene expression. For example, in the budding yeast *S. cerevisiae*, the transcription factor Ace1 activates the metallothionein gene Cup1, resulting in protection of cells against copper toxicity [50,51]. Ace1 also induces the expression of Cu/Zn-SOD and another metallothionein called CRS5 [89,90]. Conversely, at limiting copper conditions Mac1, another transcription factor of yeast, activates copper importers Ctr1 and Ctr3 [91,92]. In the fission yeast *Schizosaccharomyces pombe*, the transcription factor Cuf1 regulates copper efflux from the vacuole to the cytoplasm [93]. The promoters of Cuf1 targets, including the copper transporter Ctr6, contain one or more copies of a Cu-signaling element (CuSE) of the consensus sequence 5′-D(T/A)DDHGCTGD-3′ (D = A, G, or T; H = A, C, or T) to which Cuf1 binds under conditions of copper starvation to induce transcription [93,94]. Interestingly, the CuSE motif resembles target sequences of the Ace1 transcription factor of *S. cerevisiae*. Accordingly, a Cuf1 transgene can bring about high expression levels of metallothionein (*CUP1*) transcripts in *S. cerevisiae ace1*$^{-/-}$ cells under conditions of copper deprivation. In this way *S. pombe* Cuf1 plays a dual role: at high copper it induces the expression of copper detoxification genes, on the other hand it handles copper transporter gene expression under conditions of copper scarcity [93,94].

The yeast *Saccharomyces cerevisiae* contains two transcription activators designated Aft1 and Aft2 that activate gene expression in response to iron deprivation [95,96]. Additionally, Cuf1 in *S. pombe* also represses the expression of iron-regulated genes such as *fip1*, *fio1* and *frp1* which encode proteins that are required for iron uptake [97]. In the filamentous fungus *Podospora anserina*, a copper-regulatory transcription factor called GRISEA activates expression of the gene for mitochondrial *SOD2*, which protects the fungus against oxidative stress [98]. To date, two zinc finger transcription factors that show responsiveness to zinc have been well-characterized. These are the aforementioned MTF-1 in vertebrates and insects, and Zap1 in the yeast *S. cerevisiae* [99–101]. Zinc homeostasis in yeast is primarily controlled by Zap1: at limiting zinc conditions it increases the expression of *ZRT1*, *ZRT2*, and *FET4* genes [102–104]. In *Drosophila* the transporter ZnT35C, a target of MTF-1, is involved in cellular and organismal zinc efflux and plays a major role in zinc homeostasis [101]. In mammals, MTF-1 also counteracts zinc toxicity by inducing transcription of *ZnT1*, a zinc exporter gene [100]. Although MTF-1 in higher eukaryotes and the metalloregulatory factors of yeasts are well-characterized, the transcription factor(s) responding to heavy metals in the nematode *C. elegans*, in plants and in the fungus *Neurospora crassa*, remain to be identified.

7. CONCLUDING REMARKS AND OPEN QUESTIONS

Every organism has to cope with environmental fluctuations of essential and non-essential metals. Metallothioneins, short cysteine-rich proteins with the ability to avidly bind a variety of heavy metals, typically act as a first line of defense against heavy metal toxicity. In resting, non-stressed cells, metallothioneins are present at a basal level. Heavy metal load results in a strong transcriptional activation of metallothionein genes via metal-regulatory transcription factors, such as MTF-1 from mammals to insects, and Ace1 in yeast. While yeast contains a separate factor (Mac1) to cope with copper starvation, MTF-1 of *Drosophila* is able to regulate the appropriate target genes upon heavy metal load as well as under copper scarcity.

One major open question concerns the metal specificity of MTF-1 action: metallothionein genes are induced by several metals, though with some metal preference, while the zinc exporters Znt1 and ZnT35C in mouse and *Drosophila*, respectively, are only activated in response to zinc load. It will be most challenging to identify the molecular basis of this metal specificity. Besides, the apparently independent, convergent evolution of metal-responsive transcription factors in higher versus lower eukaryotes is an enigma in the field of gene and protein evolution.

ACKNOWLEDGMENTS

We are grateful to Alisa Davis for valuable discussions and for critical reading of the manuscript. This work was supported by the Kanton Zürich and by the Swiss National Science Foundation.

ABBREVIATIONS

aNLS	auxiliary localization signal
CuSE	copper-signaling element
LMB	leptomycin B
MREs	metal response elements
MTF-1	metal-responsive transcription factor-1
MTs	metallothioneins
NES	nuclear export signal
NLS	nuclear localization signal
PCs	phytochelatins
ROS	reactive oxygen species

REFERENCES

1. N. Ercal, H. Gurer-Orhanmand and N. Aykin-Burns, *Curr. Top. Med. Chem.*, 2001, **1**, 529–539.
2. M. Valko, H. Morris and M. T. Cronin, *Curr. Med. Chem.*, 2005, **12**, 1161–1208.
3. H. Bartsch and J. Nair, *Cancer Detect. Prev.*, 2004, **28**, 385–391.
4. E. D. Harris, *Annu. Rev. Nutr.*, 2000, **20**, 291–310.
5. K. M. Hambidge and N. F. Krebs, *J. Nutr.*, 2007, **137**, 1101–1105.
6. C. Perafan-Riveros, L. F. Franca, A. C. Alves and J. A. Sanches, Jr., *Pediatr. Dermatol.*, 2002, **19**, 426–431.
7. S. Lutsenko and M. J. Petris, *J. Membr. Biol.*, 2003, **191**, 1–12.
8. E. D. Harris, *Crit. Rev. Clin. Lab. Sci.*, 2003, **40**, 547–586.
9. J. C. Rutherford and A. J. Bird, *Eukaryot. Cell*, 2004, **3**, 1–13.
10. C. M. Moore and J. D. Helmann, *Curr. Opin. Microbiol.*, 2005, **8**, 188–195.
11. K. Balamurugan and W. Schaffner, *Biochim. Biophys. Acta*, 2006, **1763**, 737–746.
12. I. Bremner, *Prog. Food Nutr. Sci.*, 1987, **11**, 1–37.
13. J. H. R. Kägi, *Methods Enzymol.*, 1991, **205**, 613–626.
14. R. D. Palmiter, *Proc. Natl. Acad. Sci. USA*, 1998, **95**, 8428–8430.
15. P. Coyle, J. C. Philcox, L. C. Carey and A. M. Rofe, *Cell Mol. Life Sci.*, 2002, **59**, 627–647.
16. R. Nath, R. Kambadur, S. Gulati, V. K. Paliwal and M. Sharma, *CRC Crit. Rev. Food Sci. Nutr.*, 1988, **27**, 41–85.

17. G. K. Andrews, *Biometals*, 2001, **14**, 223–237.
18. M. Margoshes and B. L. Vallee, *J. Am. Chem. Soc.*, 1957, **79**, 4813–4814.
19. K. Munger, U. A. Germann and K. Lerch, *Experientia Suppl.*, 1987, **52**, 393–400.
20. G. Maroni, A. S. Ho and L. Theodore, *Environ. Health Perspect.*, 1995, **103**, 1116–1118.
21. N. J. Robinson, S. K. Whitehall and J. S. Cavet, *Adv. Microb. Physiol.*, 2001, **44**, 183–213.
22. B. Zhang, D. Egli, O. Georgiev and W. Schaffner, *Mol. Cell Biol.*, 2001, **21**, 4505–4514.
23. E. A. Peroza and E. Freisinger, *J. Biol. Inorg. Chem.*, 2007, **12**, 377–391.
24. S. Clemens, *Planta*, 2001, **212**, 475–486.
25. S. Lee, J. S. Moon, T. S. Ko, D. Petros, P. Goldsbrough and S. S. Korban, *Plant Physiol.*, 2003, **131**, 656–663.
26. C. Cobbett and P. Goldsbrough, *Annu. Rev. Plant Biol.*, 2002, **53**, 159–182.
27. M. M. Pena, K. A. Koch and D. J. Thiele, *Mol. Cell Biol.*, 1998, **18**, 2514–2523.
28. L. T. Jensen, W. R. Howard, J. J. Strain, D. R. Winge and V. C. Culotta, *J. Biol. Chem.*, 1996, **271**, 18514–18519.
29. A. Pagani, L. Villarreal, M. Capdevila and S. Atrian, *Mol. Microbiol.*, 2007, **63**, 256–269.
30. G. Maroni, J. Wise, J. E. Young and E. Otto, *Genetics*, 1987, **117**, 739–744.
31. D. Lastowski-Perry, E. Otto and G. Maroni, *J. Biol. Chem.*, 1985, **260**, 1527–1530.
32. B. Zhang, D. Egli, O. Georgiev and W. Schaffner, *Mol. Cell Biol.*, 2001, **21**, 4505–4514.
33. G. Maroni, E. Otto and D. Lastowski-Perry, *Genetics*, 1986, **112**, 493–504.
34. D. Egli, H. Yepiskoposyan, A. Selvaraj, K. Balamurugan, R. Rajaram, A. Simons, G. Multhaup, S. Mettler, A. Vardanyan, O. Georgiev and W. Schaffner, *Mol. Cell Biol.*, 2006, **26**, 2286–2296.
35. D. Egli, J. Domenech, A. Selvaraj, K. Balamurugan, H. Hua, M. Capdevila, O. Georgiev, W. Schaffner and S. Atrian, *Genes Cells*, 2006, **11**, 647–658.
36. P. Coyle, J. C. Philcox and A. M. Rofe, *Biochem. J.*, 1995, **309**, 25–31.
37. Y. Takahashi, Y. Ogra and K. T. Suzuki, *J. Cell. Physiol.*, 2005, **263**, 563–569.
38. B. A. Masters, E. J. Kelly, C. J. Quaife, R. L. Brinster and R. D. Palmiter, *Proc. Natl. Acad. Sci. USA*, 1994, **91**, 584–588.
39. J. Liu, Y. Liu, S. S. Habeebu and C. D. Klaassen, *Toxicol. Appl. Pharmacol.*, 1999, **159**, 98–108.
40. D. M. Durnam and R. D. Palmiter, *Experientia Suppl.*, 1987, **52**, 457–463.
41. M. Penkowa, M. Giralt, J. Camats and J. Hidalgo, *J. Comp. Neurol.*, 2002, **444**, 174–189.
42. M. Aschner, M. G. Cherian, C. D. Klaassen, R. D. Palmiter, J. C. Erickson and A. I. Bush, *Toxicol. Appl. Pharmacol.*, 1997, **142**, 229–242.
43. C. J. Quaife, S. D. Findley, J. C. Erickson, G. J. Froelick, E. J. Kelly, B. P. Zambrowicz and R. D. Palmiter, *Biochemistry*, 1994, **33**, 7250–7259.
44. F. Doz, N. Roosen and M. L. Rosenblum, *J. Neurooncol.*, 1993, **17**, 123–129.

45. S. E. Theocharis, A. P. Margeli and A. Koutselinis, *Int. J. Biol. Markers*, 2003, **18**, 162–169.
46. D. J. Ecker, T. R. Butt, E. J. Sternberg, M. P. Neeper, C. Debouck, J. A. Gorman and S. T. Crooke, *J. Biol. Chem.*, 1986, **261**, 16895–16890.
47. F. Bonneton, L. Theodore, P. Silar, G. Maroni and M. Wegnez, *FEBS Lett.*, 1996, **380**, 33–38.
48. G. K. Andrews, *Biochem. Pharmacol.*, 2000, **59**, 95–104.
49. A. P. Morby, J. S. Turner, J. W. Huchle and N. J. Robinson, *Nucleic Acids Res.*, 1993, **21**, 921–925.
50. D. J. Thiele, *Mol. Cell. Biol.*, 1988, **8**, 2745–2752.
51. D. J. Thiele and D. H. Hamer, *Mol. Cell. Biol.*, 1986, **6**, 1158–1163.
52. G. Keller, A. Bird and D. R. Winge, *Eukaryot. Cell*, 2005, **4**, 1863–1871.
53. J. Beaudoin and S. Labbe, *J. Biol. Chem.*, 2001, **276**, 15472–15480.
54. S. C. Swain, K. Keusekotten, R. Baumeister and S. R. Stürzenbaum, *J. Mol. Biol.*, 2004, **341**, 951–959.
55. P. Lichtlen and W. Schaffner, *BioEssays*, 2001, **23**, 1010–1017.
56. D. P. Giedroc, X. Chen and J. L. Apuy, *Antioxidants Redox Signal.*, 2001, **3**, 577–596.
57. P. F. Searle, G. W. Stuart and R. D. Palmiter, *Mol. Cell. Biol.*, 1985, **52**, 407–414.
58. X. Qi, Y. Zhang and T. Chai, *Plant Physiol.*, 2007, **143**, 50–59.
59. H. Zhou, K. M. Cadigan and D. J. Thiele, *J. Biol. Chem.*, 2003, **278**, 48210–48218.
60. A. Selvaraj, K. Balamurugan, H. Yepiskoposyan, H. Zhou, D. Egli, O. Georgiev, D. J. Thiele and W. Schaffner, *Genes & Dev.*, 2005, **19**, 891–896.
61. K. Munger, U. A. Germann and K. Lerch, *J. Biol. Chem.*, 1987, **262**, 7363–7367.
62. K. Balamurugan, D. Egli, A. Selvaraj, B. Zhang, O. Georgiev and W. Schaffner, *Biol. Chem.*, 2004, **385**, 597–603.
63. G. Westin and W. Schaffner, *EMBO J.*, 1988, **7**, 3763–3770.
64. A. Auf der Maur, T. Belser, G. Elgar, O. Georgiev and W. Schaffner, *Biol. Chem.*, 1999, **380**, 175–185.
65. E. Brugnera, O. Georgiev, F. Radtke, R. Heuchel, E. Baker, G. R. Sutherland and W. Schaffner, *Nucleic Acids Res.*, 1994, **22**, 3167–3173.
66. C. Günes, R. Heuchel, O. Georgiev, K. H. Müller, P. Lichtlen, H. Blüthmann, S. Marino, A. Aguzzi and W. Schaffner, *EMBO J.*, 1998, **17**, 2846–2854.
67. D. Egli, A. Selvaraj, H. Yepiskoposyan, B. Zhang, E. Hafen, O. Georgiev and W. Schaffner, *EMBO J.*, 2003, **22**, 100–108.
68. D. C. Bittel, I. V. Smirnova and G. K. Andrews, *J. Biol. Chem.*, 2000, **275**, 37194–77201.
69. F. Radtke, O. Georgiev, H. P. Müller, E. Brugnera and W. Schaffner, *Nucleic Acids Res.*, 1995, **23**, 2277–2286.
70. H. Jiang, P. J. Daniels and G. K. Andrews, *J. Biol. Chem.*, 2003, **278**, 30394–30402.
71. Y. Li, T. Kimura, J. H. Laity and G. K. Andrews, *Mol. Cell. Biol.*, 2006, **26**, 5580–5587.

72. J. H. Laity and G. K. Andrews, *Arch. Biochem. Biophys.*, 2006, **463**, 201–210.
73. B. Zhang, O. Georgiev, M. Hagmann, Ç. Günes, M. Cramer, P. Faller, M. Vašák and W. Schaffner, *Mol. Cell. Biol.*, 2003, **23**, 8471–8485.
74. N. Saydam, O. Georgiev, M. Y. Nakano, U. F. Greber and W. Schaffner, *J. Biol. Chem.*, 2001, **276**, 25487–25495.
75. N. Saydam, T. K. Adams, F. Steiner, W. Schaffner and J. H. Freedman, *J. Biol. Chem.*, 2002, **277**, 20438–20445.
76. P. Lichtlen, Y. Wang, T. Belser, O. Georgiev, U. Certa, R. Sack and W. Schaffner, *Nucleic Acids Res.*, 2001, **29**, 1514–1523.
77. H. P. Müller, E. Brugnera, O. Georgiev, M. Badzong, K. H. Müller and W. Schaffner, *Somat. Cell Mol. Genet.*, 1995, **21**, 289–297.
78. X. Chen, B. Zhang, P. M. Harmon, W. Schaffner, D. O. Peterson and D. P. Giedroc, *J. Biol. Chem.*, 2004, **279**, 4515–4522.
79. U. Lindert, M. Cramer, O. Georgiev and W. Schaffner, unpublished data.
80. A. Vardanyan, V. Günther, K. Balamurugan, O. Georgiev and W. Schaffner, unpublished data.
81. U. Wimmer, Y. Wang, O. Georgiev and W. Schaffner, *Nucleic Acids Res.*, 2005, **33**, 5715–5727.
82. F. Hilberg, A. Aguzzi, N. Howells and E. F. Wagner, *Nature*, 1993, **365**, 179–181.
83. A. A. Beg, W. C. Sha, R. T. Bronson and D. Baltimore, *Genes Dev.*, 1995, **9**, 2736–2746.
84. M. L. Turski and D. J. Thiele, *J. Biol. Chem.*, 2007, **282**, 24017–24026.
85. K. Balamurugan and W. Schaffner, unpublished data.
86. K. Balamurugan, D. Egli, H. Hua, R. Rajaram, G. Seisenbacher, O. Georgiev and W. Schaffner, *EMBO J.*, 2007, **26**, 1035–1044.
87. M. J. Petris, K. Smith, J. Lee and D. J. Thiele, *J. Biol. Chem.*, 2003, **278**, 9639–9646.
88. C. E. Ooi, E. Rabinovich, A. Dancis, J. S. Bonifacino and R. D. Klaisner, *EMBO J.*, 1996, **15**, 3515–3523.
89. E. B. Gralla, D. J. Thiele, P. Silar and J. S. Valentine, *Proc. Natl. Acad. Sci. USA*, 1991, **88**, 8558–8562.
90. V. C. Culotta, W. R. Howard and X. F. Liu, *J. Biol. Chem.*, 1994, **269**, 25295–25302.
91. S. Labbe, H. Zhu and D. J. Thiele, *J. Biol. Chem.*, 1997, **272**, 15951–15958.
92. Y. Yamaguchi-Iwai, M. Serpe, D. Haile, W. Yang, D. J. Kosman, R. D. Klausner and A. Dancis, *J. Biol. Chem.*, 1997, **272**, 17711–17718.
93. D. R. Bellemare, L. Shaner, K. A. Morano, J. Beaudoin, R. Langlois and S. Labbe, *J. Biol. Chem.*, 2002, **277**, 46676–46686.
94. J. Beaudoin and S. Labbe, *J. Biol. Chem.*, 2001, **276**, 15472–15480.
95. C. Casas, M. Aldea, C. Espinet, C. Gallego, R. Gil and E. Herrero, *Yeast*, 1997, **13**, 621–637.
96. P. L. Blaiseau, E. Lesuisse and J. M. Camadro, *J. Biol. Chem.*, 2001, **276**, 34221–34226.
97. S. Labbe, M. M. Pena, A. R. Fernandes and D. J. Thiele, *J. Biol. Chem.*, 1999, **274**, 36252–36260.

98. C. Borgouts, C. Q. Scheckhuber, O. Stephan and H. D. Osiewacz, *Int. J. Biochem. Cell Biol.*, 2002, **34**, 1355–1371.
99. H. Zhao and D. J. Eide, *Mol. Cell. Biol.*, 1997, **17**, 5044–5052.
100. S. J. Langmade, R. Ravindra, P. J. Daniels and G. K. Andrews, *J. Biol. Chem.*, 2000, **275**, 34803–34809.
101. H. Yepiskoposyan, D. Egli, T. Fergestad, A. Selvaraj, C. Treiber, G. Multhaup, O. Georgiev and W. Schaffner, *Nucleic Acids Res.*, 2006, **34**, 4866–4877.
102. H. Zhao and D. J. Eide, *J. Biol. Chem.*, 1996, **271**, 23203–23210.
103. B. M. Waters and D. Eide, *J. Biol. Chem.*, 1996, **277**, 33749–33757.
104. H. Zhao and D. Eide, *Proc. Natl. Acad. Sci. USA*, 1996, **93**, 2454–2458.

Met. Ions Life Sci. **2009**, *5*, 51–81

3

Bacterial Metallothioneins

Claudia A. Blindauer

Department of Chemistry, University of Warwick, Coventry CV4 7AL, UK
<c.blindauer@warwick.ac.uk>

Metal Ions in Life Sciences, Volume 5 Edited by Astrid Sigel, Helmut Sigel, and Roland K. O. Sigel

Published by the Royal Society of Chemistry, www.rsc.org DOI: 10.1039/9781847558992-00051

ABSTRACT: The bacterial metallothioneins (MT), BmtAs, constitute a well defined sub-family, sharing a number of characteristics with eukaryotic metallothioneins. They are small, cysteine-rich proteins, are inducible by metal ion stress, and contain metal-thiolate clusters with topologies similar to those of mammalian MTs. They also have a number of features that set them apart from most previously characterized metallothioneins. Most importantly, they adopt a zinc finger-like protein fold, and they contain conserved aromatic residues including histidines. The latter have been shown to modulate thermodynamics and kinetics of metal binding in various ways. The two metal-coordinating His residues in *Synechococcus* PCC 7942 SmtA enhance the stability of the zinc-thiolate cluster, compared to an all-Cys cluster generated by site-directed mutagenesis. His40 confers structural order to the entire C-terminus, whilst His49 has been shown to play a pivotal role in regulating zinc transfer reactions. The sequences of SmtA orthologues from other species show varations in the loop containing this residue, and preliminary studies have shown that this crucially affects metal binding properties of the proteins.

KEYWORDS: cadmium · cysteine · histidine · metal-thiolate cluster · zinc

1. INTRODUCTION

1.1. Bacterial Metal Homeostasis

As single-cell organisms, bacteria are constantly exposed to their environment, and have therefore developed efficient mechanisms to deal with a variety of toxic substances including metal ions such as Ag, Cd, Co, Cu, Hg, Ni, Pb, and Zn [1,2]. However, some of these metal ions are also essential, e.g., Zn, Cu, Ni, and Co, each being a constituent of a variety of enzymes and other proteins. Therefore, to ensure the survival, correct functioning, and reproduction of bacterial cells, the fluxes of essential and toxic metal ions need to be controlled exquisitely. The most intriguing mystery, not only in bacterial metal homeostasis, is the question how metal ion specificity in metal-binding proteins is achieved.

Knowledge of how proteins discriminate between different metal ions is crucial to understand (a) how cells ensure that they have the correct "quota" of each essential metal ion whilst excluding the adverse effects of closely similar toxic ions, and (b) how the incorporation of the correct metal ion into a newly synthesized protein is ensured. Addressing question (a), there has been much progress in the past two decades. In bacteria, metal concentrations are kept constant by the regulation of uptake and efflux, or, more rarely, intracellular sequestration. The majority of metal-handling genes are inducible, i.e., transcription and expression of the gene is regulated by the action of metal-dependent transcription factors – also termed metal sensors [3]. Studies of metal ion homeostasis mechanisms in bacteria have often been facilitated by the arrangement of metal-handling genes and the associated sensors in operons (see below). Question (b) has been moving into focus more recently; whilst, for example in the case of copper and zinc, the initial notion had been that new proteins simply pick their preferred metal ion from a cellular pool of labile ions, it is now clear that this is not the case, and that the concentrations of "free" Cu and Zn in *E. coli* are zepto- [4] and femtomolar [5], respectively – this corresponds to less than one ion per cell.

Breakthroughs in the understanding of intracellular metal trafficking were achieved by the identification of metallochaperones for Cu, Ni, and Co, and of chelatases for Fe, Co, and Mg, reviewed elsewhere [e.g., 6,7]. However, despite the fact that Zn is the second most abundant "trace" metal ion in both eukaryotic and prokaryotic cells, and displays possibly the widest variety in terms of biological functions, with hundreds of zinc metalloproteins occurring in all six enzyme classes [8], and despite significant advances in the understanding of prokaryotic zinc homeostasis [9] including the identification of uptake systems [10,11], efflux pumps [12], and sensors [3,13], zinc metalloprotein assembly is far from being understood in either bacteria or eukaryotes. The existence of intracellular "zinc chaperones" has frequently been evoked to fill this gap – with the metallothioneins (MT) as favorite candidates for this role.

1.2. The Discovery of Bacterial Metallothioneins

The ideas that evolution was chemically constrained [14] and that the organisms that produced the first oxygen, namely the cyanobacteria, were also the first that had to deal with the altered metal concentrations in the new aerobic environment, have led to the suggestion that cyanobacterial metal requirements (and presumably homeostatic mechanisms) and biogeochemical cycles at the archean-proterozoic boundary co-evolved [15]. Therefore, cyanobacteria "command an evolutionary transition in metal availability, toxicity, and use" [11], and it seems appropriate that the first bacterial metallothionein was also discovered in a cyanobacterium.

The earliest indication for the presence of metallothioneins in a prokaryote came from studies by Olafson et al., who isolated a protein with high cysteine (15%) and metal ion content (Cd, Zn, and Cu, 2.2 mol equiv. in total) from the marine cyanobacterium *Synechococcus* RRIMP N1 [16]. The cells had been stepwise adapted to growth in 50 μM $CdCl_2$, and sufficient material for a preliminary characterization was isolated from a 20-L culture. Expression of the novel MT was only detectable in the presence of excess metal, suggesting that gene expression was inducible. This hypothesis was confirmed in a subsequent study [17], which established that the bacterial MT was produced in response to zinc and cadmium, but not to copper, and in which regulation at the transcriptional level was suggested. Considering the minute amounts of available material, sequencing was not possible, but nearly a decade later, an almost full protein sequence was determined for the metallothionein isolated from the freshwater *Synechococcus* strain TX-20 [18]. This strain is thought to be identical to strain *Synechococcus* PCC 7942, and is also known as *Anacystis nidulans* R2 [19]. Like the material isolated from the marine strain, the freshwater MT was associated with Cd, Zn, and Cu, even though only Cd^{2+} had been added. Spectroscopic studies suggested the presence of metal-thiolate clusters, possibly in one single domain. Sequence comparisons with mammalian, crustacean, echinoderm, fungal, and plant MT sequences indicated that there was no evolutionary relationship between prokaryotic and any known eukaryotic MTs, making convergent evolution a likely scenario.

In the same decade, Sadler and colleagues discovered "pseudo-thioneins" in *Pseudomonas putida*, ubiquitous soil bacteria, that had been isolated from sewage sludge and had been adapted to growth in 3 mM Cd^{2+} [20]. Three proteins with molecular masses of 7, 7, and 3.5 kDa, cysteine contents between 12 and 23%, and different ratios of Cd, Cu, and Zn were isolated. In remarkable contrast to known eukaryotic MTs, both pseudo-thioneins and cyanobacterial MTs were shown to contain aromatic residues including histidines. ^{113}Cd NMR spectra of CdPT1, a 7 kDa pseudo-thionein expressed during the exponential phase of cell growth, displayed four peaks, with two of them (483 and 476 ppm) clearly outside the shift range expected for $Cd(Cys)_4$ [20]. Yet again, the small amounts that could be isolated precluded the determination of protein sequences.

This situation changed a few years later with recombinant technology becoming more accessible. The *smtA* gene of *Synechococcus* PCC 7942 was isolated by Robinson et al. in 1990 [21],[1] cloned into the vector pGEX3X, and expressed in *E. coli* as a glutathione-S-transferase fusion protein in the presence of either Cd^{2+}, Cu^{2+}, Hg^{2+}, or Zn^{2+}. Analysis of both fusion protein and cleaved SmtA demonstrated the binding of all metals tested [22]. In parallel, it was found that in *Synechococcus* PCC 7942, the gene *smtA* is arranged in an operon with a divergently transcribed gene which was named

[1]Note that the sequences derived from protein and DNA differ by the presence of Cys [18] instead of Ser [21] at position 33.

smtB [23,24]. The protein SmtB is the metal-responsive transcriptional regulator of SmtA expression (see Section 1.4), and became a founding member of a whole family of metal sensors [3].

In 1992, a gene encoding a protein with close sequence similarity to the PCC 7942 sequence was discovered in the thermophilic strain *Synechococcus vulcanus* (now *Thermosynechococcus vulcanus*) in the course of exploring the gene neighborhood of the *psaC* gene (a photosystem-I subunit), and was named *mtnA* [25]. Subsequently, NMR and mutagenesis studies for SmtA from *Synechococcus* PCC 7942 established the participation if His residues in zinc and cadmium binding [26], and motivated more detailed studies including the determination of the 3D structure of SmtA in 2001 [27]. SmtA orthologues from *Anabaena* PCC 7120, *Pseudomonas aeruginosa* and *putida* were identified, cloned, expressed and characterized in 2002 [28], and the term "BmtA" was suggested for the designation of bacterial proteins with features similar to SmtA. Furthermore, the presence of metallothionein in *Synechococcus* PCC 7942 was used by several authors as a model system to develop analytical methodology with possible applications in metalloproteomics, e.g., HPLC [29], capillary electrophoresis [30], and gel electrophoresis coupled with laser-ablation [31], all hyphenated to inductively coupled plasma-mass spectrometry (ICP-MS).

1.3. Functional Studies in Freshwater *Synechococcus*

Discussions of functions for metallothioneins are often fraught with words such as elusive, enigmatic, or multipurpose. To some extent, this vagueness is due to the presence of multiple isoforms in higher eukaryotes and the impossibility of generating meaningful knockout mutants that would allow the determination of a phenotype. In contrast, the majority of bacterial genomes (see Section 4.1) contain a single copy for a metallothionein. Moreover, genetic manipulation of *Synechococcus* PCC 7942 is possible, and hence a number of functional studies are available for SmtA from this strain. Several excellent reviews with an emphasis on *in vivo* studies and molecular genetics have been compiled [11,19,32], therefore the discussion of functional studies will be kept short in the present review.

SmtA transcripts are elevated when cells are challenged with Zn^{2+}, Cd^{2+}, Hg^{2+}, Cu^{2+}, Co^{2+}, Cr^{3+}, and Ni^{2+} [24], and binding of Zn^{2+}, Cd^{2+}, Hg^{2+}, and copper was demonstrated *in vivo* in *E. coli* [22]. Furthermore, as explained above, bacterial metallothioneins were discovered by adapting bacteria to growth in elevated levels of Cd^{2+} [16]. Later studies revealed that this adaptation involved disabling the gene for the sensor SmtB (see below), and increasing the copy number of the *smtA* gene [23,33]. This leads to the presence of *smtA* RNA, even in the absence of Zn^{2+} or Cd^{2+}, and increased tolerance to as well as intracellular accumulation of these metal ions. However, even though a range of metal ions elicit SmtA production, and amplification of the *smtA* gene gives cells an advantage during cadmium stress, studies using *lacZ* as a reporter gene revealed that $Zn^{2+}$is the strongest inducer of expression, followed by Cd^{2+} and Cu^{2+} [24]. In addition, deletion or disruption of the *smtA* gene leads to a 5-fold decrease in zinc tolerance, but a less-pronounced loss in tolerance against Cd^{2+}. No reduction in tolerance against Ag^{+}, Cu^{2+}, or Hg^{2+} was detected [34]. These results are consistent with a role for SmtA in dealing with elevated levels of Zn^{2+} in *Synechococcus* PCC 7942. Caution should however be applied to transfer conclusions made for this particular strain to other bacterial species, as neither metal ion specificity, nor the mechanism of regulation are necessarily identical for MTs from other (cyano)bacteria, as explained below.

1.4. Regulation of BmtA Expression: The smt Operon and Other Systems

Zinc-dependent expression of *smtA* in *Synechococcus* PCC 7942 is regulated on the transcriptional level by the action of the zinc sensor SmtB, which, in the absence of zinc, binds to the *smtA* operator-promoter region and represses transcription of *smtA* [35] as well as of itself. The operator-promoter region of the smt operon (Figure 1) consists of 100 basepairs and contains two 12-2-12 imperfect inverted repeats with the core sequence 5′-TGAAxx-xx-xxTTCA, each of which appears to be capable of binding (at least) two apo-SmtB homodimers. Only one inverted repeat, comprising SmtB binding sites S1 and S2, appears to be necessary for regulation of *smtA* transcription (depicted in Figure 1), with the other one, comprising sites S3

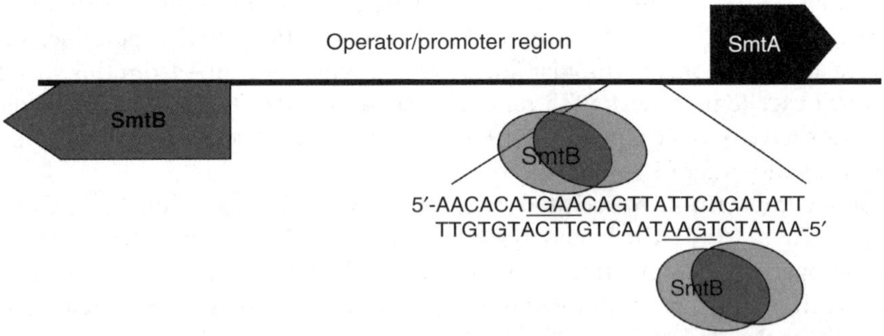

Figure 1. Simplified scheme of gene organisation and zinc-dependent regulation of *smtA* expression by the zinc sensor SmtB in *Synechococcus* PCC 7942, showing two apo-SmtB homodimers bound to one of the two 12-2-12 inverted repeats.

and S4 (not shown), suggested to play a role in *smtB* autoregulation [36]. Although SmtB homodimers are thought to contain in total four zinc binding sites, only two mol equiv. of Zn^{2+} per dimer seem to be necessary and sufficient to decrease the DNA affinity of SmtB by about 3 orders of magnitude via negative allosteric regulation [36]. The analysis of mutant *Synechococcus* PCC 7942 indicated that SmtB might not be the only factor regulating SmtA transcription [33]: cells devoid of functional *smtB* showed constitutive expression of SmtA in the absence of excess metal, but still an increase in *smtA* transcripts was observed upon exposure to 1.4 μM Cd^{2+}. The agent or mechanism for this additional metalloregulation remains unknown.

It is noteworthy that a close orthologue of SmtB, termed ZiaR, regulates the expression of a zinc efflux pump (ZiaA) in the freshwater cyanobacterium *Synechocystis* PCC 6803 [12], which is devoid of metallothionein. Further SmtB-related sensors, termed BxmR and AztR, have been characterized in *Oscillatoria brevis* and *Anabaena* PCC 7120, respectively. BxmR regulates the expression of both efflux pump Bxa1 and metallothionein BmtA [37] in response to divalent Zn and Cd, and monovalent Cu and Ag [38], whilst AztR was shown to regulate the expression of the highly zinc-specific efflux pump AztA [39]. Notably *Anabaena* also contains a metallothionein [28], but nothing about its regulation is known. Structures, mechanisms, and phylogeny of SmtB/ArsR-like sensors are discussed in recent reviews [3,13]. However, it needs to be emphasized that many sequenced cyanobacterial genomes do not contain recognizable SmtB orthologues, even if a gene for a bacterial MT is present [40]. This brief overview demonstrates that even amongst freshwater cyanobacteria there is a tremendous diversity in metal ion housekeeping mechanisms including varying specificity of metal-handling genes.

2. STRUCTURE: A HYBRID METALLOTHIONEIN/ZINC FINGER

2.1. Structure Determination by Nuclear Magnetic Resonance

The structure of zinc-loaded SmtA (residues 5–56, the N-terminal Met is not present, and three further residues are too flexible to be observed) was determined by solution NMR [27]. Evaluation of ^{1}H and ^{15}N NMR data for Zn_4SmtA allowed the determination of a low-resolution structure (1.36 ± 0.35 Å R.M.S.D. for backbone atoms) with a clearly defined backbone trace that included a β hairpin and a short α helix. Comparison of homonuclear 2D ^{1}H NMR spectra acquired for both Zn^{2+} and fully Cd^{2+} substituted SmtA (see Section 3.1.) suggested that the protein refolds in the

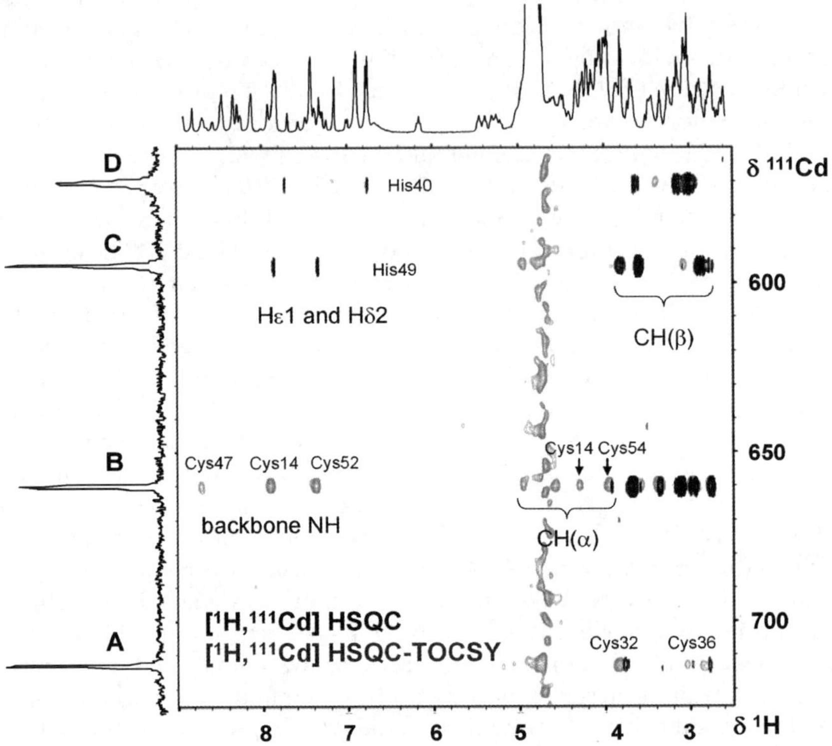

Figure 2. Heteronuclear [^{1}H,^{111}Cd] NMR correlation spectra of Cd_4SmtA. The figure shows an overlay of two spectra that provide complementary information. A HSQC spectrum (black) demonstrates histidine coordination for sites C and D through coupling between ^{111}Cd and aromatic C-H protons, and informs on several ^{111}Cd-Cys-H(β) connectivities. A HSQC-TOCSY spectrum (grey), optimized for larger coupling constants ($^{3}J^{1}$H,^{111}Cd = 45 Hz), identifies unambiguously all four Cd^{2+}-binding residues in site B through CH(α) and NH protons. The left-hand side of the 2D spectrum shows the 1D ^{111}Cd spectrum, and a partial 1D ^{1}H spectrum of Cd_4SmtA is shown on top. The resonance at *ca.* 4.7 ppm corresponds to H_2O.

presence of Cd^{2+} into an equally well-defined structure closely similar to that of Zn_4SmtA. Thus, as found for mammalian MTs [41], Cd^{2+} replaces Zn^{2+} isostructurally in SmtA, allowing to utilize heteronuclear 2D [^{1}H,^{111}Cd] NMR experiments (Figure 2) to determine the metal-to-ligand connectivities. These experiments demonstrated the presence of two $M(Cys)_4$ and two $M(Cys)_3$His sites, as judged from both ^{111}Cd NMR chemical shifts and connectivity patterns (Figure 3a). The incorporation of metal-to-ligand restraints derived from these experiments into the structure calculations

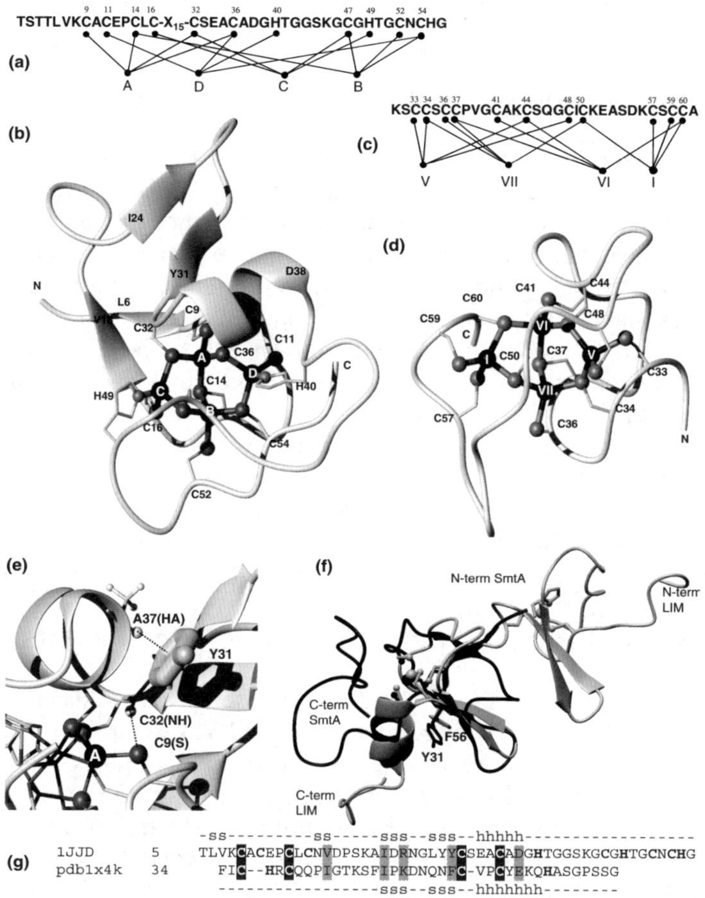

Figure 3. Structural features of Zn_4SmtA and comparison with related proteins. (a) Metal-to-ligand connectivity pattern determined by heteronuclear [^{1}H,^{111}Cd] correlation NMR spectroscopy (see Figure 2). (b) 3D solution structure of Zn_4SmtA showing the $Zn_4Cys_{11}His_2$ cluster and protein fold (pdb 1jjd; [27]). (c) Metal-to-ligand connectivity pattern in the α-domain in rat liver MT, showing no similarities with the pattern shown in (a). (d) Solution structure of the cadmium-loaded α-domain in rat liver MT [42]. The cluster structures in (b) and (d) are very similar. (e) Weak interactions around the zinc finger site A. The thiolate of Cys9 accepts a hydrogen bond from the backbone NH proton of Cys32, and the aromatic ring of Tyr31 is involved in a CH-π interaction with the CH(α) atom of Ala37. (f) Structural alignment of Zn_4SmtA (black) with model 1 of the pdb entry1x4k, i.e., the solution structure of the LIM domain in human LIM-protein 3 [46] (grey). The Phe56 residue in the LIM domain appears to have a similar role to Tyr31 in SmtA. (g) Sequential alignment corresponding to the structural alignment shown in (f).

yielded a well-defined structure (Figure 3b) with a backbone R.M.S.D. of 0.90 ± 0.11 Å.

2.2. Comparison with Other Metallothioneins

SmtA contains a $Zn_4Cys_9His_2$ cluster, with two fused six-membered rings in distorted boat conformations which bears striking similarity to the M_4Cys_{11} cluster in the α-domains of mammalian MTs, although there is no congruency in the metal-to-ligands connectivity patterns (Figures 3a and c) [42]. There are five bridging and four terminal Cys residues, one attached to each Zn^{2+}. Sites C and D, in analogy to sites I and V in mammalian MTs (Figures 3c and d), have two terminal ligands, but in SmtA one of them is a His residue. The conformations of the six-membered rings in SmtA and the α cluster in rat liver MT are very similar, with minor deviations mainly due to the difference in Cd-S and Zn-S bond lengths. We have previously noted [43] that the structures of the Zn-S or Cd-S clusters in metallothioneins are distorted versions of the minerals wurtzite (hexagonal ZnS) or Greenockite (hexagonal CdS). The distortions are likely imposed by the surrounding protein and might have significance for the dynamics of metal binding.

2.3. Related Protein Folds: Treble-Clef Zinc Fingers

In contrast to other MTs, SmtA displays extensive secondary structure, namely a β bridge, formed by residues Val7 and Val18, a β hairpin, formed by residues Ile24-Tyr31, and a short α helix formed by residues Glu34-Asp38 (Figure 3b). The mutual arrangement of helix and hairpin is stabilized by a CH-π interaction between the CH(α) proton of Ala37 and the aromatic ring of Tyr31 (Figure 3e), and an NH···S hydrogen bond (2.61 Å for S···H) between the backbone NH proton of Cys32 and thiolate sulfur of Cys9, the terminal Cys ligand for site A.

The combination of a four-coordinate zinc site with β hairpin and α helix is reminiscent of a zinc finger. Indeed, using the four Cys residues of site A as anchoring points, residues 5–39 of the SmtA structure [27] overlay quite well with both C- and N-terminal zinc finger of the eukaryotic transcription factor GATA, which also contain Cys_4 zinc sites [44], stabilized by NH···S bonds. An even better match is found with some LIM domain proteins [45,46], in which even the CH-π interaction is reproduced (Figures 3f and 3g). The protein folds of GATA fingers and LIM domains have been classified together with those of a number of further proteins as "treble-clef finger" [47]. Common to all of these domains is the ability to mediate biomolecular interactions – a fact that opens intriguing possibilities for bacterial MTs. The fact that BmtAs are single zinc finger proteins, together with an analysis of their electrostatic surfaces, argues against DNA binding.

However, a role for the zinc finger in protein-protein interactions is a definite option, although interaction partners have yet to be identified.

3. THERMODYNAMIC AND DYNAMIC PROPERTIES

3.1. Complex Stability

Native length SmtA, expressed as a non-fusion protein and purified maintaining pH >7.5, is consistently isolated with 4 molar equivalents of zinc, as determined by inductively-coupled plasma-atomic emission spectroscopy (ICP-AES) and electrospray mass spectrometry (ESI-MS) [27,28,49]. Previous preparations via a GST-fusion protein had led to the isolation of mixtures with lower zinc content (ca. 3 mol equiv.) [26], and comparison of 2D ^{1}H NMR spectra of these two different preparations indicate that the latter consists of a mixture of Zn_4SmtA and a second species, possibly identical to Zn_1SmtA (see Section 3.3.).

The zinc affinity of SmtA, as measured by hydrogen ion competition (pH of half-displacement, pH(1/2) = 4.1 [28]), is reported to be higher than that of mammalian MTs, whereas its affinity for copper (pH(1/2) = 2.35) and cadmium (pH(1/2) = 3.5) is somewhat lower [22]. The zinc affinity of Zn_4SmtA has also been measured by competition with the chelator 5F-BAPTA (1,2-bis(2-amino-5-fluorophenoxy)ethane-*N*,*N*,*N'*,*N'*-tetraacetic acid), following a procedure previously employed for mammalian MTs [48]. Under identical conditions as used for mammalian MTs (4 mM ionic strength, pH 8.1), very little zinc was transferred from SmtA to 5F-BAPTA, even if the 5F-BAPTA concentration was doubled. This allowed only the estimation of a lower limit for the conditional average stability constant of log $K \approx 12.5$–13 – at least one order of magnitude higher than that of rabbit liver MT-2a. Thus, the higher zinc affinity of SmtA is not only reflected in a lower pH(1/2), but also in its apparent stability constant. It needs to be emphasized that Zn_4SmtA is in principle able to transfer Zn^{2+} to 5F-BAPTA; this was verified by repeating the experiments under different conditions (Figure 4), which led to the observation of Zn^{2+}-loaded 5F-BAPTA, and allowed the determination of a value of log $K = 10.9 \pm 0.3$ for the conditional stability constant at pH 7.4 and $I = 93$ mM [49]. The origins of the higher absolute and relative (with respect to Cd^{2+}) zinc affinity of SmtA in comparison to other MTs are discussed in Section 3.4.

The zinc affinities of several orthologues of SmtA were explored qualitatively by equilibrating the proteins at pH 4.1 and determining the percentage of zinc bound to the proteins [28]. BmtAs from *Anabaena* PCC 7120, *Pseudomonas putida* KT 2440, and *Pseudomonas aeruginosa* (clinical isolate),

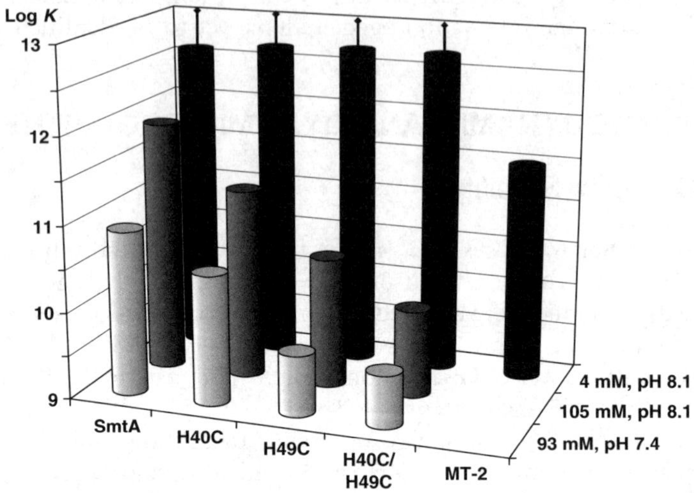

Figure 4. Estimation of zinc affinity by competition with 5F-BAPTA and ^{19}F NMR [48]. The method gives an average apparent stability constant over all available binding sites. Protein and zinc contents in the reaction mixtures were determined by ICP-AES. Note the log *K* scale, which has been restricted to log $K = 9$–13 for better recognition of differences.

expressed recombinantly in *E. coli*, retained 54, 47, and 68%, respectively, of the zinc they were originally isolated with, indicating similar pH stability of these proteins.

3.2. Metal Exchange: An Inert Site in SmtA

The preparation of homogeneous zinc-free Cd^{2+}-loaded SmtA requires generation of the apo-protein and reconstitution with Cd^{2+} [27]. The corresponding ^{111}Cd NMR spectra contained four well-defined peaks (Figure 2). Stepwise addition of $^{111}Cd^{2+}$ to folded Zn_4SmtA at neutral pH followed by ^{111}Cd NMR spectroscopy (Figure 5a) demonstrated site-specific incorporation of Cd^{2+}, with the first equivalent binding quantitatively to the Cys_4 site B. The exchange of Cd^{2+} into the Cys_3His sites D and then C follows consecutively, with several coexisting mixed-metal cluster species observed during the titration. The same species can also be observed by ESI-MS (Figure 5b). However, the zinc finger site A could not be populated by Cd^{2+} by this method, and both ESI-MS and ICP-AES confirm the $ZnCd_3$ stoichiometry. Intriguingly, site A is also a Cys_4 site, and should therefore

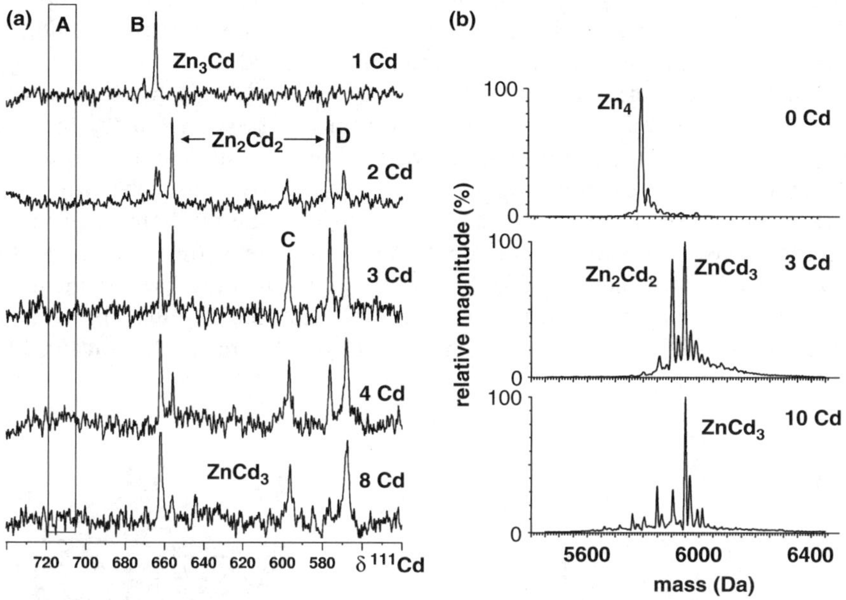

Figure 5. Multinuclear NMR and mass spectrometry give complementary information on metal exchange in SmtA. (a) Titration of Zn_4SmtA with Cd^{2+} followed by 1D ^{111}Cd NMR. Substoichiometric Cd^{2+} is incorporated preferentially in the Cys_4 site B, followed by site D, and then site C. No incorporation into site A (boxed area, see Figure 2) can be observed. Different mixed-metal species can be distinguished by slight differences in their ^{111}Cd chemical shifts. (b) MS experiments give simultaneous access to Zn^{2+} and Cd^{2+} speciation during metal exchange. Top spectrum: SmtA as isolated contains four zinc ions. Middle spectrum: Sub-stoichiometric amounts of Cd^{2+} lead to clearly defined mixed-metal clusters with four metal ions per protein. Bottom spectrum: Only three out of four Zn^{2+} ions can be exchanged by incubation of Zn_4SmtA with a 2.5-fold excess of Cd^{2+}. The $ZnCd_3$ composition of the final product was also corroborated by independent ICP-AES measurements. Note that at 2 mol equivs of Cd^{2+}, at least three co-existing species can be observed.

display a pronounced preference (see Section 3.4. for details) for Cd^{2+} over Zn^{2+}, as observed for site B.

This unusual behavior of SmtA is in stark contrast to that of previously studied metallothioneins, for which usually a stoichiometric amount of Cd^{2+} is sufficient for complete replacement of Zn^{2+} [50]. This preference of MTs for Cd^{2+} is due to the softer character of Cd^{2+} that leads to a ca. four orders of magnitude difference between the stability constants for Cd^{2+}- and Zn^{2+}-loaded mammalian metallothioneins [51]. The fact that stable, well-folded

Cd_4SmtA with a sharp ^{111}Cd NMR peak for site A can readily be prepared via the apo-protein argues against a general problem with Cd^{2+} binding to this site, therefore, we hypothesized that the end product of metal exchange was not under thermodynamic, but kinetic control. This hypothesis was verified by developing a novel high-resolution mass spectrometry method [52], in which zinc self-exchange was followed using stable isotopes. Remarkably, the reaction stopped at a ratio clearly consistent with only three exchanging sites (Figure 6a). Assuming that the mechanism of metal ion exchange involves the attack of a terminal thiolate by the incoming metal ion, an analysis of the solvent-accessibility of zinc ligands gives a clue as to how inertness might be achieved (Figure 6b): only the terminal thiolates of sites B (Cys52), C (Cys16), and D (Cys11) are surface-accessible.

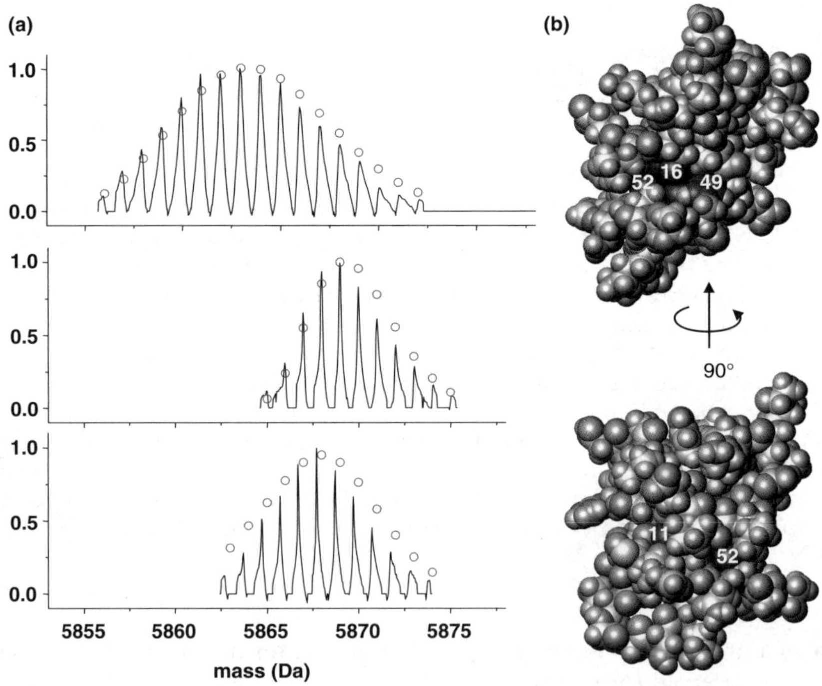

Figure 6. Zinc self-exchange in SmtA observed by high resolution FT-ICR-MS. (a) Incubation of natural abundance Zn_4SmtA (top spectrum) with a 10-fold molar excess (with respect to Zn^{2+}) of $^{67}ZnCl_2$ (93.11% pure) leads to a shift to a higher mass corresponding to three exchanging sites (bottom spectrum), that differs significantly from fully-exchanged $^{67}Zn_4SmtA$ (middle spectrum). (b) Accessibility of terminal metal ligands.

Confirmation for this hypothesis came from the study of metal-deficient BmtAs [28], and mutant SmtAs [49], which display additional, accessible terminal thiolates for site A, and all of which lack inertness.

The metal exchange behavior of Zn_4SmtA is unique amongst both metallothioneins and zinc fingers, all of which tend to exchange metal ions readily [50]. The complete lack of incorporation of external metal ion into site A also testifies to the absence of intramolecular metal exchange between sites. It is the unique combination of a zinc finger with a metallothionein that leads to the inertness of the finger site A: Cys9 is protected from attack by the steric bulk of the zinc finger portion of the protein, whilst attack on the other three Cys ligands (14, 32, 36) is not possible, as they are buried and protected by bridges to the other three metal ions.

3.3. Metal Release: Differential Reactivity of Individual Zinc Ions

The fact that SmtA expression leads to the accumulation of zinc inside bacterial cells, as well as the possibility for protein-protein interactions raises questions about the further intracellular fate of SmtA-bound zinc.

In its metal-bound form, Zn_4SmtA is an extremely stable protein that can be kept in relatively concentrated solutions (mM) for years without noticeable degradation (except for a loss of two non-zinc binding residues (TS) from the flexible N-terminus). In contrast, immediate breakdown into fragments was observed in preparations in which zinc had been stripped from the protein [53]. Although caution should be applied when extrapolating from *in vitro* experiments [54], these observations might have significance for intracellular conditions, and we suggest that zinc depletion might be a prerequisite for SmtA degradation *in vivo*. If this is the case, there should be a mechanism for removing zinc from SmtA under physiological conditions. Inter-protein metal transfer from MTs to other zinc-requiring proteins is an attractive hypothesis and has been demonstrated *in vitro* [55,56]. Since at present the physiological putative interaction partner(s) for SmtA remain unknown, the metal transfer capability of SmtA was probed by studying reactions with EDTA [52,53]. In these reactions, SmtA behaved yet again in an unexpected way. The removal of metal ions from either metallothioneins or zinc fingers usually results in the complete loss of ordered structure. However, the reaction of SmtA with a slight excess of EDTA (with respect to Zn^{2+}) results in the appearance of a structured intermediate, recognized by well-dispersed NH proton resonances in 1D (and 2D, not shown) ^{1}H NMR spectra (Figure 7a). Investigation of the same reaction by ESI-MS at pH 7.4 (Figure 7b) allowed the observation of two steps of metal loss, with the Zn_1SmtA form as a prominent intermediate.

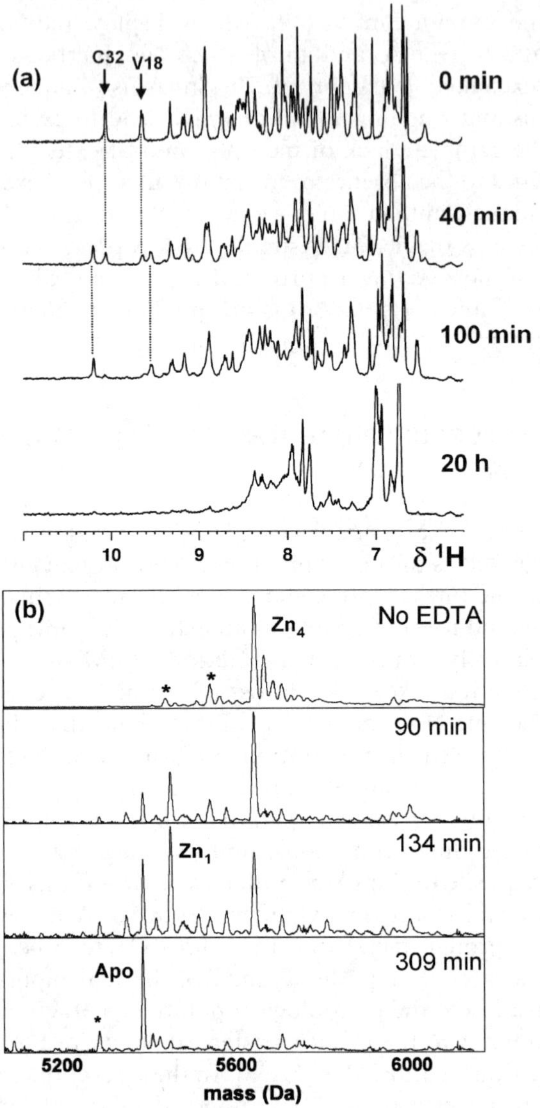

Figure 7. Zinc transfer from Zn_4SmtA to EDTA observed by NMR and ESI-MS. (a) 1D ^{1}H NMR indicates the formation of a folded intermediate that is metastable in the presence of free EDTA. (b) An intermediate corresponding to Zn_1SmtA is observed by ESI-MS. The peaks labelled with an asterisk are truncation products. Note that the reaction conditions in (a) and (b) are not identical; the MS experiment was carried out at lower concentrations and different buffer conditions, and proceeds a little more slowly.

NMR results suggest that the slower reacting zinc is again that in site A. A more complete understanding of the metal transfer reaction was achieved by extending the studies to His-to-Cys mutants (see Section 3.4.), which revealed that initial attack of EDTA occurs exclusively via site C. Thus SmtA contains several sites with different and exquisitely tuned reactivity.

The functional rationale for the resilience of site A in both metal exchange and metal transfer reactions remains unknown, but it appears likely that site A provides stabilization for a scaffold that facilitates cooperative binding of the other three metal ions. It would be desirable to see whether and under which conditions metal-depleted Zn_1SmtA occurs *in vivo*. Zinc release from intracellular SmtA expressed in *E. coli* has been observed utilizing the electrophilic attack of NO on the thiolate sulfurs, but these experiments did not allow the observation of "individual" protein species [57].

3.4. The Role of Histidine Residues: Studies on Mutant SmtAs

Besides the zinc finger fold, another striking feature that sets bacterial MTs apart from the majority of other characterized metallothioneins, is the presence of His residues which participate in metal ion binding. Confirmation for the importance of His residues in zinc binding came from the study of mutated SmtAs in which individual His residues had been converted to arginine. Mutation of His40 led to an increase of the pH of half-displacement from 4.1 to 4.62, and the His49Arg mutant had a pH(1/2) of 4.48 [26]. In contrast, the His55Arg mutant displayed, with a pH(1/2) of 3.81, a slightly increased zinc affinity. In addition, with only two molar equivalents of Zn^{2+} released in PMPS/PAR titrations, the His49 mutant and the His40/His49/His55 triple mutant displayed significantly lower zinc contents than the wild-type. These findings are in agreement with the 3D structure, in which His40 and His49 are coordinating zinc [27].

It is conceivable that the complete removal of a metal ligand, as performed in the His-to-Arg mutants, would lead to a decrease in zinc affinity. However, in order to probe the "value" of histidine residues in a metal-thiolate cluster, it is necessary to replace the His residues by other metal-binding residues, e.g., Cys. The reverse experiment, i.e., the replacement of Cys residues by histidines in mammalian MTs, has been carried out by several groups [58,59], and in summary, an increase in the *relative* Zn^{2+} affinity compared to that for Cd^{2+} has been reported. The same conclusion can be drawn from a study on a consensus zinc finger peptide [60]: whilst Cd^{2+} affinity was dramatically increased by two orders of magnitude for each His-to-Cys mutation, the effect for Zn^{2+} was, with a 2–3 fold increase in the stability constant per Cys residue, only moderate, leading to preferred binding of Cd^{2+} to Cys_4 sites and of Zn^{2+} to Cys_2His_2 sites, in accordance with Pearson's HSAB principle. This general trend explains the higher *relative* zinc affinity of SmtA, but not the increase in absolute stability. To address this, single His40Cys and His49Cys, as well as the His40Cys/His49Cys double mutant proteins were expressed in *E. coli* and characterized *in vitro* [49].

The close structural similarity of the inorganic cores of SmtA and mammalian α-domains had suggested that the replacement of both His40 and His49 could furnish a mammalian MT-like cluster in SmtA, and molecular modelling implied that only minor structural adjustments of the backbone would be necessary to accommodate the mutated residues [61]. In all cases, zinc-loaded proteins were isolated, albeit with a consistently lower zinc content (2.9–3.5 mol equivs) than the wild-type (4.4 ± 0.4 mol equivs). MS experiments demonstrated that up to four Zn^{2+} or Cd^{2+} ions could be bound to either mutant, but that in each case, there was also a clear indication for species with only three metal ions, which are completely absent from spectra of the wild-type (see Figure 5b, top). In accordance with the conclusion that the mutants possess a site with lowered zinc affinity, competition experiments with 5F-BAPTA indicated that at moderate ionic strengths ($I = 93$ mM or 105 mM), all three mutants displayed a significantly lower average zinc affinity than the wild-type (Figure 4), although at low ionic strength ($I = 4$ mM) all four proteins displayed very high, indistinguishable zinc affinities. Note that the former conditions are thought to mimic physiological conditions more closely.

The finding that complexes involving His ligands were more stable than those with pure Cys coordination was rather unexpected from a coordination chemistry point of view: the conditional stability constant at pH 7.4, calculated from absolute constants [62], for the simple 1:1 complex of Zn^{2+} with Cys is almost two orders of magnitude larger than that for Zn(His) (log $K' = 8.3$ vs 6.5, respectively), and the same trend prevails in the 2:1 complexes as well as in the zinc finger peptides mentioned above. A comparison of ^{1}H NMR data for wild-type and mutant SmtAs, and an analysis of the structural neighborhoods of the mutated residues (Figure 8) sheds some light on the origins of the lowered overall stability found in the mutants. NMR lineshapes and general appearance of the fingerprint region of 1D and 2D ^{1}H NMR spectra indicated that all three mutants were at least partially folded, but it was evident that neither mutant protein folded as well as the wild-type – a conclusion also reflected in a lower thermostability of the mutants, which precipitated at temperatures above 278–283 K [49]. Thus, each mutation had impaired the ability of SmtA to adapt a stable protein fold, and it is reasonable to expect that these more disordered structures impact on the overall metal binding affinity. In summary, the His residues are essential for a stable protein fold, and concomitantly a stable metal cluster.

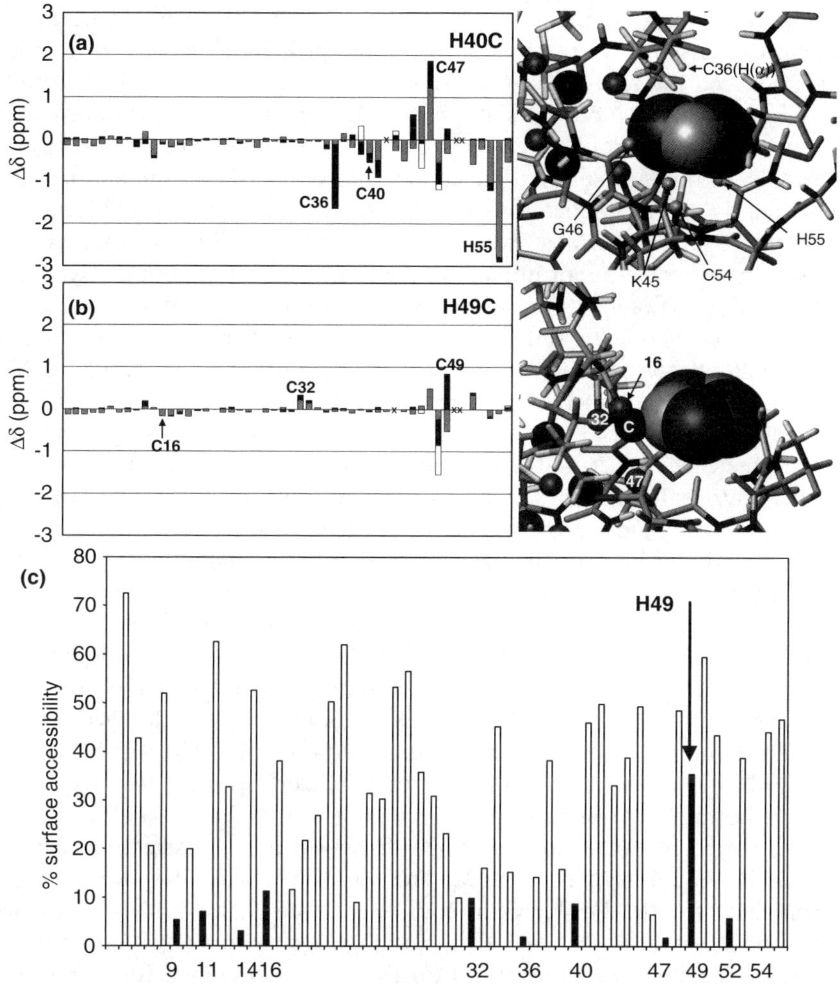

Figure 8. The two metal-binding His residues have different functions in SmtA. (a) Chemical shift perturbation analysis of NH (light grey) and CH(α) (dark grey) backbone protons of the His40Cys mutant. Major perturbations occur throughout the entire C-terminus. Inspection of the environment of His40 shows that it is buried and confers structural order to the C-terminus. The labelled atoms are backbone oxygens. (b) The effects on chemical shifts upon the mutation of His49 are restricted to neighboring and site C residues. (c) His49 is the most solvent-exposed metal-binding residue (black bars) [53].

Analysis of the chemical shift differences between wild-type and mutant allowed dissecting the impacts of the individual His residues (Figure 8). To summarize, His40 is deeply buried inside the protein, situated at the crossroads between the zinc finger (N-terminal 38 residues) and the metallothionein-like (C-terminus) portions of SmtA, and plays a crucial role to order the C-terminal region through weak interactions of its imidazole ring. Its replacement by a thiolate sulfur, that cannot reproduce these interactions, has consequently severe implications for the structure of the entire C-terminus. His49 is in contrast the most exposed zinc ligand of all (Figure 8c), and does not appear to have a major impact on the overall structure, but on dynamics and metal stoichiometry. In addition, the juxtaposition of two relatively exposed thiolates (Cys16 and Cys49) appears to create a site that is prone to disulfide bond formation [53]. Crucially, in contrast to mammalian MT domains, which contain two (α-domain) or three (β-domain) sites with two terminal Cys residues, the four sites in SmtA all have only a single terminal Cys residue each, lowering the probability for disulfide bond formation, yet another feature that helps to explain the remarkable overall stability of wild-type SmtA. The tendency of the mutated site C in the His49Cys mutant to lose its metal and become oxidized has a parallel in mammalian MT-3, the α-domain of which displays the same characteristics [63].

Unsurprisingly, the mutations also strongly influence the metal exchange behavior [49] and transfer kinetics [53] of SmtA in different ways. Regarding metal exchange, small, but significant amounts of the fully-exchanged Cd_4 or Cd_3 forms were present in all mutants, although fully-exchanged species were the dominant forms in the double mutant only. The His49 mutant displayed a strong tendency to form metal deficient species, possibly due to disulfide bond formation during the incubations. Concerning metal transfer, the Zn_4 forms of both single mutants displayed much faster zinc transfer kinetics in reactions with EDTA. The Zn_1 form was also a prominent intermediate for the His40Cys mutant, indicating that His40 has a major impact on the reactivity of sites B, C, and D, but a less pronounced influence on site A. This can be understood on the basis of the structural role that His40 plays for the entire C-terminus; all three sites have ligands derived from this portion of the protein (Cys47, 52, and 54), whereas site A has only ligands from the N-terminal, zinc finger-related part, with Cys36 being the "last" ligand (Figure 3a). The pronounced effect of the His40 mutation on sites B and C, but not A, is also demonstrated by the chemical shift changes for the ^{111}Cd NMR resonances [49]: whilst peak A hardly changed at all (1 ppm), peaks B and C shifted by 20 and 31 ppm, respectively.

The oxidized Zn_3 form of the His49Cys mutant, which is thought to contain a disulfide bond between Cys16 and Cys49, gave crucial insight into the mechanism of metal transfer: its complete inertness in reactions with EDTA demonstrated clearly that site C is the primary point of attack for

EDTA, not only in the mutants, but also the wild-type. This coincides with the observation that His49 is also the most solvent-exposed of all metal ligands (Figure 8c) [53].

4. THE BmtA FAMILY

4.1. Distribution of BmtAs in Bacterial Genomes

The adaptation to available metal ion levels in an organism's habitat has been considered as an important factor in phytoplankton evolution and speciation [64], and similar considerations may also pertain to bacteria in habitats other than water. Figure 9 shows a phylogenetic tree, prepared using ClustalW [65], of all available BmtA sequences retrieved through BLAST searches [66], of the genome sequences available at NCBI [67], and provides a summary of the occurrence of BmtAs in prokaryotes. Present in only 37 out of 940 bacterial genomes, true BmtAs are not common in bacteria, but there is a formal possibility that other genes with unrelated sequences, but similar functions have not been discovered so far. This is particularly likely because of the usually low complexity and small size of metallothioneins, which might cause them to be overlooked during the annotation process.

The majority of BmtAs have been detected in the genomes of cyanobacteria and pseudomonads (which belong to the γ-proteobacteria), but examples are also found in α-proteobacteria (e.g., *Magnetospirillum magnetotacticum,* several *Methylobacter* species, and the recently discovered pathogenic *Granulibacter bethesdensis*), other proteobacteria (*Nitrosococcus oceanii*) and firmicutes (pathogenic *Staphylococcus epidermidis* RP62A). An interesting unresolved question concerns the origin of BmtA genes in bacteria from various divisions. With the amount of available data it is not yet possible to conclusively decide whether those bacteria that do not have the gene have lost it during evolution, or whether those that do have the gene, have acquired it through horizontal gene transfer.

All sequenced *Pseudomonas aeruginosa* (a human pathogen)*, putida* and *fluorescens* strains, as well as the insect pathogen *Pseudomonas entomophila* contain a BmtA gene, but it is absent from all strains of the plant pathogen *Pseudomonas syringae*. The presence of BmtA in the soil-dwelling *P. putida* and *P. fluorescens* is not surprising, but the occurrence of BmtAs in pathogens from at least three different divisions raises the question whether the protein might aid pathogenicity or confer resistance towards antibacterials.

BmtA-containing cyanobacteria occupy a variety of habitats including freshwater (e.g., *Synechococcus* PCC 7942, *Anabaena, Oscillatoria*), salt

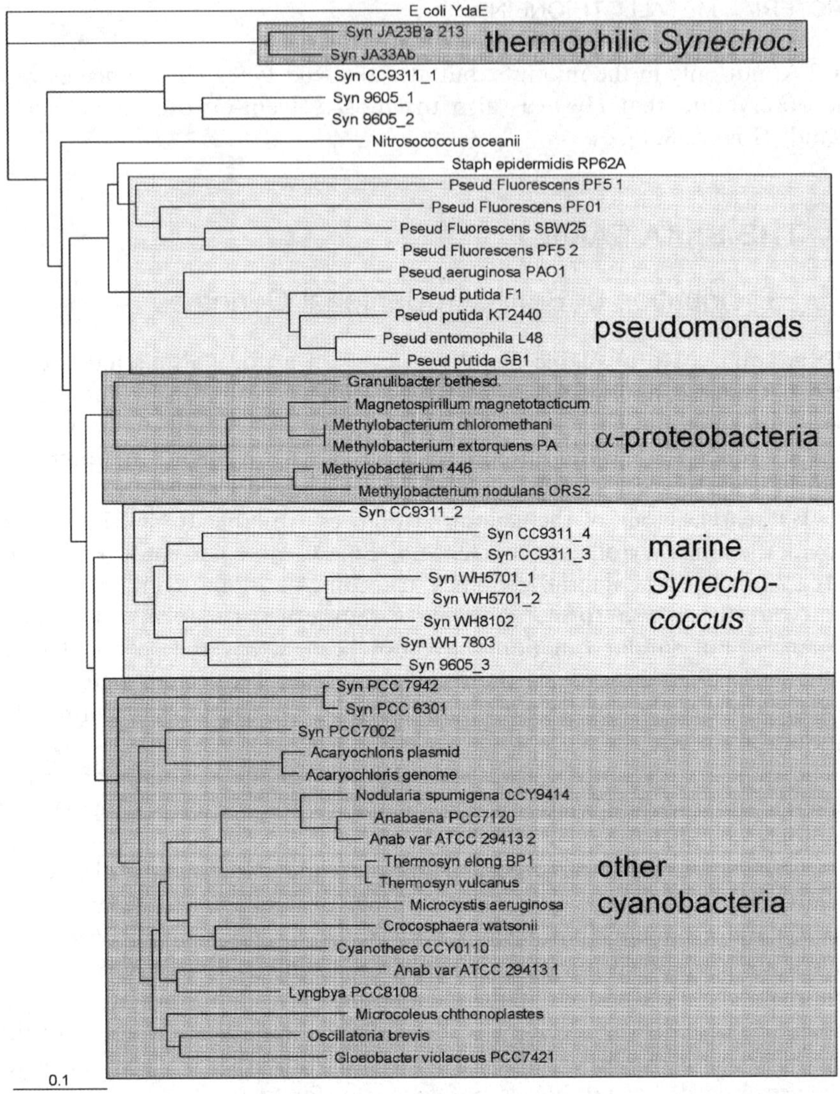

Figure 9. Phylogenetic tree of available BmtA sequences, rooted with the sequence of *E. coli* YdaE as an outgroup. This protein is predicted to have a related fold, but is not a metallothionein [28]. BmtAs are also present in the genomes of five further *Pseudomonas aeruginosa* and one further *Pseudomonas putida* strains, but their sequences are 100% identical to the ones used for the construction of the phylogenetic tree, and were therefore not included. The majority of BmtA genes are genomic, but there are also examples of plasmid-borne genes, e.g., in *Acaryochloris marina*. There is relatively high divergence between sequences, suggesting that BmtA sequences evolved rapidly, possibly under the influence of environmental conditions. Despite the shortness, and hence relatively low information content, of the sequences, the tree reproduces bacterial subdivisions, for example, all sequences of α-proteobacteria are clustered together, and so are those of all pseudomonads. Most sequences from marine *Synechococcus* also form a defined subgroup.

water (several marine *Synechococcus, Crocosphaera*), microbial mats in intertidal zones (*Lyngbya, Microcoleus, Cyanothece*), and hot springs (*Synechococcus* JA strains and *Thermosynechococcus*). The two toxic bloom forming cyanobacteria *Microcystis aeruginosa* and *Nodularia spumigena* are also represented, and *Acaryochloris marina*, a minor symbiont of an ascidian, contains genes for BmtA in its genome and on two plasmids. The presence of the BmtA gene in the rock-dwelling *Gloeobacter violacaeus* is most significant, since this cyanobacterium is thought to have diverged from others even before the evolution of chloroplasts [68] – indicating that the BmtA gene is very old.

Only about half of the 40 cyanobacterial genomes sequenced until February 2008 contain a gene for a bacterial metallothionein [40]. This balance might be somewhat skewed due to the availability of no less than 12 *Prochlorococcus* genomes, none of which contain a BmtA gene. It should be noted that marine *Synechococcus* are more closely related to *Prochlorococcus* than to freshwater *Synechococcus* (such as strain PCC 7942), and both marine genera are thought to have evolved from a common ancestor under iron limitation [69]. *Synechococcus* have a broader global distribution and are thought to be more generalists than *Prochlorococcus*, but also display more genetic diversity, a fact that points to adaptation to specific ecological niches. *Synechococcus* are also more abundant in mesotrophic waters where nutrient levels are higher, e.g., in upwelling zones and in coastal areas [70].

Six out of 13 sequenced marine *Synechococcus* strains contain at least one gene for BmtA. Interestingly, the coastal species CC 9311 [71] contains no less than four different copies of BmtA. A number of other metal-handling genes including bacterial ferritin also occur in multiple copies in this strain, and it has been suggested that this enhanced capacity for metal storage enables this strain to deal with changes in metal availability in its coastal habitat, where fluctuations between limiting and replete conditions are common. The distribution of the four different genes throughout the whole genome points towards gene acquisition through horizontal gene transfer rather than amplification and divergence.

4.2. Variations in Primary Sequences and Metal-Binding Properties

Figure 10 shows a selection of BmtA sequences, demonstrating that there is considerable divergence in bacterial metallothionein sequences, probably because of the fast mutation rate in bacteria. Nevertheless, the first six and last two cysteines, as well as His40 (SmtA notation), are nearly fully conserved, whereas more variation is seen in the stretch between His40 and Cys52, which has been termed the variable loop [49]. Cys47, Cys52, and

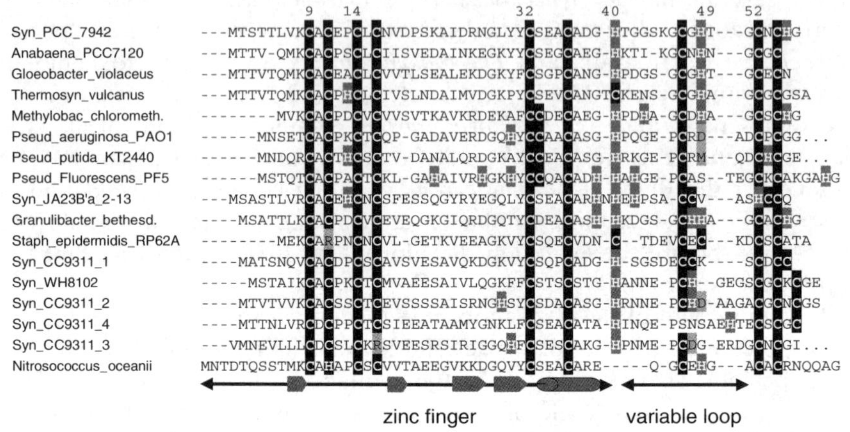

Figure 10. Conservation and variations in BmtA sequences. Representative sequences of bacterial metallothioneins are shown, and sequences are ordered to highlight variations in the presence and arrangement of metal-binding residues. Cys residues are printed in white on black, and His residues in white on dark grey. Other residues of note are highlighted in light grey. Some sequences include a hydrophobic C-terminal extension, which is not shown and replaced by dots.

Cys54 tether the C-terminus via sites C, B, and D, respectively. The residues crucial for the zinc finger fold are also highly conserved, in particular, the aromatic ring in position 31 is fully conserved, and a Gly residue in position 27 or 28 is also conserved to facilitate the tight turn of the β hairpin. Very few sequences show a replacement of His40, whereas His49 is by far the most variable metal ligand. There are examples where His49 is replaced by residues with metal-binding potential, e.g., aspartate (*P. aeruginosa*) or methionine (*P. putida*). "Natural" His-to-Cys mutants are also found; however, these are either associated with shifts in the sequence (e.g., in *Thermosynechococcus vulcanus*) or an absence of other site D ligands, e.g., in *Staphylococcus epidermidis,* where Cys11 is missing. Another variation is the presence of CH and CC motifs instead of the CxH motif (residues 47–49 in SmtA), common in many sequences from marine species, and an apparent complete loss of His49 is seen in sequences from *P. fluorescens*; however, these sequences have several additional His residues in other positions. A large number of sequences also display additional Cys residues, immediately following either Cys32 or Cys54, but it is not known whether these participate in metal coordination, or whether they behave like His55, which is not necessary for stable zinc binding.

We have only just begun to explore the properties of representative BmtA orthologues in order to understand the rationale for the divergence of the variable loop. So far, proteins from *Anabaena* PCC 7120, *P. aeruginosa*, and

P. putida KT2440 were expressed recombinantly and characterized [28]. The *Anabaena* protein, in which all 11 zinc ligands are conserved, showed high similarity to SmtA in all aspects investigated (metal content, ^{1}H and ^{111}Cd NMR chemical shifts, pH stability), whereas both proteins from *Pseudomonas* species were isolated with only *ca.* 3 molar equivalents of Zn^{2+} bound. A fourth site could be occupied in the presence of excess Zn^{2+} or Cd^{2+} [28], indicating the existence of four sites in total, with one of them having a weaker affinity, similar to the observations in the mutant SmtA proteins. The only ligand that is not conserved in the *Pseudomonas* sequences is His49 (see above), and neither Asp (*P. aeruginosa*) nor Met (*P. putida*) are able to provide the same stability to site C conferred by His. It is noteworthy that the *Pseudomonas* proteins also did not display an inert site; this is consistent with the previously formulated idea that only an intact four-metal cluster is able to protect all four Cys residues of site A from attack by an incoming metal. Considering the important role of site C in metal transfer reactions, it is conceivable that the *Pseudomonas* proteins, as well as other BmtAs, will display very different kinetics in such reactions. It is also tempting to speculate that this most variable, most exposed and kinetically most active site plays a role in specific recognition of interaction partners. Although no experimental evidence to that effect has been produced yet, we hypothesize that the alterations of site C in sequences from different species have a significant impact on the behavior of the proteins *in vivo*.

5. BIOTECHNOLOGICAL USES OF METALLOTHIONEINS IN BACTERIA

5.1. Biosensors

A biosensor encompasses the coupling of a biological material with a microelectronic system or device for the detection of an analyte in a given sample [72]. Several bacterial systems conferring metal resistance, including bacterial metallothioneins, have provided a basis for the development of sensory devices for environmental monitoring [73]. Fundamental prerogatives for the applicability of a certain protein in a biosensor are high stability, the capability to reversibly bind metal ions, and specificity. In principle, bacterial metallothioneins can fulfil at least the first two criteria, but are not expected to discriminate well between different soft metal ions.

The direct interaction between heavy metals and SmtA, immobilized on gold as a GST-fusion protein, elicited capacitance changes, and allowed the detection of femtomolar to millimolar concentrations of Hg^{2+}, Cu^{2+}, Zn^{2+}, and Cd^{2+} in pure solution [74,75]. The response at low concentrations was interpreted as non-specific interactions with unfolded protein, and at higher concentrations as metal-dependent protein folding. Stored in the presence of metal ions, the electrodes were stable for about a week, but rapidly deteriorated in the presence of EDTA. These findings concur with our own observation of rapid apo-protein degradation during Zn_4SmtA reaction with equimolar EDTA [53]. Although the responses to different metals are not identical, specific metal detection is problematic with wild-type SmtA from *Synechococcus* PCC 7942. It is an intriguing perspective to make use of the zinc finger scaffold and employ site-directed mutagenesis to improve specificity.

5.2. Bioremediation

The requirements for the remediation of heavy metals differ considerably from those of organic pollutants, which ideally can be degraded, e.g., by bacteria, to harmless chemical species [76]. For most cases of heavy metal ion contamination, remediation is only possible by immobilization and isolation in a non-bioavailable form, sometimes enabling "biomining". The observation that bacteria are capable of synthesizing metallothioneins that lead to intracellular metal accumulation, has led to the idea that such bacteria could be used for the treatment of heavy-metal polluted soil, sediments, and wastewater [76].

Indeed, cadmium-resistant pseudomonads can be isolated directly from sewage sludge (including *P. putida* and *fluorescens*) [20,77], although state-of-the-art genetic engineering plays a major role in the production of strains with a tailored repertoire [78]. Most studies aiming at the use of metallothioneins expressed in bacteria for bioremediation involve the heterologous expression of non-prokaryotic MTs, including the fusion of MTs to outer membrane proteins [79,80] to enhance bio-adsorption. Sode et al. demonstrated that insertion of the *smtA* (from *Synechococcus* PCC 7942) gene into a zinc-sensitive, *smtA*-devoid marine *Synechococcus* strain (NKBG 15041c) led to increased tolerance towards zinc, copper, and cadmium, and suggested that such strains could be useful for the bioremediation of marine environments [81].

6. CONCLUDING REMARKS

The intracellular sequestration of zinc performed by bacterial metallothioneins might be seen as an alternative to zinc detoxification provided by efflux pumps such as *Synechocystis* ZiaA. It is interesting to consider the longer term fate of the sequestered metal ions. In theory, transport to a pump for export could be envisaged and might be operating in those strains that contain both metallothionein and pump, for example *Oscillatoria brevis* and *Anabaena*. However, in *Synechococcus* PCC 7942 this is unlikely, as the amount of intracellular zinc was demonstrated to be dependent on the expression of SmtA in both permissive and lethal concentrations [33]. This suggests that SmtA may act as intracellular "sink" for zinc, and it is possible that it constitutes a store that may be mobilized under less zinc-replete conditions. It has been remarked that in order to perform such a role, SmtA would need to be able to discriminate between "adventitious" and "advantageous" zinc binding sites [19]. Our findings on structure and kinetics of metal binding of SmtA are compatible with this idea, but further studies are required to identify interaction partners for the zinc finger, and to demonstrate interprotein zinc transfer *in vitro* and *in vivo*. Equally, more studies regarding the biological function of BmtAs in different species are required to understand the diversity in zinc and other metal ion housekeeping genes.

ACKNOWLEDGMENTS

I am indebted to Professor Peter Sadler (University of Warwick) and Professor Nigel Robinson (University of Newcastle) in whose labs much of the reported work was initiated and carried out. I also wish to thank everybody with whom I have worked together on BmtAs over the years. Support from the Swiss National Science Foundation, the Novartis Foundation, the Biological and Biotechnological Research Council, the Engineering and Physical Sciences Research Council, the European Commission, the Wellcome Trust, and the Royal Society are gratefully acknowledged.

ABBREVIATIONS AND DEFINITIONS

5F-BAPTA	1,2-bis(2-amino-5-fluorophenoxy)ethane-*N*,*N*,*N*′,*N*′-tetraacetic acid
BmtA	bacterial metallothionein with sequences similarity to *Synechococcus* SmtA
EDTA	ethylenediamine-*N*,*N*,*N*′,*N*′-tetracetate
ESI-MS	electrospray mass spectrometry
FT-ICR-MS	Fourier-transform ion cyclotron resonance mass spectrometry
HPLC	high performance liquid chromatography
HSQC	heteronuclear single quantum coherence spectroscopy
ICP-AES	inductively coupled plasma-atomic emission spectroscopy
ICP-MS	inductively coupled plasma-mass spectrometry

MT	metallothionein
NCBI	National Center for Biotechnology Information
PAR	4-(pyridyl-2-azo)resorcinol
PMPS	*p*-hydroxy-mercuri-phenylsulfonate
pH(1/2)	pH of half-displacement, pH value at which 50% of bound metal is lost
R.M.S.D.	root mean square deviation
SmtA	metallothionein from *Synechococcus* PCC 7942
SmtB	zinc sensor in *Synechococcus* PCC 7942
TOCSY	total correlation spectroscopy

REFERENCES

1. S. Silver and L. T. Phung, *J. Ind. Microbiol. Biotechnol.*, 2005, **32**, 587–605.
2. D. H. Nies, *FEMS Microbiol. Rev.*, 2003, **27**, 313–339.
3. S. Tottey, D. R. Harvie and N. J. Robinson, *Acc. Chem. Res.*, 2005, **38**, 775–783.
4. A. Changela, K. Chen, Y. Xue, J. Holschen, C. E. Outten, T. V. O'Halloran and A. Mondragon, *Science*, 2003, **301**, 1383–1387.
5. C. E. Outten and T. V. O'Halloran, *Science*, 2001, **292**, 2488–2489.
6. A. C. Rosenzweig, *Chem. Biol.*, 2002, **9**, 673–677.
7. H. L. Schubert, E. Raux, M. A. Matthews, J. D. Phillips, K. S. Wilson, C. P. Hill and M. J. Warren, *Biochem. Soc. Trans.*, 2002, **30**, 595–600.
8. S. F. Sousa, P. A. Fernandes and M. J. Ramos, *J. Am. Chem. Soc.*, 2007, **129**, 1378–1985.
9. D. K. Blencowe and A. P. Morby, *FEMS Microbiol. Rev.*, 2003, **27**, 291–311.
10. K. Hantke, *Curr. Opin. Microbiol.*, 2005, **8**, 196–202.
11. J. S. Cavet, G. P. M. Borrelly and N. J. Robinson, *FEMS Micorbiol Rev.*, 2003, **27**, 165–181.
12. C. Thelwell, N. J. Robinson and J. S. Turner-Cavet, *Proc. Natl. Acad. Sci. USA*, 1998, **95**, 10728–10733.
13. M. A. Pennella and D. P. Giedroc, *Biometals*, 2005, **18**, 413–428.
14. R. J. P. Williams and J. J. R. F. Da Silva, *J. Theor. Biol.*, 2003, **220**, 323–343.
15. M. A. Saito, D. M. Sigman and F. M. M. Morel, *Inorg. Chim. Acta.*, 2003, **356**, 308–318.
16. R. W. Olafson, K. Abel and R. G. Sim, *Biochem. Biophys. Res. Commun.*, 1979, **89**, 36–43.
17. R. W. Olafson, S. Loya and R. G. Sim, *Biochem. Biophys. Res. Commun.*, 1980, **95**, 1495–1503.
18. R. W. Olafson, W. D. McCubbin and C. M. Kay, *Biochem. J.*, 1988, **251**, 691–699.
19. N. J. Robinson, S. K. Whitehall and J. S. Cavet, *Adv. Microb. Physiol.*, 2001, **44**, 183–213.
20. D. P. Higham, P. J. Sadler and M. D. Scawen, *Science*, 1984, **225**, 1043–1046.

21. N. J. Robinson, A. Gupta, A. P. Fordham-Skelton, R. R. D. Croy, B. A. Whitton and J. W. Huckle, *Proc. Biol. Sci.*, 1990, **242**, 241–247.
22. J. G. Shi, W. P. Lindsay, J. W. Huckle, A. P. Morby and N. J. Robinson, *FEBS Lett.*, 1992, **303**, 159–163.
23. A. Gupta, B. A. Whitton, A. P. Morby, J. W. Huckle and N. J. Robinson, *Proc. Biol. Sci.*, 1992, **248**, 273–281.
24. J. W. Huckle, A. P. Morby, J. S. Turner and N. J. Robinson, *Mol. Microbiol.*, 1993, **7**, 177–187.
25. T. Shimizu, T. Hiyama, M. Ikeuchi and Y. Inoue, *Plant Mol. Biol.*, 1992, **20**, 565–567.
26. M. J. Daniels, J. S. Turner-Cavet, R. Selkirk, H. Sun, J. A. Parkinson, P. J. Sadler and N. J. Robinson, *J. Biol. Chem.*, 1998, **273**, 22957–22961.
27. C. A. Blindauer, M. D. Harrison, J. A. Parkinson, A. K. Robinson, J. S. Cavet, N. J. Robinson and P. J. Sadler, *Proc. Natl. Acad. Sci. USA*, 2001, **98**, 9593–9598.
28. C. A. Blindauer, M. D. Harrison, A. K. Robinson, J. A. Parkinson, P. W. Bowness, P. J. Sadler and N. J. Robinson, *Mol. Microbiol.*, 2002, **45**, 1421–1432.
29. K. Takatera and T. Watanabe, *Anal. Sci.*, 1992, **8**, 469–474.
30. A. F. Lavorante, M. F. Gine, A. P. Gervasio, C. E. Miranda, M. F. Fiore, C. M. Bellato and E. Carrilho, *Anal. Sci.*, 2003, **19**, 1611–1616.
31. M. R. Binet, R. Ma, C. W. McLeod and R. K. Poole, *Anal. Biochem.*, 2003, **318**, 30–38.
32. J. S. Turner and N. J. Robinson, *J. Ind. Microbiol.*, 1995, **14**, 119–125.
33. J. S. Turner, N. J. Robinson and A. Gupta, *J. Ind. Microbiol.*, 1995, **14**, 259–264.
34. J. S. Turner, A. P. Morby, B. A. Whitton, A. Gupta and N. J. Robinson, *J. Biol. Chem.*, 1993, **268**, 4494–4498.
35. A. P. Morby, J. S. Turner, J. W. Huckle and N. J. Robinson, *Nucl. Acids Res.*, 1993, **21**, 921–925.
36. M. L. VanZile, X. Chen and D. P. Giedroc, *Biochemistry*, 2002, **41**, 9776–9786.
37. T. T. Liu, S. Nakashima, K. Hirose, Y. Uemura, M. Shibasaka, M. Katsuhara and K. Kasamo, *FEBS Lett.*, 2003, **542**, 159–163.
38. T. Liu, S. Nakashima, K. Hirose, M. Shibasaka, M. Katsuhara, B. Ezaki, D. P. Giedroc and K. Kasamo, *J. Biol. Chem.*, 2004, **279**, 17810–17818.
39. T. Liu, J. W. Golden and D. P. Giedroc, *Biochemistry*, 2005, **44**, 8673–868.
40. C. A. Blindauer, *Chem. Biodiv.*, in press.
41. B. A. Messerle, A. Schaeffer, M. Vašák, J. H. R. Kägi and K. Wüthrich, *J. Mol. Biol.*, 1992, **225**, 433–443.
42. P. Schultze, E. Wörgötter, W. Braun, G. Wagner, M. Vašák, J. H. R. Kägi and K. Wüthrich, *J. Mol. Biol*, 1988, **203**, 251–268.
43. C. A. Blindauer and P. J. Sadler, *Acc. Chem. Res.*, 2005, **38**, 62–69.
44. J. M. Matthews and M. Sunde, *IUBMB Life*, 2002, **54**, 351–355.
45. R. Weiskirchen and K. Gunther, *Bioessays*, 2003, **25**, 152–162.
46. F. He, M. Inoune, T. Kigawa, M. Shirouzu, T. Terada and S. Yokoyama, unpublished.
47. N. V. Grishin, *Nucleic Acids Res.*, 2001, **29**, 1703–1714.
48. D. W. Hasler, L. T. Jensen, O. Zerbe, D. R. Winge and M. Vašák, *Biochemistry*, 2000, **39**, 14567–14575.

49. C. A. Blindauer, M. T. Razi, D. J. Campopiano and P. J. Sadler, *J. Biol. Inorg. Chem.*, 2007, **12**, 393–405.
50. D. G. Nettesheim, H. R. Engeseth and J. D. Otvos, *Biochemistry*, 1985, **24**, 6744–6751.
51. M. Vašák and J. H. R. Kägi, in *Metal Ions Biol. Systems*, Vol. 15, Ed. H. Sigel, Marcel Dekker Inc., New York and Basel, 1983, pp. 213–273.
52. C. A. Blindauer, N. C. Polfer, S. E. Keiper, M. D. Harrison, N. J. Robinson, P. R. R. Langridge-Smith and P. J. Sadler, *J. Am. Chem. Soc.*, 2003, **125**, 3226–3227.
53. O. I. Leszczyszyn, C. D. Evans, S. E. Keiper, G. Z. L. Warren and C. A. Blindauer, *Inorg. Chim. Acta*, 2007, **360**, 3–13.
54. D. H. Petering, J. Zhu, S. Krezoski, J. Meeusen, C. Kiekenbush, S. Krull, T. Specher and M. Dughish, *Exp. Biol. Med.*, 2006, **231**, 1528–1534.
55. M. Huang, C. F. Shaw and D. H. Petering, *J. Inorg. Biochem.*, 2004, **98**, 639–648.
56. L. J. Jiang, M. Vašák, B. L. Vallee and W. Maret, *Proc. Natl. Acad. Sci. USA*, 2000, **97**, 2503–2508.
57. M. R. Binet, H. Cruz-Ramos, J. Laver, M. N. Hughes and R. K. Poole, *FEMS Microbiol. Lett.*, 2002, **213**, 121–126.
58. N. Romero-Isart, N. Cols, M. K. Termansen, J. L. Gelpi, P. Gonzalez-Duarte, S. Atrian and M. Capdevila, *Eur. J. Biochem.*, 1999, **259**, 519–527.
59. Y. J. Zhou, N. Zhang, L. Y. Li and B. G. Ru, *Protein Pept. Lett.*, 2000, **7**, 9–16.
60. B. A. Krizek, D. L. Merkle and J. M. Berg, *Inorg. Chem.*, 1993, **32**, 937–940.
61. C. A. Blindauer, *J. Inorg. Biochem*, 2008, **102**, 507–521.
62. IUPAC Stability Constants Database, Data Version 4.56 (compiled by L. D. Pettit and H. K. J. Powell), Academic Software, Timble, Otley, UK, 2005.
63. B. Roschitzki and M. Vašák, *Biochemistry*, 2003, **42**, 9822–9828.
64. W. G. Sunda, D. G. Swift and S. A. Huntsman, *Nature*, 1991, **6321**, 55–57.
65. J. D. Thompson, D. G. Higgins and T. J. Gibson, *Nucl. Acids Res.*, 1994, **22**, 4673–4680.
66. S. F. Altschul, T. L. Madden, A. A. Schaffer, J. Zhang, Z. Zhang, W. Miller and D. J. Lipman, *Nucleic Acids Res.*, 1997, **25**, 3389–3402.
67. http://www.ncbi.nlm.nih.gov/genomes/lproks.cgi, accessed 16/02/2008.
68. B. Nelissen, Y. Van de Peer, A. Wilmotte and R. De Wachter, *Mol. Biol. Evol.*, 1995, **12**, 1166–1173.
69. C. S. Ting, G. Rocap, J. King and S. W. Chisholm, *Trends Microbiol.*, 2002, **10**, 134–142.
70. K. Zwirglmaier, J. L. Heywood, K. Chamberlain, E. M. S. Woodward, M. V. Zubkov and D. J. Scanlan, *Env. Microbiol.*, 2007, **9**, 1278–1290.
71. B. Palenik, Q. Ren, C. L. Dupont, G. S. Myers, J. F. Heidelberg, J. H. Badger, R. Madupu, W. C. Nelson, L. M. Brinkac, R. J. Dodson, A. S. Durkin, S. C. Daugherty, S. A. Sullivan, H. Khouri, Y. Mohamoud, R. Halpin and I. T. Paulsen, *Proc. Natl. Acad. Sci. USA*, 2006, **103**, 13555–13559.
72. S. Belkin, *Curr. Opin. Microbiol.*, 2003, **6**, 206–212.
73. N. L. Brown, J. R. Lloyd, K. Jakeman, J. L. Hobman, I. Bontidean, B. Mattiasson and E. Csöregi, *Biochem. Soc. Trans.*, 1998, **26**, 662–665.

74. P. Corbisier, D. van der Lelie, B. Borremans, A. Provoost, V. de Lorenzo, N. L. Brown, J. R. Lloyd, J. L. Hobman, E. Csöregi, G. Johansson and B. Mattiasson, *Anal. Chim. Acta.*, 1999, **387**, 235–244.
75. I. Bontidean, J. R. Lloyd, J. L. Hobman, J. R. Wilson, E. Csöregi, B. Mattiasson and N. L. Brown, *J. Inorg. Biochem.*, 2000, **79**, 225–229.
76. M. Valls and V. de Lorenzo, *FEMS Microbiol. Rev.*, 2002, **26**, 327–338.
77. K. Chovanova, D. Sladekova, V. Kmet, M. Proksova, J. Harichova, A. Puskarova, B. Polek and P. Ferianc, *Biologia*, 2004, **59**, 817–827.
78. M. Mejare and L. Bülow, *Trends Biotech.*, 2001, **19**, 67–73.
79. M. Valls, R. Gonzalez-Duarte, S. Atrian and V. De Lorenzo, *Biochimie*, 1998, **80**, 855–861.
80. M. Valls, V. de Lorenzo, R. Gonzalez-Duarte and S. Atrian, *J. Inorg. Biochem.*, 2000, **79**, 219–223.
81. K. Sode, Y. Yamamoto and N. Hatano, *J. Mar. Biotechnol.*, 1998, **6**, 174–177.

Met. Ions Life Sci. **2009**, *5*, 83–105

4

Metallothioneins in Yeast and Fungi

Benedikt Dolderer, [1] ***Hans-Jürgen Hartmann,*** [1] ***and Ulrich Weser*** [1,2,*]

[1]Anorganische-Biochemie, Interfakultäres Institut für Biochemie, University of Tübingen, Hoppe-Seyler-Str. 4, D-72076 Tübingen, Germany
<ulrich.weser@uni-tuebingen.de>

[2]Centro di Risonanze Magnetiche, University of Florence, Via Luigi Sacconi 6, I-50019 Sesto Fiorentino (Firenze), Italy

Metal Ions in Life Sciences, Volume 5 Edited by Astrid Sigel, Helmut Sigel, and Roland K. O. Sigel

Published by the Royal Society of Chemistry, www.rsc.org DOI: 10.1039/9781847558992-00083

ABSTRACT: Small cysteine-rich proteins sharing most if not all of the general features used to define the metallothionein (MT) superfamily are found in yeast and fungi. Unlike MTs from mammalian sources, most of the known yeast and fungal MTs are Cu(I) rather than Zn(II) or Cd(II) binding proteins. The sequences of fungal MTs reported so far are quite diverse, in such a way that fungal MTs are assigned to six different families. Family 8 contains the MTs with the highest similarity to the N-terminal domains of mammalian MTs. The best characterized member of this family is isolated from the ascomycete *Neurospora crassa*. It represents a copper-induced polypeptide of only about 25 amino acid residues and harbors a single cluster made up of six Cu(I) that are bound to its seven cysteine residues. The MTs assigned to families 9 and 10 are MT-1 and MT-2 found in the human pathogenic yeast *Candida glabrata*. The regulation of these proteins employing a copper sensitive transcription factor shares the same principle as were described for the MTs found in *Saccharomyces cerevisiae*, Cu-MT and Crs5, that are assigned to families 12 and 13. *S. cerevisiae* Cu-MT is the only MT, of which the structure including its $Cu(I)_8$-thiolate core has been revealed. It should be emphasized that this is the largest known Cu cluster in biological systems. Besides the presentation of these well studied aspects, the open questions of Cd(II) and Zn(II) binding in yeasts and fungi are addressed and future directions of the MT research are discussed.

KEYWORDS: *Candida glabrata* · copper · fungus · metallothionein · *Neurospora crassa* · *Saccharomyces cerevisiae* · yeast

1. INTRODUCTION

Metallothioneins (MTs) are generally described as cysteine-rich polypeptides sharing particular features of equine MT, the first MT to be isolated (see also Chapter 1). The first fungal MT to be identified was Cu-MT from the yeast *Saccharomyces cerevisiae*. Although it exhibits most of the common MT characteristics, the differences observed between this MT and horse MT were larger than those between two different vertebrate MTs. Therefore, in an early attempt to classify the family of MTs, *S. cerevisiae* MT was assigned to the class II of a three-class-system [1]. The primary structure of these proteins being only very distantly or not at all related with the sequence of the equine prototype MT was used as the criterion for the assignment to class II. Based on phylogenetic analyses a more recent proposal for the MT classification divided the known MT molecules into 15 families, these altogether forming the superfamily of MTs [2,3]. The different MTs of fungal origin known so far are quite diverse such that they have been assigned to six different families, numbered 8 to 13 (Figure 1).

Besides their sequential diversity, the exclusive coordination of copper ions rather than divalent Zn(II) or Cd(II) is a common characteristic of most fungal MTs. Family 8 constitutes the subgroup with the largest number of known MT sequences. The most striking feature of this group of very small MTs of about 25 residues each is their sequence similarity with the N-terminal domains of mammalian MTs. The most prominent and best

Family 8	
Neurospora crassa MT	--GDCGCSG-ASSCNCGSG-CSCSNCGSK
Agaricus bisporus MT	--GDCGCSG-ASSCTCASGQCTCSGCG-K
Podospora anserina MT	MGGSCNCSGSASSCSCGSD-CSCSGCK--
C. gloeosporioides MT-1	MSG-CGCAS-TGTCHCGKD-CTCAGCPHK
C. gloeosporioides MT-2	-MAPCSCKSCGTSCAGSCTSCSCGSCSH-
Heliscus lugdunensis MT	--SPCTCSTC--NCAGACNSCSCTSCSH-
Family 9	
Candida glabrata MT-1	ANDCKCPNGCSCPNCANGGCQCGDKCECKKQSCHGCGEQCKCGSHGSSCHGSCGCGDKCECK
Family 10	
Candida glabrata MT-2	PEQVNCQYDCHCSNCACENTCNCCAKPACACTNSASNECSCQTCKCQTCKC
Family 11	
Yarrowia lipolytica MT-1	MEFTTAMFGTSLIFTTS-TQSKHNLVNNCCCSSSTSESSMPASCACTKCGCKTCKC
Yarrowia lipolytica MT-3	MEFTTAMLGASLISTTS-TQSKHNLVNNCCCSSSTSESSMPASCACTKCGCKTCKC
Yarrowia lipolytica MT-2	MEFTSALFGASLVQSKHKTTKKHNLVDSCCCSKPTEKPTN--SCTCSKCACDSCKC
Yarrowia lipolytica MT-4	MEFLNANFGASLIQSKHKTTKKHNLVNSCCCSKPAEKPTN--SCTCSKCACDSCKC
Family 12	
Saccharomyces cerevisiae MT	-QNEGHECQCQCGSCKNNEQCQKSCSCPTGCNSDDKCPCGNKSEETKKSCCSGK
P_{-1-40}	YQNEGHECQCQCGSCKNNEQCQKSCSCPTGCNSDDKCPCGN-------------
P_{5-40}	-----HECQCQCGSCKNNEQCQKSCSCPTGCNSDDKCPCGN-------------
Family 13	
Saccharomyces cerevisiae Crs5	MTVKICDCEGECCKDSCHCGSTCLPSCSGGEKCKCDHSTGSPQCKSCGEKCKCETTCTCEKSKCNCEKC

Figure 1. Amino acid sequences of all MTs assigned to families 8–13. "–" indicates gaps in the alignment of family 8, 11, and 12 MTs.

studied member of this group represents the MT from the ascomycete *Neurospora crassa*. The first subsection therefore deals with this protein followed by a brief introduction of the other members of family 8. The following sections cover the members of families 12, 13 (both MTs from *Saccharomyces cerevisiae*), 9 and 10 (both MTs from *Candida glabrata*), which contain only one member each.

The sequence of the characterization of these families was changed as the MTs from *S. cerevisiae* are studied in more detail and as the general principles presented for *S. cerevisiae* Cu-MT also apply to the *C. glabrata* MTs. The four members of family 11 were discovered when the complete assembled genomes of four yeast species were compared and their genes were classified according to *S. cerevisiae* proteins [4]. No further studies dealt with these proteins. But a closer inspection reveals large differences with respect to known MT sequences. All cysteine residues in these putative proteins are restricted to the C-terminal halves, starting with an unusual sequence of three following cysteine residues. Typical cysteine spacings are only seen in the last 13 residues, that also contain the remaining cysteine residues, having the structure CXCXXCXCXXCXC. Although this sequence contains typical MT motifs, it is too compact compared with other known MT sequences. In our opinion it seems questionable whether these genes really represent members of the MT superfamily.

To date only limited knowledge exists concerning Zn(II) and Cd(II) induced MTs. Therefore, a section has been added to briefly mention both their existence and the synthesis of phytochelatins in fungi.

2. FAMILY 8 METALLOTHIONEINS

2.1. *Neurospora crassa* Metallothionein

The MT isolated from the ascomycete *Neurospora crassa* represents the prototype for the family 8 of the MT superfamily. This small polypeptide contains only 25 amino acid residues and shares a striking similarity with the N-terminal domain of mammalian MTs [5]. In particular, its seven cysteine residues are positioned as the first seven cysteine residues of mammalian MTs. The lack of aromatic and the degeneracy of the occurring amino acid residues are well in line with the definition of MTs.

Like the genes of mammalian MTs, that contain two introns, the coding region of the *N. crassa* MT gene is interrupted by an 94-bp intron. The exon-intron boundaries, however, are not found at similar positions of the genes [6]. Sequences related to the mammalian or *S. cerevisiae* MT metal regulatory sequences were not found in the promoter region of *N. crassa* MT. Unlike mammalian MTs the *N. crassa* protein is induced only by copper ions. Cd(II), Zn(II), Co(II), and Ni(II) turned out to be ineffective [7]. The exact regulatory mechanism is not yet known.

Isolated *N. crassa* MT contains exclusively six copper ions per mole of protein. The occurrence as Cu(I) and the coordination by cysteine thiolate, thereby forming a $Cu(I)_6S_7$ cluster, was deduced from circular dichroism, electronic absorption and electron paramagnetic resonance spectroscopic results [8,9]. An orange-red luminescence emission upon excitation in the UV region ($\lambda \sim 300$ nm), which is attributed to charge-transfer transitions within the Cu(I)-thiolate chromophore, is a characteristic feature of Cu-MTs and also seen using the *N. crassa* protein [10]. Unlike Cu-MTs, low-molecular Cu(I) complexes are not luminescent in solution at room temperature, but only in the dried state or at very low temperatures as 77 K, the temperature of liquid nitrogen. This difference is explained with solvent interactions of low-molecular luminophores which leads to quenching of their luminescence. The luminophores of Cu-MTs are partially shielded from such interactions by the protein backbone. This explanation is further emphasized by the fact that Cu-MT luminescence in solution can be additionally increased by replacement of H_2O with D_2O, removal of dissolved O_2, or by the use of more viscous media [8]. The solvent shielding capability of such a small 25 residue polypeptide was indicative for a compact tertiary structure of the protein, the Cu(I)-thiolate complex being located in its center.

Despite its sequence homology to the N-terminal domain of mammalian MTs, *N. crassa* MT is exclusively isolated as Cu_6-MT [5]. Therefore, it was of interest to study its behavior in the presence of other metal ions. *In vitro* titration revealed that this MT was capable of binding up to three molar equivalents of the divalent metal ions Zn(II), Cd(II), Co(II), and Hg(II). Deduced from the electronic absorption and electron paramagnetic resonance features of the Co(II) containing species a tetrahedral coordination geometry was suggested [11].

Interestingly, the *in vitro* reconstitution of apo-MT to yield Cu_6-MT was dependent upon prior formation of Zn_3-MT (see below) [12]. This shows that intermediate binding of at least Zn(II) to *N. crassa* MT seems to be necessary in the course of its folding pathway, although the protein's physiological function seems to be more related to Cu(I) binding, as Cu(I) is the only metal ion capable of inducing the expression of this protein.

2.2. Structure of *Neurospora crassa* Metallothionein

Using two-dimensional homonuclear ^1H NMR the polypeptide fold of *N. crassa* Cu_6-MT was determined [12]. The structure determination relied solely on proton-proton interactions, as the copper isotopes ^{63}Cu and ^{65}Cu are NMR-insensitive quadrupolar nuclei and the substitution of NMR-insensitive copper isotopes with the NMR active spin $\frac{1}{2}$ isotope ^{109}Ag did only result in mixed-metal species, but never in Ag_6-MT. From other studies dealing with *S. cerevisiae* and murine MT it was also concluded before that Ag(I) might not always be suitable to isomorphically replace Cu(I) [13,14]. Interestingly, the *in vitro* reconstitution of apo-MT was dependent on a prior formation of Zn_3-MT and subsequent titration with Cu(I) [12]. Only if carried out like that, Cu_6-MT, which is the native form isolated from *N. crassa*, was obtained. Direct titration of apo-MT led to the formation of a Cu_4-species. In contrast to the well defined one-dimensional ^1H NMR spectrum of native Cu_6-MT [15] the Cu_4-MT species resulted in a very broad unresolved spectrum, being indicative for a mixture of fast interconverting structures or structural instabilities [12].

Like the known structures of other MTs, this protein lacked any typical secondary structural elements [15]. The N-terminal half of its polypeptide chain is wound around an empty space, which is supposed to contain the Cu(I)-thiolate center, in a left-handed, its C-terminal half in a right-handed fashion (Figure 2). Such a change in the sense of rotation with which the polypeptide of a MT wraps around its metal-thiolate cluster was also observed in the structures of the C-terminal domain of blue crab MT and the $Zn_xCu(I)_3$-containing α-domain of mouse MT-1 [16,17]. Not all of the cysteine sulfur atoms of the structure presented by Cobine et al. [12] pointed towards the center of the protein and were therefore not capable of

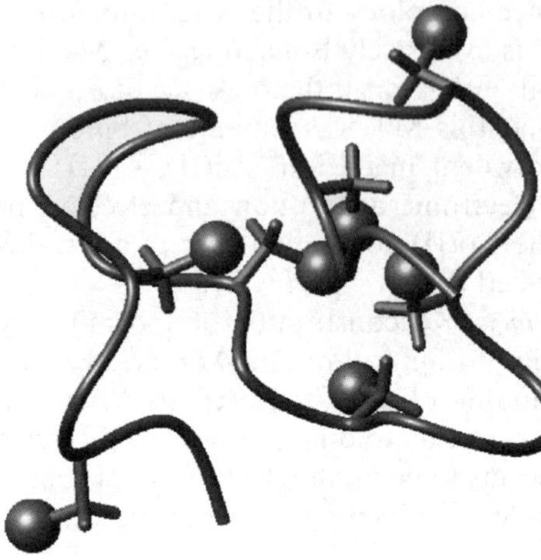

Figure 2. Ribbon display of the *Neurospora crassa* MT structure [12]. The cysteine side chains and sulfur atoms are shown. The figure was prepared using the data deposited in the Protein Data Bank under accession number 1T2Y.

contributing to the formation of a single Cu(I)-thiolate cluster. This was probably due to the lack of structural restraints which might have led to a wrong orientation of the N- and C-terminal cysteine side chains. It was tried to model six Cu(I) into the free space within the polypeptide portion of *N. crassa* MT forcing both terminal cysteine sulfur atoms to point towards the center of the whole structure. One possible structure is shown in [12]. But it should be mentioned that in an earlier study dealing with the *S. cerevisiae* Cu-MT structure the modelling approach did not lead to an unambiguous result [13]. Another problem complicating the situation is that Cu(I) can adopt variable coordination numbers. Thus, in *S. cerevisiae* Cu-MT Cu(I) is found both in linear and in trigonal geometries surrounded by thiolate sulfurs (see below). In this light the presented model might indeed be chemically plausible on the basis of the present structural knowledge about this protein, but the true structure of the metal-thiolate cluster might well be still different from it.

2.3. Other Members of the Family 8

A metallothionein sharing almost 80% sequence identity with *N. crassa* MT was isolated from mushroom (*Agaricus bisporus*) [18]. As already known from the *N. crassa* protein, *A. bisporus* MT contains six Cu(I) per protein molecule. Electronic absorption, circular dichroism and luminescence emission of the two MTs are very similar arguing for a similar structure of their metal-thiolate chromophores.

The ascomycete *Podospora anserina* possesses a Cu(I)-inducible MT the protein of which is expressed at higher levels also in the course of aging [19]. Its primary structure shows 62% identity with the *N. crassa* protein. As a long-living mutant of this fungus was associated with a loss-of-function mutation in the transcription factor GRISEA, which resulted in reduced copper uptake, copper-mediated oxidative stress was discussed to have an effect on aging of this fungus [20,21]. Therefore, it was interesting to study how levels of *P. anserina* MT, another player controlling the intracellular availability of copper ions, change during aging and in response to copper levels in the environment [19]. It turned out that *P. anserina* MT is expressed already at trace amounts of copper ions in the growth medium and that further expression can be induced by an increase of copper concentrations. Levels of this MT were also found to be significantly increased during aging of wild-type *P. anserina*. As mitochondria lose their functionality in the course of aging, disturbances in the turnover of their respiratory complex proteins lead to the release of copper ions previously bound to cytochrome oxidase. This in turn would cause the induction of *P. anserina* MT.

The facultative parasite *Colletotrichum gloeosporioides* expresses two proteins that are homologous to other family 8 MTs [22]. Interestingly, these two proteins are expressed mainly during appressorium formation. In this context they are induced by the wax of this parasite's hosts. The induction by copper and cadmium ions is weak. Therefore, these two MTs are suggested to have a specialized function related to the process of appressorium formation. The nature of the metals bound to theses proteins is yet unknown. But as both copper and cadmium ions are capable of inducing these MTs, it is likely that both can be found coordinated to them. Another interesting point to mention is that with these two members family 8 contains not only Cu-MTs, but also Cd- and therefore most likely also Zn-MTs.

Yet another Cd-MT belonging to family 8 MTs was identified in the aquatic hyphomycete *Heliscus lugdunensis* under conditions of Cd(II) stress [23]. Under these conditions not only MT was expressed, but also elevated levels of glutathione and phytochelatin were found. *Heliscus lugdunensis* MT binds at least two Cd(II), as demonstrated by mass spectrometry.

3. Cu-METALLOTHIONEIN IN *SACCHAROMYCES CEREVISIAE*

3.1. General Aspects

On the search of possible low molecular mass precursors of Cu_2Zn_2-superoxide dismutase and/or to elucidate how copper is delivered into the apo-protein, an exclusively Cu(I)-loaded MT was found and isolated from baker's yeast [24]. The mature protein consisting of 53 amino acid residues is obtained with a cysteine content of only 20%, which is lower compared to the 30% cysteine content known from mammalian MTs [25]. Unlike vertebrate MT, the yeast protein contains an aromatic histidine residue. Furthermore, the amino acid composition of yeast MT differs from its vertebrate counterparts in a higher content of glutamic acid and asparagines [26].

This protein is encoded by the *CUP1* gene, which is regulated by Cu(I) and Ag(I), but not induced by metal ions like Cd(II) or Zn(II). In the presence of Cu(I) the transcription factor CUP2 (also termed ACE1) undergoes structural changes within its cysteine-rich DNA-binding domain. Following this action it binds to the regulatory sequence of the *CUP1* gene thereby activating its transcription [26–28]. The N-terminal eight amino acid residues predicted to form the open reading frame of *CUP1* are not found in the isolated Cu-MT and are most probably cleaved off from the premature polypeptide chain [29].

Under physiological conditions yeast MT is found containing Cu(I) within its metal-thiolate cluster. The exclusive existence of Cu(I) bound to cysteine thiolate sulfur atoms was proven using X-ray photoelectron spectrometry [30]. *In vitro*, Ag(I) binds to yeast MT with the same stoichiometry of up to eight ions/molecule [25]. Within the Cu-thiolate cluster of yeast MT two different Cu(I) binding modes were suggested, as a biphasic Cu(I) replacement kinetic was observed, when bathocuproine sulfonate, a Cu(I) chelating agent, was added to Cu-MT. In this reaction 25% of Cu(I) bound to Cu-MT was released within seconds, while it took another 20 hours until the remaining Cu(I) was set free [31]. Another binding configuration was suggested for divalent metal ions including Zn(II) and Cd(II), as maximal stoichiometries of four ions/molecule were found in *in vitro* substitution experiments [25]. Reconstitution studies employing Co(II) [32] and Fe(II) [33] revealed the incorporation of four metal ions each being tetrahedrally coordinated. The fact that changes in the concentrations of divalent metal ions are not involved in the regulation of yeast MT indicates that binding of such metals, if at all, is only a physiological side reaction of this protein.

Like *N. crassa* MT Cu-MT from *S. cerevisiae* shows a characteristic orange-red luminescence exerting emission and excitation maxima at 609 nm and 277 nm, respectively [34]. The quantum yield of its luminescence was determined at 0.0058, which is relatively large compared to values of low-molecular weight Cu(I) complexes. This phenomenon is explained by a protective function of the polypeptide backbone shielding its metal-thiolate core from contact with water. Thereby, not only the quenching effect of water is reduced, but also the thiolate-coordinated Cu(I) is protected from contact with hard ligands, which would cause Cu(I) disproportionation into Cu(II) and elemental Cu^0 [35].

The sequence of yeast MT contains twelve cysteine residues. But only the N-terminal ten cysteine residues turned out to be involved in Cu(I) binding, which was shown in experiments using truncated forms lacking parts of the C-terminus including either the C-terminal two or four cysteine residues [34]. This finding displays another important difference compared to mammalian MTs, whose cysteine residues all participate in metal binding. Based on data of EXAFS (extended X-ray absorption fine structure) measurements a trigonal coordination of Cu(I) by cysteine thiolate sulfur atoms was suggested. Cu-S distances were determined at 2.23 Å. The data was also consistent with thiolate sulfur atoms bridging each of the Cu(I) to another Cu(I). Cu-Cu distances were identified at 2.7 Å and 3.9 Å [36]. Surprisingly, EXAFS measurements employing both intact Cu-supplemented yeast cells and the purified Cu-MT revealed essentially the same Cu-S distances [37]. An unexpected observation was the release of substantial amounts of intact Cu-MT from living yeast cells which were clearly not attributable to uncontrolled cell rupture [38].

3.2. NMR Studies

In a first one-dimensional ^{1}H NMR study His5 of yeast MT, which is an untypical residue for the class of MT molecules, was excluded from playing a role in metal coordination. The characteristic C(4) and C(2) resonances of His in the regions of 6.5–7.5 ppm and 8 ppm of apo- and Cu-MT were compared with each other and no significant differences were found [39].

Structural studies employing NMR techniques were impeded by the fact that ^{63}Cu and ^{65}Cu are NMR-insensitive quadrupolar nuclei, which made it impossible to obtain direct information about the Cu(I) positions within the protein. In order to circumvent the problem NMR-active ^{109}Ag(I) ($I = 1/2$) was used as an isomorphic replacement of Cu(I) [40,41]. Employing ^{1}H-^{109}Ag HMQC (heteronuclear multiple quantum coherence) experiments earlier findings that His5, Cys49, and Cys50 do not participate in metal binding were confirmed. Interestingly, seven instead of eight different frequencies of Ag(I) were found in the NMR spectra, their chemical shifts being spread from 790 ppm to 1250 ppm (Table 1). This large range of chemical shifts was explained by the occurrence of different coordination numbers for the different Ag(I). The analysis of the ^{1}H-^{109}Ag HMQC data revealed the coordination scheme depicted in Figure 3. According to this scheme a linear coordination mode is expected for three Ag(I), whereas four Ag(I) are bound in a trigonal fashion. The deep-field shifted resonances of Ag(I) I and II were explained with their linear coordination modes. The resonance of Ag(I) VI, that according to Figure 3 is supposed to be coordinated by two cysteine residues, would be more in line with trigonal coordination. This discrepancy was explained by a Ag-S coupling of Ag(I) VI with one of the cysteine

Table 1. ^{109}Ag chemical shift data for silver-substituted *S. cerevisiae* MT.

Metal resonance	Ag-MT
I	790
II	890
III	1000
IV	1037
V	1183
VI	1220
VII	1250

All values are given in parts per million.
Source: Data taken from [41] with permission from the American Chemical Society (copyright 1993).

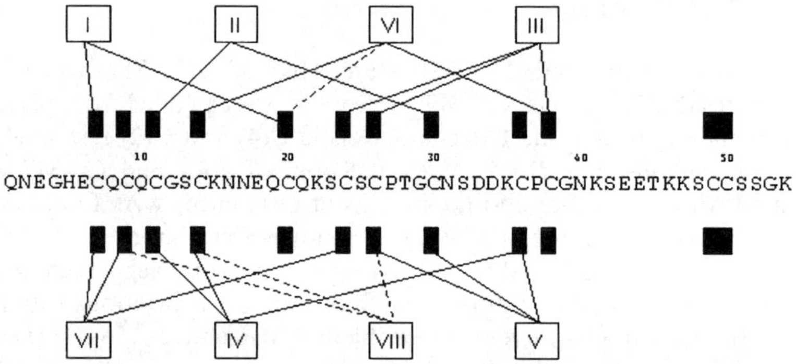

Figure 3. Schematic representation of the metal-thiolate bonds. Metal ions are indicated by roman numerals. The scheme was adopted from [41] with permission from the American Chemical Society (copyright 1993). The dashed lines and metal$_{VIII}$ were added according to the X-ray data [44].

residues which was below the detection limit. If this were the case, the metal-thiolate cluster of Ag_7-MT would be composed of two bi- and five three-coordinated Ag(I) bound to ten cysteinyl thiolate sulfur atoms [40].

An isomorphic replacement of Cu(I) by Ag(I) in yeast MT was deduced by the same metal-to-protein stoichiometry observed for both metal ions [25] and confirmed by similar patterns of the ^{1}H-^{1}H NOESY spectra of MT loaded with either metal ion [41]. The higher number of NOE crosspeaks seen for the Cu_7-MT was interpreted with a more compact protein structure in this derivative. This is in line with a smaller ionic radius of Cu(I) at 0.95 Å compared to 1.15 Å of Ag(I). The analysis of the sequential assignments of the NOESY spectra revealed that except for type I β-turns no typical secondary structure elements were present in yeast MT.

A doubling of resonances attributable to residues Lys41 to Lys53 was indicative for a higher flexibility of the C-terminus and its presence in more than one conformation. This effect was not seen using a truncated mutant lacking the C-terminal five residues. As Cys49 and Cys50 were absent in this mutant, it might well be that these two residues are responsible for the flexibility of the native protein by interacting with other cysteine residues or metal ions [41].

Exploiting the available NMR data, Peterson, Narula, and Armitage performed structure calculations for both the Cu(I) and the Ag(I) containing form of yeast MT [42]. Under the assumption of an isomorphic replacement of Cu(I) by Ag(I) metal to cysteine connectivities of the Ag(I) derivative were included in the structure calculations of Cu(I)-MT. As expected, the two structures shared great similarity. The metal-thiolate cluster was embedded in a pocket of two parallel loops being solvent-accessible only to one side. Again as expected from the NMR raw data the cluster pocket of Cu-MT was smaller than that of its Ag(I) counterpart due to the smaller ionic radius of Cu(I). In the metal-thiolate cluster obtained the Cu(I) were positioned on seven corners of a distorted cube. Seven of nine neighboring Cu(I) were found to be at the same distance of 3.31 Å apart from each other.

The hypothesis of an isomorphical replacement of Cu(I) by Ag(I), however, was challenged for the known differences of these two metal ions with respect to their coordination chemistry and ionic radii [13]. Therefore, another structure determination of Cu(I)-MT was performed only relying on 2D ^{1}H-^{1}H TOCSY and NOESY data obtained on a 800 MHz NMR spectrometer. Having high-resolution NMR spectra at hand 1048 NOEs could be used as input for structure calculations, which resulted in a highly resolved conformation of the polypeptide part of this protein [43]. The overall polypeptide fold was similar, but not identical to the one presented by Peterson et al. [42]. As before, a higher flexibility was seen for the N- and C-terminal end, residues 1–5 and 41–53. The structure of the remaining part including all ten metal coordinating cysteine residues fulfilled the authors' expectations in that it was of a higher quality compared to the one previously available. Therefore, it was tried to model the Cu(I) into the exactly defined free spaces between the highly resolved cysteinyl sulfur atoms to find the correct architecture of the metal-thiolate cluster. Unfortunately, rather than one, a whole set of possibilities of filling yeast apo-MT with seven Cu(I) were found, all of them being chemically plausible [13]. The arrangement proposed by Peterson et al. [42] was among these possibilities, not necessarily meaning that this was the correct one. To explain the result of more than one cluster structure employing the modelling approach, a certain flexibility of Cu(I) bound to yeast MT was proposed including the possibility of changing coordination partners for the metal ions. Subtle changes in the overall protein structures due to such flexibility were suggested to be the reason why earlier attempts to crystallize the protein had failed.

3.3. Crystal Structure

Based on the NMR structural result the flexibility of the C- and N-terminal ends were discussed to prevent the crystallization of the full-length protein [43]. In order to tackle that problem two truncated mutants, P_{5-40} and P_{-1-40}, were designed lacking either the C-terminal or both flexible ends (Figure 1). An additional tyrosine residue was added to the longer fragment as an aromatic residue in the exposed first position was expected to facilitate crystallization.

The two truncated mutants were synthesized employing solid phase peptide synthesis and reconstituted with Cu(I). The integrity of the Cu(I)-thiolate cluster in the shortened polypeptide matrix was demonstrated using electronic absorption, circular dichroism, luminescence, and ^{1}H NMR. The shortened fragments displayed virtually identical spectroscopic properties and even a structure determination of the longer fragment exploiting its ^{1}H NMR data revealed a structure identical to that of the native protein. The shorter fragment was also suggested to possess an identical structure as the native protein due to the similarity of its spectra with those obtained for the other two species. A separate structure determination was omitted.

Fortunately, crystallization of the shortest yeast MT fragment P_{5-40} was successful [44]. Attempts to crystallize the native protein or the truncated fragment P_{-1-40} under the same or similar conditions all resulted in amorphous precipitates unsuitable for X-ray crystallography [45]. Using the small cubic crystals obtained for P_{5-40} the structure of the residues 5 to 40 of native yeast MT including its metal-thiolate core was determined (Figure 4). The fold of the polypeptide backbone was essentially the same as proposed in earlier studies using NMR [13,42,43]. This structure, however, was the first to show the exact architecture of the metal-thiolate core of this protein (Figure 5). Intriguingly, a stoichiometry of eight rather than seven Cu(I) per molecule MT and two different binding modes for Cu(I) were found. Six of the Cu(I) ions, Cu37 and Cu39–43, are coordinated trigonally by cysteine thiolate sulfur atoms, whereas two, Cu38 and Cu44, are bound in a linear fashion. The eight Cu(I) are ordered in two groups of four Cu(I) each, thereby forming two distorted tetrahedrons. The one comprising Cu39, Cu40, Cu41, and Cu44 displays shorter Cu-Cu distances ranging from 2.59 to 2.91 Å, whereas in the other tetrahedron formed by Cu37, Cu38, Cu42, and Cu43 slightly longer

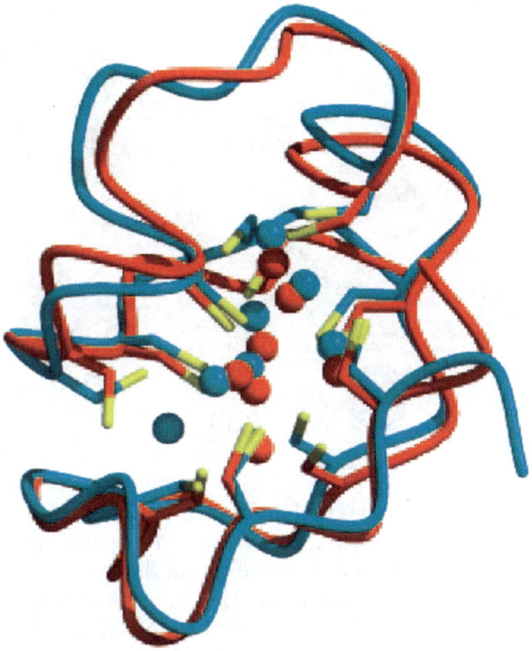

Figure 4. Superposition of the X-ray Cu_8-MT (cyan tube) and the NMR Cu_7-MT (red tube) structures from *S. cerevisiae*. The Cu(I) of each structure are represented in the same color as the respective polypeptide backbone. The eighth cyan Cu sphere is overlapped by the lowest red Cu ball. The cysteine side chains are also shown. Reproduced from [44] with permission from the National Academy of Sciences, USA (copyright 2005).

Cu-Cu distances of 2.78–3.12 Å are seen. For Cu-Cu distances below the sum of their Van-der-Waals radii, which is 2.8 Å, the interactions between the respective pair of Cu(I) ions include some metal-metal bonding character.

All ten cysteine residues participate in the formation of the metal-thiolate cluster. Eight of the thiolate sulfur atoms each coordinate two Cu(I). The other two cysteine sulfur atoms of Cys10 and Cys22 each bind three Cu(I) and are thus surrounded by 4 atoms. These two cysteine residues are located in the center of the whole metal-thiolate cluster both connecting the two aforementioned two groups of four Cu(I) each.

The X-ray structure of yeast MT delivers the explanation for its characteristic luminescence features. As mentioned above, a strikingly high quantum yield of 0.0058 was determined for this Cu(I)-binding protein, which was explained by an effective shielding of the Cu-thiolate cluster from interaction with solvent molecules. As was already shown by the NMR

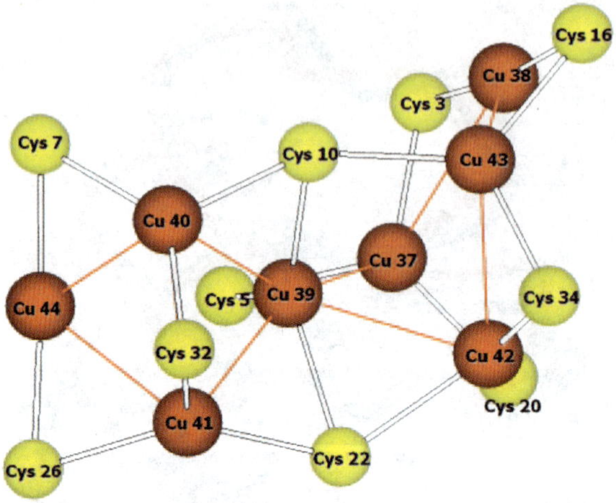

Figure 5. Structure of the Cu(I)-thiolate cluster of *S. cerevisiae*. Cu(I) are represented by orange-brown and cysteine sulfur atoms by yellow spheres. The cysteine residue numbering refers to the cysteine positions in the truncated polypeptide P_{5-40} used for the X-ray structure determination. The orange-brown bonds connect Cu(I) with interatomic distances <3.0 Å. Reproduced from [44] with permission from the National Academy of Sciences, USA (copyright 2005).

structures and again approved by the X-ray structure, the cysteine sulfur atoms are all located in the inside of the polypeptide, such that the Cu(I) bound to the large hydrophobic thiolate sulfur atoms are inaccessible for solvent molecules. Consequently, the luminescence quenching due to interactions with solvent molecules is diminished, which in turn leads to this relatively high quantum yield.

An in-depth inspection of the superposition of the Cu_7-MT NMR structure [43] and the Cu_8-MT X-ray structure [44] revealed another intriguing finding. All cysteine residues of Cu_7-MT were found in similar conformations as their counterparts in Cu_8-MT, with the only exception of Cys11 (NMR nomenclature). The close-up view on that region is seen in Figure 6. The cysteine thiolate sulfur of this Cys11 points into the opposite direction with respect to the X-ray structure towards the outside of the protein. In this conformation Cys11 is incapable of participating in the formation of the Cu(I)-thiolate cluster.

As mentioned above, Cu_8-MT contains two Cu(I) that are more easily removed than the other six [31]. Thus, Cu-MT in solution consists of a mixture with stoichiometries of six to eight Cu(I) per molecule MT. The more labile Cu(I) are very likely the ones that are coordinated only by two

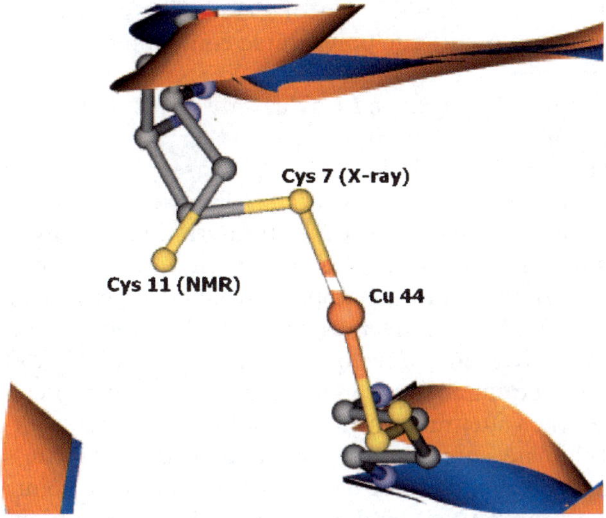

Figure 6. High magnification display of the Cys7(X-ray)-Cu44 region of the Cu_8-MT, superimposed to the same region of the Cu_7-MT NMR structure. A different orientation of the corresponding cysteine residues Cys11 (NMR) and Cys7 (X-ray) is seen. The different cysteine numbering comes from the use of the full length protein for the NMR structure determination and the truncated form P_{5-40} for the X-ray structure determination. Reproduced from [44] with permission from the National Academy of Sciences, USA (copyright 2005).

cysteine residues, namely Cu44 and Cu38. Those two Cu(I) are bound at the opposite ends of the cluster. Interestingly, Cu44 is coordinated by the abovementioned Cys11 (NMR nomenclature). Therefore, it was suggested that Cu44 was the Cu(I) missing in the structure of Cu_7-MT and that the cysteine residues coordinating the diagonally bound Cu(I) play a role during the uptake or release of Cu(I) by switching their conformations.

4. METALLOTHIONEIN-LIKE PROTEIN Crs5 IN *SACCHAROMYCES CEREVISIAE*

CRS5 (copper-resistant suppressor 5) was identified as a novel gene conferring copper resistance in yeast cells lacking the *CUP1* gene, that encodes for the yeast Cu-MT covered in Section 3 [46]. The sequence of the polypeptide encoded by *CRS5* exhibits similarities to mammalian and invertebrate MTs and shows all features to be classified as a MT. Only very little homology, however, to Cu-MT was observed. The open reading frame of *CRS5* encodes for a 69 residue polypeptide containing 19 cysteine residues, many of them ordered in Cys-X-Cys, Cys-X-Y-Cys, or Cys-Cys configurations. Other frequently occurring residues are lysine and serine, while aromatic residues are totally absent.

The regulation of transcription of *CRS5* shares the same mechanisms as shown for *CUP1* employing the Cu(I) binding transcription factor CUP2 [46]. Besides the up-regulation of *CRS5* transcription as a response towards higher levels of Cu(I), a constitutive expression of its protein was observed [47]. *CUP1* contributes more to copper resistance of yeast than does *CRS5* [47]. Part of the higher effectiveness is due to the action of the *CUP1* promoter, another part being due to a higher avidity of Cu-MT for Cu(I). Comparison of the luminescence emission of the two proteins revealed a lower luminescence intensity of Crs5 indicating a higher solvent accessibility of its metal-thiolate core. Crs5 is found to bind 11–12 Cu(I). Exchange reactions with the Cu(I) chelator bathocuproine sulfonate pointed towards a kinetically more labile Cu(I) binding in Crs5. Moreover, a biphasic reaction with bathocuproine sulfonate was observed, in which approximately 35% of the Cu(I) bound by Crs5 are transferred to the Cu(I) chelator within 5 minutes.

As deletion of Crs5 resulted in higher sensitivity towards Cu(I) toxicity, this protein was first suggested to play a role in Cu(I) homeostasis and detoxification [46]. The protein, however, is also capable of binding divalent metal ions such as Zn(II) and Cd(II). *In vitro* Crs5 Cd(II) binding capacity was determined at six Cd(II) per Crs5 molecule [47]. The results of a more recent study suggested that Crs5 confers resistance towards Zn(II) overload [48]. From yeast cells cultivated in Cu-supplemented media, heterometallic Zn, Cu-Crs5 was isolated. Only under certain physiological conditions, i.e., under anaerobic conditions, homometallic Cu-Crs5 was obtained. Therefore, and also from its primary structure Crs5 was suggested to be the evolutionary ancestor of the eukaryotic Zn-MT lineage.

5. METALLOTHIONEIN-1 AND METALLOTHIONEIN-2 IN *CANDIDA GLABRATA*

Two distinct families of Cu(I) inducible MTs, designated MT-1 and MT-2, were identified in the human pathogenic yeast *Candida glabrata* [49]. Whereas the MT-1 family appears to consist of only one member, two isoforms of MT-2 were isolated. MT-1 and MT-2 consist of 62 and 51 amino acid residues, respectively [50]. Besides the full-length MT-2, a processed shorter mutant lacking the first six amino acid residues at the N-terminus was observed under certain environmental conditions. Both MTs lack any homology between each other and with vertebrate MTs. As members of the MT superfamily they both have a high content of cysteine residues in their primary structures. Out of the 18 cysteine residues of MT-1 14 are found in Cys-X-Cys and two in a Cys-X-X-Cys sequence, motives typical for MTs. MT-2 contains 16 cysteine residues out of which 14 are found in a Cys-X-Cys motif. As Cys6 of MT-2 is absent in the aforementioned truncated isoform, this residue is likely not to be involved in metal-thiolate cluster formation. Deviating from the general definition of MTs, unusual aromatic tyrosine and histidine residues are found in the sequences of MT-1 and MT-2 from *C. glabrata*. Another characteristic feature is the occurrence of internal sequence repeats. In MT-1 the sequences Cys-X-Cys-Pro-Asn and Cys-Gly-Asp-Lys-Cys-Glu-Cys-Lys are both found twice, while the pentapeptide Gln-Thr-Cys-Lys-Cys is repeated in MT-2. This indicates that the two polypeptides were elongated in the course of evolution by gene duplications.

The metal-binding properties of MT-1 and MT-2 from *C. glabrata* have not yet been examined in such detail as for example that of Cu-MT from *S. cerevisiae*. Based on luminescence and UV absorption properties, copper ions were suggested to be bound as Cu(I) by cysteine thiolate sulfur atoms. The stoichiometries were determined to be 11–12 and 10 for MT-1 and MT-2, respectively [50]. The binding of other metal ions was not studied, but most probably does not play any physiological role, as indicated by the sole regulation of their expression by Cu(I) and Ag(I) (see below). Thus far the Cu(I)-thiolate cluster of Cu_8-MT from *S. cerevisiae* is the largest oligonuclear Cu(I)-thiolate cluster of which a structure is available [44]. For the two MTs from *C. glabrata* even higher Cu(I):protein stoichiometries are suggested. In this view, it would be interesting to know whether they are all ordered to form a single cluster in these proteins or whether more than one domain is formed upon Cu(I) binding. In *S. cerevisiae* Cu-MT eight Cu(I) are coordinated to ten cysteine thiolates. Two cysteine residues do not participate in the cluster formation. For *C. glabrata* MT-1 and MT-2 Cu(I):Cys ratios of 11–12:18 and 10:16 were determined, which are lower than the 8:10 ratio of the *S. cerevisiae* Cu-MT cluster. One possible explanation might be that not all cysteine residues of the two *C. glabrata* MTs participate in Cu(I) binding, as was suggested above for Cys6 of MT-2. A lower Cu(I):Cys ratio might also indicate a lower number of bridging thiolate sulfur atoms, the absence of thiolate sulfur atoms coordinating three Cu(I) and/or less or no Cu(I) that are coordinated linearly.

C. glabrata MT-1 and MT-2 are encoded by three different genes, *MT-1, MT-2a and MT-2b* [51]. Whereas *MT-1 and MT-2b* occur as single-copy genes, various copies of tandemly amplified *MT-2a* are found in the genomes of different strains of *C. glabrata* [52]. The coding sequences of *MT-2a* and *MT-2b* are identical. Slight differences in the 5′ regions, substantial differences in the 3′ regions of the two genes are observed. *MT-2a* has autonomously replicating sequences in its 3′ region that are missing in *MT-2b* [51]. Like in the case of Cu-MT from *S. cerevisiae* the expression of MT-1 and MT-2 in *C. glabrata* is regulated by Cu(I) and Ag(I), but not by divalent metal ions like Zn(II) and Cd(II) [50]. Comparable with the regulation of expression of *S. cerevisiae* Cu-MT, a Cu(I) sensing transcription factor, termed Amt1, was identified [53]. The DNA-binding domain of Amt1 is located at the N-terminus. In its activated state it harbors a $Cu(I)_4$ cluster and is structurally different from its inactive apo-form. Additionally, a binding site for Zn(II) was found nearby the tetra-copper cluster. DNA binding, however, only depends on the presence of Cu(I), but not of Zn(II) [54]. Amt1 regulates all three *C. glabrata* MT promoters. Moreover, it even works with the *CUP1* promoter from *S. cerevisiae*, whereas *S. cerevisiae* CUP2 is capable of activating only *C. glabrata MT-1*, but not *MT-2a* and *MT-2b* [55].

6. ZINC AND CADMIUM BUFFERING SYSTEMS

Unlike mammalian MTs fungal MTs described above appear solely induced by copper ions and mainly function in the sequestration of copper ions within their hosts. *C. glabrata* was the first fungus for which the parallel existence of both MTs and γ-glutamyl peptides were reported [49]. Whereas the expression of the first is stimulated by copper, the latter is produced in response to cadmium stress. Such phytochelatins were also identified in *S. cerevisiae*, synthesized in the presence of Cd, Zn, and Cu ions, and *N. crassa*, where the synthesis was only proven in response to Cd ions [56]. For further information about phytochelatins see Chapter 15 of this book.

MTs that are expressed in response to Cd(II) have already been mentioned in Section 2. Zym1 represents a MT involved in the sequestration of Zn(II) in *Schizosaccharomyces pombe* [57]. This MT is induced by Zn but not Cu ions. The protein shares some similarity with mammalian MTs, consisting of only 51 amino acid residues and contains only twelve cysteine residues which are capable of binding up to four equivalents of Zn(II). The cysteine spacing differs from that of the mammalian MTs. Zym1 has not yet been assigned to a family of the MT superfamily system.

7. CONCLUDING REMARKS

As demonstrated above, most of the fungal MTs are induced by copper and are isolated as Cu-MTs. Despite their sequential diversity the mechanisms of their copper induced regulation follow general principles. In the presence of elevated intracellular copper concentrations copper-sensing transcription factors bind Cu(I) via cysteine thiolates. Thereby structural changes within the transcription factors occur, thus enabling them to bind to certain DNA motifs and to induce the transcription of the respective MT genes.

Unlike the well studied mammalian Zn/Cd-MTs only limited knowledge is available on the structures of both mammalian and fungal Cu-MTs. *S. cerevisiae* Cu-MT is the only MT of which the protein structure including its metal-thiolate core is known [44]. The characteristics of its metal binding and the exact Cu(I) to MT stoichiometry have been clarified unambiguously from its crystal structure. As known from mammalian Zn/Cd-MTs the formation of a metal-thiolate cluster located in the center of the protein determines the fold of the polypeptide chain, which is wrapped around the cluster and fixed to it via its cysteine residues. Out of the eight Cu(I) two are coordinated by only two thiolate sulfur ligands, whereas the remaining six are tri-coordinated. On the other side eight of the ten thiolate ligands have been found to bridge two Cu(I) each, while two cysteine residues at central positions of the cluster bind even three Cu(I). These variations regarding the Cu(I) coordination chemistry have to be kept in mind, when trying to deal with structures of other Cu-MTs.

Mammalian MTs are usually isolated as mixed Zn,Cu-MTs. In fetal liver the copper concentration is much more distinct [58,59]. Recently, in a study about the NMR structures of Cu(I)-loaded domains of murine MT-1 [17] the stoichiometries of Cu(I) binding to the single α- and β-domains was three and four, respectively. They adopt detectable protein folds distinct from their Zn(II)-loaded structures. The NMR structures of the two Cu(I)-loaded domains, however, lack information about the spacial positioning of the Cu(I). Knowledge gathered during the elucidation of the *S. cerevisiae* Cu-MT structure might facilitate the complement of this information and therefore help to understand the function of Cu-MT in mammals.

An interesting issue in MT research is the question about what dictates the metal specificity of different members of the MT superfamily. One might divide MTs belonging either to the Zn or the Cu lineage depending on the metal with which they are usually isolated [48]. The preference for either metal can be explained by the sole availability of the respective metal due to presence of excess binding capacity for the other metal. This would mean that MTs do not have a certain specificity for certain metal ions and that their loading is determined by the whole network of proteins and low molecular compounds regulating the intracellular availability of Zn and Cu ions. The other possibility is that the specificity of MTs is determined by their primary structure. Interestingly, family 8 contains highly homologous MTs both of the Cu-type as well as of the Zn/Cd-type. Therefore, it would be interesting to analyze the sequences of the different members of this family in order to find out whether any differences in the primary structure account for the different metal specificities.

ACKNOWLEDGMENTS

The fruitful collaboration and stimulating discussions with many colleagues including I. Bertini, C. Del Bianco, G. Liu, and C. Luchinat (Florence), V. Calderone, S. Mangani (Siena), H. Echner, T. Klein, and W. Voelter (Tübingen) are warmly acknowledged. Support from the German National Research Foundation (DFG) and of the European Community Access to Research Infrastructure action of the Improving Human Potential (LSF user project) is gratefully acknowledged. B. Dolderer was a recipient of a predoctoral fellowship of the Konrad-Adenauer-Stiftung. This work was partly supported by EC programs SPINE II (n° LSHG-CT-2006-031220) and UPMAN (n° LSHG-CT-2004-512052), and by Ente CR Firenze.

ABBREVIATIONS

Crs5	copper-resistant suppressor 5
EXAFS	extended X-ray absorption fine structure
HMQC	heteronuclear quantum coherence
MT	metallothionein
NMR	nuclear magnetic resonance
NOESY	nuclear Overhauser effect spectroscopy
P_{-1-40}	truncated fragment of yeast Cu-MT comprising residues 1–40 of native yeast MT plus an additional tyrosine residue at the N-terminus
P_{5-40}	truncated fragment of yeast Cu-MT including residues 5–40 of native yeast MT
TOCSY	total correlation spectroscopy

REFERENCES

1. Y. Kojima, *Methods Enzymol.*, 1991, **205**, 8–10.
2. Y. Kojima, P.-A. Binz and J. H. R. Kägi, in *Metallothionein IV*, Ed. C. Klaassen, Birkhäuser Verlag, Basel, 1999, pp. 3–6.
3. P.-A. Binz and J. H. R. Kägi, in *Metallothionein IV*, Ed. C. Klaassen, Birkhäuser Verlag, Basel, 1999, pp. 7–13.
4. B. Dujon, D. Sherman, G. Fischer, P. Durrens, S. Casaregola, I. Lafontaine, J. De Montigny, C. Marck, C. Neuveglise, E. Talla, N. Goffard, L. Frangeul, M. Aigle, V. Anthouard, A. Babour, V. Barbe, S. Barnay, S. Blanchin, J. M. Beckerich, E. Beyne, C. Bleykasten, A. Boisrame, J. Boyer, L. Cattolico, F. Confanioleri, A. De Daruvar, L. Despons, E. Fabre, C. Fairhead, H.

Ferry-Dumazet, A. Groppi, F. Hantraye, C. Hennequin, N. Jauniaux, P. Joyet, R. Kachouri, A. Kerrest, R. Koszul, M. Lemaire, I. Lesur, L. Ma, H. Muller, J. M. Nicaud, M. Nikolski, S. Oztas, O. Ozier-Kalogeropoulos, S. Pellenz, S. Potier, G. F. Richard, M. L. Straub, A. Suleau, D. Swennen, F. Tekaia, M. Wesolowski-Louvel, E. Westhof, B. Wirth, M. Zeniou-Meyer, I. Zivanovic, M. Bolotin-Fukuhara, A. Thierry, C. Bouchier, B. Caudron, C. Scarpelli, C. Gaillardin, J. Weissenbach, P. Wincker and J. L. Souciet, *Nature*, 2004, **430**, 35–44.

5. K. Lerch, *Nature*, 1980, **284**, 368–370.
6. K. Munger, U. A. Germann and K. Lerch, *EMBO J.*, 1985, **4**, 2665–2668.
7. K. Munger, U. A. Germann and K. Lerch, *J. Biol. Chem.*, 1987, **262**, 7363–7367.
8. M. Beltramini and K. Lerch, *Biochemistry*, 1983, **22**, 2043–2048.
9. M. Beltramini and K. Lerch, *FEBS Lett.*, 1982, **142**, 219–222.
10. M. Beltramini and K. Lerch, *FEBS Lett.*, 1981, **127**, 201–203.
11. M. Beltramini, K. Lerch and M. Vašák, *Biochemistry*, 1984, **23**, 3422–3427.
12. P. A. Cobine, R. T. McKay, K. Zangger, C. T. Dameron and I. M. Armitage, *Eur. J. Biochem.*, 2004, **271**, 4213–4221.
13. I. Bertini, H. J. Hartmann, T. Klein, G. Liu, C. Luchinat and U. Weser, *Eur. J. Biochem.*, 2000, **267**, 1008–1018.
14. R. Bofill, O. Palacios, M. Capdevila, N. Cols, R. Gonzalez-Duarte, S. Atrian and P. Gonzalez-Duarte, *J. Inorg. Biochem.*, 1999, **73**, 57–64.
15. J. A. Malikayil, K. Lerch and I. M. Armitage, *Biochemistry*, 1989, **28**, 2991–2995.
16. S. S. Narula, M. Brouwer, Y. Hua and I. M. Armitage, *Biochemistry*, 1995, **34**, 620–631.
17. B. Dolderer, H. Echner, A. Beck, H. J. Hartmann, U. Weser, C. Luchinat and C. Del Bianco, *FEBS J.*, 2007, **274**, 2349–2362.
18. K. Munger and K. Lerch, *Biochemistry*, 1985, **24**, 6751–6756.
19. N. B. Averbeck, C. Borghouts, A. Hamann, V. Specke and H. D. Osiewacz, *Mol. Gen. Genet.*, 2001, **264**, 604–612.
20. K. Marbach, J. Fernandez-Larrea and U. Stahl, *Curr. Genet.*, 1994, **26**, 184–186.
21. H. D. Osiewacz and U. Nuber, *Mol. Gen. Genet.*, 1996, **252**, 115–124.
22. C. S. Hwang and P. E. Kolattukudy, *Mol. Gen. Genet.*, 1995, **247**, 282–294.
23. P. Jaeckel, G. Krauss, S. Menge, A. Schierhorn, P. Rucknagel and G. J. Krauss, *Biochem. Biophys. Res. Commun.*, 2005, **333**, 150–155.
24. R. Prinz and U. Weser, *Hoppe Seylers Z. Physiol. Chem.*, 1975, **356**, 767–776.
25. D. R. Winge, K. B. Nielson, W. R. Gray and D. H. Hamer, *J. Biol. Chem.*, 1985, **260**, 14464–14470.
26. T. R. Butt, E. J. Sternberg, J. A. Gorman, P. Clark, D. Hamer, M. Rosenberg and S. T. Crooke, *Proc. Natl. Acad. Sci. USA*, 1984, **81**, 3332–3336.
27. C. Buchman, P. Skroch, J. Welch, S. Fogel and M. Karin, *Mol. Cell Biol.*, 1989, **9**, 4091–4095.
28. M. Karin, R. Najarian, A. Haslinger, P. Valenzuela, J. Welch and S. Fogel, *Proc. Natl. Acad. Sci. USA*, 1984, **81**, 337–341.

29. C. F. Wright, K. McKenney, D. H. Hamer, J. Byrd and D. R. Winge, *J. Biol. Chem.*, 1987, **262**, 12912–12919.
30. U. Weser, H. J. Hartmann, A. Fretzdorff and G. J. Strobel, *Biochim. Biophys. Acta*, 1977, **493**, 465–477.
31. U. Weser and H. J. Hartmann, *Biochim. Biophys. Acta*, 1988, **953**, 1–5.
32. H. J. Hartmann and U. Weser, *Biochem. Biophys. Res. Commun.*, 1985, **132**, 277–283.
33. X. Q. Ding, E. Bill, A. X. Trautwein, H. J. Hartmann and U. Weser, *Eur. J. Biochem.*, 1994, **223**, 841–845.
34. J. Byrd, R. M. Berger, D. R. McMillin, C. F. Wright, D. Hamer and D. R. Winge, *J. Biol. Chem.*, 1988, **263**, 6688–6694.
35. H. Rupp, R. Cammack, H. J. Hartmann and U. Weser, *Biochim. Biophys. Acta*, 1979, **578**, 462–475.
36. G. N. George, J. Byrd and D. R. Winge, *J. Biol. Chem.*, 1988, **263**, 8199–8203.
37. A. Desideri, H. J. Hartmann, S. Morante and U. Weser, *Biol. Met.*, 1990, **3**, 45–47.
38. K. Felix, H. J. Hartmann and U. Weser, *Biol. Met.*, 1989, **2**, 50–54.
39. J. Bordas, M. H. J. Koch, H. J. Hartmann and U. Weser, *Inorg. Chim. Acta*, 1983, **78**, 113–120.
40. S. S. Narula, R. K. Mehra, D. R. Winge and I. M. Armitage, *J. Am. Chem. Soc.*, 1991, **113**, 9354–9358.
41. S. S. Narula, D. R. Winge and I. M. Armitage, *Biochemistry*, 1993, **32**, 6773–6787.
42. C. W. Peterson, S. S. Narula and I. M. Armitage, *FEBS Lett.*, 1996, **379**, 85–93.
43. C. Luchinat, B. Dolderer, C. Del Bianco, H. Echner, H. J. Hartmann, W. Voelter and U. Weser, *J. Biol. Inorg. Chem.*, 2003, **8**, 353–359.
44. V. Calderone, B. Dolderer, H. J. Hartmann, H. Echner, C. Luchinat, C. Del Bianco, S. Mangani and U. Weser, *Proc. Natl. Acad. Sci. USA*, 2005, **102**, 51–56.
45. B. Dolderer, *On the Molecular Architecture of Cu(I) Thiolate Centres in Cu-Thioneins*, Ph.D. Thesis, University of Tübingen, 2004.
46. V. C. Culotta, W. R. Howard and X. F. Liu, *J. Biol. Chem.*, 1994, **269**, 25295–25302.
47. L. T. Jensen, W. R. Howard, J. J. Strain, D. R. Winge and V. C. Culotta, *J. Biol. Chem*, 1996, **271**, 18514–18519.
48. A. Pagani, L. Villarreal, M. Capdevila and S. Atrian, *Mol. Microbiol.*, 2007, **63**, 256–269.
49. R. K. Mehra, E. B. Tarbet, W. R. Gray and D. R. Winge, *Proc. Natl. Acad. Sci. USA*, 1988, **85**, 8815–8819.
50. R. K. Mehra, J. R. Garey, T. R. Butt, W. R. Gray and D. R. Winge, *J. Biol. Chem.*, 1989, **264**, 19747–19753.
51. R. K. Mehra, J. L. Thorvaldsen, I. G. Macreadie and D. R. Winge, *Gene*, 1992, **114**, 75–80.
52. R. K. Mehra, J. R. Garey and D. R. Winge, *J. Biol. Chem.*, 1990, **265**, 6369–6375.
53. P. B. Zhou and D. J. Thiele, *Proc. Natl. Acad. Sci. USA*, 1991, **88**, 6112–6116.
54. J. L. Thorvaldsen, A. K. Sewell, A. M. Tanner, J. M. Peltier, I. J. Pickering, G. N. George and D. R. Winge, *Biochemistry*, 1994, **33**, 9566–9577.

55. J. L. Thorvaldsen, A. K. Sewell, C. L. McCowen and D. R. Winge, *J. Biol. Chem.*, 1993, **268**, 12512–12518.
56. R. Kneer, T. M. Kutchan, A. Hochberger and M. H. Zenk, *Arch. Microbiol.*, 1992, **157**, 305–310.
57. G. P. Borrelly, M. D. Harrison, A. K. Robinson, S. G. Cox, N. J. Robinson and S. K. Whitehall, *J. Biol. Chem.*, 2002, **277**, 30394–30400.
58. H. J. Hartmann and U. Weser, *Biochim. Biophys. Acta.*, 1977, **491**, 211–222.
59. H. J. Hartmann and U. Weser, in *Biological Inorganic Chemistry, Structure and Reactivity*, Eds. I. Bertini, H. B. Gray, E. I. Stiefel and J. S. Valentine, University Science Books, Sausalito, California, 2007, pp. 156–162.

Met. Ions Life Sci. **2009**, *5*, 107–153

5

Metallothioneins in Plants

Eva Freisinger

Institute of Inorganic Chemistry, University of Zürich, Winterthurerstrasse 190, CH-8057 Zürich, Switzerland

<freisinger@aci.uzh.ch>

Metal Ions in Life Sciences, Volume 5 Edited by Astrid Sigel, Helmut Sigel, and Roland K. O. Sigel

Published by the Royal Society of Chemistry, www.rsc.org DOI: 10.1039/9781847558992-00107

ABSTRACT: The earliest perception restricted the occurrence of metallothioneins to animals, fungi, and certain bacteria and assigned the corresponding functions in plants to the enzymatically synthesized phytochelatins. This picture has now clearly changed, and the existence of plant metallothioneins is generally accepted. Compared to the vertebrate forms, plant metallothioneins show an intriguing sequence diversity, which is contrasted by an enormous lack of information about their possible functions, properties, and structures. The majority of functions proposed is derived from gene expression studies, which are reviewed with a not exhaustive but representative list of examples. However, it becomes evident that so far it is not possible to draw a clear-cut picture that ascribes specific functions to a given metallothionein. Rather, proposed functions overlap or are complementary to each other; sometimes even contradictory results are obtained. Regarding the structures and properties of the plant metallothionein proteins the amount of literature available is, compared to that of other MTs, considerably less comprehensive illustrating the early stages, in which this research area resides. Informations are at hand concerning the metal ion content of the different plant metallothionein species and metal-thiolate cluster stabilities, the number of clusters formed is proposed and predictions regarding the secondary structure are made. Three-dimensional data is so far restricted to one domain of a metallothionein from wheat providing fascinating insights into a cluster arrangement never reported for any metallothionein so far.

KEYWORDS: apparent pK_a values · gene expression · His residues · metal-thiolate clusters · plant metallothioneins · recombinant metallothioneins · sulfide · yeast complementation studies

1. INTRODUCTION

While the discovery of the first mammalian metallothionein (MT) dates back more than 50 years [1], the first cysteine-rich small protein from plant material was isolated only 25 years ago from wheat germs (*Triticum aestivum*) [2]. This protein is solely encoded by the mRNA conserved in dry mature wheat embryos showing declining levels already around 5 hours past seed imbibition. Accounting for 20–25% of the total cysteine (Cys) incorporation into nascent proteins, the protein was denoted "early cysteine-labeled protein" or E_c. Initially, E_c was isolated in form of the S-alkylated protein following iodoacetamide treatment of the soluble protein fraction obtained from wheat germs in a yield of 5–10 mg per 100 g of plant material.

Solely based on the peculiar amino acid composition without any knowledge of the sequence or information concerning the potential metal ion binding abilities, a relationship of E_c with the likewise Cys-rich mammalian MTs was already suggested at that time [2]. However, it was not until the determination of the partial amino acid sequence, the metal ion content, and spectrophotometric measurements on the Cd^{II}-substituted form that the designation of *T. aestivum* E_c as the first metallothionein isolated from a higher plant was substantiated by sufficient evidence [3]. Apart from the presence of histidine (His) residues, E_c actually meets all criteria used to designate Cys- and metal ion-rich proteins as metallothioneins: A low molecular weight (M_r 5 000–10 000 Da), the absence of aromatic amino acids (other than His), a characteristic distribution of Cys residues along the amino acid chain, e.g. the presence of CysXaaCys motifs with Xaa marking any amino acid other than Cys, and the characteristic optical features of metal-thiolate complexes such as a shoulder of the peptide backbone transitions around 250 nm for the Cd^{II}-substituted forms [4,5]. Since then, the number of newly discovered plant MT sequences, mostly on the mRNA level, has increased significantly. For example, a search in the Swiss-Prot protein sequence database for the term "metallothionein" currently lists 259 entries [6]. 64 of these sequences originate from the plant kingdom.

Nevertheless, only for *T. aestivum* E_c-1 evidence is available for its existence on the protein level. No other plant MT was isolated directly from plant material so far. The existence of 23 sequences listed in the database was only confirmed on the transcript level, all other sequences are designated with the protein existence criteria "inferred by homology". This constriction is based on the fact that small RNAs can be both, actual transcripts for small proteins such as MTs or regulatory, not translated RNAs. Therefore, the sole presence of *mt* mRNA in a plant tissue is not sufficient for a protein confirmation on the transcript level in the sense used by the Swiss-Prot database. As additional differentiating criteria such as the inducibility of genes are often just not listed in the database due to unknown reasons although this inducibility is described in the corresponding literature, it can be assumed that the majority of MT sequences listed have indeed been identified as such based on gene expression trials and do encode for MT proteins.

The number of mRNA transcripts putatively encoding plant MTs gets even larger when additionally considering a total of 570 MT sequences from all different phylae of life deposited in the TrEMBL database, a computer-annotated supplement of the Swiss-Prot database [6]. So far, investigations based on gene expression trials represent the major source of informations about plant MTs. Actual knowledge about the translational products of these mRNAs, the MT proteins, however, is scarce. Not only is *T. aestivum* E_c-1 the only plant MT available from its native source, moreover, also the number of recombinant plant MT forms obtained by overexpression in a bacterial host is limited to currently ten [3,7–18]. This number even has to be cut down to seven, when disregarding the three MT forms that were solely studied in form of their glutathione S-transferase (GST) fusion proteins [7,14,18]. The GST tag consists of 220 amino acids and is commonly used for purification and detection purposes. The tag alone accounts thus for approximately 75% of total residues in the GST-MT fusion protein, and extra caution has to be applied when trying to confer properties determined for the fusion protein onto the relatively small MT part.

With respect to the proposed functions of plant MTs we also have to rely mainly on conclusions drawn from gene expression studies conducted on plants at different developmental stages and under certain stress conditions such as exposure to heavy metal ions. The use of *mt*-knockout plants for functional studies is presumably a rather tedious endeavor due to the multiple MT subfamilies found in plants, the members of which additionally most likely have overlapping functions [19]. A different approach to obtain informations about the function of plant MTs makes use of complementation studies with MT-deficient or metal ion-sensitive *Saccharomyces cerevisiae* cell lines [20–23]. Due to the lack of three-dimensional structural data from NMR spectroscopic or crystallographic studies, hypotheses about the structure of the plant MTs are currently based on methods such as theoretical calculations [14,24], IR and Raman spectroscopy [9,13], as well as proteolytic digestion experiments [8,12,25]. Only recently, the three-dimensional structure of the larger domain of the *T. aestivum* E_c-1 MT was determined by NMR spectroscopy [26]. It reveals interesting insights into an unprecedented combination of a metal-thiolate cluster similar to the β-domain of the vertebrate isoforms engulfing a single metal ion binding site with a Cys_2His_2 coordination motif as known from zinc-finger proteins.

2. CLASSIFICATION

2.1. Nomenclature

Historically, MTs were subdivided into three different classes [27,28]: Class I MTs all matched exactly the positions of Cys residues found in horse kidney cortex MT, so to speak the prototype of a metallothionein [1]. Class II MTs shared the general characteristics for metallothioneins, but their Cys distribution pattern along the peptide chain was distinctively different to equine MT. Class III was made up by the phytochelatins (PCs), non-gene encoded polypeptides of the general formula $(Glu\text{-}Cys)_n$-Gly ($n = 2-11$) synthesized by the specific γ-Glu-Cys dipeptidyl transpeptidase phytochelatin synthase using glutathione as a substrate [29]. While *T. aestivum* E_c-1 was assigned to class II [30], more uncertainty prevailed when trying to classify other

Cys-rich proteins originating from plants. These proteins, such as the *Pisum sativum* $PsMT_A$, were reported to be homologs of MTs or MT-like [31]. However, the increasing number of MT sequences discovered and their diversity demanded a revision of the constricted three-classes nomenclature system. Currently, the MT superfamily is made up by all proteinaceous MTs and divided into 15 families based on sequence similarities and phylogenetic relationships [32]. For example, the mammalian MTs are part of family 1, the vertebrate MTs, while all MTs of plant origin constitute now family 15, enclosing also the proteins formerly only designated MT-like. The non-gene encoded PCs are no longer regarded as metallothioneins and now form the PC superfamily (see also Chapter 15). The plant MT isoforms are additionally subdivided into essentially four subfamilies called p1, p2, p3, and pec, based on the number of cysteine residues per protein and their distribution along the peptide chain (Figure 1 and Table 1).

With few exceptions, a typical plant MT constitutes 60–85 amino acids, depending on the subfamily it belongs to, and is thus generally slightly larger than members of the mammalian MT family. A major feature of members of the plant p1, p2, and p3 subfamilies is the clustering of Cys residues in the amino- (N-) and the carboxy- (C-) terminal regions. The Cys pattern of the N-terminal region is strictly conserved within a subfamily and one of the major distinguishing features between subfamilies. In contrast, the distribution of Cys residues within the C-terminal part is distinctively shared by all three subfamilies. The overall Cys content amounts to 12 (p1), 14 (p2), and 10 (p3) residues, respectively, and is thus lower than in mammalian MTs

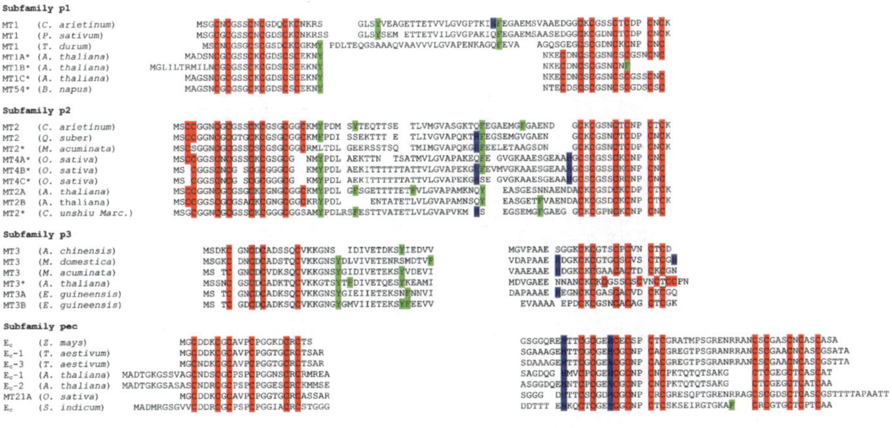

Figure 1. Amino acid sequence alignment of representative members of the four subfamilies p1, p2, p3, and pec of plant MTs. Cys residues are highlighted in red, His in blue, and the aromatic amino acids Phe and Tyr in green. See also Table 1.

Table 1. Assignment to subfamilies, general properties of the amino acid sequences, and accession numbers of selected plant MTs.

Subfamily[a]	Source			aa	Cys (%)	Aromatic amino acids			Swiss-Prot [6] Accession no.
						His	Phe	Tyr	
p1	*Cicer arietinum*	MT1	(chickpea)	75	12 (16.0)	1	1	1	Q39458
p1	*Pisum sativum*	MT1	(garden pea)	75	12 (16.0)	0	1	1	P20830
p1	*Triticum durum*	MT1	(durum wheat)	75	12 (16.0)	0	1	1	Q68HV7
p1v	*Arabidopsis thaliana*	MT1A	(mouse-ear cress)	45	13 (28.9)	0	0	1	P43392
		MT1B		45	10 (22.2)	0	1	1	Q38803
		MT1C		45	13 (28.9)	0	0	1	Q38804
p1v	*Brassica napus*	MT54	(rape)	45	13 (28.9)	0	0	1	P43402
p2	*Cicer arietinum*	MT2	(chickpea)	79	14 (17.7)	0	2	2	Q39459
p2	*Quercus suber*	MT2	(cork oak)	77	14 (18.2)	1	2	0	Q93X22
p2v	*Musa acuminata*	MT2	(banana)	78	13 (16.7)	1	1	0	O22319
p2v	*Oryza sativa*	MT4A	(rice)	78	13 (16.7)	1	1	1	Q0IMG5
		MT4B	(rice)	79	12 (15.2)	2	1	1	Q2QNE8
		MT4C	(rice)	78	12 (15.4)	2	0	1	Q2QNC3
p2	*Arabidopsis thaliana*	MT2A	(mouse-ear cress)	81	14 (17.3)	0	2	2	P25860
		MT2B		77	14 (18.2)	0	1	2	Q38805
p2v	*Citrus unshiu* Marc.	MT2	(satsuma orange)	79	12 (15.2)	1	2	1	O80335

p3	*Actinidia chinensis*	MT3	(kiwi)	63	10 (15.9)	0	0	1	P43389
p3	*Malus domestica*	MT3	(apple)	66	10 (15.2)	2	0	1	O24059
p3	*Musa acuminata*	MT3	(banana)	65	10 (15.4)	1	0	2	Q40256
p3v	*Arabidopsis thaliana*	MT3	(mouse-ear cress)	69	12 (17.4)	0	1	2	O22433
p3	*Elaeis guineensis*	MT3-A	(oil palm)	65	10 (15.4)	1	1	1	Q9STC4
		MT3-B		63	10 (15.9)	0	1	2	Q9STC3
pec	*Zea mays*	E_c	(maize)	76	17 (22.4)	2	0	0	P43401
pec	*Triticum aestivum*	E_c-1	(bread wheat)	80	17 (21.2)	2	0	0	P30569
		E_c-3		81	17 (21.0)	2	0	0	P30570
pec	*Arabidopsis thaliana*	E_c-1	(mouse-ear cress)	84	17 (20.2)	2	0	0	P93746
		E_c-2		85	17 (20.0)	2	0	0	Q42377
pec	*Oryza sativa*	MT21a	(rice)	87	17 (19.5)	2	0	0	Q109B0
pec	*Sesamum indicum*	E_c	(sesame)	83	17 (20.5)	2	1	0	Q9FUJ8

[a] Amino acid sequences of plant MTs varying from the consensus sequence (Figure 1) or displaying other peculiarities (see text) are denoted with v.

(20 Cys). Both Cys-rich regions are separated by a long Cys-free linker region with a typical length between ~30–45 amino acids. In contrast to all other MT forms known so far, a distinctive feature is the presence of the aromatic amino acids phenylalanine and/or tyrosine as well as of an occasional histidine in the linker regions of p1, p2, and p3. These extra residues even reveal a certain degree of conservation within a subgroup (compare also Figure 1 and Table 1). The importance of the central linker region for folding or its targeting was hypothesized [8,33]. With a total of 17 Cys residues, members of the pec subfamily, such as the *T. aestivum* E_c-1, show a higher cysteine content and percentage than the plant MTs from the other three subfamilies, and are thus more comparable to the well characterized mammalian iso-forms. Members of the pec subfamily further differ from other plant MTs in that they contain a third Cys-rich region and two considerably shorter linker regions of ~11–15 amino acids in length, which are additionally devoid of any aromatic amino acids, but feature two highly conserved histidine residues as part of the central cysteine-rich region. The Cys-pattern of this central region is identical to the one found in the C-terminal Cys-rich part of plant MTs from the other three subfamilies.

However, a few examples of plant MT sequences have been identified so far, which do not exactly match the consensus sequence for either one of the four subfamilies. These sequences are marked with an asterisk in the sequence alignment in Figure 1. For example, the putative MT1A, MT1B, and MT1C proteins from *Arabidopsis thaliana* share the typical Cys distribution pattern in the N-terminal region with other plant MT1 proteins, however, the linker region between the two Cys-rich regions is reduced to only 10 amino acids and also contains an additional Cys residue just prior to the C-terminal region [34]. The consensus sequence of the otherwise highly conserved C-terminal region is altered to CxxCxCxxxCxCxxxCxC (Figure 1) and thus all three CxC motifs are separated by three amino acids in contrast to three and two amino acids for all other members of the p1 subfamily. The same variance is displayed by the metallothionein denoted MT54 from *Brassica napus*, which otherwise can be clearly assigned to the p1 subfamily as well [35].

Mt1b from *A. thaliana* shows an even greater number of peculiarities [34]. In the putative MT1b protein an 11 amino acids long Cys-free stretch precedes the first Cys residue of the N-terminal domain, the length of which as well as its hydrophobic character being highly unusual. Most common are 2–4 amino acids. While featuring the additional Cys residue located just prior to the C-terminal Cys-rich region analogously as encoded by *mt1a* and *mt1c*, a deletion encompassing six codons as well as a single point mutation limit the total number of conserved Cys from 6 to 3. It is not clear, whether these alterations in the gene sequence generate a protein with new specific functionalities – or result in a nonfunctional translation product.

Variations from the consensus sequence can also be found within the p2 subfamily. In MT2 from *Musa acuminata* the second Cys residue is altered to a Ser [36]. However, comparing the respective codon it appears likely that this conservative mutation was brought about by a simple t→a nucleotide transversion on the gene level further justifying assignment of the protein to the p2 subfamily. Less unambiguous is the classification of three MTs from *Oryza sativa* denoted MT4A, MT4B, and MT4C [37]. While in MT4A only the last Cys residue of the N-terminal Cys-rich region is missing, MT4B and MT4C also lack the first Cys residue of the N-terminal consensus sequence. However, a certain conservation of residues with other members of the p2 subfamiliy cannot be denied and might warrant the assignment of the three proteins from *O. sativa* as p2 variants.

With 60–65 amino acids, 10 of which are Cys residues, members of the p3 subfamily represent the smallest plant MTs (not regarding the p1 variants mentioned above). The 10 sequences currently listed in the Swiss-Prot protein sequence database all closely match the consensus sequence for this subfamily. But also for this subclass of plant MTs variants are known, e.g., a mRNA transcript from *A. thaliana* listed in the TrEMBL database, which encodes a protein with high similarity to other members of the p3 subfamily. Specifically, the Cys distribution pattern in the N-terminal domain is conserved as well as the first amino acids of the Cys-free linker region. What is striking, however, is the total number of eight Cys residues found in the C-terminal region. While the "inner" six Cys residues satisfy the highly conserved pattern of six Cys residues prevailing throughout the entire plant MT family, the two additional Cys residues flanking this motif distinguish the putative *A. thaliana* MT3 protein uniquely from all other members currently assigned to the p3 subfamily. The distant possibility that these two Cys residues evolved from genetic mutation originating from the relative conserved Gly residues generally found at these positions, i.e., that two ggc→tgc transversions occurred, remains to be investigated further. One general peculiarity of proteins from the p3 subfamily is the relative high conservation of a His residue N-terminal of the C-terminal Cys-rich region as well as the occasional occurrence of a His residue at or close to the end of the protein.

Even less examples for plant MTs are known, which can be assigned to the pec subfamily. A search in the Swiss-Prot protein sequence database for "metallothionein" returns four hits, namely the mRNA sequences of *T. aestivum* E_c-1 and E_c-3, *Zea mays* E_c, as well as *O. sativa* MT21A. Two more sequences, this time on the gene level, can be obtained when using the search term "Ec protein homolog": *A. thaliana* E_c-1 and E_c-2. The *Sesamum indicum* mRNA encoding for a member of the pec subfamily is listed only in the TrEMBL database. Among all members of the pec subfamily a high degree of amino acid sequence conservation is observed without any variants. As in the p3 subfamily, also for the pec proteins the high conservation of two His residues is striking, one located directly before the central Cys-rich region and another one after its first CxC pair.

2.2. Gene Structure

The majority of data available about plant MTs is based on gene expression studies, thus on the analysis of cDNA libraries obtained from mRNA isolated from different plant tissues. Knowledge about the encoding genes themselves, however, is scarce.

A. thaliana is the only plant species so far, where isoforms of all four plant MT subfamilies were identified. The MT gene family features a total of eight different members. Five of them, *mt1a* (GenBank [38] accession number U11253), *mt1b* (U11254), *mt1c* (U11255), *mt2a* (D11394), and *mt2b* (U11256), contain a single intron with the intron splice site being conserved to one base pair (bp) before the last Cys codon of the N-terminal Cys-rich region [34]. Thus, all five genes are supposed to originate from a single ancestor. Interestingly, also analysis of the two gene sequences encoding proteins of the pec subfamily, E_c-1 (U90439) and E_c-2 (AC002391), reveals the presence of one intron each, and the position of the intron splice site is identical to the one of the other five MT genes.

The only member of the *A. thaliana* gene family containing two intron sequences, one located 2 bp after the last Cys codon of the N-terminal Cys-rich region and the other one in approximately the center of the sequence encoding for the Cys-free linker region, is *mt3* (AF013959). Three of the eight genes, *mt1a*, *mt1b*, and *mt1c*, encode for MT1-like proteins based on the Cys distribution pattern in the N-terminal region, but show certain variations in the putative amino acid sequence as discussed in Section 2.1 including the reduction of the Cys-free linker region to only seven amino acids as the most peculiar characteristic. In addition, actual transcription of the *mt1b* gene is questionable due to certain deletions near the TATA box when compared to the highly homologous promoter region in *mt1a*. The other two genes, *mt2a* and *mt2b*, encode for proteins from the p2 subclass. No metal responsive element (MRE) could be identified in the promoter region of any of the five *mt* genes from the p1 or p2 subfamily although inducibility of *mt2* gene expression by copper was shown [39]. The two genes encoding for the two proteins from the pec subfamily, E_c-1 and E_c-2, contain a promoter sequence with homology to abscisic acid (ABA) responsive elements [40]. ABA is a plant hormone and a key regulator of gene expression during seed maturation [41].

In contrast to the two genes from *A. thaliana*, the gene sequence of E_c-3 (X68288) from *T. aestivum* encodes an intron-less mRNA [30]. However, alike is the presence of an ABA responsive element in the 5′ flanking region of the coding sequence as well as the absence of any MREs of the kind found in mammalian MTs. This is in accordance to data showing that E_c-1 is not induced by elevated Zn^{II} ion concentrations [30]. Gene expression studies actually identified three different cDNA clones in *T. aestivum*: cDNA-I, -II, and -III. cDNA-I and -II encode the same protein, named E_c-1. cDNA-III encodes a protein, E_c-3, of the same size and with the same Cys distribution pattern as E_c-1 but with three amino acid alterations, all originating from single g→a transitions in the respective codons. While mammalian MTs are part of multigene families [42], the three cDNA sequences identified are probably derived from single-copy genes in each of the A, B, and D genomes of the hexaploid wheat *T. aestivum*. Specifically, the cDNA-III clone was probably transcribed from the D genome, which originates from the weed progenitor *Triticum* (or *Aegilops*) *tauschii* [30].

A more elaborate study of the promoter region was performed with a gene encoding a MT3 isoform, MT3-B (AJ236914), from *Elaeis guineensis* [18]. In the open reading frame (ORF) two introns of 1432 and 306 bp length, respectively, divide the coding region into three exons of 47, 48, and 97 bp. The promoter region contains several DNA motifs with similarities to known responsive elements. Among them are a putative MRE as found in vertebrate MTs, a putative root-specific element (RSE), and three putative ethylene responsive elements (ERE). Two of these ERE are similar to a ripening-specific *E4* gene from *Solanum lycopersicum* (tomato) with homology to the oxidative-stress-repair enzyme peptide methionine sulfoxide reductase of *B. napus* and *A. thaliana* and to a senescence-specific glutathione-S-transferase gene from *Dianthus caryophyllus* (clove). The third ERE resembles a number of pathogenesis-related genes. Only one copy of the gene seems to be present in the *E. guineensis* genome based on Southern blot analyses next to another single copy of a gene encoding for a second MT3 isoform, MT3-A.

3. FUNCTION

A number of hypotheses have been made concerning the functions of plant MTs and MTs in general, which are largely based on gene expression studies. Thus an exemplary selection of such studies for members of the four different plant MT subfamilies is presented in the following Section 3.1. While gene expression studies solely monitor mRNA levels, yeast complementation studies, as elaborated in more detail in Section 3.2, are more suited to draw conclusions on the protein level. Re-establishment of, e.g., metal ion tolerance in a yeast mutant by a foreign protein is a rather explicit hint that this protein indeed is important in metal ion homeostasis or detoxification. Manipulation of *mt* gene expression leads to so-called *mt*-knockout plants as described in Section 3.3. Monitoring of metal ion concentrations in certain tissues and other phenomenological properties in comparison with unmodified wild-type plants directly visualize the effect of MT deficiency, unless the function of the protein encoded by the knocked out gene is taken over or compensated by another protein. Possible functions derived from the previous sections will be discussed in Section 3.4.

3.1. Differential Gene Expression

The majority of information available about plant MTs is based on gene expression studies. The visualization of mRNA levels in tissues from different natural sources is a well established sensitive method applicable for a broad range of applications. Due to the difficulties associated with direct measurements of MT protein levels, such gene expression studies are often the only method to gain knowledge about the possible functions of MTs *in vivo*. However, it should be one of the future goals to verify that alterations in *mt* mRNA levels observed are indeed directly reflected in MT protein expression.

An early investigation on different tissues from *P. sativum* showed that *mt1* transcript levels are most abundant in roots, clearly discernible in etiolated leaves, weakly detectable in seed leaves, but absent in tissues from green leaves [31]. A similar experiment, performed with *Z. mays* shows the same high root *mt1* transcript levels, but additionally also, though lower, abundance in green leaves and the pith as well as residual levels in seeds [43]. Finally, also *mt1* genes of *A. thaliana* are strongly transcribed in the root tissue of both, young and mature, plants, but are less abundant in leaves at the same developmental stage and barely detectable in flowers and pods [34]. Fusion of the 5′ promotor regions of the *A. thaliana mt1a, mt2a, mt2b,* and *mt3* genes to the β-glucuronidase (GUS) reporter allows direct visualization of *mt* gene expression by staining (Figure 2) [40]. In this way, expression of *mt1a::GUS* could be more precisely allocated to the vascular tissue of all organs and predominantly to the mature region of roots, but it is absent in the elongation zone of roots. *Mt1* mRNA was also found in the leaf tissue of *B. napus* although in strong dependence of the respective developmental stage [35]. Transcripts of the *mt1* gene did not appear until the flowering stage, increasing further concomitantly with pod development reaching highest abundance in yellow leaves of the still developing plant. Interestingly and in contrast to results from other plants, no *mt1* transcripts were observed in the root tissue of *B. napus*.

To test for a probable role of MT1 in metal ion tolerance or detoxification, mRNA levels in the whole tissue of germinating seedlings from different *A. thaliana* ecotypes were analyzed in response to increasing Cu^{2+} concentrations [44]. However, the *mt1* gene was found to be equally

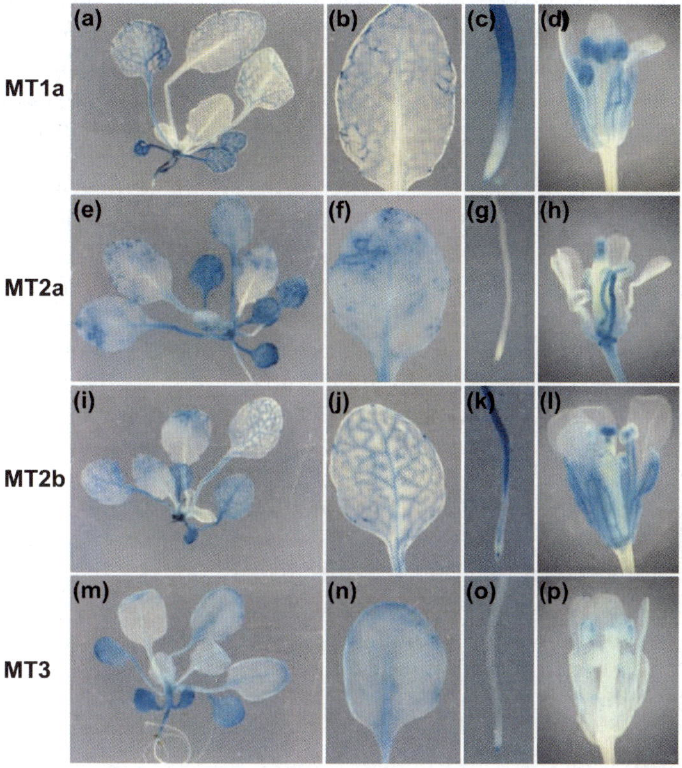

Figure 2. Visualization of *mt1a* (a–d), *mt2a* (e–h), *mt2b* (i–l), and *mt3* (m–p) gene expression patterns in whole plants (a,e,i,m), mature leaves (b,f,j,n), and lateral roots (c,g,k,o) from three-week-old as well as flowers (d,h,l,p) from six-week-old *A. thaliana* plants upon fusion to the β-glucuronidase (GUS) reporter gene. Reproduced from [40] by permission of Wiley-Blackwell; copyright (2003).

transcribed in all samples even up to 40 μM $CuCl_2$. In contrast, expression of the *mt1a::GUS* gene was shown to increase in roots after a 2 days treatment of three-week-old *A. thaliana* plants with 25 μM $CuSO_4$ [40]. In addition, the same conditions increased the expression of all four *mt::GUS* genes in trichomes.

Mt2 gene transcripts are generally more abundant in leaves than *mt1* mRNA, especially in the early developmental stages. For example, in *B. napus mt2* mRNA is already present in leaves of young plants increasing in abundance during flowering and keeping a nearly constant level thereafter [35]. However, transcription of *mt2* can be further observed in mature roots, young and mature leaves as well as in flowers of *A. thaliana* [34]. Lower levels of *mt2b* mRNA are also detectable in pods. The more recent study with the GUS reporter mentioned above allocates *mt2a::GUS* gene expression more specifically to mesophyll cells of the leaves, while *mt2b::GUS* transcripts are found in the vascular tissue of all organs and in mature roots in analogy to *mt1a::GUS* [40]. Mature root tissue is low in *mt2a::GUS* mRNA, however, strong expression is evident in the root tips. Screening of cDNA libraries of the satsuma orange (*Citrus unshiu* Marc.) obtained from mature leafs, flowers, and developing fruits in a time frame spanning 26 to 179 days after flowering revealed expression of a gene encoding for a protein with sequence similarity to the p2 subfamily of MTs in all samples at almost equal levels [45]. The putative MT2-like protein is peculiar as it is two Cys residues short with respect to the consensus sequence of the N-terminal domain (Figure 1). Thus the total number of Cys residues renders it similar to the p1 subfamily, while the Cys distribution pattern clearly justifies assignment as an MT2-like protein. In contrast, Northern blot analyses on tissues from *Malus domestica* shows that *mt2* RNA is virtually undetectable in leaves, while it is abundant in flowers and in the early stages of fruit development with decreasing levels approaching harvest age [46]. Additionally, it was shown that *mt2* is up-regulated rapidly in cool-stored fruit (0.5 °C) as well as in apples left to ripening at 20 °C after harvest. In contrast, abundance of *mt2* mRNA in bananas (*M. acuminata*) is highest directly after harvest, when the fruit peel still has a very green color, and declines concomitantly with post-harvest ripening [36].

Inducibility of *mt2* gene transcription by external stimulants was studied in detail with leaves from *Nicotiana glutinosa* (tobacco) [47]. Strong induction was detected after mechanical wounding of leaves and upon infection with the tobacco mosaic virus. A 10 mM solution of $CuSO_4$, sprayed onto the leave surface, resulted in twofold increased *mt2* mRNA levels, while no effect was noticed after treatment with salicylic acid, an inducer of pathogenesis related protein gene expression, methyl jasmonate, a signal molecule for wound-inducible gene expression, and ethylene, which is known to be produced during fruit ripening and in senescing tissues in general. A possible *mt2* gene transcription in response to elevated Cu^{2+} was also investigated on seven-day-old *A. thaliana* seedlings, which were incubated with increasing metal ion concentrations (0–50 μM) for 30 h [34]. While *mt2b* mRNA was present at high levels in all samples increasing only slightly upon metal ion treatment, strong induction of *mt2a* was evident.

Another study, published in the same year, analyzed *mt1* and *mt2* mRNA levels in the whole tissue of germinating seedlings from different *A. thaliana* ecotypes (see above) [44]. In contrast to the *mt1* gene, expression of the *mt2* gene was clearly Cu^{2+}-inducible, although transcript levels were highly variable among the different ecotypes. Similar results were obtained for 1 mM $AgNO_3$, 40 μM $CdSO_4$, 500 μM $ZnSO_4$, 180 μM $NiCl_2$, as well as in heat shock treated seedlings. 200 μM $AlCl_3$, however, had no effect. *Mt2* mRNA levels further seem to be linked to root extension in the seedlings analyzed. After inhibition of root growth by 40 μM $CuSO_4$ increasing levels of *mt2* mRNA paralleled the onset of the recovery phase approximately 4 h after exposure. In contrast to the results obtained with *A. thaliana*, no elevated transcript levels for *mt2* were observed when fully-expanded leaves of *C. unshiu* Marc. were placed into solutions containing 0–1 mM $ZnCl_2$ or $CuSO_4$ for 2 days [45].

During the investigation of *C. unshiu* Marc. mentioned above, transcripts of an MT3 isoform were nearly exclusively detected in the developing fruits at 70 to 179 days after flowering [45]. In mature fruits (179 days after flowering) *mt3* was with 20.9% even the most abundant of the transcribed genes. Also in *M. domestica, mt3* RNA accumulated with fruit development (ca 56 to 154 days after flowering) and was absent in flowers [46]. Additionally, *mt3* expression is up-regulated in mature leaves and in cool-stored fruit in analogy to *mt2* (see above), while post-harvest ripening at 20 °C causes down-regulation of *mt3*. *M. acuminata mt3* mRNA first increases in abundance until the banana peel reaches a ripening stage marked by a more green than yellow color followed by declining levels upon further ripening as found for *M. domestica*. The peak in *mt3* transcription coincides with ethylene biosynthesis by the fruit. Also in analogy to results obtained from *M. domestica, mt3* transcripts in leaves of *M. acuminata* are more abundant than *mt2* transcripts, while the reverse is true in samples taken from the corm.

Finally, mRNA encoding for two different MT3 isoforms, MT3-A and MT3-B, were identified in *E. guineensis* [18]. *Mt3-a* mRNA was abundant in developing fruits with a maximum value around 100 days after flowering accounting for around 1% of the total mRNA. Low levels were also detected in kernels at ca. 100 days after flowering as well as in root tissue. There is no *mt3-a* transcription in normal leaves, but distinct up-regulation during leaf senescence. Only trace levels of *mt3-b* mRNA were found in the tissues investigated, namely in developing fruits 100–120 days after flowering as well as in the root. While *mt3* gene expression is generally fruit-specific, transcripts can be also observed in mesophyll cells in plants that do not produce fleshy fruits such as *A. thaliana* [40]. In analogy to *mt2a::GUS*, *mt3::GUS* transcripts are also found in root tips.

Treatment of fully expanded leaves of *C. unshiu* Marc. with 0–1 mM $ZnCl_2$ or $CuSO_4$ (see above) revealed no elevated transcript levels for *mt3* in analogy to the results obtained for *mt2* expression [45]. In contrast, the *mt3::GUS* gene is increasingly expressed in young leaves and root tips upon supplementation of the growth substrate with 25 μM $CuSO_4$ (see above) [40].

Expression of the E_c proteins is restricted to developing seeds, as well as to embryogenic microspores and pollen embryoids [48]. For example, the highest level of *T. aestivum* E_c-1 mRNA is reached in immature wheat embryos approximately 15 days post anthesis and declines rapidly post imbibition of the dried seed prior to germination.

Transcription of E_c-1 genes in *Z. mays* is stimulated by ABA or when maize embryos are subjected to osmotic stress by supplementation of growth media with 20% (w/v) sucrose [49]. mRNA accumulation was not observed for *T. aestivum* E_c-1 upon supplementation with 45 or 450 μM $ZnCl_2$ [30].

While expression of *mt1* is high in root tissue, of *mt2* in leaves, and of *mt3* in the tissue of fleshy fruits, transcription is not resticted to the respective tissues, complicating a clear-cut designation to functions. In addition, mRNA levels vary with developmental stages, but do not necessarily show consistency across different plant species. The same is true for the inducibility of *mt* gene expression by certain enviromental factors. For example, whereas *mt1* expression in *A. thaliana* was shown to increase in response to treatment with Cu^{2+} in one case, it was unaffected in another study.

3.2. Yeast Complementation Studies

The manipulation of gene expression in order to investigate possible functions of proteins is a comparatively tedious task when working with genetically complex higher eukaryotes. An indirect way to determine the function or functions of a MT of interest is to perform complementation experiments with organisms that can be more easily genetically manipulated so that they are incapable of MT synthesis. One such organism is *S. cerevisiae*, budding or baker's yeast. *S. cerevisiae* expresses two metallothioneins, CUP1 and CRS5, which are involved in copper resistance and sensitivity and show properties analogous to MTs studied from higher eukaryotes [50–52]. Most laboratory strains of yeast contain multiple copies of the CUPl locus, designated $cup1^R$, and are resistant to 500 μM $CuSO_4$, while resistance is reduced to 75 μM in the CUP1 null mutant, $cup1^{\Delta}$ [53].

Recently, CRS5 has been suspected to play a certain role in Zn^{2+} tolerance [54]. This becomes especially evident under mixed-metal growth conditions (15 mM Zn^{2+} and 5 μM Cu^{2+}) with CRS5 expression providing a small advantage over CUP1 expression. Intermediate growth of yeast cells under these conditions can also be achieved upon complementation with mammalian MT1. Yeast complementation studies with the plant MT2 from *Quercus suber* show that overexpression of this protein restores viability of $cup1^{\Delta}$ cells in 75 μM Cu^{2+} [20]. 7 mM $ZnSO_4$ has no negative effects on the growth of $cup1^{\Delta}$ compared to the mutant featuring just a single copy of the CUPl locus, $cup1^S$, or $cup1^{\Delta}$, expressing *Q. suber* MT2. Nevertheless, a superior protective role of *Q. suber* MT2 becomes evident at 2.5 or 3.5 mM Cd^{2+} [21]. Expression of MT1a, MT2a, MT2b, MT3, E_c-1, or E_c-2 from *A. thaliana* in the $cup1^{\Delta}$ yeast strain re-establishes copper tolerance (50 μM

$CuSO_4$) in all instances [39,55]. To investigate sensitivity against Zn^{II}, Cd^{II}, and Co^{II} a yeast mutant, *zrc1*$^{\Delta}$ *cot1*$^{\Delta}$, was chosen that is hypersensitive to zinc due to a deficiency for overexpression of two vacuolar Zn^{II} transporters [55]. All six *A. thaliana* MTs were able to restore zinc tolerance of the mutant to the wild-type level with the E_c proteins being the most effective. Interestingly, E_c-1 also restores cobalt tolerance, but was ineffective against cadmium, while the reverse is true for the other five MTs.

Yeast complementation studies consistently show that copper sensitivity of the *cup1*$^{\Delta}$ mutant can be efficiently overcome with MT2 from *Q. suber* as well as with *A. thaliana* MTs from all four subfamilies. With the exception of *A. thaliana* E_c-1, all afore mentioned proteins were also able to confer a certain protection against elevated cadmium levels to yeast mutants, while E_c-1 was most effective in zinc and cobalt tolerance.

3.3. *mt*-Knockout Plants

Recently, two *A. thaliana* knockout plants, *mt1a*$^{\Delta}$ and *mt2b*$^{\Delta}$, have been engineered carrying *mt1a* and *mt2b* genes that have been made inoperative by insertion of transferred DNA (T-DNA) into the respective exons [55]. The *mt1a*$^{\Delta}$ *mt2b*$^{\Delta}$ double mutant as well as different combinations with a phytochelatin synthase deficient mutant, *cad1-3*, were obtained by crossing. These mutants provide a deeper insight into the functions of MT1a and MT2b *in vivo*. The three *mt* deficient mutants, *mt1a*$^{\Delta}$, *mt2b*$^{\Delta}$, and *mt1a*$^{\Delta}$ *mt2b*$^{\Delta}$, are viable under normal growth conditions and in media supplemented with 30 μM $CuSO_4$. While copper accumulation in shoots is comparable in the wild-type and in the mutants, reduced concentrations are found in the roots of the *mt1a*$^{\Delta}$ and *mt1a*$^{\Delta}$ *mt2b*$^{\Delta}$ mutants. To investigate a possible interaction of MTs and PCs, the response of a number of cross-mutants with a PC synthase mutant, *cad1-3*$^{\Delta}$, to elevated $CuSO_4$ and $CdSO_4$ levels was analyzed [55].

Interestingly, the *mt1a*$^{\Delta}$ *mt2b*$^{\Delta}$ double mutant showed normal metal tolerance, while growth of the *mt1a*$^{\Delta}$ *mt2b*$^{\Delta}$ *cad1-3*$^{\Delta}$ triple mutant experienced stronger inhibition than the *cad1-3*$^{\Delta}$ single mutant. It is thus obvious that a strict functional differentiation of MTs and PCs in plants is not possible as roles may overlap and be complementary to each other.

3.4. Functional Insights

Generally, the high Cys content of plant MTs implies that these proteins function as ligands for metal ions. However, heavy metal inducibility of the *mt* genes is not necessarily given, an example being the induction of *T. aestivum* E_c-1 by the phytohormone ABA but not by elevated zinc levels. Moreover, heavy metal response in plants seems to be essentially regulated by the afore mentioned PCs (see also Chapter 15). Nevertheless, a homeostatic function for metal ions is feasible under normal growth conditions in analogy to MT members from the other phyla of life, and was also suggested for the seed-specific zinc-thionein E_c-1. In addition, the developmental and spatial expression of the different MTs in response to environmental stresses points to additional more specific functions. The differential expression pattern observed for the different plant MT genes suggests a tissue- and development-specific regulation. For some genes, even distinct controlling elements in the promotor regions of the genes have been identified (see Section 2.2).

With respect to the fruit specific MT3 isoforms it seems unlikely to assume a role in metal ion tolerance as fruit development generally is not a period of excessive metal ion stress [18]. Thus, a function in Zn^{2+} or Cu^{+} homeostasis seems more feasible. However, even the possibility that *mt3* has no direct function in the ripening process, but is rather a post-harvest senescence-related gene, should not be neglected. Along this line, the up-regulation of *mt1* and *mt3* gene transcription during leaf senescence in *B. napus* and *E. guineensis* allows a number of speculations regarding a possible function [18,35].

Leaf senescence is accompanied by organized biochemical processes aimed at sequestering valuable nutrients such as metal ions that are discharged during protein degradation and cellular death. It would thus be a valuable investment of the plant to synthesize specific proteins for the storage of these nutrients for subsequent remobilisation and redistribution to newly developing tissues including young leaves and seeds. Hence, overexpression of *mt1* and *mt3* in senescing leaf tissue points to a function in sequestration of Zn^{2+} or Cu^{+} released from other degraded metalloproteins, where it can be stored for future use. Alternatively, sequestration of excessively released metal ions might also be important in light of a detoxifying function [35]. In addition, oxidative processes play a role in leaf senescence, and the degradation of, e.g., cell membranes and chlorophyll can produce free radicals [56]. Therefore, MTs expressed at this stage might also function as reactive oxygen species (ROS) scavengers as analogously proposed for the mammalian isoforms [57], and prevent oxidative damage of nuclear DNA to allow transcription of senescence-specific genes.

Studies with *A. thaliana mt*-knockout mutants show that accumulation of copper is reduced in roots and unaffected in shoots, while the general copper tolerance remains unchanged compared to the wild-type plants [55]. These observations led to the conclusion that MT1a and MT2b are not important for copper storage in shoots. On a speculative basis and in accordance with the high *mt1a* transcript levels in senescing leaves, MT1a might be involved in the transport of copper from shoots to roots or other sink tissues via the phloem. Investigations of cross-mutants with a PC synthase mutant suggest overlapping functions in copper tolerance and cadmium detoxification for MTs and PCs.

4. ISOLATION AND PURIFICATION

4.1. Native Metallothionein

The only plant MT isolated directly from its host so far is the *T. aestivum* E_c-1 protein. It can be purified from wheat germ with a procedure derived from the isolation of mammalian MTs and involving a series of precipitation and size-exclusion chromatography steps [3]. Initially, components with hydrophobic characteristics such as phospholipids are precipitated from a water soluble wheat germ extract with a 95% ethanol/chloroform mixture using an extract:ethanol:chloroform ratio of 1:1.21:0.095 (v/v). Subsequently, E_c-1 as well as other contaminating proteins are precipitated from the supernatant with a twofold volume excess of 95% ethanol.

After resolubilization, proteins are fractionated with fast protein liquid chromatography (FPLC) using a size exclusion column containing preferentially Sephadex G-75 resin. E_c-1 containing fractions are identified via the zinc content of the protein using flame atomic absorption spectroscopy (F-AAS) in connection with protein quantification via the Bradford assay, a staining method, which makes use of the formation of a protein-dye complex with Coomassie Brilliant Blue G-250 and its absorption at 595 nm [58]. Classically, bovine serum albumin (BSA) is used as a concentration calibration standard, but hen egg lysozyme was found to perform equally well for the plant isoforms. A more MT-specific method makes use of the absorption at 235 nm arising from the presence of the zinc-thiolate clusters (see also Section 5) monitored directly during the FPLC procedure. Finally, a very accurate method is the quantification of thiolate groups with 2,2′-dithiodipyridine (2-PDS) at pH 4.0 following the stoichiometric formation of thiopyridinone via its absorption maximum at 343 nm (absorption coefficient 7600 M^{-1} cm^{-1}) [59]. The protein concentration is then calculated assuming all 17 cysteine residues to be present in the reduced state.

4.2. Recombinant Metallothionein

A number of methods are available for the production of recombinant plant MTs, which are generally also applicable to MTs from the other families. As an expression host, *Escherichia coli* was used exclusively so far. Two general approaches were followed, tag-free expression as well as expression in form of a large fusion protein. Expression without any purification tag was performed with *T. aestivum* E_c-1 and *P. sativum* MT1 [8,60]. In the case of *T. aestivum* E_c-1, *E. coli* cells, incubated for 6 h past induction of protein overexpression at 30 °C in the presence of 1 mM $CdCl_2$, were harvested by centrifugation and ruptured by sonification. Two subsequent precipitation steps were performed in analogy to the isolation of native E_c-1 described above. Prior to a final purification with size-exclusion chromatography the resolubilized pellet from the precipitation with 95% ethanol was treated with DNaseI in order to remove any DNA contaminants still present. Typical yields were 15 mg of purified E_c-1 per liter of cell culture.

Attempts in our lab to purify members of the p1, p2, or p3 subfamily with the same strategy were not successful. We ascribe this failure to either proteolytic degradation of the small MTs *in vivo* [61] or considerably altered solubilities of these MTs in the ethanol/chloroform mixtures with respect to the other soluble cell components from *E. coli*. Nevertheless, it was possible to purify *P. sativum* MT1 from *E. coli* cells that were allowed to grow for 4.5 h past heat induction of protein expression in the presence of 300 μM Cd^{II} and at an incubation temperature of 37 °C. After cell lysis and centrifugation, the supernatant was heated to 65 °C for 15 min and centrifuged to remove denatured proteins, while *P. sativum* remained in solution. Subsequently, purification by size exclusion chromatography on Sephadex G-75 and ion-exchange on DEAE-cellulose was carried out. However, it should be noted that partial proteolytic cleavage of the linker region was observed. In the course of temperature-dependent measurements performed with dynamic light scattering we observed aggregation of plant MTs from the p1, p2, and p3 subfamily above approximately 50°C (unpublished results). Hence, the purification procedure described for *P. sativum* is probably not generally applicable.

To circumvent the severe drawback of proteolytic degradation and difficulties of protein detection during purification, e.g., the unusual migration behavior of MTs in sodium dodecyl sulfate-polyacrylamide gel electrophoresis (SDS-PAGE) described below, larger purification tags in form of fusion proteins are increasingly used for small proteins or peptides. As design of these tags usually allows their removal after purification, recombinant proteins without additional vector derived amino acids can be obtained. An exeption is, for example, the addition of the two amino acids proline and glycine when using the XmaI restriction site for cloning of a C-terminal intein tag (see below) [9,12,15,16,62]. However, these are residues without considerable metal ion binding abilities, in contrast to, e.g., the His purification tags commonly used for protein purification. Two larger purification tags have been applied so far for the purification of plant MTs: The glutathione S-transferase, a protein consisting of 220 amino acids and having a mass of 25.7 kDa, as well as the intein tag coupled to a chitin-binding domain encompassing a total of 517 residues and a mass of 57.5 kDa. For expression and purification of the GST-tagged proteins two vectors from the pGEX family (GE Healthcare, Amersham, Buckinghamshire, UK) were used, both encoding an N-terminal GST tag but one with a factor Xa (pGEX-3X) and the other one with a thrombin cleavage site (pGEX-4T-2) C-terminal of the tag, which in principle allows the cleavage of the tag after purification.

Expression and purification of all six proteins, *P. sativum* and *T. durum* MT1, *Q. suber* MT2, *A. thaliana* MT2A, *E. guineensis* MT3-B, and *S. indicum* E_c, were performed using standard procedures [7,10,14,17,23,63]. The vectors containing the respective MT encoding sequences were transformed into *E. coli* expression hosts (JM101, BL21(DE3) or C41(DE3)). After induction of protein synthesis, cultures were incubated for additional 3–4 h at 37 °C in the presence of Zn^{II} (300–500 μM), Cd^{II} (50–500 μM), or Cu^{II} (300–500 μM). Cells were harvested by centrifugation and lyzed by sonication, and the GST fusion proteins purified with affinity chromatography using glutathione Sepharose-4B (GE Healthcare, Amersham, Buckinghamshire, UK). Recombinant plant MTs were either eluted from the column in form of the fusion proteins with glutathione of cleaved from the column-bound tag with the factor Xa or thrombin proteases.

The second tag that has been used to recombinantly express plant MTs in *E. coli* is a C-terminal intein tag from the pTYB2 vector of the IMPACT-CN system (New England Biolabs, Ipswich, MA, USA). Inteins are protein-splicing elements, which undergo self-cleavage in presence of low molecular weight thiols such as, e.g., 1,4-dithiothreitol (DTT). C-terminal fusion of a chitin-binding domain to the intein tag allows affinity purification of the fusion-protein on a chitin-column. This purification system was used for *Cicer arietinum* MT1 and MT2, *M. acuminata* MT3, and *T. aestivum* E_c-1 [9,12,15,16,62]. Briefly, the pTYB2 vector was modified with the respective coding sequence and transformed into the *E. coli* cell line BL21(DE3). After induction of protein overexpression growth media were supplemented with usually 100 μM $ZnCl_2$ or $CdCl_2$, and cells were harvested after 6 h at 30 °C. Following cell lysis by sonication the soluble cell extract was passed over a chitin column and intein self-splicing was induced with 50 mM DTT at 4 °C overnight and an additional 2–4 h at room temperature. A beneficial side-effect of this procedure is the reducing environment preventing – or reversing – oxidation of the thiolate groups and in this way ensuring the fully reduced state of the protein and thus its maximum metal ion binding capacity.

However, even at slightly alkaline conditions with pH values around 8.5, the intact Zn^{II}- or Cd^{II}-thiolate clusters were found to be sufficiently stable in air to handle the proteins outside an anaerobic chamber. Naturally, the situation changes when performing experiments with oxidation-sensitive metal ions such as Cu^{I} and Co^{II} or apo-MT. After elution from the affinity column a polishing step with size-exclusion chromatography over a Sephadex G-75 column was performed, from which each purified plant MT eluted in form of a single sharp peak (Figure 3A). Noteworthy in this context is a special peculiarity of the intein tag purification system that often goes unnoticed when working with larger enzymes but increases in relevance for small metal ion binding proteins such as MTs: Thiol-mediated cleavage of the intein tag proceeds via a protein-thioester intermediate, which

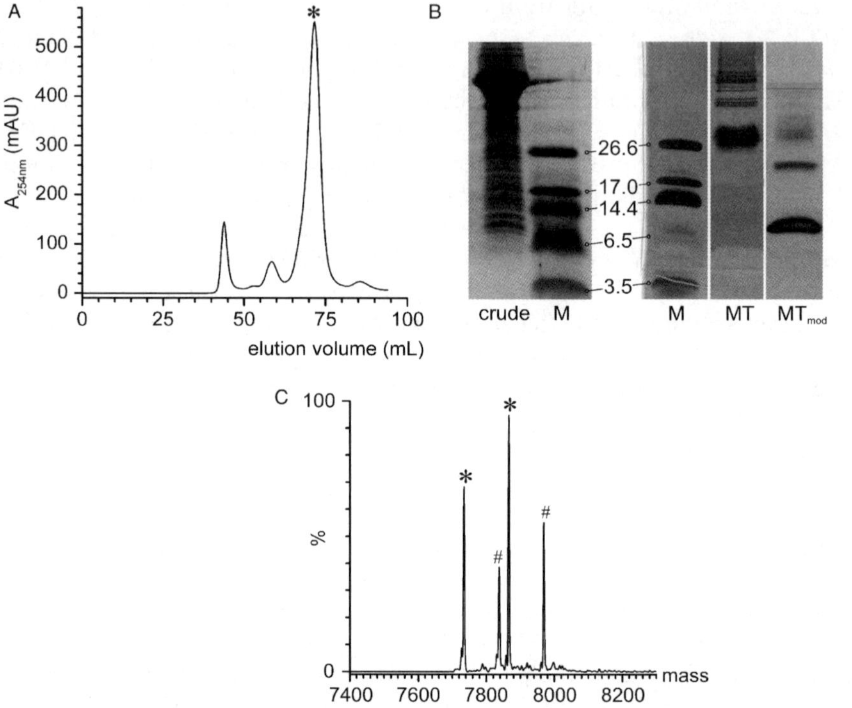

Figure 3. Purification of recombinant plant MTs overexpressed in *E. coli* in the form of an intein-fusion protein. (**A**) Typical elution profile of the final purification step with size exclusion chromatography (Sephadex G-75). The main peak, denoted with a *, contains the monomeric protein. (**B**) SDS-PAGE of the crude soluble *E. coli* cell extract (crude), the MT after size exclusion chromatography (MT), and the same protein after modification with mBBr (MT_{mod}). Two lanes with protein marker (M) allow assignment of apparent molecular weights given in kDa. The purified MT has a molecular weight around 8 kDa, but shows an apparent molecular weight above 26 kDa. After modification of thiol groups with mBBr, the apparent molecular weight around 10 kDa is identical to the expected molecular weight of the mBBr-modified protein. (**C**) Typical ESI-MS spectrum of a metal ion loaded MT isoform acquired at pH 7.5. Peaks denoted with * originate from the MT with and without the N-terminal translation initiation methionine, which frequently gets cleaved off in *E. coli*. Peaks denoted with # are the same two species bearing a C-terminal Tris-adduct from the purification procedure. See text for details.

subsequently undergoes hydrolysis to the free protein carboxy group. These adducts can be observed with ESI-MS spectroscopy, e.g., in the case of DTT as low molecular weight thiol the adduct gives rise to an additional mass peak increased by 136.2 Da relative to the mass of the protein. As complete hydrolysis at pH ~8.5 can take up to two days depending on the protein, unnoticed presence of this thioester can seriously influence the metal ion binding abilities of MTs. Hence meticulous monitoring of the hydrolysis process to exclude contributions of undesired adducts is crucial.

An additional adduct formation, though with negligible influence on the metal ion binding abilities, is the aminolysis reaction of the thioester intermediate with the commonly used buffer substance tris-(hydroxymethyl)-aminomethane hydrochloride (Tris-HCl) to yield a protein-Tris adduct with a stable amide bond increasing the mass of the protein by 103.1 Da [62]. However, if desired, the latter adduct formation can be easily avoided by switching to a buffer without a primary amino function, e.g. 4-(2-hydroxyethyl)-1-piperazineethanesulfonic acid (HEPES), during the self-splicing reaction and all subsequent dialysis steps.

Following protein purity with SDS-PAGE we persistently observed anomalous migration behaviors leading to increased apparent molecular masses of the plant MTs under investigation (Figure 3B) [9,12,16]. Due to the reducing conditions applied during sample preparation oxidative MT di- or even oligomerization seems improbable. A more likely explanation is sub-stoichiometric loading of the MT with SDS, i.e., binding of less than 1.4 g SDS per g of protein [64]. Reduced SDS-to-protein ratios are generally observed for incompletely denatured proteins, e.g., when compact structures are preserved by disulfide bridges [65,66]. For MTs, – partially – preserved metal-thiolate clusters could bring about the same effect, limiting SDS-binding to the solvent accessible outer protein surface and leading to higher apparent molecular masses in the gel experiment due to decreased charge-to-protein mass ratios. In corroboration with this interpretation, disruption of the metal-thiolate clusters upon either modification of the thiolate groups with monobromobimane (mBBr) [67] or treatment with EDTA lowers the apparent molecular weight in the SDS-PAGE experiment (Figure 3B) [9,12,16,17].

5. SPECTROSCOPIC CHARACTERIZATION

5.1. Metal Ion Content

Such basic information as the metal ion content of a yet uncharacterized MT requires knowledge of the protein as well as the metal ion concentration. Concentrations of proteins are classically determined via their absorption at 280 nm. Although most plant MTs do contain aromatic amino acids, their small number decreases the sensitivity of the method, which can be additionally hampered by a certain degree of spectroscopic overlap with charge transfer bands from the thiolate clusters of some metal ions. Provided the preparation of an adequate calibration curve, application of the Bradford assay is actually a true alternative yielding surprisingly exact concentration values. However, if the MT under investigation is present in the fully reduced state, quantification of cysteine thiol groups of the Zn^{II} or Cd^{II} forms with the 2-PDS assay in analogy to methods described for the mammalian MTs is a fast and reliable method to determine accurate protein concentrations.

For the quantification of metal ions F-AAS was shown to be the method of choice. For low concentrations and metal ions with a low natural abundance also a graphite furnace can be routinely applied. The metal ion-to-protein ratio can further be directly determined with ESI-MS at neutral pH (Figure 3C). Nevertheless, one drawback of this method is the possibility of metal ion loss during the ionization process depending on the stability of the metal-thiolate clusters and the parameters used for the measurement. As a consequence, MT species loaded with substoichiometric amounts of metal ions could be observed. Another difficulty is the concomitant presence of species with different stoichiometries, because the intensity ratios of the peaks observed in MS do not necessarily reflect the molar ratio of species in solution.

Compared to the mammalian MTs, plant MTs feature a lower content of the amino acid cysteine and hence, also a lower metal ion content can be expected. Vertebrate MTs coordinate divalent metal ions in form of two different metal-thiolate clusters, the α-domain with the stoichiometry M_4Cys_{11} and a metal ion-to-Cys ratio of 1:2.75, and the β-domain forming an M_3Cys_9 cluster and a ratio of 1:3. Assuming the presence of similar metal-thiolate clusters and thus similar metal ion-to-Cys ratios in plant MTs, the following theoretical ranges of divalent metal ion contents can be calculated: 4.0–4.4 for MT1 forms featuring 12 Cys residues, 4.7–5.1 for MT2 with 14 Cys, 3.3–3.6 for MT3 with 10 Cys, and 5.7–6.2 for the E_c proteins with 17 Cys residues. Experimentally found metal ion contents are summarized in Table 2.

Some values in Table 2 show considerable deviations from the theoretical ratios. Most obvious is the high Zn^{II} content (11.5) of *P. sativum* MT1 determined after cleavage of the MT1-GST fusion protein with factor Xa [10]. Also Cd^{II} contents are higher than expected, both, for the Zn^{II}MT1-GST fusion protein after incubation with an excess of Cd^{II} (5.8) [10] as well as for the protein expressed without any tag (5.6–6.1 Cd^{II}) [8]. Considerably lower metal ion contents are observed for the Zn^{II} forms of *Q. suber* MT2 (3.5 Zn^{II}), determined by inductively coupled plasma optic emission spectroscopy (ICP-OES) with the acidified protein after thrombin cleavage of the GST tag [23], of the *E. guineensis* MT3-GST fusion protein, which binds 1.7 equivalents Zn^{II} more than the GST protein alone [18] as well as of *S. indicum* E_c (2.4 Zn^{II}), determined after cleavage of the GST tag with factor Xa.

To corroborate the initial metal ion-to-protein stoichiometries determined, additional experiments were performed for some of the plant MTs listed in Table 2. ESI-MS spectra at pH values above ~7 showed that *C. arietinum* Zn_4MT1, *Q. suber* Zn_4MT1 and Cd_6MT1, *C. arietinum* Cd_5MT2, as well as *T. aestivum* Zn_6E_c-1 are the predominant species in the respective

Table 2. Metal ion contents of plant MTs calculated based on metal ion-to-protein ratios in vertebrate MTs and determined experimentally on the purified proteins.

Species	Zn^{II} Form	Cd^{II} Form	Cu^{I} Form	Ref.
MT1 (12 Cys)[a]	4.0–4.4	4.0–4.4	6.6–8.0	
C. arietinum MT1[b1,c1]	4, 5	4, 5	6, 9[e]	[98]
P. sativum MT1[b2,c1]	5.6[d]	3.9[d]	2.3[d]	[10]
	11.5	5.8[d,e]	6.2[d,e]	[10]
		5.6–6.1[b3,c1]		[8]
T. durum MT1[d,f1]		4 (±1)		[14]
MT2 (14 Cys)[a]	4.7–5.1	4.7–5.1	7.7–9.3	
C. arietinum MT2[b1,c1]	5	5	8	[15]
Q. suber MT2[f2]	3.5	6.3	5.5 (+ 1.7 Zn^{II})	[23]
MT3 (10 Cys)[a]	3.3–3.6	3.3–3.6	5.5–6.7	
M. acuminata MT3[b1,c1]	3, 4	4		[16]
E. guineensis MT3[b2,c1,d]	1.7			[18]
E_c (17 Cys)[a]	5.7–6.2	5.7–6.2	9.3–11.3	
T. aestivum E_c-1[c1]	5[b2]			[3]
	6[b1]	6[b1]		[12]
S. indicum E_c[b2,c2]	2.4 (±0.6)		2.2 (±0.5)	[17]

[a] Range calculated based on the metal ion-to-protein ratios in the α-domain ($M^{II}_4Cys_{11}$, 1:2.75; $M^{I}_6Cys_{11}$, 1:1.83) and β-domain ($M^{II}_3Cys_9$1:3; $M^{I}_6Cys_9$, 1:1.5) of vertebrate MTs.
[b] Protein concentration determined with: [1]2-PDS assay, [2]Bradford assay, [3]amino acid analysis.
[c] Metal ion concentration determined with: [1]F-AAS, [2]ICP-AES.
[d] GST-fusion protein.
[e] Prepared by substitution of Zn^{II} ions with the respective metal ions.
[f] Metal ion-to-protein ratio determined by: [1]MALDI-TOF MS, [2]ICP-OES.

MT preparations [9,12,15,23]. Another method for the determination of the metal ion binding capacity of MTs is to follow metal ion exchange reactions with UV spectroscopy. Zn^{II}-thiolate clusters give rise to a slight shoulder of the peptide backbone amide bands at ~230 nm due to transitions mainly originating from thiolate ligand-to-metal charge transfer (LMCT) transitions into the Zn^{II} 4 s orbital (Figure 4) [68]. The characteristic transition of Cd^{II}-thiolate clusters is shifted to lower energy forming a pronounced shoulder of the amide bands at ~250 nm (Figure 4). Also this energy band has been described to originate from transitions with LMCT character involving thiolate p_π and Cd^{2+} 5 s orbitals, as well as additional contributions from lower energy requiring ligand to ligand – or intraligand – charge transfer (LLCT) transitions, namely from sulfur to antibonding p_π ($p_{\pi*}$) orbitals within thiolates [69]. The transitions of the Cd^{II}-thiolate cluster are sufficiently separated from the position of the $S \rightarrow Zn^{II}$ LMCT band, and

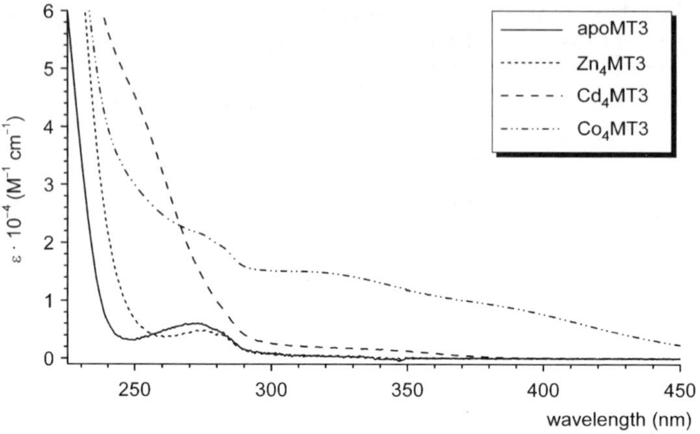

Figure 4. UV absorption spectra of a plant MT from the p3 subfamily showing the typical LMCT bands in the range of ~230–400 nm originating from the metal-thiolate clusters of the indicated divalent metal ions.

therefore the substitution of Zn^{II} ions in a Zn^{II} form MT by Cd^{II} and thus the increase in intensity ~250 nm with each substituted Zn^{II} ion can be conveniently followed with UV spectroscopy. As Cd^{II} usually adopts the same tetrahedral tetrathiolate coordination environment as Zn^{II}, such a titration experiment is a direct and very exact measure for the number of Zn^{II} ions originally bound to the MT. Such titration experiments performed with *C. arietinum* MT1 showed the formation of a Cd_4 or Cd_5 species depending on the protein batch used; the reason for this could not be determined so far [9]. Additionally, the existence of *C. arietinum* Cd_5MT2 [15], *Q. suber* Cd_4MT2 [21], *M. acuminata* Cd_4MT3 [16], and *T. aestivum* Cd_6E_c-1 [12] was confirmed in this way suggesting also the existence of the corresponding Zn^{II} forms.

5.2. Formation of Metal-Thiolate Clusters: Co^{II} Titration Studies

Co^{II} is used as a paramagnetic probe for Zn^{II} coordination sites in metallothioneins and other Zn^{II}-containing proteins [70,71]. UV absorption spectra of Co^{II} in tetrahedral tetrathiolate coordination spheres are characterized by charge transfer transitions in the near UV and in the visible range of the electromagnetic spectrum. The near UV transitions form an absorption envelope with a minimum around 290 nm followed by a maximum positioned around 320 nm with a pronounced shoulder around 380 nm (Figure 4). As for Zn^{II}- and Cd^{II}-thiolate clusters, these charge transfer transitions have their origin in S→Co^{II} LMCTs or in internal ligand excitations affected by interactions with the metal ion [72,73]. The visible range of the spectra features the *d-d* transitions of the ligand field of tetrahedrally coordinated high-spin Co^{II}-tetrathiolate complexes and is dominated by three broad bands around approximately 620, 690, and 740 nm originating from spin-orbit coupling [70,74]. Incremental reconstitution of apoMTs with Co^{II} under strictly anaerobic conditions first leads to a linear increase of the absorption profile without appreciable alteration in shape. This is indicative for Co^{II} binding to independent binding sites inside the protein, i.e., exclusive coordination to terminal thiolate groups. With increasing equivalents of Co^{II} hypso- and bathochromic shifts of absorption bands are observed indicating the onset of cluster formation. For example, the transition around 320 nm experiences a red shift due to the polarizing effect on the thiolate groups when originally terminal ligands are bridging two Co^{II} ions in the cluster [12,70].

So far, Co^{II} titration studies have been only performed with *T. aestivum* E_c-1 [12]. The absorption features observed in the near UV range are virtually identical to spectra of rabbit liver and crab MT [70,74]. The molar extinction of the maximum around 320 nm shows a value of $\sim 1000\,M^{-1}cm^{-1}$ per thiolate group and of $\sim 3000\,M^{-1}cm^{-1}$ per Co^{II} ion for all three MT isoforms calculated based on the extinction of the species fully loaded with metal ions. Cluster formation in *T. aestivum* E_c-1 starts with the addition of the fourth equivalent of Co^{II}, characterized by a red shift of the absorption band from initially 306 nm to a final wavelength of 318 nm for Co_6E_c-1.

However, the *d-d* transitions differ in shape and intensity. While the maximum around 690 nm has molar extinction values of $\sim 425\,M^{-1}cm^{-1}$ per Co^{II} ion in the case of rabbit liver and crab MT, the value for *T. aestivum* E_c-1 amounts only to $\sim 280\,M^{-1}cm^{-1}$ per Co^{II} ion. The relative intensity of the three bands to each other, 620:690:720 nm, is comparable in rabbit liver (0.7:1.0:1.0) and crab MT (0.7:1.0:0.9) but clearly different in *T. aestivum* E_c-1 (0.9:1.0:0.7) (Figure 5). It remains to be investigated whether these differences point to a different cluster structure in E_c-1, both with respect to stoichiometries as well as distortion from regular tetrahedral coordination geometry, or whether the change in absorption actually results from contributions of the highly conserved His residues in metal ion coordination in E_c-1. A corroboration of the latter can be found in spectra of Co^{II} complexes containing both Cys and His residues in their coordination spheres. The intensity of the highest energy *d-d* transition is increased relative to the transition ~690 nm for a $CoCys_3His$ [75,76] or $CoCys_2His(H_2O)$ environment [77], and the lowest energy *d-d* transition ~750 nm has completely disappeared (Figure 5) [75,77]. The spectrum of a peptide containing a $CoCys_2His_2$ coordination site shows two bands at 568 and 645 nm in a 0.3 : 1.0 ratio [76], while the spectrum of a second peptide is even reduced to just a single band in the visible range at 635 nm [78].

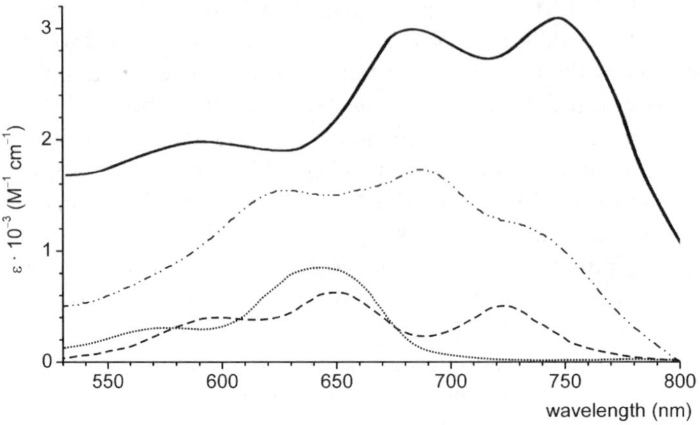

Figure 5. Spectra of the visible *d-d* transition range of different Co^{II}-containing MTs and peptides: rabbit liver $Co_{6.8}MT$ (solid line, [70]), *T. aestivum* Co_6E_c-1 (dash-dot-dotted line, [12]), and two model peptides with a $CoCys_3His$ (dashed line) and a $CoCys_2His_2$ site (dotted line) [76].

5.3. Stability of Metal-Thiolate Cluster

Both, the Zn^{II}- and Cd^{II}-thiolate clusters of MTs, are generally relatively stable against metal ion loss at neutral to slightly basic pH even under limited exposure to oxygen. However, complete release of metal ions occurs below pH ~3.5 for the Zn^{II} form and ~2.5 for the Cd^{II} form. A pH titration of the proteins whilst recording the decrease of the LMCT bands at 230 or 250 nm, respectively, allows the determination of the apparent pK_a values of the thiolate groups in presence of the respective metal ions, which is the equivalent of the pH value of half-maximum absorbance, $\frac{1}{2}(A_{max} + A_{min})$, if just a single pK_a value is considered. This pH value of half-dissociation of metal ions (in general ~4.5 for Zn- and ~3 for Cd-MTs) is one of the operational characteristics used to differentiate MTs from non-MT metal ion binding proteins [79]. This value can be read off more or less precisely directly from the pH titration curve, but mainly depending on the data quality, a more accurate measure is its determination via a curve fitting procedure with the following equation as described in detail in [16]:

$$A_{tot} = \frac{A_{MT} + A_{MTH_n} 10^{n(pK-pH)}}{1 + 10^{n(pK-pH)}} \tag{1}$$

A_{MT} is the absorptivity of the fully metal-loaded protein ($=A_{max}$), and A_{MTH_n} is the value obtained for apoMT after acidification ($=A_{min}$). The variable n was introduced into equation (1) solely to vary the slope of the titration step. This is necessary because the majority of thiolate groups in MTs have similar apparent pK_a values increasing the slope of the observed pH titration plot generally beyond the slope for a single protonation site. However, n does not reflect the actual number of protonation sites with (nearly) equal pK_a values. Whether n can be seen as a measure for cooperative effects in analogy to the coefficient in the corresponding Hill equation remains to be investigated [80]. The pH values of half-dissociation are reported for a number of MTs in the literature, although mostly given without any standard deviation, and thus offer the best way so far to compare cluster properties across the different MT families. Table 3 lists pH values of half-maximum absorbance as reported in the literature.

However, when evaluating the corresponding data plots, some of these values are derived from a relatively limited number of data points, which also show some spreading, and thus knowledge of the error range would be advantageous in order to allow comparisons. Hence, these data plots as well as all other published plots were re-evaluated with equation (1). For the Cd^{II} forms from mouse, monkey, and human a distinct two step behavior of the titration curve was observed. Therefore these data plots were also re-evaluated with an equation considering two different pK_a values:

$$A_{\text{tot}} = \frac{A_{\text{MT}} + A_{\text{MTH}_\text{m}} 10^{m(\text{p}K_2-\text{pH})} + A_{\text{MTH}_{\text{n+m}}} 10^{n\cdot\text{p}K_1+m\cdot\text{p}K_2-(n+m)\text{pH}}}{1 + 10^{m(\text{p}K_2-\text{pH})} + 10^{n\cdot\text{p}K_1+m\cdot\text{p}K_2-(n+m)\text{pH}}} \tag{2}$$

Figure 6 shows a graphical presentation of increasing pH values of half-maximum absorbance and the pK_a values determined with equations (1) or (2) as well as error bars for the standard deviations of the pK_a values on the 3σ level. One of the most surprising results is the extraordinary high pH stability of the Zn^{II}-thiolate clusters formed by *T. aestivum* E_c-1 having even a lower apparent pK_a value of the thiolate groups than equine MT, although the values overlap within the 3σ range. The pH stability of the human Zn^{II}-MT2 form is slightly higher but still in the same range around 4.3 pK_a units. This finding actually underlines the proposed function of E_c-1 as a seed zinc storage protein, high affinity for zinc being one of its characteristics. The next three pK_a values are clustered around 4.6. Again, a plant MT, *C. arietinum* MT2, shows a relatively high pH stability of its Zn^{II} cluster, which is comparable to values for the equine mixed $Zn^{II}Cd^{II}$-MT species as well as for the α-domain of human MT2. The highest pK_a values around 4.9 and thus the lowest pH stabilities in this row are found for the plant MTs from the p1 and p3 subfamilies and the β-domain of human MT2. With a value of 5.4 *P. sativum* MT1 marks the top of the chart.

The comparison of pH stabilities of the different Cd^{II} forms again features highest values for the plant MTs from the p1 and p3 subfamilies, and also *C.*

Table 3. pH values of half-maximum absorbance for different Zn- and Cd-MT forms as read off from the plot of absorbance against pH value (½ ($A_{min} + A_{max}$)) or determined by a curve fitting procedure with equation (1) and (2), respectively.

Species	Form	pH at ½ ($A_{min} + A_{max}$)	Eq. (1)	Eq. (2)[b]	Ref.
C. arietinum	Zn_5MT1		4.94(3)		[9]
P. sativum	$Zn_{5.6}MT1$[a]	5.25	5.4(1)[b]		[10]
C. arietinum	Zn_5MT2		4.51(1)		[15]
M. acuminata	Zn_4MT3		4.90(3)		[16]
T. aestivum	$Zn_{5.6}E_c$-1		4.26(1)		[12]
	$Zn_{5.8}Cu_{0.6}E_c$-1	4.7	4.68(10)[b,c]		[11]
Equine	$Cd_{4.5}Zn_{1.6}MT$[a]	4.6	4.6(4)[b,c]		[99]
	$Cd_{0.3}Zn_{5.7}MT$	4.4	4.31(1)[b,d]		[100]
Human	$Cd_{0.5}Zn_7MT2$	4.3	4.36(2)[b]		[101]
	α-Zn_4MT2	4.6	4.66(2)[b]		[101]
	β-Zn_3MT2	4.9	4.94(1)[b]		[101]
C. arietinum	Cd_5MT1		4.47(7)		[9]
P. sativum	$Cd_{5.8}MT1$[a]	3.95	4.03(7)[b]		[10]
C. arietinum	Cd_5MT2		3.73(3)		[15]
M. acuminata	$Cd_{4.3}MT3$		4.18(3)		[16]
T. aestivum	Cd_6E_c-1		3.35(3)		[12]
Equine	$Cd_{4.5}Zn_{1.6}MT$[a]	3.5	3.57(14)[b,c]		[99]
	$Cd_{6.9}Zn_{0.2}MT$		3.07(2)[b]		[102]
Mouse	Cd_7MT2		3.38(4)[b]	3.02(1) 3.81(2)	[103]
Monkey	$Cd_{6.9}MT2$		3.65(2)[b]	3.372(3) 4.069(8)	[104]
Human	Cd_7MT2		3.03(3)[b]	2.84(1) 3.50(9)	[101]
	α-Cd_4MT2	3.1	3.20(1)[b]		[105]
	β-Cd_3MT2	3.6	3.65(1)[b]		[105]
Rat	$Cd_{4.7}Zn_{2.0}MT$		3.34(1)[b,c]		[106]
	α-Cd_4MT[a]		3.30(3)[b]		[106]

[a] Limited number of data points in published plot.
[b] Re-evaluated with data taken from the literature.
[c] pH at which 50% of the Zn^{2+} or Cd^{2+} ions, respectively, of the mixed-metal species were displaced.
[d] pH of half-maximum absorption at 220 nm.

arietinum MT2 is only slightly more stable. The pH stability of Cd^{II} clusters formed by *T. aestivum* E_c-1 shares the position around 3.3 pK_a units with the mixed $Zn^{II}Cd^{II}$-MT species and the α-domain from rat, as well as the mouse MT2. However, considerably lower pK_a values around 3.1 are characteristic for the Cd^{II} forms of equine as well as the full-length human MT2 and its α-domain.

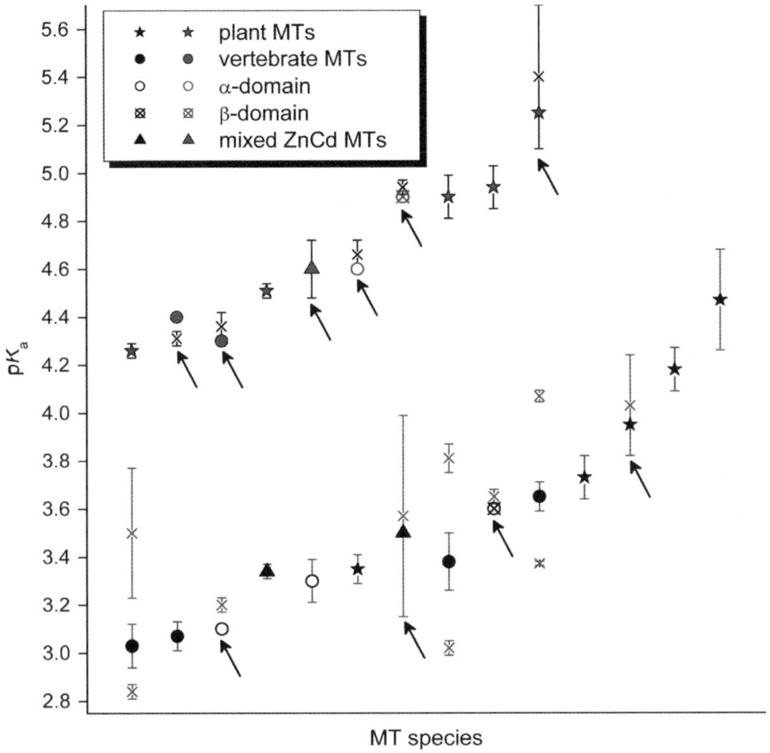

Figure 6. pH values of half-dissociation of metal ions from different MTs as calculated from published data with equation (1) (see Table 3). The error bars mark three times the standard deviation (3σ). For some species, pH values of half-maximum absorption are given in the literature (indicated with arrows). In these cases, the published values are marked with the appropriate data point symbol, while the pK_a value calculated from the published data plot with equation (1) is marked with a cross. In three cases, two additional pK_a values and the corresponding error bars are given above and below the data point symbol resulting from a second fitting using equation (2) (see Table 3). Top row: Zn^{II}MTs (from left to right: *T. aestivum* $Zn_{5.6}E_c$-1, equine $Zn_{5.7}Cd_{0.3}$MT, human $Zn_{7.0}Cd_{0.5}$MT2, *C. arietinum* Zn_5MT2, equine $Zn_{1.6}Cd_{4.5}$MT, human α-Zn_4MT2, human β-Zn_3MT2, *M. acuminata* Zn_4MT3, *C. arietinum* Zn_5MT1, *P. sativum* $Zn_{5.6}$MT1). Bottom row: Cd^{II}MTs (from left to right: human Cd_7MT2, equine $Zn_{0.2}Cd_{6.9}$MT, human α-Cd_4MT2, rat $Zn_{0.2}Cd_{4.7}$MT, rat α-Cd_4MT, *T. aestivum* Cd_6E_c-1, equine $Zn_{1.6}Cd_{4.5}$MT, mouse Cd_7MT, human β-Cd_3MT2, monkey $Cd_{6.9}$MT, *C. arietinum* Cd_5MT2, *P. sativum* $Cd_{5.8}$MT1, *M. acuminata* $Cd_{4.3}$MT3, *C. arietinum* Cd_5MT1). See text for details.

Due to the limited number of data it is so far not possible to make any predictions whether the order of pK_a values depicted in Figure 6 actually represents a trend with general applicability for the members of the four plant MT subfamilies. Intriguing, however, is the finding that not all plant MTs formerly classified as MT-like proteins are actually found at the bottom of the list of pH stabilities as is obvious from the values listed for *C. arietinum* MT2. Finally, it is also not clear yet, how the different pH stabilities of the different forms and species translate into the possible functions *in vivo*.

Stability constants of metal-thiolate clusters can be determined in a number of ways, one possibility being performance of a competition experiment with a metal ion chelator. For the study of Zn^{II}MTs 4-(2-pyridylazo)resorcinol (PAR) and 1,2-bis(2-amino-5-fluorophenoxy)ethane-*N,N,N′,N′*-tetraacetic acid (5F-BAPTA) have been used to name just two. Competition experiments with 100 μM PAR in 200 mM Tris-HCl pH 7.4 resulted in conditional stability constants of $2.9 \times 10^{10}\,M^{-1}$ for *C. arietinum* Zn_5MT2 and of $1.6 \times 10^{10}\,M^{-1}$ for *M. acuminata* Zn_4MT3 [25]. This is in between values determined for the amyloid precursor protein ($3 \times 10^9\,M^{-1}$, pH 7.5), which forms a Zn_2Cys_6 cluster in the protein dimer [81], and human MT3 ($6.2 \times 10^{10}\,M^{-1}$, pH 8.0) as well as rabbit liver MT2a ($3.1 \times 10^{11}\,M^{-1}$, pH 8.0) [82]. The competition reaction between *T. aestivum* E_c-1 and 5F-BAPTA in 10 mM Tris pH 8.1 yields apparent binding constants of $5 \times 10^{10}\,M^{-1}$ (no salt added) and $4.3 \times 10^8\,M^{-1}$ in 10 mM NaCl [11]. Note that all values listed are mean values for the coordination of one zinc ion.

6. INCORPORATION OF SULFIDE IONS

Sulfide ions increase the stability of Cd^{II}-phytochelatin complexes and additionally also the amount of Cd^{II} ions that can be bound per PC molecule. Even aggregates consisting of a CdS crystallite core coated with PCs have been reported [19]. Also inorganic coordination compounds in form of oligonuclear benzylthiolate complexes of zinc or cadmium are known that contain sulfide ions in their inner core, the clusters being shielded from their surroundings by the phenyl rings of the ligands [83,84]. Recently, also incorporation of sulfide ions into a number of MT isoforms from the MT families 1 (vertebrate MTs), 4 (echinodermata), 5 (diptera), 8–12 (fungi), and 15 (plant) was observed, all of them were expressed recombinantly in *E. coli* using the GST fusion tag system [85]. The number of sulfide ions incorporated into the Zn^{II} forms during expression was with 0.4–3.8 equivalents per protein generally lower than in the Cd^{II} forms with 1.9–4.7 equivalents. However, the sulfide content was not found to increase the metal ion binding ability of the vertebrate MTs beyond 7 equivalents Zn^{II} or Cd^{II}. ESI-MS

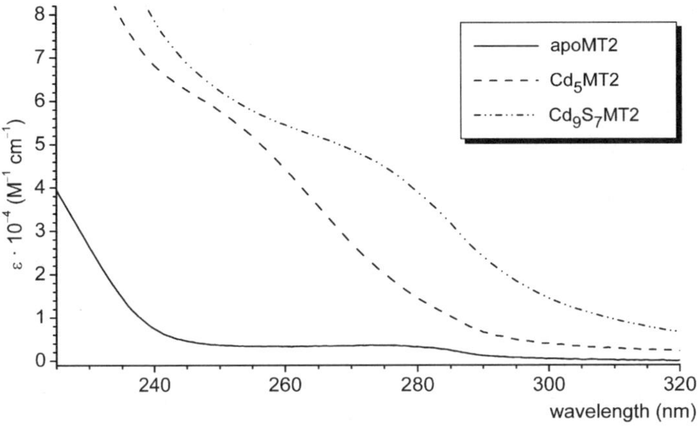

Figure 7. UV absorption spectra of different *C. arietinum* MT2 species as indicated showing the characteristic red-shift of the LMCT band upon incorporation of sulfide into the cadmium-thiolate cluster.

studies on the plant MT *Q. suber* MT2 reveal for the Zn^{II} form Zn_4MT2 as the major species next to some Zn_4S_2MT2 and for the Cd^{II} form Cd_6S_4MT2 next to some Cd_5MT2 [13]. Plant MTs, expressed recombinantly in *E. coli* using the intein fusion tag system, were generally not found to contain sulfide ions [9,12,15,16], which – at least for the Cd^{II} forms – can be easily detected as sulfide incorporation into Cd^{II}-thiolate clusters shifts the characteristic LMCT band from 250 nm to approximately 275 nm (Figure 7).

Nevertheless, a cluster with the stoichiometry Cd_9S_7MT2 is indeed formed upon titration of Zn_5MT2 with increasing equivalents of Cd^{II} in presence of Na_2S [86]. It remains to be investigated, whether the observed sulfide content of MTs is a general feature long overlooked or an artefact of recombinant protein overexpression in *E. coli*, and possibly even somehow linked to the GST fusion tag system. Also, the use of dialysis membranes has to be considered carefully, as certain commercial products, especially with low molecular weight cut-offs, contain considerable amounts of residual sulfide. Finally, the function and mechanism of such sulfide incorporation into MTs *in vivo* would require further research.

7. STRUCTURE

7.1. Secondary Structure of the Peptide Backbone

We recently investigated the secondary structure of *C. arietinum* Cd_5MT1 with CD and IR spectroscopy as well as with the aid of a structural prediction server

[9]. According to the results from structure prediction, the sequence of MT2 does not contain α-helical elements but shows contribution from β-sheets located in the linker region [87–89]. In detail, 40% of the linker region have a 50% probability of β-sheet formation (=23% of the total protein); using the lower threshold of 40% probability even 64% of the amino acids of the linker might participate in a β-sheet structure (=36% total). CD spectroscopy corroborates these results. The CD spectrum of Cd_5MT2 shows no indication for α-helices, and the processing of the spectrum using the k2D server and program [90–92] yields 58% random coil, 32% β-sheets as well as 10% helices.

As discussed [9], the k2D as well as the programs used for the evaluation of the IR data use a set of known regular protein structures without peculiarities like metal-thiolate clusters. Thus, the calculation of 10% helical content might as well be an artefact, possibly arising from spectral contributions of the cadmium-thiolate chromophore, which overlap with the contributions of α-helices. The analysis of secondary structural elements by IR spectroscopy was performed using the second derivative of the deconvoluted amide I band, which originates from the C=O stretching vibration of the peptide amide group. Also IR spectroscopy confirms with a value of 30% the percentage of β-sheets predicted and derived with CD spectroscopy. What was assigned as random coil with the CD analysis program can be further differentiated into actual random coil (34%) and β-turns (32%), which are major components of secondary structure in metal-thiolate clusters.

A similar investigation of apoMT2 was performed with IR spectroscopy. Again, 29% of β-sheets are found next to 30% random coil, 11% β-turns, and, not surprisingly, 29% of aggregated strands. It is known from the mammalian MTs that the apo-forms lack secondary structural features in the absence of metal ions and are also extremely prone to oxidation. Thus, while the β-sheet structure of *C. arietinum* MT2 is preserved in the apo-form, the β-turns of the cluster disappear to a large extend.

The only other plant MT, for which secondary structural elements were determined experimentally, is *Q. suber* MT2 [13]. Analysis of the amide I band from Raman spectroscopy and the amide III band from IR spectroscopy both, for the Zn^{II} and Cd^{II} forms, revealed a low α-helical content (0–4%), a β-turn contribution of 22–26%, 55–64% β-sheets, and 13–16% random coil. Hence, the amount of α-helices and β-turns is in the same range both, for *C. arietinum* MT1 and *Q. suber* MT2. Although being from two different subfamilies, p1 and p2, this similarity is not too surprising as the length of the Cys-rich regions is comparable and so should be the β-turn content of the formed clusters. The amount of β-sheets, however, is much higher in *Q. suber* MT2 and even higher than the percentage of amino acids making up the linker region (51%). In contrast, the amount of random coil structures in *C. arietinum* MT1 is with 34% approximately twice as high as reported for *Q. suber* MT2.

For two more plant MTs the secondary structure of the linker regions was determined based on theoretical calculations. *Ab initio* calculations performed with the sequence of *T. durum* MT1 yield two short α-helices separated by an antiparallel β-sheet [14]. For *Actinidia chinensis* MT3 a homology search combined with secondary structure prediction results in a hairpin structure with two β-sheets [24].

7.2. Arrangement of Metal-Thiolate Cluster(s)

One of the basic questions regarding the structure of plant MTs is the number of metal-thiolate clusters formed, i.e., do the Cys-rich domains separated by the linker regions form separate clusters or do they join forces to result in a single larger cluster? In theory, the members of the pec subfamily might even be able to arrange their coordinated metal ions in three separate clusters. Even in the absence of three-dimensional structural data several efforts were made to answer this question.

The N- and C-terminal Cys-rich regions in *P. sativum* MT1 were proposed to form a joint metal-thiolate cluster based on proteolytic digestion experiments with proteinase K [8]. The two regions coelute from a Sephadex G-75 column at neutral pH even after cleavage of the peptide backbone of the connecting linker region. After dissociation of metal ions at pH 2 and thus the breakdown of cluster(s), two different peptides can be resolved with a reverse phase column. This allows the conclusion that the N- and the C-terminal Cys-rich regions were originally joined by a metal-thiolate cluster. A molecular modelling study of another MT1 form, in this case from *T. durum*, presents a structure, in which the two Cys-rich regions form separate clusters with three Cd^{II} ions each [14]. However, one has to be aware that this separate cluster formation was arbitrarily chosen by the authors as a starting set for calculations lacking any experimental backup.

For *C. arietinum* MT1 we proposed the possibility for both cluster arrangements, a single cluster or two separate ones, based on the observation of a M_4^{II} as well as a M_5^{II} species, although interconversion of the forms was not possible so far [9]. Binding of four divalent metal ions can be accomplished in two separate cluster with two metal ions each resulting in a metal ion-to-Cys ratio of 2:6, and thus forming a type of cluster proposed for the γ-domain of *T. aestivum* E_c-1 (Figure 8A). The two separate clusters could be joined by coordination of a fifth metal ion resulting in a structure with similarity to a known mercury-selenolate cluster, $[Hg_5(SePh)_{12}]^{2-}$ [93].

For *Q. suber* MT2, experiments point towards a single metal-thiolate cluster based on comparative metal ion binding studies with the separate N- and C-terminal domains [21]. In analogy to results obtained with *P. sativum* MT1, also limited proteolytic digestion of *C. arietinum* Cd_5MT2 causes peptide backbone cleavage in the linker region with the resulting

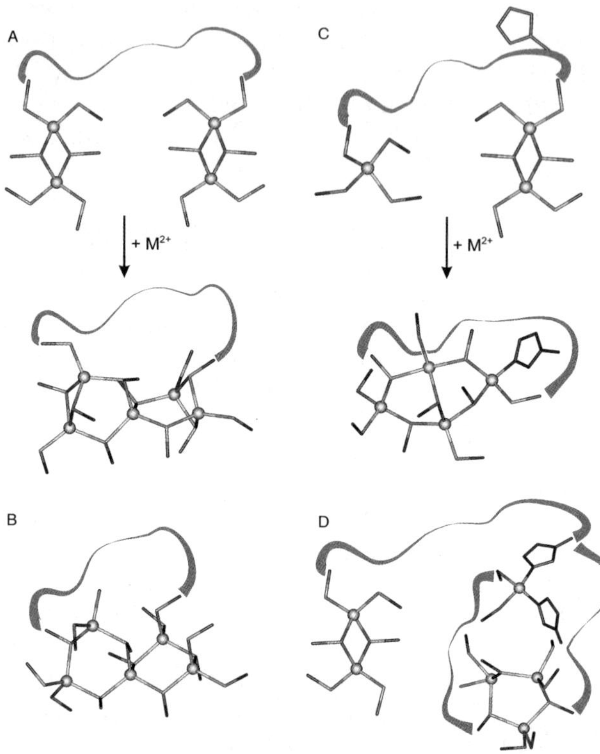

Figure 8. Hypothetical metal-thiolate cluster structures and arrangements of the members of the p1, p2, p3, and pec subfamilies of plant MTs based on the assumption that only Cys and possibly also His residues participate in the tetrahedral coordination of divalent metal ions. Metal ions are depicted as spheres, from the Cys and His residues only the thiolate groups and the imidazole rings, respectively, as well as the β-carbon atoms are drawn as sticks, while the other amino acids are schematically represented by grey ribbons. (**A**) MTs from the p1 subfamily can coordinate four divalent metal ions in two separate clusters. A fifth metal ion would require a joint single cluster. (**B**) For the coordination of five metal ions in MTs from the p2 subfamily a similar single cluster structure as in (A) must be formed, although with considerably less steric strain as two neighboring metal ions are connected by a single thiolate bridge, respectively, and not by two. (**C**) Proteins from the p3 subfamily can accomodate three metal ions in two separate clusters using exclusively thiolate ligands as depicted or additionally the conserved His residue. Assuming no other ligands, four metal ions require the formation of a joint single cluster. (**D**) Separate metal-thiolate clusters in MTs from the pec subfamily are corroborated by proteolytic digestion experiments and a solution structure from NMR spectroscopy as described in the text.

fragments coeluting from a Sephadex G-75 column at slightly basic pH [15]. MALDI-TOF analysis of the peak fraction showed peptide fragments belonging to the C- and N-terminal Cys-rich regions and the absence of fragments from the linker, which was confirmed by an amino acid analysis. Results from dynamic light scattering (DLS) agree with the described cluster arrangement. A hydrodynamic radius R_h of 1.5 nm was observed for Cd_5MT2, which results in a molecular mass of 9 kDa based on a globular protein model, which is close to the expected molecular mass of 8.5 kDa [15]. An approximately globular protein shape would agree well with a single metal-binding domains joining the two ends of the protein (Figure 8B). In contrast, for a structure consisting of two separate clusters one would expect a more elongated structure and thus, a larger value for R_h. For example, DLS measurements with *T. aestivum* Cd_6E_c-1, which consists of two different metal-binding domains (see below) and has the same molecular weight as Cd_5MT2, results in a R_h value of 1.8 nm (unpublished results). The apparent molecular mass calculated from this radius based on a globular structural model is accordingly higher (14 kDa) and a clear hint that E_c-1 indeed adopts an elongated conformation.

Although the modeling study performed on *A. chinensis* MT3 predicted the linker region to adopt a hairpin structure (see above), modeling of the Cys-rich regions was performed presuming formation of separate clusters with one divalent metal ion being coordinated by the four Cys residues of the N-terminal domain and another two divalent metal ions bound to the six Cys residues of the C-terminal domain [24]. Experimental data showed that another member of the p3 subfamily, *M. acuminata* MT3, can form two species: one coordinating 3 Zn^{2+} ions and the other 4 Zn^{2+} or Cd^{2+} ions [16]. In agreement with the theoretical study but purely based on stoichiometric considerations, a two-cluster structure would be the most likely arrangement for the three metal ion species. However, a single cluster connecting both Cys-rich domains is more probable for the species with four divalent metal ions. In this respect two cluster stoichiometries are possible. One arrangement, $M^{II}_4Cys_{10}$, would consider only Cys residues as ligands, while in the other, $M^{II}_4Cys_{10}His$, the steric strain is reduced by assuming participation of the His residue in metal ion coordination (Figure 8C). DLS measurements performed with Zn_3- and Zn_4MT3 corroborate the model of two different cluster arrangements even further, as the hydrodynamic radius obtained for Zn_3MT3 is with 1.9 nm significantly larger than the one for Zn_4MT3 ($\sim$1.5 nm).

Finally, for the pec subfamily the question regarding the number and arrangement of clusters seems resolved. The N-terminal Cys-rich domain, featuring six Cys residues as potential metal ion coordinating ligands, should form one cluster consisting of two divalent metal ions and has been denoted the γ-domain. A second metal-thiolate arrangement connects the central and C-terminal Cys-rich regions containing a total of 11 Cys residues next to two conserved His residues (Figure 8D). This conclusion can be drawn from limited proteolytic digestion experiments of *T. aestivum* Cd_6E_c-1 with proteinase K [12]. A subsequently performed size exclusion chromatography step revealed the presence of two well-resolved peaks, both with longer retention times than the undigested protein. Assignment of peaks to peptide fragments was performed with ESI-MS and amino acid analysis. Of special interest was the finding, that although the peptide backbone of the linker between the central and C-terminal Cys-rich region had been cleaved, the two fragments were still coeluting in one peak with a smaller elution time than the γ-domain. Hence, they must have been connected somehow, most likely by a joined metal-thiolate cluster arrangement. This finding of two separate metal-thiolate arrangements in *T. aestivum* E_c-1 was further corroborated with experiments performed on the separately synthesized or expressed domains showing their independence to a large extent [60]. The three-dimensional structure of the metal-thiolate arrangement formed by the central and C-terminal Cys-rich region is discussed in Section 7.4.

7.3. Coordination of Metal Ions by Histidine Residues

A recent review on His versus Cys coordination of Zn^{II} in MTs has been given in [94] (see also Chapter 3). Generally, the tetrahedral 1:2 complexes of Zn^{II} and Cd^{II} with the isolated amino acids Cys and His have stability constants of $\sim 10^{18}\,M^{-2}$ ($ZnCys_2$), $\sim 10^{12}\,M^{-2}$ ($ZnHis_2$), $\sim 10^{17}\,M^{-2}$ ($CdCys_2$), and $\sim 10^{10}\,M^{-2}$ ($CdHis_2$) [95], and thus the complexes with thiolate ligands have a greater stability than those with imidazole ligands both, for Zn^{II} and Cd^{II}. The situation changes somewhat when considering binding sites in proteins or peptides. A systematic study on three zinc finger peptides with a single metal ion binding site each, CCHH, CCHC, and CCCC, showed that gradual replacement of His residues by Cys led to a slight increase of the affinity for Zn^{II} ($1.6 \times 10^{11}\,M^{-1} \rightarrow 9.1 \times 10^{11}\,M^{-1}$), while a considerably larger stabilizing effect was observed for Cd^{II} ($5.0 \times 10^{8}\,M^{-1} \rightarrow 2.5 \times 10^{13}\,M^{-1}$) [76], This survey also showed that the CCHH zinc finger peptide prefers Zn^{II} over Cd^{II} while the reverse is true for the CCCC peptide. Although more stable in a thermodynamic sense, Cys_4 sites can be kinetically more labile than Cys_2His_2 sites [94]. This has been explained by electrostatic repulsion between deprotonated thiolate ligands leading to a temporary breaking of the individual coordinative bonds. *Vice versa*, an overall charge reducing and thus stabilizing effect on metal-thiolate clusters in MTs is expected upon partial exchange of Cys residues by His.

Currently, three three-dimensional structures of MTs featuring His residues in their amino acid sequences are available. In the solid state structure of the truncated Cu_8CUP1 protein from *S. cerevisiae* (PDB entry code 1RJU) His is not involved in metal ion coordination but seems to have an important role in stabilization of a loop structure via two hydrogen bonds to carbonyl oxygen atoms of the backbone (Figure 9A) [96]. The solution structure of Zn_4SmtA from the cyanobacterium *Synechococcus* PCC 7942 (PDB entry code 1JJD) shows the replacement of two terminal Cys residues by His in a cluster that resembles the Zn_4Cys_{11} cluster found in the α-domain of vertebrate MTs [97]. Overall, two Cys_3His and two Cys_4 sites are formed (Figure 9B). Finally, the solution structure of the C-terminal part of *T. aestivum* E_c-1 featuring the central and C-terminal Cys-rich regions including the two highly conserved His residues shows an isolated $ZnCys_2His_2$ site next to a Zn_3Cys_9 cluster similar to the β-domain of the vertebrate MTs (Figure 9C, see also Section 7.4) [26].

The latter structure provides thus the first unequivocal evidence that the conserved His residues in plant MTs, more specifically in the pec subfamily, indeed participate in metal ion coordination. Evidence in favor of an active role of His residues in Zn^{II} binding in *T. aestivum* E_c-1 was already presented in the form of ^{1}H NMR NOESY spectra showing the close proximity of the two His residues in space as well as to β-protons of Cys residues [11]. As elaborated above in Section 5.2, the increased absorption of the highest and the decreased absorption of the lowest energy band of the *d-d* transitions observed during the titration of apoE_c-1 with Co^{II} might also give a hint towards His coordination [12]. Finally, EXAFS studies, performed on both, the full-length Zn_6E_c-1 protein and the Zn_4E_c-1 species obtained after truncation of the N-terminal γ-domain as depicted in Figure 9C, show contribution of nitrogen and/or oxygen ligands in the coordination spheres of the Zn^{II} ions [60]. EXAFS studies have the drawback of only providing average values for the coordination spheres of all Zn^{II} ions present in the protein, nevertheless, results agree within the error range with values calculated based on the three dimensional structure of Zn_4E_c-1 (see Section 7.4). For Zn_6E_c-1 the EXAFS data was fitted with a model showing contribution of 3.8 ± 0.1 S and 0.2 ± 0.1 light ligands (O or N) per binding site (calc. $Cys_{3.67}His_{0.33}$), while for Zn_4E_c-1 values of 3.5 ± 0.1 S and 0.5 ± 0.1 light ligands were obtained (calc. $Cys_{3.5}His_{0.5}$). For the Cd^{II} form of *Q. suber* MT2 the pH-dependent appearance of a positive ellipticity band around 250 nm in the CD spectra was associated with metal ion coordination by His residues [21], as well as the appearance of a vibration at 1598 cm^{-1} in the Raman spectra, which was assigned to the $C_4{=}C_5$ stretching vibration of the imidazole ring in the N2 tautomer [13]. The impact of His ligands on the metal ion binding properties has yet to be investigated but promises interesting new insights into the functions of plant MTs.

A

B

C

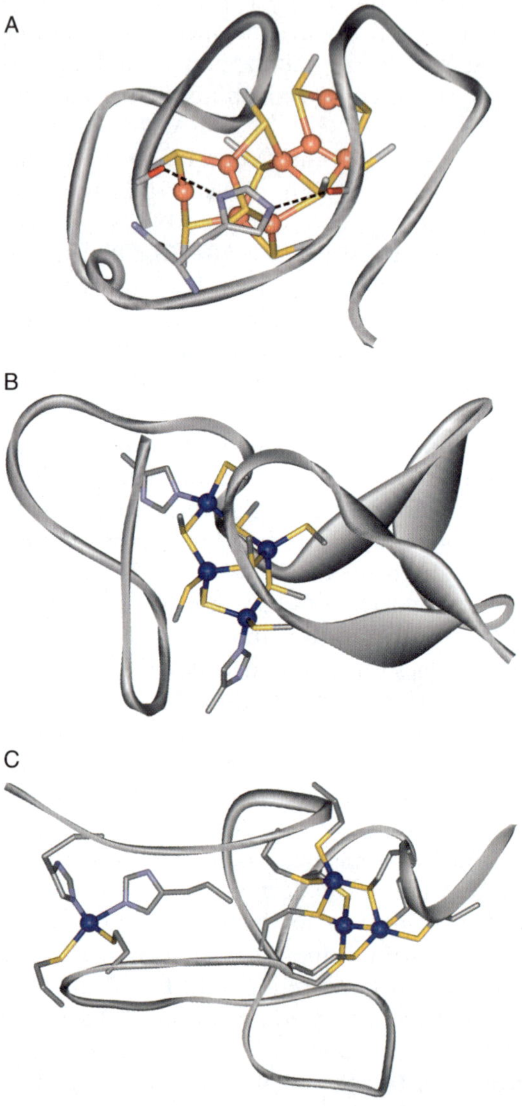

Figure 9. Solution structures of three His-containing MTs determined by NMR spectroscopy. (**A**) The His residues in *S. cerevisiae* Cu_8CUP1 (PDB entry code 1RJU) are not involved in coordination of Cu^I ions but rather form hydrogen bonds to carbonyl oxygens of the peptide backbone [96]. (**B**) In *Synechococcus* PCC 7942 Zn_4SmtA (PDB entry code 1JJD) two terminal His residues participate in the formation of a cluster with a certain similarity to the all-Cys cluster of the α-domain found in vertebrate MTs [97]. (**C**) The C-terminal part of *T. aestivum* Zn_4E_c-1 contains an isolated $ZnCys_2His_2$ site next to a Zn_3Cys_9 cluster similar to the β-domain of vertebrate MTs.

7.4. Three-dimensional Structure

Knowledge about the three-dimensional structures of plant MTs is currently limited to a single solution structure of the C-terminal part of *T. aestivum* E_c-1, Zn_4E_c-1, featuring the central and C-terminal Cys-rich regions determined by NMR spectroscopy (Figure 9C) [26]. A large degree of independence of Zn_4E_c-1 to the rest of the protein, i.e., the N-terminal γ-domain, was demonstrated by comparing metal ion binding abilities and NMR spectra of the truncated form with those of the full-length protein [60]. An earlier study suggested the formation of a single cluster with similarity to the α-domain of the vertebrate MTs based on the identical number of Cys residues in both forms and the observation that the central and the C-terminal Cys rich regions coelute from the size exclusion chromatography column after cleavage of the connecting linker region [12].

The present structure however, reveals a rather surprising and unexpected metal-thiolate cluster structure consisting of a Zn_3Cys_9 cluster similar to the β-domain found, e.g., in mammalian MTs, as well as an isolated $ZnCys_2His_2$ site as known, e.g., from zinc finger proteins. The latter motif is unprecedented in any MT studied so far. Interestingly, the amino acids forming the isolated binding site are not solely located at the N-terminus of the truncated protein, but their positions in the amino acid chain are rather interwoven with the positions of the Cys ligands constituting the Zn_3Cys_9 cluster.

The novel cluster structure also sheds light onto two curiosities described before. One-dimensional $^{111/113}$Cd NMR spectra show two relatively sharp and intense signals [12,60] next to three broader ones with lower intensity [60]. In light of the determined structure, assignment of the two strong signals to the γ-domain and the three weaker signals to the β-domain-like cluster seems reasonable, especially as the arrangement in the β-domain is known to exhibit an increased flexibility leading to broadened signals in the NMR experiment. Also the difficulties in replacing the Zn^{II} ions in Zn_6E_c-1 completely by Cd^{II} [11] can be partially explained with the preference of Cys_2His_2 sites in zinc finger peptides for Zn^{II} compared to Cd^{II} (see Section 7.3) [76].

8. CONCLUDING REMARKS AND FUTURE DIRECTIONS

The investigation of plant MTs has received considerably little interest when compared to the vertebrate isoforms. An increase in research activities has only become apparent in the last few years. With respect to the possible functions of plant MTs *in vivo*, the majority of informations is the result of gene expression studies. Only recently, results from investigations with gene-knockout plants have become available und provide, at least to some extent, intriguing new insights into the differential properties of, on the first glance, closely related proteins.

Little attention has been paid so far to the properties and structures of the translated proteins. In part this is due to the difficulties associated with direct isolation of MTs from plants and to the drawback of proteolytic digestion of recombinantly expressed proteins in *E. coli* reported in earlier studies. At least the latter problem is resolved with the application of fusion proteins for protein overexpression. Hence, informations on the basic properties of plant MTs have become available. However, there is still an enormous lack of three-dimensional data, let alone knowledge concerning metal-thiolate cluster formation and protein folding, and it is one of the eminent future goals to add more pieces to the structural picture, which will certainly smoothen the way to understand the functions of MTs *in vivo* and their interplay with potential binding partners in the plant organism.

ACKNOWLEDGMENTS

Financial support from the Swiss National Science Foundation is gratefully acknowledged: SNF grant 20-113728/1 and SNF-Förderungsprofessur PP002-119106/1.

ABBREVIATIONS

2-PDS	2,2′-dithiodipyridine
5F-BAPTA	1,2-bis(2-amino-5-fluorophenoxy)ethane-*N*,*N*,*N*′,*N*′-tetraacetic acid
aa	amino acids
ABA	abscisic acid
bp	base pair
BSA	bovine serum albumin
CD	circular dichroism
cDNA	complementary DNA, synthesized from a mature mRNA template
DLS	dynamic light scattering
DTT	1,4-dithiothreitol
E_c	early cysteine labeled protein
EDTA	ethylenediamine-*N*,*N*,*N*′,*N*′-tetraacetate
ERE	ethylene responsive element
ESI-MS	electrospray ionization mass spectroscopy
EXAFS	extended X-ray absorption fine structure

F-AAS	flame atomic absorption spectroscopy
FPLC	fast protein liquid chromatography
GST	glutathione S-transferase
GUS	β-glucuronidase
HEPES	4-(2-hydroxyethyl)-1-piperazineethanesulfonic acid
ICP-AES	inductively coupled plasma-atomic emission spectroscopy
ICP-OES	inductively coupled plasma optic emission spectroscopy
IR	infrared
LLCT	ligand-to-ligand charge transfer
LMCT	ligand-to-metal charge transfer
MALDI-TOF	matrix-assisted laser desorption/ionization – time of flight
mBBr	monobromobimane, 4-bromomethyl-3,6,7-trimethyl-1,5-diazabicyclo[3.3.0]octa-3,6-diene-2,8-dione
MT	metallothionein
mt	metallothionein gene
MRE	metal responsive element
NOESY	nuclear Overhauser effect spectroscopy
ORF	open reading frame
PAGE	polyacrylamide gel electrophoresis
p1, p2, p3	subfamilies 1, 2, and 3 of plants metallothioneins
PAR	4-(2-pyridylazo)resorcinol
PC	phytochelatin
pec	subfamily of the E_c plant metallothioneins
ROS	reactive oxygen species
RSE	root specific element
SDS	sodium dodecyl sulfate
T-DNA	transferred (foreign) DNA into a host's nuclear DNA genome
Tris-HCl	tris-(hydroxymethyl)-aminoethane hydrochloride

REFERENCES

1. M. Margoshes and B. L. Vallee, *J. Am. Chem. Soc.*, 1957, **79**, 4813–4814.
2. L. Hanley-Bowdoin and B. G. Lane, *Eur. J. Biochem.*, 1983, **135**, 9–15.
3. B. G. Lane, R. Kajioka and T. D. Kennedy, *Biochem. Cell Biol.*, 1987, **65**, 1001–1005.
4. J. H. R. Kägi, S. R. Himmelhoch, P. D. Whanger, J. L. Bethune and B. L. Vallee, *J. Biol. Chem.*, 1974, **249**, 3537–3542.

5. Y. Kojima, C. Berger, B. L. Vallee and J. H. R. Kägi, *Proc. Natl. Acad. Sci. USA*, 1976, **73**, 3413–3417.
6. http://expasy.org/sprot/.
7. N. J. Robinson, J. R. Wilson and J. S. Turner, *Plant Mol. Biol.*, 1996, **30**, 1169–1179.
8. P. Kille, D. R. Winge, J. L. Harwood and J. Kay, *FEBS Lett.*, 1991, **295**, 171–175.
9. O. Schicht and E. Freisinger, *Inorg. Chim. Acta*, 2008, accepted.
10. A. M. Tommey, J. Shi, W. P. Lindsay, P. E. Urwin and N. J. Robinson, *FEBS Lett.*, 1991, **292**, 48–52.
11. O. I. Leszczyszyn, R. Schmid and C. A. Blindauer, *Proteins: Structure, Function, and Bioinformatics*, 2007, **68**, 922–935.
12. E. A. Peroza and E. Freisinger, *J. Biol. Inorg. Chem.*, 2007, **12**, 377–391.
13. J. Domènech, A. Tinti, M. Capdevila, S. Atrian and A. Torreggiani, *Biopolymers*, 2007, **86**, 240–248.
14. K. Bilecen, U. H. Ozturk, A. D. Duru, T. Sutlu, M. V. Petoukhov, D. I. Svergun, M. H. J. Koch, U. O. Sezerman, I. Cakmak and Z. Sayers, *J. Biol. Chem.*, 2005, **280**, 13701–13711.
15. X. Wan and E. Freisinger, 2008, *to be submitted.*
16. E. Freisinger, *Inorg. Chim. Acta.*, 2007, **360**, 369–380.
17. C. L. Chyan, T. T. Lee, C. P. Liu, Y. C. Yang, J. T. Tzen and W. M. Chou, *Biosci. Biotechnol. Biochem.*, 2005, **69**, 2319–2325.
18. S. N. A. Abdullah, S. C. Cheah and D. J. Murphy, *Plant Physiol. Biochem.*, 2002, **40**, 255–263.
19. C. Cobbett and P. Goldsbrough, *Annu. Rev. Plant Biol.*, 2002, **53**, 159–182.
20. J. Domènech, G. Mir, G. Huguet, M. Capdevila, M. Molinas and S. Atrian, *Biochimie*, 2006, **88**, 583–593.
21. J. Domènech, R. Orihuela, G. Mir, M. Molinas, S. Atrian and M. Capdevila, *J. Biol. Inorg. Chem.*, 2007, **12**, 867–882.
22. J. Lee, D. Shim, W. Y. Song, I. Hwang and Y. Lee, *Plant Mol. Biol.*, 2004, **54**, 805–815.
23. G. Mir, J. Domènech, G. Huguet, W.-J. Guo, P. Goldsbrough, S. Atrian and M. Molinas, *J. Exp. Bot.*, 2004, **55**, 2483–2493.
24. Z. Chunming, L. Tun, Z. Riqing, Z. Nanming and L. Jinyuan, *Chinese Sci. Bull.*, 2000, **45**, 1413–1417.
25. A. Cabral and E. Freisinger, 2008, *work to be published.*
26. E. A. Peroza, O. Zerbe and E. Freisinger, 2008, *in preparation.*
27. B. A. Fowler, C. E. Hildebrand, Y. Kojima and M. Webb, *Experientia Suppl.*, 1987, **52**, 19–22.
28. Y. Kojima, in *Metallobiochemistry, Part B: Metallothionein and Related Molecules*, Vol. **205** *of Methods in Enzymology*, Ed. J. F. Riordan and B. L. Vallee, Academic Press, New York, 1991, pp. 8–10.
29. M. Oven, J. E. Page, M. H. Zenk and T. M. Kutchan, *J. Biol. Chem.*, 2002, **277**, 4747–4754.
30. I. Kawashima, T. D. Kennedy, M. Chino and B. G. Lane, *Eur. J. Biochem.*, 1992, **209**, 971–976.

31. I. M. Evans, L. N. Gatehouse, J. A. Gatehouse, N. J. Robinson and R. R. D. Croy, *FEBS Lett.*, 1990, **262**, 29–32.
32. Y. Kojima, P.-A. Binz and J. H. R. Kägi, in *Metallothionein IV*, Ed. C. Klaassen, Birkhäuser Verlag, Basel, 1999, pp. 3–6.
33. N. J. Robinson, A. M. Tommey, C. Kuske and P. J. Jackson, *Biochem. J.*, 1993, **295**, 1–10.
34. J. M. Zhou and P. B. Goldsbrough, *Mol. Gen. Genet.*, 1995, **248**, 318–328.
35. V. Buchanan-Wollaston, *Plant Physiol.*, 1994, **105**, 83846.
36. S. K. Clendennen and G. D. May, *Plant Physiol.*, 1997, **115**, 463–469.
37. G. K. Zhou, Y. F. Xu, J. Li, L. Y. Yang and J.-Y. Liu, *J. Biochem. Mol. Biol.*, 2006, **39**, 595–606.
38. http://www.ncbi.nlm.nih.gov/Genbank/index.html.
39. J. M. Zhou and P. B. Goldsbrough, *Plant Cell*, 1994, **6**, 875–884.
40. W. J. Guo, W. Bundithya and P. B. Goldsbrough, *New Phytol.*, 2003, **159**, 369–381.
41. A. Marion-Poll, *Trends Plant Sci.*, 1997, **2**, 447–448.
42. M. Karin and R. I. Richards, *Nature*, 1982, **299**, 797–802.
43. A. J. de Framond, *FEBS Lett.*, 1991, **290**, 103–106.
44. A. Murphy and L. Taiz, *Plant Physiol.*, 1995, **109**, 945–954.
45. T. Moriguchi, M. Kita, S. Hisada, T. Endo-Inagaki and M. Omura, *Gene*, 1998, **211**, 221–227.
46. S. J. Reid and G. S. Ross, *Physiol. Plant.*, 1997, **100**, 183–189.
47. D. Choi, H. M. Kim, H. K. Yun, J.-A. Park, W. T. Kim and S. H. Bok, *Plant Physiol.*, 1996, **112**, 353–359.
48. T. L. Reynolds and R. L. Crawford, *Plant Mol. Biol.*, 1996, **32**, 823–829.
49. C. N. White and C. J. Rivin, *Plant Physiol.*, 1995, **108**, 831–832.
50. A. Brenes-Pomales, G. Lindegren and C. C. Lindegren, *Nature*, 1955, **176**, 841–842.
51. T. R. Butt, E. Sternberg, J. Herd and S. T. Crooke, *Gene*, 1984, **27**, 23–33.
52. V. Cizewski Culotta, W. R. Howard and X. F. Liu, *J. Biol. Chem.*, 1994, **269**, 25295–25302.
53. D. H. Hamer, D. J. Thiele and J. E. Lemontt, *Science*, 1985, **228**, 685–690.
54. A. Pagani, L. Villarreal, M. Capdevila and S. Atrian, *Mol. Microbiol.*, 2007, **63**, 256–269.
55. W.-J. Guo, M. Meetam and P. Goldsbrough, *Plant Physiol.*, 2008, **146**, 1697–1706.
56. M. N. Merzlyak and G. A. F. Hendry, *P. Roy. Soc. Edinb. B*, 1994, **102**, 459–471.
57. L. S. Chubatsu and R. Meneghini, *Biochem. J.*, 1993, **291**, 193–198.
58. M. M. Bradford, *Anal. Biochem.*, 1976, **72**, 248–254.
59. A. O. Pedersen and J. Jacobsen, *Eur. J. Biochem.*, 1980, **106**, 291–295.
60. E. A. Peroza, A. Al Kaabi and E. Freisinger, 2008, *in preparation.*
61. S. Hong, M. Toyama, W. Maret and Y. Murooka, *Protein Expres. Purif.*, 2001, **21**, 243–250.
62. E. A. Peroza and E. Freisinger, *Protein Expres. Purif.*, 2008, **57**, 217–225.

63. A. Siti Nor Akmar, S. C. Cheah and D. J. Murphy, *Plant Physiol. Bioch.*, 2002, **40**, 255–263.
64. J. A. Reynolds and C. Tanford, *Proc. Natl. Acad. Sci. USA*, 1970, **66**, 1002–1007.
65. R. Pitt-Rivers and F. S. A. Impiomba, *Biochem. J.*, 1968, **109**, 825–830.
66. J. G. Valenzuela, I. M. B. Francischetti, V. M. Pham, M. K. Garfield, T. N. Mather and J. M. C. Ribeiro, *J. Exp. Biol.*, 2002, **205**, 2843–2864.
67. G. Meloni, M. Knipp and M. Vašák, *J. Biochem. Bioph. Meth.*, 2005, **64**, 76–81.
68. E. I. Solomon, S. I. Gorelsky and A. Dey, *J. Comp. Chem.*, 2006, **27**, 1415–1428.
69. H.-J. Liu, J. T. Hupp and M. A. Ratner, *J. Phys. Chem.*, 1996, **100**, 12204–12213.
70. M. Vašák and J. H. R. Kägi, *Proc. Natl. Acad. Sci. USA*, 1981, **78**, 6709–6713.
71. M. Vašák, *J. Am. Chem. Soc.*, 1980, **102**, 3953–3955.
72. M. Vašák, J. H. R. Kägi, B. Holmquist and B. L. Vallee, *Biochemistry*, 1981, **20**, 6659–6664.
73. D. Mastropaolo, J. A. Thich, J. A. Potenza and H. J. Schugar, *J. Am. Chem. Soc.*, 1977, **99**, 424–429.
74. J. Overnell, M. Good and M. Vašák, *Eur. J. Biochem.*, 1988, **172**, 171–177.
75. L. M. Green and J. M. Berg, *Proc. Natl. Acad. Sci. USA*, 1989, **86**, 4047–4051.
76. B. A. Krizek, D. L. Merkle and J. M. Berg, *Inorg. Chem.*, 1993, **32**, 937–940.
77. A. J. Sytkowski and B. L. Vallee, *Proc. Natl. Acad. Sci. USA*, 1976, **73**, 344–348.
78. A. D. Frankel, J. M. Berg and C. O. Pabo, *Proc. Natl. Acad. Sci. USA*, 1987, **84**, 4841–4845.
79. M. Vašák and I. M. Armitage, *Environ. Health Perspect.*, 1986, **65**, 215–216.
80. A. V. Hill, *J. Physiology-London*, 1910, **40**, 4–7.
81. E.-D. Ciuculescu, Y. Mekmouche and P. Faller, *Chem. Eur. J.*, 2005, **11**, 903–909.
82. D. W. Hasler, L. T. Jensen, O. Zerbe, D. R. Winge and M. Vašák, *Biochemistry*, 2000, **39**, 14567–14575.
83. M. Gelinsky and H. Vahrenkamp, *Z. Anorg. Allg. Chem.*, 2002, **628**, 1017–1021.
84. G. S. H. Lee, K. J. Fisher, A. M. Vassallo, J. V. Hanna and I. G. Dance, *Inorg. Chem.*, 1993, **32**, 66–72.
85. M. Capdevila, J. Domenech, A. Pagani, L. Tio, L. Villarreal and S. Atrian, *Angew. Chem. Int. Edit.*, 2005, **44**, 4618–4622.
86. X. Wan and E. Freisinger, *in preparation*, 2008.
87. http://www.predictprotein.org.
88. B. Rost, P. Fariselli and R. Casadio, *Protein Sci.*, 1996, **5**, 1704–1718.
89. B. Rost and C. Sander, *J. Mol. Biol.*, 1993, **232**, 584–599.
90. M. A. Andrade, P. Chacón, J. J. Merelo and F. Morán, *Prot. Eng.*, 1993, **6**, 383–390.
91. http://www.embl-heidelberg.de/~andrade/k2d.html.
92. J. J. Merelo, M. A. Andrade, A. Prieto and F. Morán, *Neurocomputing*, 1994, **6**, 443–454.
93. M. Berardini, T. J. Emge and J. G. Brennan, *Inorg. Chem.*, 1995, **34**, 5327–5334.
94. C. A. Blindauer, *J. Inorg. Biochem.*, 2008, **102**, 507–521.

95. *NIST Critically Selected Stability Constants of Metal Complexes*, Reference Database 46, Version 7.0 (data collected and selected by R. M. Smith and A. E. Martell), U. S. Department of Commerce, National Institute of Standards and Technology, Gaithersburg, MD, USA, 2003.
96. V. Calderone, B. Dolderer, H. J. Hartmann, H. Echner, C. Luchinat, C. Del Bianco, S. Mangani and U. Weser, *Proc. Natl. Acad. Sci. USA*, 2005, **102**, 51–56.
97. C. A. Blindauer, M. D. Harrison, J. A. Parkinson, A. K. Robinson, J. S. Cavet, N. J. Robinson and P. J. Sadler, *Proc. Natl. Acad. Sci. USA*, 2001, **98**, 9593–9598.
98. O. Schicht and E. Freisinger, *results to be published.*
99. J. H. R. Kägi and B. L. Vallee, *J. Biol. Chem.*, 1960, **235**, 3460–3465.
100. M. Vašák and J. H. R. Kägi, in *Metal Ions in Biological Systems*, Vol. **15**, Ed. A. Sigel and H. Sigel, Marcel Dekker Inc, New York, 1983, pp. 213–273.
101. L. J. Jiang, M. Vašák, B. L. Vallee and W. Maret, *Proc. Natl. Acad. Sci. USA*, 2000, **97**, 2503–2508.
102. J. H. R. Kägi and B. L. Vallee, *J. Biol. Chem.*, 1961, **236**, 2435–2442.
103. B. Cai, Q. Zheng and Z. X. Huang, *Protein J.*, 2005, **24**, 327–336.
104. W.-H. Yu, B. Cai, Y. Gao, Y. Xie and Z.-X. Huang, *J. Protein Chem.*, 2002, **21**, 177–185.
105. P. K. Y. Pan, Z. F. Zheng, P. C. Lyu and P. C. Huang, *Eur. J. Biochem.*, 1999, **266**, 33–39.
106. D. R. Winge and K.-A. Miklossy, *J. Biol. Chem.*, 1982, **257**, 3471–3476.

Met. Ions Life Sci. **2009**, *5*, 155–181

6

Metallothioneins in Diptera

Sílvia Atrian

Department of Genetics, Faculty of Biology, Universitat de Barcelona, Av. Diagonal 645, E-08028 Barcelona, Spain

<satrian@ub.edu>

Metal Ions in Life Sciences, Volume 5 Edited by Astrid Sigel, Helmut Sigel, and Roland K. O. Sigel

Published by the Royal Society of Chemistry, www.rsc.org DOI: 10.1039/9781847558992-00155

ABSTRACT: Dipterean insecta owe their relevance to the well-known model organism *D. melanogaster*, and extensively to the *Drosophila* genus. In the frame of the study of metallothioneins (MTs), they constitute a bizarre exception, since *Drosophila* is the only metazoan organism synthesizing only MTs similar to the yeast (*S. cerevisiae*) Cup1 paradigmatic copper-thionein. *D. melanogaster* MTs are optimized for copper binding and metabolism, and no MT analogous to the mammalian MTs has been identified, unlike in the cases of other Arthropoda and invertebrates analyzed so far (Chapters 7 and 8 of this book). The *D. melanogaster* genome contains four MT genes (*MtnA*, *MtnB*, *MtnC*, and *MtnD*), which are clustered in the same genomic region, probably arising from amplifying duplications. The main isoforms are *MtnA* and *MtnB*, encoding peptides 40- and 43-amino acids long, comprising 10 and 12 cysteines, respectively. MtnC and MtnD probably arise from recent MtnB duplications and play a minor role in flies. Expression of the four genes is dependent on dMTF-1, a transcriptional activator homologous to the mammalian MTF-1. Major MT synthesis in larvae and adults is observed in the digestive tract, mainly in the midgut, although with a different distribution between *MtnA* and *MtnB*. Other tissues of *Mtn* gene expression are salivary glands, ventricula, Malpighian tubules, and hemocytes. Metal induction increases the rate of MT synthesis in these tissues, but not normally in MT gene-silent organs, and invariably, the best response is obtained for copper, followed by cadmium, and very poorly for zinc. In conclusion, all the available data suggests a definitive role of *D. melanogaster* MT in copper metabolism (from ingestion to distribution, storage, delivery, and detoxification in the organism) as well as in cadmium tolerance (by digestive assimilation blockage). This is readily confirmed by studies of population genetics that show that both spontaneous *MtnA* genomic duplications, independently isolated in worldwide locations, and fly strains containing supernumerary Mtn genes constructed in the laboratory are able to tolerate increased copper and cadmium concentrations; while *Mtn* and *dMTF-1* knock-outs are extremely sensitive to these conditions.

KEYWORDS: cadmium · copper · diptera · *Drosophila* · insects · metallothionein · zinc

1. INTRODUCTION

Despite the well-established role of *Drosophila* as one of the model organisms in classic and molecular genetics, until recent years the information available on the *Drosophila* metallothionein (MT) system was appallingly scarce in comparison to mammalian MTs, or even to other invertebrate MTs, such as Crustacea or Echinodermata. Fortunately, two lines of research have come to compensate for this paradox, one focused on the molecular mechanisms regulating the expression of the *Drosophila* MT genes and their relation with MT function and metal homeostasis, and another devoted to the characterization of *Drosophila* MT proteins and their metal-binding abilities, made possible by their recombinant synthesis. Tackling the issue from these two approaches has provided a much better and comprehensive understanding of the function played by MTs in fruit fly physiology and evolution, although we are still a long way from fully comprehending these subjects.

2. METALLOTHIONEIN GENES IN *DROSOPHILA MELANOGASTER:* GENOMIC AND CHROMOSOMAL ARCHITECTURE. GENE AMPLIFICATION

Among *Drosophila* species, major research has obviously been focused and carried out in *Drosophila melanogaster*. Work performed on other *Drosophila* species, mainly concerning evolutionary aspects, will be commented in Section 6 of this chapter.

Molecular genetics techniques based on the identification of transcripts whose levels significantly increased in organisms and/or cells exposed to excess metal (copper or cadmium) made it possible, in the 1980s, to identify two MT genes in *Drosophila*: *Mtn* [1,2] and *Mto* [3], nowadays denominated *MtnA* and *MtnB* respectively, according to the FlyBase Consortium gene nomenclature (*cf.* http://flybase.bio.indiana.edu). Surprisingly, these encode distant paralog proteins, sharing only 25% amino acid identity. More recently, the exhaustive analysis of the complete *D. melanogaster* genome revealed the existence of two additional MT genes, therein named *MtnC* and *MtnD*, close in structure, sequence and localization to *MtnB* [4]. Although the four genes map in the right arm of the third chromosome of *D. melanogaster* (3R), *MtnB*, *MtnC*, and *MtnD* cluster together in the 92E-F region (*MtnB:* 92E12-E13; *MtnC:* 92E4; and *MtnD:* 92F1) [4,5] while *MtnA* is located farther away, in the 85E10-15 segment of the polytene chromosomes [1] (Figure 1). The fact that *MtnC* and *MtnD* encode 67% similar peptides, while mapping so close in the genome, greatly supports the hypothesis that they constitute a subfamily (the *MtnB*-like genes) arisen from a much more recent duplication than that which is hypothesized to have generated the two genes encoding MtnA and MtnB, the main MT isoforms [6].

2.1. *MtnA (Mtn)*

The isolation of a cDNA from copper-fed *D. melanogaster* larvae that cross-hybridized with a mouse MT1 cDNA probe confirmed that the encoded

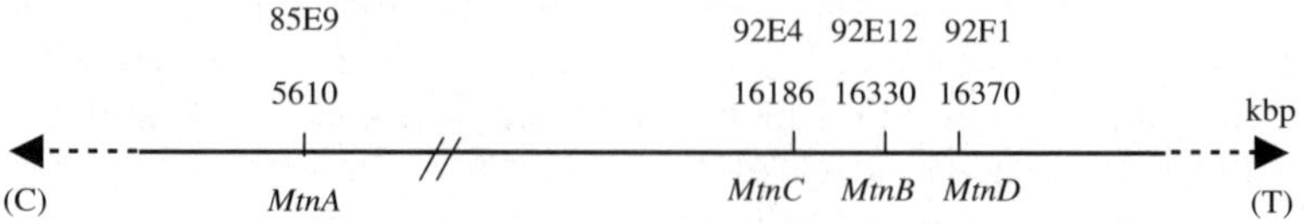

Figure 1. Schematic representation of the localization of the four MT genes in the right arm of the third chromosome (3R) of *D. melanogaster*. The relative positions of the centromer (C) and the telomer (T) are indicated, as well as the bands of the polytene chromosomes for each MT locus. Their sequence location according to the *D. melanogaster* genome version FB2007_02 (http://flybase.bio.indiana.edu) is: *MtnA*, 56094787-5610093; *MtnB,* 16328960-16329321; *MtnC,* 16186068-16186258; *MtnD,* 16360727-16361007 bp.

peptide, then called Mtn and now MtnA, could be indeed a metallothionein-like protein, although considerably shorter than all the known mammalian forms. MtnA is made up of 40 amino acids, 10 of which are cysteines distributed in C-X-C motives [7] (Table 1). The corresponding gene is interrupted by a 265-bp intron, thus belonging to the subtype of long *Drosophila* introns [1]. *MtnA* is the *D. melanogaster* MT gene exhibiting the highest degree of genomic plasticity, in view of the fact that, as explained in detail below, it is the only *Drosophila* MT gene for which natural alleles have been reported and, furthermore, it has been shown to be extremely prone to spontaneous genomic duplication.

Two allelomorfs are known for *MtnA,* the first one, corresponding to the first isolated version of this gene [3], is also designated as Mtn^{1} when the differentiation between alleles has to be specifically stated. The subsequently discovered allele was named $Mtn^{\bullet 3}$ and has a clearly lower frequency in populations [8]. Four single nucleotide substitutions distinguish both allelic sequences: two in their promoter regions and one in their 3′ UTR, and only one is located in the coding region, yielding an amino acid substitution at the C-terminal fragment of the protein: the 40th residue in Mtn^{1} being a Glu, while a Lys in $\mathrm{Mtn}^{\bullet 3}$. Also, the 3′ UTR of the $Mtn^{\bullet 3}$ mRNA is 49 bp longer than that of the mature Mtn^{1} messenger, as detected by Northern blot analysis [8].

In 1986, two distinct duplications of the *MtnA* gene were identified in a cadmium resistant *D. melanogaster* strain [3]. One genomic duplication encompassed a direct repeat of the entire *MtnA* transcriptional unit in 2.2 kb of DNA flanked by no definite signal sequences. Thus, its molecular origin remained unclear, once homologous recombination was discarded, and no clear evidence of transposition events could be established. Contrarily, a very recent origin for this duplication could be envisaged, since both copies of *MtnA* share the same DNA sequence, including their intronic regions,

Table 1. Amino acid sequence alignment of the four *Drosophila melanogaster* MT isoforms. The MtnA sequence indicated corresponds to the Mtn^{-1} allele, while the $Mtn^{\bullet 3}$ allele includes a lysine at position 40 [8]. Also, a OregonR strain has been isolated encoding for a truncated MtnD peptide due to the presence of a codon stop instead of a Lys-codon at position 32 [6]. These are the only two natural polymorphisms reported at present for *D. melanogaster* MTs. The codon interrupted by the presence of an intron in the gene is indicated. A partial sequence of the yeast Cup1 metallothionein is also included.

```
                ↓
MtnA 1 MPCP-CGSGCKCASQATKGSCNCGSDCKC---GGDKKSACGCSE  40
MtnB 1 MVCKGCGTNCQCSAQKCGDNCACNKDCQCVCKNGPKDQCCSNK   43
MtnC 1 MVCKGCGTNCKCQDTKCGDNCACNQDCKCVCKNGPKDQCCKSK   43
MtnD 1 MGCKACGTNCQCSATKCGDNCACSQQCQCSCKNGPKDKCCSTKN  44

Cup1 1 ---QNEGHECQCQCGSCKNNEQCQKSCSCPTGCNSDDKCPCGNK---
```

which are not constrained by selective restrictions. The other isolated duplication involved a longer DNA genomic segment (4.6 kb) with flanking regions that showed no relation with those of the 2.2-kb duplication, thus all data suggests independent duplication events in the genome. Hence, the *Drosophila* genome region comprising the *MtnA* gene shows an enhanced plasticity, enabling gene duplications that have been generally related to increased metal tolerance, and which have been reported to occur not only spontaneously (i.e. detected in natural populations) but also as a result of artificial induction (laboratory strains).

In a considerable effort to analyze the distribution of the spontaneous occurrence of *MtnA* duplications, Maroni et al. analyzed 302 lines of *D. melanogaster* obtained from flies caught in natural populations in all five continents [9]. *MtnA* was duplicated in 20 of these, of worldwide origins, apart from tropical Africa, and with the highest frequency among French populations. Molecular analyses of the *MtnA* region in these populations revealed different genomic structures, which could be classified into 43 direct tandem duplications and 2 transposition-like duplications, although the exact molecular mechanism that might have caused them remained undetermined. Since all the duplications yielded *MtnA* mRNA levels that were approximately twice those detected in wild-type strains (values between 1.7 × and 2 ×), and both the copper and cadmium LC50 values were clearly higher in these flies, enhanced metal resistance was hypothesized as the main force responsible for keeping the *MtnA* duplication at such high frequencies in natural populations. Selective pressure in geographically separate areas could originate from diverse sources, such as industrial pollution by heavy metals or the extensive use of copper-containing antibacterial and antifungal agents in orchards and vineyards.

More recently, in 2004, *MtnA* has been identified as a gene that is currently submitted to adaptive selection because the abundance of its transcripts shows a high variation within a given *Drosophila* species, while there are relatively low differences between them. These features suggest that active selection processes are currently occurring because they suggest that gene-environment interactions are responsible for the stabilization of the segregating populations carrying genetic variants that determine dissimilar levels of *MtnA* gene expression [10]. These natural adaptive events have been reproduced on a laboratory scale by growing *D. melanogaster* populations in cadmium-enriched media (80 μL/mL) for 20 generations and comparing their ability to tolerate high cadmium with several parameters involved in fitness, such as the fecundity rate and body weight. The results clearly show that metal resistance is acquired at the expense of fitness reduction, this providing a convincing explanation of why in the absence of environmental selection, high metal resistance variants are not fixed in populations [11].

Finally, the fact that the number of *MtnA* gene copies is directly responsible for augmented metal tolerance has been demonstrated by the high cadmium tolerance exhibited by transgenic flies where a second copy of *MtnA* has been introduced artificially [12], or even by the enhanced tolerance to copper of transgenic flies incorporating a yeast MT (*S. cerevisiae*, Cup1) [13]. In this latter case, transgenic flies exhibited also a fitness reduction comparable to that detected in the flies with the *MtnA* genomic duplication [11].

2.2. *MtnB (Mto)*

One year later than *MtnA*, a second MT gene, *Mto* or *MtnB*, was identified in *D. melanogaster*, through isolation of its cDNA from a cell line showing enhanced resistance to cadmium [14]. DNA sequencing showed that the encoded metallothionein, MtnB, was longer than MtnA (43 versus 40 amino acids), also comprising two additional Cys residues, thus 12 Cys (versus 10 Cys in MtnA). Other gene features were also shown to be divergent between the two genes, especially the length of the interrupting intron, here belonging to the "*Drosophila* short intron type" (61 bp), which supported the hypothesis that *MtnA* and *MtnB* were created by an old duplication/divergence event in the *D. melanogaster's* evolutionary history [5].

Unlike for *MtnA*, strains carrying *MtnB* duplications have neither been isolated from nature, nor induced in flies by metal treatment. The only known *MtnB* duplication has been reported in a cadmium-resistant cell line (CdR200) [9]. In recent years, transgenic flies carrying a *MtnB* duplication have been constructed, and it has been confirmed that they exhibited greater cadmium tolerance [15]. Therefore, the absence of spontaneous duplications involving *MtnB* may be related to some structural restriction in the corresponding genomic region, rather than to the functional properties of the encoded peptide.

2.3. *MtnC* and *MtnD*

The existence of these two genes was ascertained by *in silico* search of the *D. melanogaster* genome [4]. Their presence has never been suspected, probably due to a very low expression level even in the presence of metals. Their relatedness to *MtnB*, both at gene structure level (short-intron type) and in gene sequence similarity, as well as their chromosomal situation, strongly suggests that they originated from successive tandem duplications of *MtnB*, which became fixed in the genome as secondary forms. They cannot be considered *MtnB* pseudogenes because they are actively expressed genes, and the corresponding peptides exhibit proficient metal-binding activity, although at lower rates than the two main isoforms (cf. Section 4 of this chapter).

3. METALLOTHIONEIN TRANSCRIPTS: TISSULAR AND DEVELOPMENTAL DIFFERENTIAL EXPRESSION

Regulation of the transcription of the *D. melanogaster* MT genes should provide the basis for understanding not only the molecular mechanism by which they respond to inducer agents (mainly metals), but also their tissular and developmental expression pattern, as well as the physiological function of the encoded peptides. Several components have to be considered to depict a complete frame for these events: the *cis* regulatory elements of the *Mtn* genes (namely promoters and/or enhancers), the *trans* regulatory elements acting upon them (mainly the transcription factor dMTF-1) and the situations capable of activating Mtn synthesis in flies. The overview given in this section is fully complemented by the deeper molecular analysis provided in Chapter 2 of this book by Balamurugan and Schaffner.

3.1. Overview of *D. melanogaster* Metallothionein Gene Transcription

The transcripts of the two main *D. melanogaster* MT genes were characterized when they served to isolate the corresponding genes. The *MtnA* (Mtn^1 allele) transcription unit is 600-nt long, its long-type intron interrupts its 8th codon and it harbours short 5′ (124 nt) and 3′ (91 nt) UTR regions, the latter including a canonical polyadenylation signal [1]. The 3′ UTR of the mRNA from the $Mtn^{\bullet 3}$ allelic variant was reported to be 49 nt longer [8]. Allelic polymorphism also leads to quantitative variation in the expression levels of the two forms, as it has been shown that the $Mtn^{\bullet 3}$ strains accumulate up to 70% less mRNA than Mtn^1 strains, this reduction mainly affecting the synthesis of MT in the midgut and not in other organs [16]. The *MtnB* transcription unit encloses a short-type intron (61 nt) also interrupting the 8th codon [5]. Flanking regions are also relatively small: 44 nt and 100 nt for the 5′ and 3′ UTRs, respectively. The *MtnA* transcribed sequence is preceded at short distance by a canonical TATA box (TATAAA, starting at –28 bp from the transcriptional start site, + 1) and several putative MRE elements, in a –200 bp upstream region, identified by their high similarity to the mammalian counterparts [1]. Although *MtnB* was first identified as a TATA-less gene [5], this has been contradicted in recent studies [4] that have localized a canonical TATA box, but more distant than usual at the upstream gene region, and also several putative MREs, besides an AP1 site at –223/–218 bp, probably acting as an enhancer binding element, as has been reported in mammals. The structure of the *MtnC* and *MtnD* transcription units is equivalent to that of *MtnB*, and both a TATA signal and MREs are present in their 5′ boundaries, according to their active, metal regulated transcription [4].

The only specific trans-acting element controlling the transcription of *Mtn* genes identified so far is dMTF-1, the fly homolog to the mammalian MTF-1 factor [17]. dMTF-1 is a 791-aa long activator that binds to the MREs through zinc-finger mediated DNA interaction, in a Zn-dependent manner. In agreement with this, the highest degree in similarity between the mammalian and *Drosophila* homologs expands the zinc-finger region, located at their N-terminal moiety [18]. Transcription of the four *Mtn* genes is completely dependent on dMTF-1, both in basal and metal-induced conditions [4,19]. dMTF-1 null flies, which are totally viable and fertile when reared in normal food, are consequently extremely sensible to copper, cadmium, and zinc intoxication, and, unexpectedly, also to copper depletion [19]. This is due to the fact that dMTF-1 also regulates the synthesis of a copper importer (Ctr1B), through interaction with MREs sharing their consensus sequence with those of the *Mtn* genes [20]. However, and for some as yet unexplained reason, the Ctr1B-MREs do not respond to induction on copper overload. Thus, the dMTF-1 role in metal homeostasis can be considered broader in *Drosophila* than in mammals, since at least for copper, it controls both responses to excess and scarcity stresses.

Beyond the parallelisms of the MTF-1 transcription factor in mammals and *Drosophila*, there are significant differences in the capacity of the different metal ions to induce the transcription of MT genes in flies and in mammals. Hence, mammalian MT genes mainly respond to zinc and cadmium, while the foremost effect on MT transcription in *D. melanogaster* is observed after copper and cadmium treatment. Paradoxically, when the capacity of induction is analyzed in dMTF-1-transfected mammalian cells, the transcription factor responds to zinc, in a similar way that the endogenous MTF-1 does, but MT genes in *Drosophila* are poorly induced by this metal, this having been attributed to global metal metabolism conditions rather than to a differential molecular behavior of dMTF-1 [4].

3.2. Localization and Quantification of *Drosophila* Metallothionein Gene Transcripts

Extensive data on the ontogenic and tissular distribution of gene expression has been extensively characterized for the *MtnA* isoform [16]. Constitutive *MtnA* mRNA is detected as early as in oogenesis/vitellogenesis (stage 10 of *Drosophila* embryo development), in a set of follicular cells on the boundary between oocyte and nurse cells. These *MtnA* mRNA levels significantly increase in females fed with a copper supplement, thus this expression is metal-regulated and is probably responsible for maternal inheritance of Cu(I)-MtnA complexes, to accomplish some unsuspected role in very early *Drosophila* zygotes.

Progressively with the advance of the embryogenesis, *MtnA* expression also becomes evident in the Malpighian tubules and hemocytes (stage 13) and midgut (stage 16). In fact, these three organs, together with the fat body, can be considered the main sites of larval MtnA synthesis, both for barrier, storage and metabolic purposes. Especially Cu(I)-MtnA complexes have been postulated as copper donors for hemocyanin synthesis, precisely in hemocytes. Overall, once *MtnA* expression starts, it continues throughout all the larval and adult life in the same described tissues [5]. Constitutive *MtnB* expression could not be detected during embryogenesis [16], but quantitative life-span comparison of the constitutive expression of *MtnA* and *MtnB* mRNA showed that constitutive *MtnA* mRNA levels are invariably higher than those of *MtnB*, i.e., ten and twenty fold times higher for larvae and adults, respectively [21]. Even though in the middle and posterior midgut both isoforms are actively synthesized, there is also some degree of specificity: *MtnA* is more abundant in the proventriculus and Malpighian larval tubules, while *MtnB* is predominant in the salivary glands. During pupal tissular reformation, expression of both genes is readily detected in the imago midgut as soon as this structure is reorganized. Finally, midgut also persists as the main adult expression site for the two MTs, with significant expression again in the Malpighian tubules, besides the cardia, for *MtnA*, and ventricula for *MtnB* [21].

Induction with metal ions causes dramatic, though differential, effects in the transcription rates of the *Mtn* genes (Figure 2). *MtnA* expression is induced in the alimentary tract, mainly midgut, cells mostly by copper, but also significantly by cadmium, and very poorly by mercury, silver or zinc [1]. mRNA quantification by Northern blot densitometry revealed that as much as 12% of the total polyadenylated midgut mRNA could be attributed to *MtnA* after copper or cadmium induction. Consequently, copper supplement in fly food produces higher and developmentally anticipated *MtnA* transcription ratios, but as a rule it can be assumed that the strong induction capacity of copper only increases *MtnA* expression in those tissues where the gene is constitutively transcribed, and never in constitutively-silent tissues [16]. Cadmium and copper greatly increase *MtnB* mRNA levels in salivary glands, but not those of *MtnA*. However, in adult flies *MtnB* exhibits a higher response to cadmium than to copper. In this case, the most pronounced effect can be appreciated in the middle midgut, precisely in the cell type that accumulates the highest MT levels in *Drosophila*: the cuprophilic cells [21]. In early studies (1986) it had already been shown that in copper-fed flies this metal is accumulated in the lysosomes of midgut cells, usually as mineralized, insoluble concretions [22]. Cadmium also accumulated in this cell type, but no "cadmium-lysosomes" have ever been detected. For a full description of the significance of the presence of copper and Cu(I)-Mtn complexes in this cell type, see Section 5.

Zinc induction assayed for the same conditions as cadmium and copper is very poor, but increased metal doses and longer treatment times also rendered increased *Mtn* mRNA accumulation. In third instar larvae, exposure to 5 mM zinc for 16 h resulted in a five fold *MtnA* and a twelve fold *MtnB* mRNA increase. In adult flies, the induction is weaker than in larvae, only reaching 1.6× and 2.4× levels, respectively, after 3 days. This induction capacity of zinc was also found to be isoform- and organ-dependent. Hence, while zinc, cadmium or copper yield increased *MtnA* expression in the Malpighian tubules, *MtnB* expression in this organ is only induced by zinc [21].

3.3. Regulation of the Differential Expression of the *D. melanogaster Mtn* Genes

The functionality of the MREs present in the promoters of the *Mtn* genes was first assayed by the construction of P-mediated transgenic flies, in which

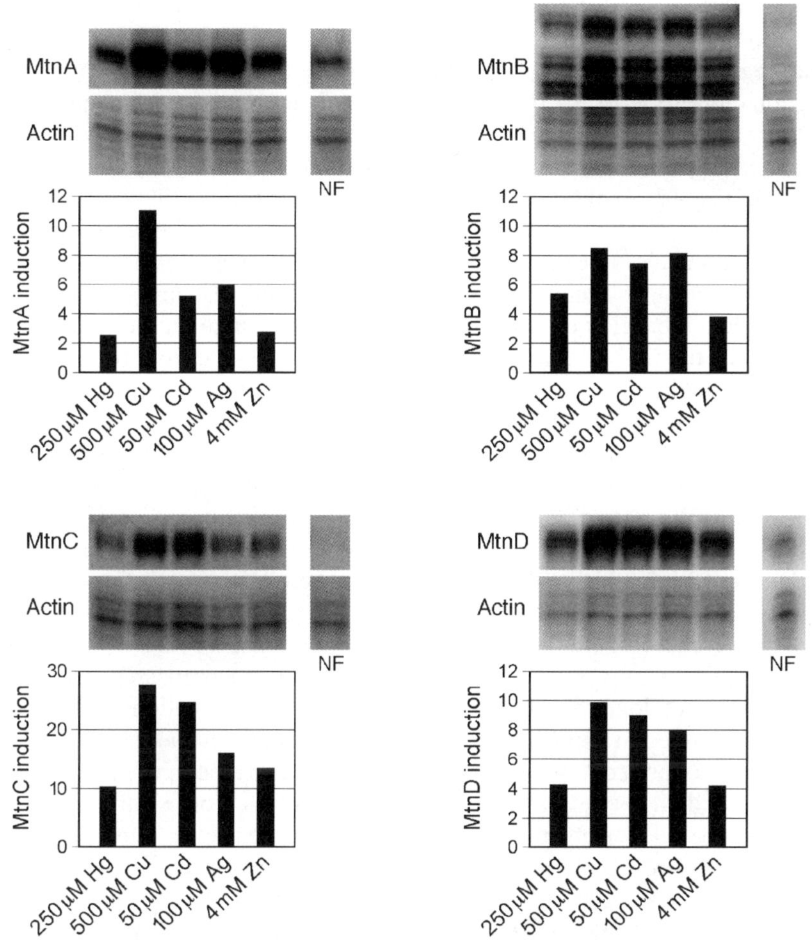

Figure 2. Induction of the *Mtn* genes by different heavy metals. Cu, Cd, Ag, Hg and Zn are all inducers of *MtnA*, *MtnB*, *MtnC*, and *MtnD*, but at different molar concentrations. Larvae were transferred to the indicated type of food for 6 h and transcript levels assayed by S1 mapping. Signal intensities were normalized using the *actin* signal and divided by the basal expression on normal food. Note that the basal expression level of *MtnC* is very low, which results in a high induction rate. NF = normal food. Reproduced from [6] with permission from Wiley-Blackwell, copyright (2006).

different portions of the *MtnA* upstream region controlled the expression of an exogenous additional *MtnA* gene [12]. The efficiency of transcription was measured by the increased cadmium tolerance exhibited by the transgenic organisms, and it was clearly stated that no more than 373 bp upstream the transcription initiation site (+1) was sufficient for *MtnA* expression. Advanced analyses using a viral thymidine kinase (TK) reporter in hamster kidney cells (see also [12]) showed that the –130 to –6 fragment was sufficient to confer metal-regulated expression of TK, this experiment also serving to confirm the functionality of the *MtnA* promoter in a mammalian cell environment.

Other systems have been used to characterize both the *MtnA* and *MtnB* promoters, and it is worth noting the construction of P-mediated transgenics including the *Adh* or the *lacZ* coding regions as reporters [23], which enabled the evaluation of the capacity of each promoter to control gene expression in response to metal and other stresses, in different *D. melanogaster* tissues, and in several developmental stages. Hence, the proximal –373 bp *MtnA* region is able to direct basal (i.e., constitutive) gene expression in the midgut and Malpighian tubules of third instar larvae, while a region essential for *MtnB* expression was located between –1220 and –221, with transcription-inducing capacity only in the midgut, and precisely in the iron-cell region. The induction capacity of several metals on these promoter regions was analyzed with the same reporter constructs. Results clearly confirmed the highest response of the *MtnA* controlling region to copper, leading to increased transcriptional activity on the tissues where the constitutive function had been detected (midgut and Malpighian tubules) and to new activation in the fat body cells. On the contrary, zinc and cadmium only caused activation in the digestive tract. Differences could be detected in the response of the *MtnB*-derived constructs depending on the promoter region they encompassed (short promoter, –221 bp; long promoter, –1220 bp). Copper induction of the short promoter resulted in expression in the middle midgut, coinciding with an area enriched in copper cells, while the long promoter expanded gene activation to the whole midgut, as well as to the Malpighian tubules. Comparably, cadmium induction on the short promoter was able to direct gene expression only in the posterior, iron-cell rich midgut portion and gastric caeca, while the long promoter expanded expression to Malpighian tubules and proventricula.

Thus, in the complex *D. melanogaster* MT gene expression pattern, different *cis* elements probably account for metal, developmental and tissular specific responses. Relatively little information has been gathered about *MtnA* and *MtnB* promoter response to non-metal stresses. Both promoters were activated by heat-shock, but exhibiting differential response, and only the latter responded to oxidative (paraquat) stress tests [23].

4. THE METALLOTHIONEIN PROTEINS IN *DROSOPHILA MELANOGASTER*

Purification of the two canonical *D. melanogaster* MTs (MtnA and MtnB) from metal-treated flies was repeatedly assayed after identification of the respective genes, but for MtnA this remained absolutely unsuccessful, while for MtnB only small amounts could be recovered. Lysosomal import and degradation after metal chelation or deposit as metal-MT insoluble concretions have been alluded as being responsible for this drawback. Recombinant synthesis was envisaged as the alternative approach to obtain enough MT proteins to perform a full characterization of the clusters they could afford with different metals, as had been extensively carried out with the mammalian and yeast MTs (cf. the respective chapters of this book). The first attempts used yeast as heterologous host [24], attempting *MtnA* or *MtnB* transcription under inducible (Cup1) and constitutive (PGK) promoters. Both MtnA and MtnB were shown to be present in the whole yeast protein extract of the corresponding transformants, and their presence rescued the copper hypersensitivity of the host strain, but no attempts to purify the recombinant proteins for further characterization were assayed.

It was as late as the 2000s when, applying recombinant expression technology in *E. coli*, our lab was able to prepare a large enough quantity of all the *Drosophila* MT proteins, as metal-MT clusters, to make the first study of their metal binding properties [6,25,26]. MTs were synthesized as fusion proteins with a stabilizing tag (GST), which were subsequently digested to release and purify the MT portion. In fact, by growing the corresponding bacterial cultures in metal (zinc, cadmium or copper)-supplemented media, MTs are directly recovered as metal complexes that have consequently been conformed inside the cell, thus in an *in vivo* environment. This methodology enables the purification of enough metal-MT complexes to either be directly characterized or to serve as the starting point for further procedures, such as metal replacement on Zn-MT forms, or denaturalization/renaturalization studies of the Cd-MT complexes.

Surprisingly, and unexpectedly among metazoans, it was clearly shown that all four *Drosophila* MTs exhibit metal coordinating properties closer to those of *Ascomyceta* (fungal, *N. crassa* and *A. bisporus*; and yeast Cup1) monodominial MTs than to those of their *Arthropoda* counterparts, and markedly to those of the vertebrate proteins. Hence, in summary, *Drosophila* MTs yield unique (i.e., single species), homometallic, highly compact copper-containing complexes when recombinantly produced in copper-supplemented media, an ability that only the so-called *copper-thioneins* are able to exhibit [27]. Contrarily, syntheses of zinc or cadmium complexes produce poorly folded species with variable stoichiometry, revealing a suboptimal aptitude for chelation of divalent metals [25,26].

Although this general behavior was attributed to the four *Drosophila* MTs, slight differential features could be identified for each of the isoforms (Table 2). In reference to zinc binding, the zinc content of the Zn-MtnA complexes was slightly lower than those obtained for all the MtnB-like forms (MtnB, MtnC, and MtnD): Zn_4-MtnA was the major form, but was

Table 2. Metal content and major species of the MtnA-, MtnB-, MtnC-, and MtnD-metal complexes recovered from recombinant synthesis in culture media supplemented with a unique metal ion, at the final concentration. Reproduced from [6] with permission from Wiley-Blackwell, copyright (2006).

Supplemented Metal	Zn (300 μM)				Cd (300 μM)				Cu (500 μM)			
Mtn	A	B	C	D	A	B	C	D	A	B	C	D
Metal/Mtn content[a]	3.5 Zn	3.7 Zn	3.9 Zn	3.8 Zn	3.8 Cd	4.1 Cd	2.3 Cd	2.0 Cd	7.0 Cu	8.9 Cu	4.2 Cu	5.3 Cu
Metal-Mtn species[b]	$Zn_4 > Zn_3$	$Zn_4 > Zn_5$	$Zn_4 > Zn_5$	$Zn_4 > Zn_5$	$Cd_4 > Cd_3 \approx Cd_4S_2$	$Cd_4 \approx Cd_4S > Cd_5 \approx Cd_5S$	$Cd_4 > Cd_5$	Cd_5	$Cu_8 > Cu_7$	$Cu_9 > Cu_8$	$Cu_5 > Cu_6$	$Cu_5 > Cu_6$
Presence of S^{2-} ligands[c]	+	+	+	+	+ +	+ +	+ +	+ +	–	–	–	–

[a] Metal per Mtn molar ratio calculated from the zinc, cadmium, copper, and sulfur content measured by acid ICP-AES [28].

[b] Metal per Mtn molar ratio calculated from the difference between holo- and apoprotein molecular masses, obtained from ESI-MS.

[c] Absence (–), moderate (+) or substantial (+ +) presence of S^{2-} in the metal-metallothionein complexes, evaluated according the methods described in [28].

followed in abundance by Zn_3-MtnA, in contrast to Zn_5-MtnB, Zn_5-MtnC or Zn_5-MtnD. This lower global binding ability can be attributed to the lower cysteine content (10 Cys versus 12 Cys, respectively). But it is worth noting that the Zn-Mtn complexes exhibit dissimilar degrees of folding, as shown by a comparative consideration of their CD spectra (Figure 3), which shows that the MtnB-like forms yield more chiral zinc complexes than MtnA.

The differences between *Drosophila* MTs are more pronounced when these polypeptides bind cadmium. Hence, the two main isoforms, MtnA and MtnB exhibit a similar cadmium binding capacity, in contrast with the much lower cadmium content of the MtnC and MtnD recombinant preparations (Table 2). Since the major species detected by ESI-MS were invariably Cd_4 and Cd_5 for the three MtnB-like forms, the diminished total cadmium content should be attributed to the presence of undermetallated molecules in the preparation or, equivalently, to an intrinsic tendency to oxidation and/or partial instability of the corresponding Cd-Mtn complexes. This would also explain why the Cd-MtnB complexes exhibit a higher chirality than Cd-MtnC and Cd-MtnD, in accordance with a less compact structure of the latter. The presence of sulfide ligands in the cadmium complexes of *Drosophila* MTs was a constant [28], and this was clearly revealed by the GC-FPD analysis of acid-labile sulfide, CD data (Figure 3), and ESI-MS peak assignment (Figure 4). Precisely, CD absorbance in the 270–280 nm range has been attributed to the chromophores participated by the sulfide anions and cadmium.

More significant was the analysis of the features of the copper complexes conformed by the *Drosophila* MTs, due to their copper-thionein character. As mentioned earlier, the four isoforms yielded only homometallic copper complexes when recombinantly synthesized in copper-supplemented bacterial cultures, although again with clear differences between them. The copper content of the MtnB preparations was higher than for MtnA, and again MtnC and MtnD exhibited a lower capacity for binding this metal ion (Table 2). The CD spectra of these Cu-Mtn complexes suggested similar but not equal numbers of Cu-SCys chromophores for MtnA and MtnB, and a significantly decreased chirality for the secondary isoforms, Cu-MtnC and Cu-MtnD.

In summary, all four *Drosophila* MT isoforms exhibit overlapping, yet distinct properties for the coordination of different heavy metals, as the stoichiometric, spectrometric and spectroscopic features of the copper and cadmium complexes with MtnA and MtnB correlate well with a greater stability of copper-MtnA and cadmium-MtnB. Finally, MtnC and MtnD, both of which are highly similar to MtnB, display lower copper and cadmium binding capabilities compared to either MtnA or MtnB, which can mainly be explained by the intrinsic instability of the complexes.

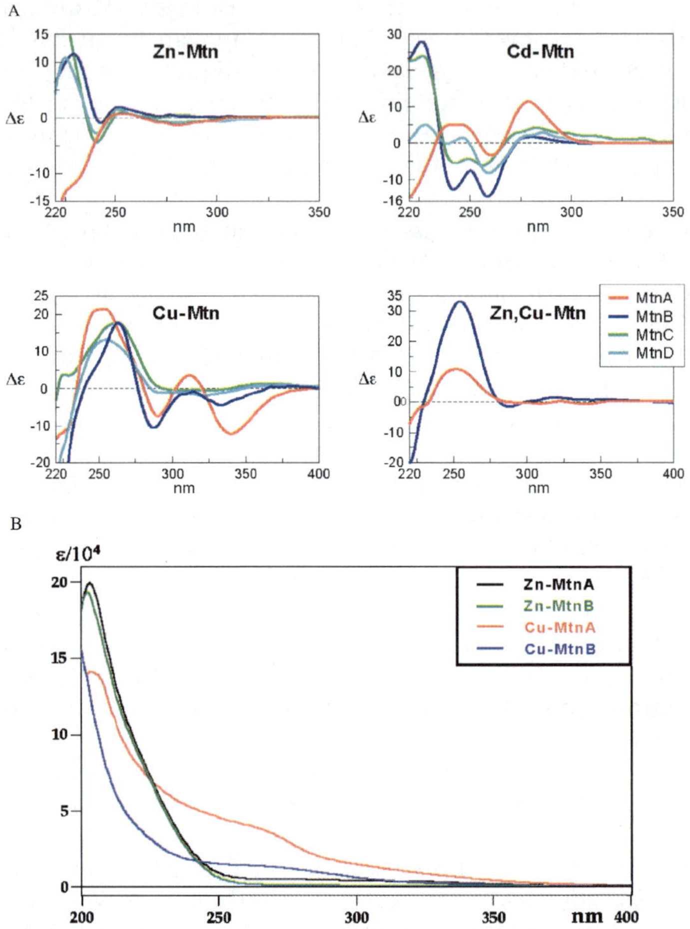

Figure 3. (**A**) CD spectra of the recombinant *Drosophila* MT preparations, obtained from bacterial cultures supplemented with the metals indicated. The differences in the region from 220 to 240 nm of the Zn and Cd complexes of MtnA spectra *versus* those of MtnB, MtnC, and MtnD are consistent with the participation of the chloride anions in metal clusters of the latter three peptides [25]. The absorptions in the 270–290 nm region of the divalent metal complexes, especially relevant in the Cd(II) species, are due to the S^{2-} ligands present in these samples [28]. Overall, MtnA and MtnB-like MTs exhibit distinct types of spectra, indicative of their differential binding behavior. Δε indicates the chiral intensity of the sample measured as the differential absorption of the left and right circularly polarized light. (**B**) UV-Vis spectra of the Zn and Cu complexes of the two main MT forms, MtnA and MtnB. Reproduced from [6] with permission from Wiley-Blackwell, copyright (2006).

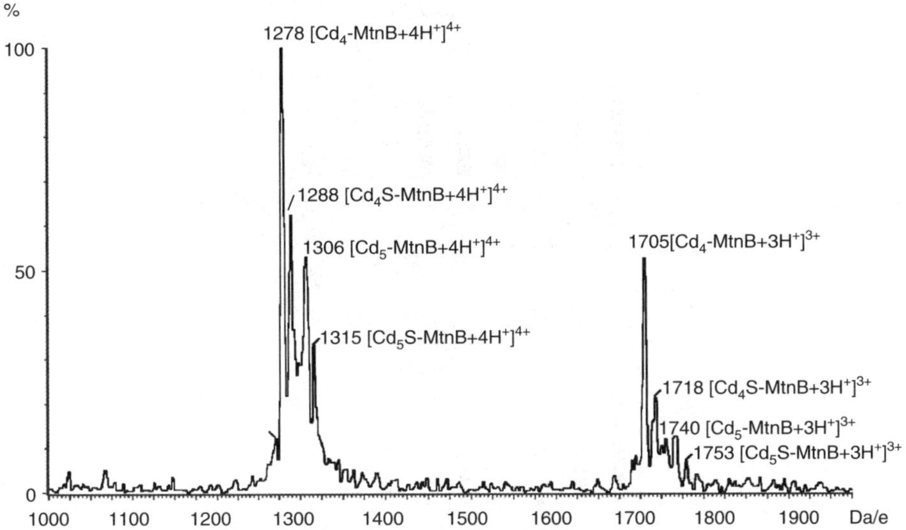

Figure 4. ESI-MS spectrum of the recombinant preparation of the MtnB peptide recovered from *E.coli* supplemented cultures. The two sets of peaks correspond to two contiguous charge states (+4 and +3), in order to illustrate the coexistence of sulfide-loaded and sulfide-devoid Cd(II)-MtnB complexes. Reproduced from [28] with permission from Wiley-VCH Verlag GmbH & Co. KGaA, copyright (2005).

5. METALLOTHIONEINS IN FLY PHYSIOLOGY: METAL HOMEOSTASIS

The physiological functions of MTs in *Drosophila* cast the same certainties and doubts that MT offers in all other organisms, with a general acceptance that these proteins serve both to ensure essential metal (here basically copper) homeostasis and to provide a sequestering strategy for preventing the free circulation of toxic heavy metals in fly larvae or adults. A graphical estimate of the effects of the different metals is clearly afforded by the results illustrated in Figure 5.

5.1. Cadmium Poisoning

D. melanogaster resistance to metal poisoning, especially to cadmium, was identified in 1983 as an inherited trait [29]. Nevertheless, at that time neither any relationship between cadmium pre-treatment and protection from toxic metal doses, nor any relation to a metallothionein-like protein, was detected

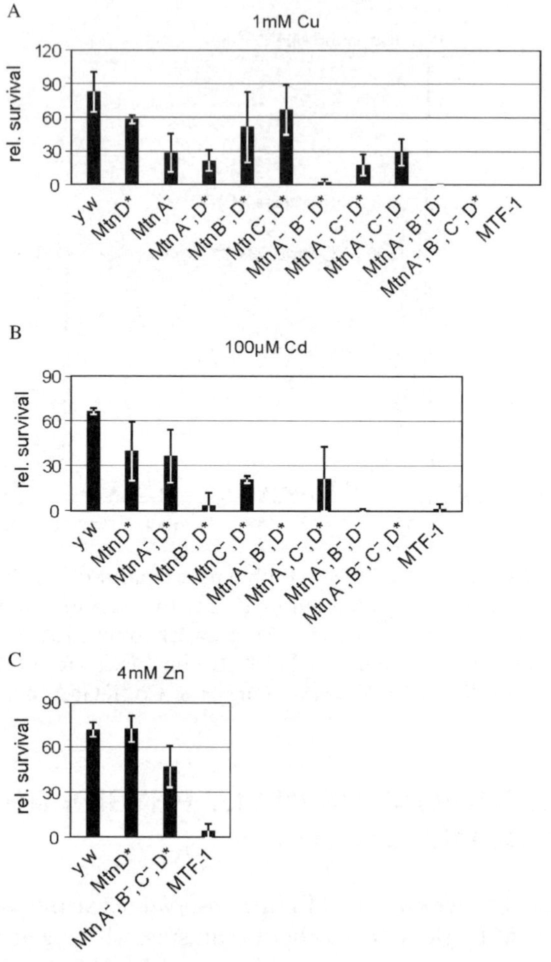

Figure 5. Viability of *Mtn* and *MTF-1* mutant versus wild-type flies at different metal concentrations. The bar diagrams depict the percentage of survival of mutant and wild-type (*yw*) embryos to adulthood, with normal food set to 100%. Flies were allowed to deposit 150-300 eggs on food containing the indicated concentrations of metal, and eclosing adults were counted. Error bars represent standard deviations of several independent experiments, calculated from the number of flies in a total of 3 to 10 different tubes. Metal concentrations (A, copper; B, cadmium; C, zinc) supplemented to the food are indicated above each panel. Reproduced from [6] with permission from Wiley-Blackwell, copyright (2006).

in adult flies. It was two years later [30] when this connection was firmly established when working with cell cultures of the R200-*Drosophila* cell type, and increased putative-MT levels were identified as being responsible for enhanced cadmium tolerance. Localization of the labelled cadmium ingested by larvae revealed that up to 95% of the radioactivity was bound to a protein exhibiting MT-like features [31].

In the same study, it remained clear that neither was zinc a good inducer of this polypeptide, nor was any protein-bound zinc detected in larvae fed with excess of the metal. Contrarily, copper, if supplied in high amounts, ended up bound to the same MT-like product and also with a considerable percentage remaining in the insoluble cell fraction. This readily suggested different paths of assimilation and MT-interaction for these three metals in *Drosophila*, as exhaustive molecular studies have later confirmed [6].

Resistance to cadmium was also measured in different laboratory strains [31], which confirmed that determinants other than MT levels could exist in *D. melanogaster*, although these studies should be considered with caution as only one of the *Drosophila* MT proteins was taken into account [32]. Then, the identification of the two *MtnA* alleles (Mtn^{1} and $Mtn^{\bullet 3}$) made it possible to compare the cadmium tolerance exhibited by strains with different allelic compositions [33]. Although the transcript levels for Mtn^{1} were reported as being three times higher than those of $\mathrm{Mtn}^{\bullet 3}$, both homozygous strains revealed equivalent cadmium tolerance levels. It was then proposed that the polymorphism at position 40 of these alleloproteins (Glu in Mtn^{1} and Lys in $\mathrm{Mtn}^{\bullet 3}$) could be responsible for the lower cadmium binding efficiency of the former, which would compensate for the increased gene expression, something that has never been confirmed by biochemical studies.

5.2. Copper Homeostasis

Copper has been identified as the best inducer of the *Drosophila Mtn* gene expression, and metal binding analysis of the 4 Mtn peptides clearly demonstrated that all the isoforms could be considered copper-thioneins (cf. Section 4). Furthermore, *MtnA* genomic duplication confers a higher resistance to copper than that exhibited by one-copy gene flies. Thus, a major role of *Drosophila* MTs in the homeostasis, or globally speaking, in the metabolism of copper in flies can be postulated. Copper metabolism in *Drosophila* has been described as a very complex interaction of import, storage, tolerance and supply mechanisms [34]. In this scenario, MT has been proposed as a sensitive indicator of total heavy metal concentration and redistribution, taking into account the fact that copper needs in insect organisms may be significantly different to those assumed for vertebrates (for example in relation to the respiratory pigments). Nowadays *Drosophila*

MTs have been described to be involved in stress-free metabolism, and both in copper-excess and copper-depletion conditions.

Regarding regular fly metabolism, the most significant finding was the correlation between MT accumulation and fluorescence in the so-called *copper* or *cuprophilic* cells in the digestive tract of flies [35]. These are acid-secreting cells located in the midgut epithelium of *Drosophila*, which exhibit an outstanding orange fluorescence only few minutes after the ingestion of copper-containing food (1 mM $CuCl_2$). S2 cells in *in vitro* tissue cultures behave similarly. Several lines of evidence suggest that copper-MT complexes in the cytoplasm of these cells are mainly responsible for fluorescence emission. First, this phenomenon was also reported in yeast cells on copper accumulation, and Cup1, the main yeast MT, is the paradigm of copper-thioneins, the kind of MT to which *Drosophila* MTs clearly belong [25,26]. Second, the two main *Drosophila* MTs exhibit expression patterns (both temporal and spatial) that are totally coincident with copper cell distribution. Along this line of reasoning, MT-mediated accumulation in intestinal cells can serve as a means for storage for periods of copper shortage, a capacity that would even benefit later generations. This capacity has been shown by the ability of *Drosophila* larvae to successfully grow under copper depletion provided that their parents were reared in copper-supplemented food, but the exact mechanism for this *inheritance* has yet to be described [34].

That *Drosophila* MTs have a determining role in the handling of copper excess was firmly evidenced by research at different levels: from cell and full organism studies to molecular experiments. Hence, three of the four *D. melanogaster* MT isoforms (*MtnA, MtnB,* and *MtnD*) are up-regulated in S2 cells in response to increased Cu levels, in a dose-dependent manner, and transcriptional impairment of the expression of these genes significantly reduced the tolerance of the cultured cells [36].

The generation of a collection of *Mtn* knock-out flies, for all, each one, and different combinations of the four *Drosophila* MTs [37] constituted a giant's step towards determining their role not only in copper, but in zinc and cadmium homeostasis. Similarly to what was reported for MT1 & MT2-KO mice [37,38], MT deficiency is not lethal, even noticeable for organisms grown on normal food, i.e., food without metal supplementation. On the contrary, MT-null flies are highly sensible to excess copper or cadmium during their development, and if exposed in their adult life, they show a markedly reduced lifespan, especially males. Zinc has a similar, although less severe, effect. The phenotypic features exhibited by the flies null for the four *Mtn* genes regarding tolerance to excess copper and cadmium were totally coincident with those of dMTF-1 defective flies, thus confirming that no significant expression of any *Mtn* gene is achieved in the absence of this transcriptional activator, and that virtually the only elements responsible for the surplus copper and cadmium handling are the Mtn proteins. On the other hand, *Mtn*-null flies were not particularly sensitive to zinc, and not sensitive at all to silver, mercury or to copper depletion. These observations suggested that a great deal of the sensitivity to these conditions exhibited by dMTF-1 nulls is not mediated by mechanisms that involve Mtn proteins [37]. Significantly, one of the features clearly shared by flies lacking all MTs and dMTF-1 null flies was the practically absent fluorescence in the cuprophilic midgut cells. Therefore, it was confirmed that this orange emission is dependent on three events: copper ingestion, acid secretion, and Mtn availability. We have yet to firmly establish the physiological relation between them, but from a molecular point of view it has been proposed that the Cu(I) ions sheltered from the aqueous environment by the MT peptide would account for the emission of orange fluorescence [37].

Finally, *D. melanogaster* strains deficient for only one of the four genes have made it possible to evaluate the contribution of each isoform to the handling of the three considered metals (copper, zinc, and cadmium) [6]. First, from the enhanced sensitivity to metals of the organisms lacking both *MtnA* and *MtnB*, it was confirmed that these two isoforms play a major role in *D. melanogaster*, while *MtnC* and *MtnD* were secondary isoforms, whose loss did not significantly alter the capacity to cope with metals if compared to that exhibited by wild-type flies. In fact their functionality was only worth considering in the absence of the major forms. Comparison between the capacities of the individual null strains to resist different metals led to the conclusion that MtnA played a major role in the defence against copper, while MtnB did likewise for cadmium, although both isoforms could perfectly serve both purposes. This was in total concordance with the metal-binding behaviors defined for each of the Mtn isoforms, and also coincident with the best inducer for each *Mtn* gene promoter: copper for *MtnA* and cadmium for *MtnB*. However, no correspondence could be detected between the *Mtn* tissue-specific expression pattern and any functional region of their gene promoters because an artificial construct of MREs derived exclusively from a non-MT gene promoter (*Ctr1B*) was able to control the distribution of *Mtn* expression in exactly the same way as that observed in wild-type flies. Therefore, metal distribution in the organisms was suggested to be solely responsible for the differential *Mtn* expression pattern [37].

6. METALLOTHIONEIN MOLECULAR DIFFERENTIATION IN THE *DROSOPHILA* GENUS

The little data available on the MTs of *Drosophila* species other than *D. melanogaster* will be briefly accounted for in this section. In particular, the *Sophophora* subgenus, which comprises the species closest to *D.*

melanogaster, is one of the fundamental sources of data for population genetics and natural selection studies. As the MT system can be clearly associated to positive traits, i.e., resistance to heavy metals in natural environments, it has been analyzed in different genera from a molecular evolutionary point of view.

Mtn in *D. simulans*, a *D. melanogaster* sibling species, is present in its $Mtn^{\bullet 3}$ allelic form as was evidenced by gene sequencing [8], and restriction site analyses of the corresponding genomic region revealed that among other *melanogaster group* species, the most common pattern also corresponded to that of the $Mtn^{\bullet 3}$ region of *D. melanogaster*. Hence, *D. yakuba*, *D. tessieri*, *D. sechellia*, and *D. mauritiana* restriction maps resembled $Mtn^{\bullet 3}$ rather than Mtn^1, and *D. erecta* and *D. orena* showed an intermediate restriction pattern. This data led to consider $Mtn^{\bullet 3}$ to be the ancestral Mtn allele in the *Drosophila* species group, an assumption that was further supported by the analysis of the *D. ananasse Mtn* gene. This species belongs to the *Sophophora* subgenus, but not to the *melanogaster* subgroup, and it is therefore more distant from *D. melanogaster* [39]. Only the $Mtn^{\bullet 3}$ allele has been isolated in this species, and in no case does it seem to have undergone duplications or other changes introducing any kind of molecular variability (intron length, non-synonymous nucleotide replacements, etc). Sequence characterization of the *D. ananasse* $Mtn^{\bullet 3}$ gene showed a total conservation of the coding region in relation to the $Mtn^{\bullet 3}$ alleles of other species, but the comprised intron was revealed to be much shorter (69 bp) than its *D. melanogaster* homolog. As this short length was also shared by *D. pseudoobscura* (another *Sophophora* species, but even more distant from *D. melanogaster* than *D. ananasse*), it was hypothesized than the short-intron *Mtn* gene was the ancestral form.

Overall, a $Mtn^{\bullet 3}$ allele comprising a short-type intron is currently considered to be the most ancestral form of the MT gene in the *Drosophila* lineage. The transition to the *D. melanogaster* group is associated with the generation of a large-size intron, but the $Mtn^{\bullet 3}$ coding region remained unaltered, including the lysine-codon in the 40th position, so that it is the only *Mtn* allele detected so far in all the *Drosophila* species studied, except *D. melanogaster*. In this species, the Mtn^1 form was generated and rapidly spread among its populations, being at present the most common *Mtn* allele in *D. melanogaster*. The other typical features of the MT system in *D. melanogaster*, i.e., the *MtnB*-like subfamily and the *MtnA* variability regarding its genomic copy number, are absent from any other *Drosophila* species analyzed to date, and thus confirm the unprecedented plasticity of this system in *D. melanogaster* [40].

The generalization of conclusions about the physiological role of *Drosophila* MTs (Section 3), drawn from their development and tissue expression patterns is uncertain because these are not conserved among different species, even among close species belonging to the same taxonomic subdivision. For example, expression distribution of the *MtnA* gene has been extensively characterized during development in seven species of the melanogaster group (*D. melanogaster*, *D. mauritiana*, *D. sechellia*, *D. simulans*, *D. teissieri*, *D. yakuba*, and *D. erecta*) [16]. Discrepancies affect either the tissue-specific pattern or temporal and even quantitative variability, and in fact, the only common trait among the species is the *MtnA* expression in the midgut. Further, *MtnA* expression in the Malpighian tubules is exclusive to *D. melanogaster*. *D. mauritiana* and *D. sechellia* exhibit a very strong *MtnA* expression in the amnioserosa cell layer, at the early stage 13 of embryo development and in hemocytes and midgut. In *D. simulans*, *D. teissieri*, and *D. yakuba* the expression is detected only in hemocytes and midgut, and finally, in *D. erecta*, only in the midgut. If the temporal switch of the *MtnA* gene is considered, *D. teissieri* stands out for its early and strong *MtnA* embryonic expression. Finally, these unexpected variations were also detected when quantifying the *MtnA* level of expression, with both intraspecific (allelic) and interspecifically significant quantitative differences. This diversity of developmental gene expression patterns suggests a rapid evolution of the regulatory system of this gene, maybe in relation to species differentiation events and/or to specific metal handling needs in each case.

7. METALLOTHIONEINS IN OTHER (NON-*DROSOPHILIDAE*) *DIPTERA* AND INSECTA

Although *Drosophila* remains the only dipterean insect whose MTs have been thoroughly characterized on a molecular (gene/protein) and physiological level, some environmental pollution studies have proposed the use of MTs from other diptera as biomarkers for heavy metal contamination and therefore some data is also available. Precisely, different species of mosquitoes have attracted researchers' attention mainly due to the extent of their habitats and also the health-related interest in controlling them. Hence, *Culex quinquefasciatus*, a year-round mosquito species that is highly common in the Calcutta (India) area, has at least one MT gene that is sufficiently similar to those of *D. melanogaster* to cross-hybridize in Southern analysis. Furthermore, it confers resistance to industrial and domestic waste containing high concentrations of heavy metals [41]. Cultured cells from another mosquito, *Aedes albopictus*, have served as host for molecular analysis of the functionality of the *D. melanogaster* *Mtn*A promoter. These studies [42] showed how the expression of the reporter gene cloned downstream of this promoter responded to cadmium and copper induction. However, the level of induction in the mosquito cells was lower than that in homologous *D. melanogaster* cells (Schneider2 line), indicating some significant divergence in the trans-acting transcription factors controlling *Mtn* expression in both diptera.

Different *Chironomus* species (*C. thummi* [43], *C. yoshimatsui* [44], and *C. riparius* [45]) also synthesize metallothionein-like products in response to cadmium exposure. Their concentration increases proportionally to tissue-accumulated cadmium, thus, MT was proposed as both individual and population biomarker of heavy metal exposure. On the other hand, larvae of phantom midges (genus *Chaorobus* [46]) are able to accumulate cadmium when growing in contaminated lakes by synthesis of high MT levels, and it has been suggested that they are able to exploit this ability to colonize such types of ecological niches. High levels of metallothionein-like, cadmium-accumulating proteins have also been described in other non-diptera aquatic insects, such as mayflies (*Baetis thermicus*, *Ephenoptera* [47]) and stoneflies (*Eustheia spectabilis*, *Plecoptera* [48]).

Among the few non dipterean insects analyzed so far, the case of the springtail *Orchesella cincta* (Collembola) must be highlighted, as the genetic and biochemical characterization of its MT system has provided the only example of MT posttranslational processing reported up to now, other than the removal of the eight N-terminal amino acids of the yeast Cup1 primary polypeptide. Like other insects, springtails synthesize greater amounts of MT to cope with cadmium pollution, in this case, of forest soils, so that their populations develop an adaptive cadmium tolerance. Subsequent studies with cadmium-intoxicated organisms have shown that all the ingested cadmium was bound to MT in the springtail gut [49]. Surprisingly, isolation of these Cd-MT complexes and analysis of their protein component revealed that two different peptides were present, corresponding to two proteolytic derivates of the primary product of the *O. cinca* MT mRNA translation [50]. The molecular mechanism and significance of this phenomenon have yet to be described.

Finally, there is only one report in the literature of the interaction of a dipterean MT with a metal other than cadmium, and surprisingly, this is not copper but zinc. It refers to the capacity of the housefly larvae (*Musca domestica*) to synthesize antibacterial, metallothionein-like peptides in response to zinc overload [51].

8. CONCLUDING OVERVIEW

Although knowledge of the *D. melanogaster* MT system has greatly improved in the last decade, there are still considerable areas of ignorance, which undoubtedly deserve further research. Among the most appealing, the following are undoubtedly worth mentioning. First, solving the 3D structure of the Cu(I) and Cd(II) complexes conformed by MtnA and MtnB would provide an insight for assigning the preference for copper binding to some polypeptide features. Second, the establishment of the fate of the Cu(I) and Cd(II) complexes inside the organism would complete our understanding of their physiological significance. Third, the complete picture of the molecular mechanisms controlling *Mtn* gene expression in response to copper and cadmium, along with similarities to and differences from those operating in mammals, would enhance our comprehension of both of them.

Finally, we should consider that *D. melanogaster* is just one of the hundreds of *Drosophila* species, and a global view of the MT system in the genus could help to disregard specific peculiarities that are maybe not of general importance. Also a complete evolutionary pattern would be most desirable. In this context, the computational analysis of MT and MTF-1 genes and corresponding regulating regions in the recently released genomes of twelve *Drosophila* species seems to be the most fascinating immediate challenge [52].

ABBREVIATIONS AND DEFINITIONS

aa	amino acid(s)
bp	base pair(s)
CD	circular dichroism
Cup1	the copper-thionein of baker's yeast (*S. cerevisiae*)
ESI-MS	electrospray ionization mass spectrometry
GC-FPD	gas chromatography with flame photometric detection
GST	glutathione-S-transferase
ICPAES	inductively-coupled plasma atomic emission spectroscopy
LC50	concentration of a chemical which kills 50% of a sample population
MRE	metal response element
MT	metallothionein
Mtn/Mtn	*D. melanogaster* MT gene/protein without allusion to any specific isoform
nt	nucleotide(s)
PGK	phosphoglycerate kinase
TK	thymidine kinase
UTR	untranslated region

REFERENCES

1. G. Maroni, E. Otto and D. Lastowski-Perry, *Genetics*, 1986, **112**, 493–504.
2. G. Maroni, E. Otto, D. Lastowski-Perry and D. H. Proce, *Experientia Suppl.*, 1987, **52**, 385–392.

3. E. Otto, J. E. Young and G. Maroni, *Proc. Natl. Acad. Sci. USA*, 1986, **83**, 6025–6029.
4. D. Egli, A. Selvaraj, H. Yepiskoposyan, B. Zhang, E. Hafen, O. Georgiev and W. Schaffner, *EMBO J.*, 2003, **22**, 100–108.
5. P. Silar, L. Theodore, R. Mokdad, N. Erraïss, A. Cadic and M. Wegnez, *J. Mol. Biol.*, 1990, **215**, 217–224.
6. D. Egli, J. Domenech, A. Selvaraj, K. Balamurugan, H. Hua, M. Capdevila, O. Georgiev, W. Schaffner and S. Atrian, *Genes to Cells*, 2006, **11**, 647–658.
7. D. Latowski-Perry, E. Otto and G. Maroni, *J. Biol. Chem.*, 1985, **260**, 1527–1530.
8. L. Theodore, A. Ho and G. Maroni, *Genet. Res.*, 1991, **58**, 203–210.
9. G. Maroni, J. Wise, J. E. Young and E. Otto, *Genetics*, 1987, **117**, 739–744.
10. S. Nuzhdin, M. L. Wayne, K. L. Harmon and L. M. McIntyre, *Mol. Biol. Evol.*, 2004, **21**, 1308–1317.
11. D. F. Shirley and R. M. Sibly, *Evolution*, 1999, **53**, 826–836.
12. E. Otto, J. M. Allen, J. E. Young, R. Palmiter and G. Maroni, *Mol. and Cell. Biol.*, 1987, **7**, 1710–1715.
13. J. L. Meyer, M. A. Hoy and A. Jeyaprakash, *Biol. Control*, 2006, **36**, 129–138.
14. R. Mokdad, A. Debec and M. Wegnez, *Proc. Natl. Acad. Sci. USA*, 1987, **84**, 2658–2662.
15. F. Bonneton, Ph. D. Thesis, Université Paris XI, 1994.
16. F. Bonneton and M. Wegnez, *Dev. Genet.*, 1995, **16**, 253–263.
17. B. Zhang, D. Egli, O. Georgiev and W. Schaffner, *Mol. Cell. Biol.*, 2001, **21**, 4505–4514.
18. X. Chen, B. Zhang, P. M. Harmon, W. Schaffner, D. O. Peterson and D. P. Giedroc, *J. Biol. Chem.*, 2004, **279**, 4515–4522.
19. K. Balamurugan, D. Egli, A. Selvaraj, B. Zhang, O. Georgiev and W. Schaffner, *Biol. Chem.*, 2004, **385**, 597–603.
20. A. Selvaraj, K. Balamurugan, H. Yepiskoposyan, H. Zhou, D. Egli, O. Georgiev, D. Thiele and W. Schaffner, *Gene Dev.*, 2005, **19**, 891–896.
21. M. Durliat, F. Bonneton, E. Boissonneau, M. Andre and M. Wegnez, *Biometals*, 1995, **8**, 339–351.
22. G. Maroni, D. Lastowski-Perry, E. Otto and D. Watson, *Environ. Health Persp.*, 1986, **65**, 107–116.
23. F. Bonneton, L. Theodore, P. Silar, G. Maroni and M. Wegnez, *FEBS Lett.*, 1996, **380**, 33–38.
24. P. Silar and M. Wegnez, *FEBS Lett.*, 1990, **269**, 273–276.
25. M. Valls, R. Bofill, I. N. Romero, D. R. Gonzalez, J. Abian, M. Carrascal, D. P. Gonzalez, M. Capdevila and S. Atrian, *FEBS Lett.*, 2000, **467**, 189–194.
26. J. Domenech, O. Palacios, L. Villarreal, P. Gonzalez-Duarte, M. Capdevila and S. Atrian, *FEBS Lett.*, 2003, **533**, 72–78.
27. M. Valls, R. Bofill, R. Gonzalez-Duarte, P. Gonzalez-Duarte, M. Capdevila and S. Atrian, *J. Biol. Chem.*, 2001, **276**, 32835–32843.
28. M. Capdevila, J. Domenech, A. Pagani, L. Tio, L. Villarreal and S. Atrian, *Angew. Chem. Int. Ed.*, 2005, **44**, 4618–4622.
29. K. B. Jacobson, J. E. Turner, N. T. Christie and R. K. Qwenby, *Sci. Total Environ.*, 1983, **28**, 355–366.

30. A. Debec, R. Mokdad and M. Wegnez, *Biochem. Bioph. Res. Com.*, 1985, **127**, 143–152.
31. G. Maroni and D. Watson, *Insect Biochem.*, 1985, **15**, 55–63.
32. H. J. Gill, D. L. Nida, D. A. Dean, M. W. England and K. B. Jacobson, *Toxicology*, 1989, **56**, 315–321.
33. G. Maroni, A. Ho and L. Theodore, *Environ. Health Persp.*, 1995, **103**, 1116–1118.
34. K. Balamurugan, D. Egli, H. Hua, R. Rajaram, G. Seisenbacher, O. Georgiev and W. Schaffner, *EMBO J.*, 2007, **26**, 1035–1044.
35. M. McNulty, M. Puljung, G. Jefford and R. R. Dubreuil, *Cell Tissue Res.*, 2001, **304**, 383–389.
36. A. Southon, R. Burke, M. Norgate, P. Batterman and J. Camarakis, *Biochem. J.*, 2004, **383**, 303–309.
37. D. Egli, H. Yepiskoposyan, A. Selvaraj, K. Balamurugan, R. Rajaram, A. Simons, G. Multhaup, S. Mettler, A. Vardanyan, O. Georgiev and W. Schaffner, *Mol. Cell. Biol.*, 2006, **26**, 2286–2296.
38. B. A. Masters, E. J. Kelly, C. J. Quaife, R. L. Brinster and R. D. Palmiter, *Proc. Natl. Acad. Sci. USA*, 1994, **91**, 584–588.
39. W. Stephan, V. S. Rodriguez, B. Zhou and J. Parsch, *Genetics*, 1994, **138**, 135–143.
40. B. W. Lange, C. H. Langley and W. Stephan, *Genetics*, 1990, **126**, 921–932.
41. S. Sarkar, A. K. Duttagupta and T. K. Mal, *Environ. Pollut.*, 2004, **127**, 183–193.
42. C. C. N. Wu and A. M. Fallon, *Comp. Biochem. Physiol. B*, 1997, **116**, 353–358.
43. L. A. Seidman, G. Bergtrom, D. J. Gingrich and C. C. Reisen, *Tissue Cell*, 1986, **18**, 407–418.
44. M. Yamarura, K. T. Suzuki, S. Hatakeyama and K. Kubota, *Comp. Biochem. Physiol. C*, 1983, **75**, 21–24.
45. P. L. Gillis, L. C. Diener, T. B. Reynoldson and D. G. Dixon, *Environ. Toxicol. Chem.*, 2004, **21**, 1836–1844.
46. M. Croteau, L. Hare, P. G. G. Campbell and Y. Coiullard, *Environ. Toxicol. Chem.*, 2002, **21**, 737–741.
47. Y. Aoki, S. Hatayema, N. Kobayashi, Y. Sumi, T. Suzuki and K. T. Suzuki, *Comp. Biochem. Physiol. C*, 1989, **93**, 345–347.
48. L. Everard and R. Swain, *Comp. Biochem. Physiol. C*, 1983, **75**, 275–280.
49. P. J. Hensbergen, M. J. van Velzen, R. A. Nugrono, M. H. Donker and N. M. van Straalen, *Comp. Biochem. Physiol. C*, 2000, **125**, 17–24.
50. P. J. Hensbergen, M. H. Donker, M. J. van Velzen, D. Roelofs, R. C. van der Schors, P. E. Hunziker and N. M. van Straalen, *Eur. J. Biochem.*, 1999, **259**, 197–203.
51. H. Y. Jin, Q. Hui, J. Y. Jun, A. C. Ju, L. D. Sen, D. R. Qian and Q. R. Lin, *Acta Biol. Hung.*, 2005, **56**, 283–295.
52. *Drosophila* 12 Genomes Consortium, *Nature*, 2007, **450** 203–218.

Met. Ions Life Sci. **2009**, *5*, 183–197

7

Earthworm and Nematode Metallothioneins

Stephen R. Stürzenbaum
School of Biomedical and Health Sciences, Department of Biochemistry, Pharmaceutical Sciences Division, King's College London, London, SE1 9NH, UK
<stephen.sturzenbaum@kcl.ac.uk>

ABSTRACT: The earthworm and the nematode are primary soil sentinel and model invertebrate organisms. Their metallothioneins (MT) display striking structural and functional features. Firstly, the primary sequences, though remarkably conserved within species belonging to the same genus, have notable divergencies containing additional potentially metal-binding residues. Secondly, cadmium is the major transcriptional activator of MT expression in the intestinal cells. Thirdly, the MT promoters seemingly

Metal Ions in Life Sciences, Volume 5 Edited by Astrid Sigel, Helmut Sigel, and Roland K. O. Sigel

Published by the Royal Society of Chemistry, www.rsc.org DOI: 10.1039/9781847558992-00183

lack functional metal response elements which underlines the finding that MTF-1 is absent from the nematode and possibly the earthworm genome. This chapter highlights past, present, and future advances aiming to dissect the synergies and discrepancies of the metallothioneins from two evolutionary very distant relatives of soil dwelling worms.

KEYWORDS: *C. briggsae* · *C. elegans* · earthworm · *L. rubellus* · metallothionein · nematode

1. INTRODUCTION

The Oxford English Reference Dictionary defines a worm as being a slender elongated invertebrate animal with a soft moist body, inhabiting marine or fresh water, burrowing in the soil, or parasitic in the bodies of vertebrates. The entry continues with the following in-depth description:

> *"The term worm was formerly used for anything that creeps or crawls, and even zoologists applied it to a wide range of soft-bodied invertebrates. Three important animal phyla include worms – the annelids, which comprise segmented worms such as earthworms, leeches, and ragworms; the nematodes or roundworms; and the platyhelminths, which include the flatworms, tapeworms, and flukes."* [1].

This chapter will focus on the earthworm (mainly *Lumbricus rubellus*) and nematode (*Caenorhabditis elegans* and *Caenorhabditis briggsae*), both of which fit the generalist definition of a worm, but are indeed very distant relatives (Figure 1).

The first scholarly authority on earthworms was Charles Darwin, FRS (1809–1882) who, although may be remembered most for his description of the *The Origin of Species*, had a life-long interest in earthworm biology. Indeed, one of his first public presentations and publications covered the subject of earthworm casts and soil formation by earthworms [2]. Similarly, in his last book (published in 1881), entitled "*The formation of vegetable mould through the action of worms with observations on their habits*", he covered nearly 40 years of meticulous record keeping on the biology of earthworms. At the time, this book was a best-seller and can be considered a turning point in the perception of the importance of earthworms by farmers, gardeners, and ecologists. Since then, earthworms have been hailed as a critical component to environment health due to their involvement in the creation of topsoil, the improvement of soil structure, water infiltration, and gas exchange, retention of nutrients, improvement of soil permeability and the reduction of erosion. More recently, the macroscopic earthworm has been selected as a keystone organism for the assessment of the toxicity of

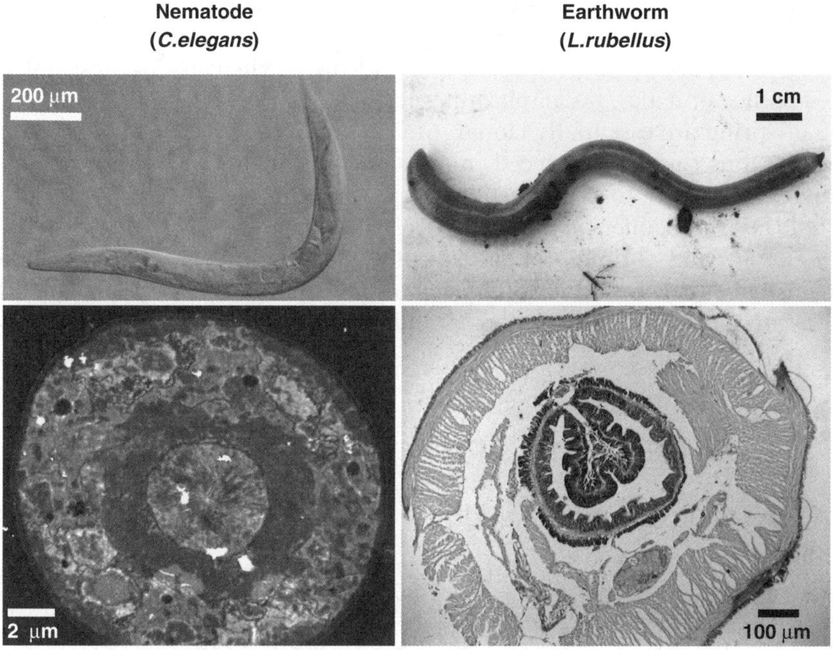

Figure 1. The microscopic nematode *Caenorhabditis elegans* (left panels) and the macroscopic earthworm *Lumbricus rubellus* (right panels) in full body view (top panels) and as a transection (bottom panels). Note the simple, transparent body plan of the nematode in comparison to the relative complexity of the segmented earthworm.

chemicals, herbicides, and pesticides. The obligate and facultative use of earthworms in respective EU and US toxicity tests is undoubtedly the second turning point highlighting the importance of the earthworm in providing an early warning of deterioration in soil quality and thus indirectly linked to the protection of human health.

The microscopic non-parsitic bacterivorous nematode *C. elegans* was first utilized in the 1970s by Sydney Brenner to study developmental biology and genetics [3] and has since progressed rapidly to becoming one of the premier invertebrate genetic model systems. Numerous advantages include the availability of a fully sequenced genome, detailed knowledge of the lineage, composition and functioning of cells, or the many experimental tools (knockdown by RNAi, chromosomal knockout, knockin, GFP tagging, etc.). It can be cryo-preserved and has a short life cycle (2 days from egg to sexual maturity, a 3 day reproductive period, a brood size of 200–300 offspring and an average lifespan of 14 days). Unlike the simultaneous (or synchronous) hermaphroditic earthworm, which relies on the reciprocal insemination of two individuals, *C. elegans* is a protandrous hermaphrodite, where the gonads first produce sperm and then switch to oogenesis [3]. In the absence of males, hermaphrodites fertilize their own eggs and as a result the offspring are essentially clonal. Males, which are rare, occur due to the spontaneous (or stress induced) nondisjunction of the X chromosome and thus can be used to cross-fertilize hermaphrodites in genetic crosses [4]. Whole genome sequencing has now been completed for *C. elegans* [5] and *C. briggsae* [6] and although phenotypically and geographically almost indistinguishable they are evolutionary some 80–110 million years apart, and are therefore more distant than human and mouse [7].

2. FROM GENES TO PROTEINS

2.1. Genomic Organisation

Compared to other Lumbricidae, the genome of *L. rubellus* is relatively compact at 420 Mbp spread over 36 chromosomes [8] but more than four times the size of *C. elegans* or *C. briggsae* (both about 100 Mbp over 6 chromosomes). This implies contrasting complexities of the genome architecture, a notion that is reflected in the genomic organisation of their respective metallothioneins. As the entire genome has been deciphered for both nematode species, it is possible to state that each of the nematode genomes contains precisely two metallothioneins (*MT-1*: WormBase identifier *mtl-1*, GenBank accession number X53244 and *MT-2*: WormBase identifier *mtl-2*, GenBank accession number X53245), both located on chromosome 5. The general intron-exon structure of the two nematode metallothioneins is highly similar and contains one short exon and one larger exon interrupted by a single intron. In comparison, the genomic organization of *L. rubellus MT-2* is more complex, with 4 short exons spanning a genomic region that is nearly 10× larger that the nematode loci (Figure 2).

All three earthworm *MT-2* genome sequences isolated to date (GenBank accession numbers: AJ299434-AJ299436) display the same general structure, but significant differences in the promoter and intronic segments [9]. As very limited information is available regarding the earthworm genome sequence, it is, at present, not possible to predict if the three sequences represent true isoforms, are a result of a gene duplication event or highlight the occurrence of intra-species variation (the three *MT-2s* were isolated from a library generated from a pool of earthworms sampled from a field site). Future sequencing efforts using different isolates and possibly whole genome sequencing will solve this enigma.

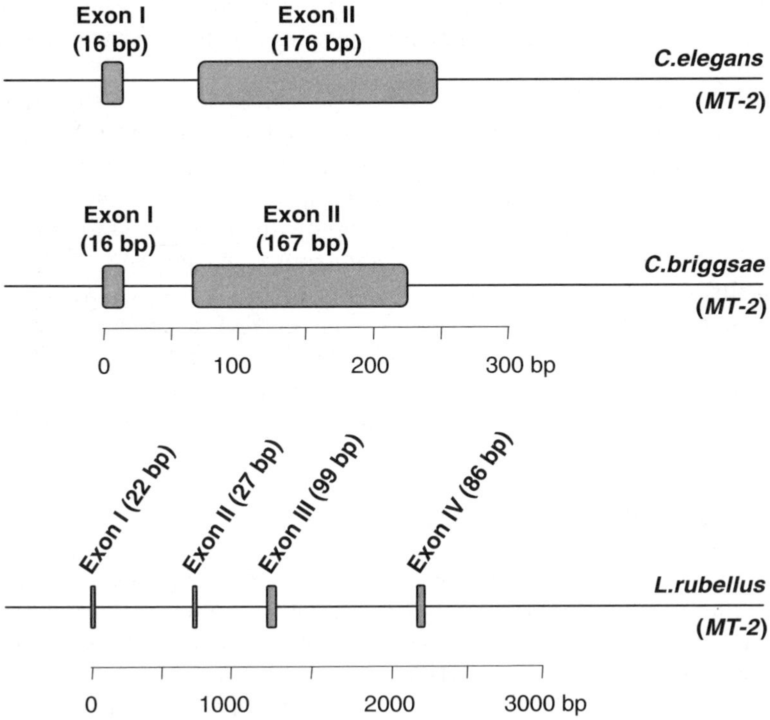

Figure 2. Schematic representation of the genomic organisation of nematode (*C. elegans* and *C. briggsae*) and earthworm *L. rubellus MT-2*. Note that although the respective coding regions are comparable in size, the genomic locus of *L. rubellus* is larger by one order of magnitude.

2.2. Expressed Sequence Tags

Expressed sequence tags (ESTs) are single pass sequences isolated at random from cDNA libraries. Though short (400–800 bps) and by definition of lower quality, they provide a rapid insight into gene splicing and expression characteristics. The current version of dbEST (release 011108), the GenBank database dedicated to ESTs, contains over 43 million sequences from 1470 different species (http://www.ncbi.nlm.nih.gov/projects/dbEST). In total, 346,107 ESTs originate from *C. elegans*, mapped onto the genome sequence and can be accessed via WormBase (www.wormbase.org). As EST libraries are typically synthesized from organisms raised under standard conditions (rather than metal exposed), this explains why only 25 MT sequences (20× *MT-1* and 5× *MT-2*) have been isolated. If the expression of every gene was equal (which obviously is not the case) and given that the *C. elegans* genome contains 20,000 genes, this number is just above the expected stochastic probability.

In contrast, the "MT hit rate" was significantly higher in the far more modest *L. rubellus* EST project (currently standing at 19,934 ESTs) as several different libraries were synthesized each utilizing earthworms from different developmental stages exposed to specific stressors including cadmium and copper. In detail, 45 ESTs were identified as metallothioneins, of which 31 were isolated from heavy metal exposed earthworms (29 from Cd and a further 2 from Cu exposures). Interestingly, and this shows the true value of random sequence generation via an EST approach, a hitherto unidentified MT, namely *MT-3*, was identified which was shown to be enriched in a library synthesized from late cocoons. This has led to the notion that the expression of earthworm *MT-3* may be induced during embryonic development [9].

2.3. Metallothionein Protein

In the early 1980s, classical biochemical techniques confirmed the presence of metallothioneins in earthworms [10] and nematodes [11]. Partial protein sequences were obtained from the amino terminal and the full length protein sequence deducted following cDNA cloning [12–16]. Both, earthworm and nematode MTs, bear the typical hallmarks of a metallothionein: they are small, cysteine-rich with no or few aromatic residues (Figure 3, top panel). Although the primary sequences are highly conserved in species belonging to the same genus or family, there is an apparent divergence in sequence conservation – at least in the two taxonomic invertebrate lineages investigated here (Figure 3, bottom panel). Whilst *C. elegans* MT-1 and *C. briggsae* MT-1 are 92% identical over the entire protein, alignments between, say, *C. elegans* MT-1 and *L. rubellus* MT-1 can only be achieved by artificially aligning the many cysteine residues which requires the introduction of multiple gaps (Figure 4). Given the importance of the cysteines and histidines in metallobiology, it is worthy to note the C-terminal extensions of *C. elegans* and *C. briggsae* MT-1 which both contain a single cysteine and three histidine residues. Similarly, it is striking that earthworm MT-3 displays two extra cysteines, both located in alignment gaps (Figure 3, top panel). It is likely that these subtle differences are the key to explaining possible differences in structure, differential metal handling, and function.

Recombinant expression of earthworm and nematode metallothioneins allowed detailed studies on pH sensitivity and metalation characteristics [17]. Electrospray ionization mass spectrometry (ESI-MS), UV absorption and circular dichroism (CD) spectroscopy applied to recombinant *L. rubellus* MT-2 showed that the 20 cysteines are capable of binding 7 Cd^{2+} ions.

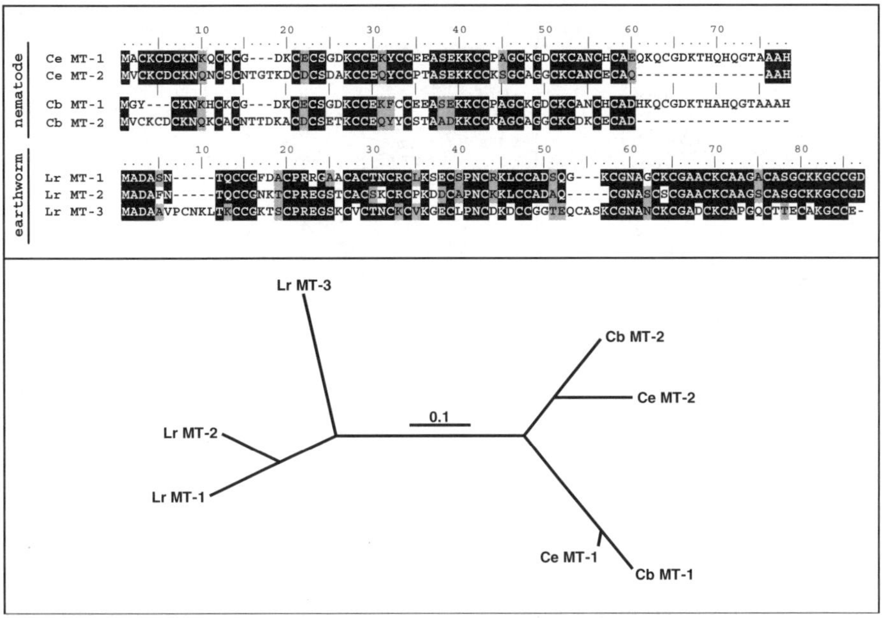

Figure 3. Protein alignment of nematode (Ce: *C. elegans*; Cb: *C. briggsae*) and earthworm (Lr: *L. rubellus*) metallothioneins (top panel) and corresponding phylogenetic tree (bottom panel). The alignment was constructed using ClustalW algorithms and visualised using the biological sequence alignment editor (BioEdit, Ibis Biosciences, Carlsbad, USA). The unrooted tree was computed using the Neighbor-Joining (NJ) method providing the genetic distance between sequences based on the number of amino acid substitutions.

Interestingly, pH titrations identified a biphasic demetallation reaction, which provides strong evidence that the protein consists of two metal-thiolate binding domains with stoichiometries of Cd3Cys9 and Cd4Cys11 with 11 cysteines at the N-terminal and 9 cysteines at the C-terminal. This is notably the reverse of mammalian MTs. Likewise, CD spectroscopy showed that the peptide wraps around Cd in a manner that is distinct from mammalian MTs [17]. Ion-spray mass spectrometry on heterologously expressed *C. elegans* MT-2 found that the nematode MT is capable of binding less Cd than its annelid counterpart, namely it has a metal-to-protein stoichiometry of 6 moles per mole with clusters of tetrahedrally coordinated cadmium-thiolate complexes. Unlike mammalian MTs, the molecular weight determined by comparative gel exclusion chromatography matches the data obtained by mass spectrometry data and therefore it has been hypothesized that the molecular shape approaches that of globular marker proteins [18].

		Ce MT-1	Ce MT-2	Cb MT-1	Cb MT-2	Lr MT-1	Lr MT-2	Lr MT-3
Ce MT-1	Identity %	100						
	Similarity %	100						
	Gaps	0						
Ce MT-2	Identity %	66	100					
	Similarity %	75	100					
	Gaps	5	0					
Cb MT-1	Identity %	92	64	100				
	Similarity %	95	75	100				
	Gaps	0	5	0				
Cb MT-2	Identity %	60	76	57	100			
	Similarity %	71	84	68	100			
	Gaps	5	0	5	0			
Lr MT-1	Identity %	33	28	43	33	100		
	Similarity %	37	41	53	39	100		
	Gaps	10	10	14	22	0		
Lr MT-2	Identity %	34	32	36	37	74	100	
	Similarity %	42	39	45	40	83	100	
	Gaps	14	18	23	7	2	0	
Lr MT-3	Identity %	37	37	38	35	56	55	100
	Similarity %	43	47	41	37	67	67	100
	Gaps	17	6	30	18	9	6	0

Figure 4. Pairwise comparison of nematode (Ce: *C. elegans*; Cb: *C. briggsae*) and earthworm (Lr: *L. rubellus*) metallothionein amino acid sequences. In addition to identity and similarity scores (expressed as %), note the number of gaps that were required to maximize the identity/similarity scores.

Although many physical, chemical, and spectroscopic features are conserved between earthworm, nematode, and higher vertebrate metallothioneins, many invertebrate specific features clearly still await discovery.

3. TRANSCRIPTIONAL REGULATION

3.1. Metal Specificity

Cd is a potent inducer of MT transcription in nematodes [19] and earthworms [20]. In laboratory exposures, transcriptional activation of *C. elegans MT-1* and *MT-2* is rapid (<1 hour) and sensitive ($<3\,\mu$M Cd). In contrast, at least 100 and 250 μM Cu are required to activate *C. elegans MT-2* and *MT-1*, respectively. A chronic exposure to 500 μM Zn only marginally induces MT, indicating that Zn is only a mild transcriptional activator of nematode MTs [21]. Likewise, using fully quantitative PCR, it was shown that earthworm *MT-2* is more responsive to an LC50 dose of Cd (6.4–7.0 μM Cd g^{-1} soil) than an LC50 of Cu (9–10 μM Cu g^{-1} soil) when exposed to metal spiked loam soil amended with 3% of composted bark [22,23]. Of course, it may be argued that laboratory exposures do not reflect the "real world", an argument that could be refuted, as earthworms exposed to Cu or Cd in a semi-natural mesocosm setup had an *MT-2* expression response that was comparable to the laboratory setup [24]. Furthermore, *MT-2* expression was found to be unaffected by diurnal, weekly or seasonal temperature fluctuations using soils sampled from a Cd, Pb, and Zn smelter [25]. Of course, *in vitro*, it is possible to fill recombinant apoMT with most heavy metals, including cadmium, copper, zinc, or arsenic. Although this specifies the affinity of MTs towards metal binding in general, metal responsive transcriptional activation and the binding of metals to the protein is dependent on the actual bioavailability of metals within the cell. In *C. elegans* for example, it was shown that the native protein of Cd-induced MT can contain between 2–20% Zn [26].

3.2. The Elusive Metal-Responsive Transcription Factor 1

Unlike Drosophila [27] or higher eukaryotes the genome of *C. elegans* does not contain a protein that, at least by sequence similarity, is orthologous to the metal-responsive transcription factor 1 (MTF-1). Metal-response elements (MREs) are absent in the promoter region of *C. elegans MT-1* and only one single MRE is located in *MT-2* [28]. Although the promoter of earthworm *MT-2* contains three MREs, there is indirect evidence that they are non-functional. Likewise, though only based on a 20,000 EST project, a classical earthworm MTF-1 has remained elusive [9].

Taken together, this suggests that the transcriptional regulation of nematode and earthworm MT is distinct from other invertebrates and vertebrates. At least for *C. elegans*, it was shown that the regulation of MTs requires ELT-2 (an erythroid-like GATA-type transcription factor) to bind to specific GATA transcription factor-binding sites located in the promoters of *MT-1* and *MT-2*. However, ELT-2 is not a metal sensor as it constitutively activates metallothioneins, thus suggesting the presence of a yet to be identified repressor-mediated heavy metal responsive gene that drives the transcriptional activation of nematode (and earthworm) MT [29].

4. CELLULAR AND SUBCELLULAR LOCALIZATION

In earthworms, X-ray distribution maps of the hepatocyte-resembling chloragocytes revealed that Ca, Pb, and Zn bioaccumulate in organelles that are distinct from Cd- and S-rich organelles which, due to their affinity to the neutral-red dye, are likely to be of lysosomal origin [9,30,31]. This established that the chloragogenous tissue surrounding the coelomic surfaces of the gut (Figure 5) is the major Cd sequestering tissue.

Antibodies raised against earthworm and nematode MT-2 independently determined that the predominant location of MT protein is in the intestine

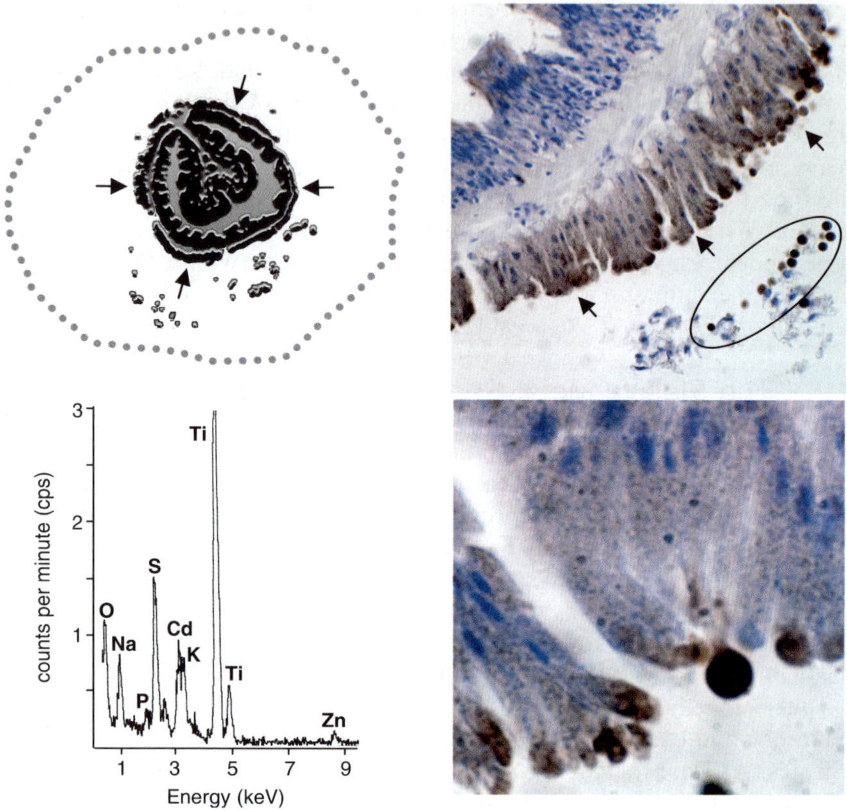

Figure 5. Localization of earthworm MT-2. The majority of Cd accumulates in the chloragogenous tissue surrounding the posterior alimentary canal (arrows, top left panel). A polyclonal antibody specific to earthworm MT-2 [16] identifies the apical ends of the chloragogenous tissue as the premier location of MT-2 (arrows, top right panel). Once saturated, the chloragogenous tissue is released into the coelomic cavity and targeted by coelomocytes (circled area, top right panel). Static electron probe X-ray microanalysis applied to the MT-2 antibody positive area contained large amounts of Cd and S demonstrating that MT protein and Cd are co-localized [30].

[16,32], a finding that has been dissected into fine detail at least for earthworm MT-2. Not only Cd, but also metallothionein levels are highest in the chloragogenous tissue which accumulate in the most apical region (Figure 5). X-ray absorption spectroscopy (XAS) confirmed that the Cd and S are coordinated with a bond distance typical of a metallothionein-type ligand [33]. The co-localization of Cd and MT was further confirmed by immunogold cytochemistry on thin cryosectioned chloragogenous tissue [16]

followed by static electron probe X-ray microanalysis. Once saturated, the MT accumulating apical ends eventually break off into the coelomic cavity where they are targeted by the earthworms' immune cells, the coelomocytes (Figure 5). This process is rapid as MT positive coelomocytes have been isolated from earthworms following a 3 days dermal exposure to Cd [34,35]. It is still unclear how long the MT laden coelomocytes remain within the body or if they are excreted via the MT immuno positive nephridia or the dorsal pores.

5. TRANSGENIC WORMS

To date, no transgenic earthworms have been made, for reasons not limited to technical reasons, but mainly due to the prohibitively long live cycle. In *C. elegans* it is common practice to inject into the gonads either β-galactosidase reporter (*lacZ*) or green fluorescent protein expressing constructs that contain either a gene specific promoter or full length coding sequences. Successful F1 lines contain the extrachromosomal array which, if sufficiently penetrant, is passed on to following generations. If required, the extrachromosomal array can be integrated into the chromosome by UV irradiation, or exposure to chemical mutagens such as 4,5′,8-trimethylpsoralen or ethane methyl sulfonate. Using the *lacZ* reporter system, it was shown that *MT-1* is constitutively expressed in the pharynx in the absence of cadmium. Upon exposure to cadmium or heat stress, however, *MT-1* and *MT-2* are both strongly induced in all intestinal cells [36]. This finding was later confirmed using highly penetrant GFP expressing lines that are under the control of the respective *MT-1* or *MT-2* promoter (Figure 6 and [21]). Intriguing is not that metallothionein is responsive to Cd, but that *MT-1* is expressed in three cells of the lower pharyngeal bulb even in the absence of Cd. Located at the entry point to the intestine, *MT-1* may function as a sensor of toxic metals or regulator of essential metal homeostasis.

Besides the ability to track MT synthesis via transcriptional and translational transgenes, it has been possible to isolate *C. elegans* strains with chromosomal deletions of *MT-1* or *MT-2* as well as a *MT-1;MT-2* double knockout [21,37]. Total brood size was reduced when one or both MTs were deleted and volumetric growth impaired in the double knockout. This finding is of significance, as these effects were observed in the absence of Cd. Although neither of the knockouts was lethal even in the presence of moderate Cd concentrations (2.5–12 μM), the dose dependent reduction in total brood size and volumetric growth was (compared to wild-type) more significant in single MT knockouts and most significant in the double mutant. This reveals that nematode MT modulates cadmium-dependent and -independent toxic effects on life history traits.

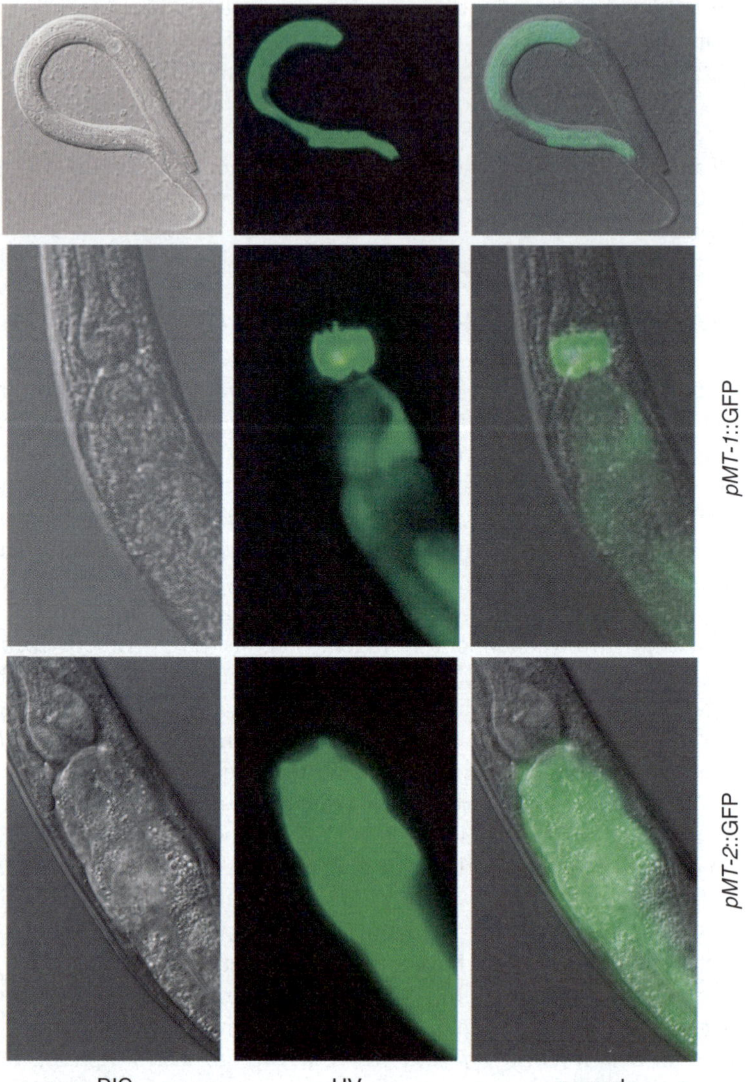

Figure 6. Localization of *C. elegans MT-1* and *MT-2*. Using GFP tagged MT promoter expressing transgenes (*pMT-1*::GFP and *pMT-2*::GFP) demonstrates that, upon exposure to Cd, the major location of MT transcription are the intestinal cells genes (top panel). In addition, note that MT-1 is constitutively expressed in the lower pharyngeal bulb, even in the absence of Cd (bottom two panels). Left panels are images taken using a microscope with Nomarski DIC optics, middle panels are the same samples under UV light visualised using a standard 'FITC' filter block and the right panel a merged image.

6. CONCLUDING REMARKS AND FUTURE DIRECTIONS

Earthworm and nematode metallothioneins certainly share many features with metallothioneins from other (higher) organisms and substantiate the notion that they play a pivotal role in the protection from cadmium toxicosis. Although metallothionein has been hailed as a major component in cadmium detoxification, the *C. elegans* double knockout is viable in the presence of cadmium. Most striking, however, is the observation that the double knockout displays a significantly impaired brood size and volumetric growth in the *absence* of cadmium, suggesting an, as yet to be discovered, alternative role independent of cadmium detoxification.

Although details concerning the general function of invertebrate MTs have been forthcoming, comprehensive studies investigating the isoform specificity and transcriptional activation have been neglected. This shortcoming should be redressed by exploiting molecular, genetic, whole animal life-cycle analysis as well as structural, biochemical, and biophysical methodologies to unravel the intrinsic nature that underpins the functional differentiation of MTs. This will help to understand the mode of action of the MTs under investigation, and contribute towards building a knowledge base on fundamental processes in metal ion-protein interactions. This area is still far from being well understood, but undoubtedly of interest to the field of metallobiology, protein evolution, toxicology, and medical and environmental genomics.

ACKNOWLEDGMENTS

I wish to thank the Natural Environmental Research Council (NERC), the Royal Society, and (more recently) the Biotechnology and Biological Sciences Research Council for financial assistance. Furthermore, I am grateful to Dr. Swain and Ms. Hughes (King's College London, UK), Prof. Baumeister (Freiburg University, Germany) and Dr. Georgiev (Zurich University, Switzerland) and in particular to Drs. Kille and Morgan (Cardiff University, UK) for their continuous support and seemingly endless enthusiasm.

ABBREVIATIONS

CD	circular dichroism
DIC	differential interference contrast
ELT-2	erythroid-like GATA-type transcription factor

ESI-MS	electrospray ionization mass spectrometry
EST	expressed sequence tag
GFP	green fluorescent protein
LC50	lethal concentration (that kills 50% of the population)
MRE	metal response element
MT	metallothionein
MTF	metal-responsive transcription factor
PCR	polymerase chain reaction
RNAi	ribonucleic acid interference
UV	ultraviolet
XAS	X-ray absorption spectroscopy

REFERENCES

1. *The Oxford English Reference Dictionary*, Ed. J. Pearsall and B. Trumble, 2nd edn., Oxford University Press, 1996, p. 1667.
2. C. Darwin, *Proc. Geolog. Soc. London*, 1838, **2**, 52: 574–576.
3. S. Brenner, in *The Nematode Caenorhabditis elegans*, Ed. W. B. Wood, Cold Spring Harbor Lab. Press, Cold Spring Harbor, NY, 1988, pp. 9–13.
4. A. D. Stewart and P. C. Phillips, *Genetics*, 2002, **160**, 975–982.
5. *C. elegans* Sequencing Consortium, *Science*, 1998, **282**, 2012–2018.
6. L. D. Stein, Z. Bao, D. Blasiar, T. Blumenthal, M. R. Brent, N. Chen, A. Chinwalla, L. Clarke, C. Clee, A. Coghlan, A. Coulson, P. D'Eustachio, D. H. Fitch, L. A. Fulton, R. E. Fulton, S. Griffiths-Jones, T. W. Harris, L. W. Hillier, R. Kamath, P. E. Kuwabara, E. R. Mardis, M. A. Marra, T. L. Miner, P. Minx, J. C. Mullikin, R. W. Plumb, J. Rogers, J. E. Schein, M. Sohrmann, J. Spieth, J. E. Stajich, C. Wei, D. Willey, R. K. Wilson, R. Durbin and R. H. Waterston, *PLoS Biol.*, 2003, **1**, e45.
7. L. W. Hillier, R. D. Miller, S. E. Baird, A. Chinwalla, L. A. Fulton, D. C. Koboldt and R. H. Waterston, *PLoS Biol.*, 2007, **5**, e167.
8. T. R. Gregory and P. D. N. Hebert, *Can. J. Zool.*, 2002, **80**, 1485–1489.
9. S. R. Stürzenbaum, O. Georgiev, A. J. Morgan and P. Kille, *Environ. Sci. Technol.*, 2004, **38**, 6283–6289.
10. K. T. Suzuki, M. Yamamura and T. Mori, *Arch. Environ. Contam. Toxicol.*, 1980, **9**, 415–424.
11. K. Maruyama, R. Hori, T. Nishihara and M. Kondo, *Jap. J. Toxicol. Environ. Health*, 1986, **32**, 22–27.
12. S. R. Stürzenbaum, P. Kille and A. J. Morgan, *FEBS Letters*, 1998, **431**, 437–442.
13. C. Gruber, S. R. Stürzenbaum, P. Gehrig, R. Sack, P. Hunziker, B. Berger and R. Dallinger, *Eur. J. Biochem.*, 2000, **267**, 573–582.
14. L. W. Slice, J. H. Freedman and C. S. Rubin, *J. Biol. Chem.*, 1990, **265**, 256–263.
15. M. Imagawa, T. Onozawa, K. Okumura, S. Osada, T. Nishihara and M. Kondo, *Biochem. J.*, 1990, **268**, 237–240.

16. S. R. Stürzenbaum, C. Winters, M. Galay, A. J. Morgan and P. Kille, *J. Biol. Chem.*, 2001, **276**, 34013–34018.
17. T. T. Ngu, S. R. Stürzenbaum and M. J. Stillman, *Biochem. Biophys. Res. Commun.*, 2006, **351**, 229–233.
18. C. You, E. A. Mackay, P. M. Gehrig, P. E. Hunziker and J. H. R. Kägi, *Arch. Biochem. Biophys.*, 1999, **372**, 44–52.
19. V. H. Liao and J. H. Freedman, *J. Biol. Chem.*, 1998, **273**, 31962–31970.
20. F. Brulle, G. Mitta, C. Cocquerelle, D. Vieau, S. Lemière, A. Leprêtre and F. Vandenbulcke, *Environ. Sci. Technol.*, 2006, **40**, 2844–2850.
21. S. C. Swain, K. Keusekotten, R. Baumeister and S. R. Stürzenbaum, *J. Mol. Biol.*, 2004, **341**, 951–959.
22. M. Galay, D. J. Spurgeon, J. M. Weeks, S. R. Stürzenbaum, A. J. Morgan and P. Kille, *Biomarkers*, 2003, **8**, 229–239.
23. D. J. Spurgeon, S. R. Stürzenbaum, C. Svendsen, P. K. Hankard, A. J. Morgan, J. M. Weeks and P. Kille, *Comp. Biochem. Physiol. C*, 2004, **138**, 11–21.
24. D. J. Spurgeon, C. Svendsen, L. J. Lister, P. K. Hankard and P. Kille, *Environ. Pollut.*, 2005, **136**, 443–452.
25. C. Svendsen, P. K. Hankard, L. J. Lister, S. K. Fishwick, M. J. Jonker and D. J. Spurgeon, *Environ. Pollut.*, 2007, **147**, 83–93.
26. K. Maruyama, R. Hori and T. Nishihara, *Eisei Kagaku*, 1986, **32**, 22–27.
27. D. Egli, A. Selvaraj, H. Yepiskoposyan, B. Zhang, E. Hafen, O. Georgiev and W. Schaffner, *EMBO J.*, 2003, **22**(1), 100–108.
28. F. Kugawa, H. Yamamoto, S. Osada, M. Aoki, M. Imagawa and T. Nishihara, *Biomed. Environ. Sci.*, 1994, **3**, 222–231.
29. L. H. Moilanen, T. Fukushige and J. H. Freedman, *J. Biol. Chem.*, 1999, **274**, 29655–29665.
30. C. A. Morris, S. R. Stürzenbaum, B. Nicolaus, A. J. Morgan, J. L. Harwood and P. Kille, in *Metallothionein IV*, Ed. C. D. Klaassen, Birkhäuser Verlag, Basel, Switzerland, 1999, pp. 621–627.
31. A. J. Morgan, S. R. Stürzenbaum, C. Winters, G. W. Grime, N. A. AbdAziz and P. Kille, *Ecotox. Environ. Safety*, 2004, **57**, 11–19.
32. M. Kondo, M. Imagawa, K. Maruyama, Y. Okada, S. Tsunasawa and T. Nishihara, *Biomed. Environ. Sci.*, 1990, **3**, 315–325.
33. J. Cotter-Howells, J. M. Charnock, C. Winters, P. Kille, J. C. Fry and A. J. Morgan, *Environ. Sci. Technol.*, 2005, **39**, 7731–7740.
34. J. Homa, E. Olchawa, S. R. Stürzenbaum, A. J. Morgan and B. Plytycz, *Environ. Pollut.*, 2005, **135**, 275–280.
35. J. Homa, S. R. Stürzenbaum, A. J. Morgan and B. Plytycz, *Eur. J. Soil Biol.*, 2007, **43**, S273–280.
36. J. H. Freedman, L. W. Slice, D. Dixon, A. Fire and C. S. Rubin, *J. Biol. Chem.*, 1993, **268**, 2554–2564.
37. S. Hughes and S. R. Stürzenbaum, *Environ. Pollut.*, 2007, **145**, 395–400.

Met. Ions Life Sci. **2009**, *5*, 199–237

8

Metallothioneins in Aquatic Organisms: Fish, Crustaceans, Molluscs, and Echinoderms

Laura Vergani

Department of Biology, University of Genova, I-16132 Genova, Italy
<laura.vergani@unige.it>

Metal Ions in Life Sciences, Volume 5 Edited by Astrid Sigel, Helmut Sigel, and Roland K. O. Sigel

Published by the Royal Society of Chemistry, www.rsc.org DOI: 10.1039/9781847558992-00199

ABSTRACT: Metallothioneins (MTs) have been described in a wide range of organisms, from bacteria to mammals, thus representing an interesting example of evolutionary molecular adaptation. If the moderate variability of MTs across phylogenetically distant organisms reflects their highly conserved function, the specific environmental requirements may explain the multiplicity of isoforms also in the same organism. The MT polymorphism is particularly important in invertebrates with respect to vertebrates. This review is an attempt to summarize the knowledge about MTs from aquatic animals, both vertebrates and invertebrates, to gain new insights into the structure-function relationship of this class of proteins. The large and increasing literature on MTs indicates that MTs from aquatic vertebrates are rather similar to mammalian counterparts, whereas a variety of structures have been described in invertebrates. Although the prototypical αβ-domain organization of vertebrate MTs has been observed in most invertebrate isoforms, some invertebrate MTs display alternative structures in which the canonical organization has been modified, such as the ββ-domain, the αββ-domain, and the multiple α-domain structures of oyster MTs, and the inverted βα-domain organization of sea urchin MTs. In this review we emphasize three major taxa of aquatic invertebrates, the molluscs, the crustaceans and the echinoderms, although some data have been reported for other invertebrates.

KEYWORDS: aquatic animals · crustaceans · echinoderms · fish · heavy metals · metallothioneins · molluscs

1. INTRODUCTION

Metallothioneins (MTs) have been found in almost all organisms, including vertebrates, invertebrates, plants, and bacteria. The evolution of new MT isoforms and variants occurred with the preservation of the positions of those amino acids which are essential for the formation and stability of the metal cluster architecture. MTs are usually classified in three classes [1]. Class I comprises all the MTs with clear sequence similarity to horse liver MT-1, including the MTs of mammals, fish, molluscs, and crustaceans. Class II comprises the MTs of plants, yeasts, and sea urchin. Class III includes atypical, non-translationally synthesized metal-thiolate proteins, such as phytochelatins and vegetal γ-glutamyl peptides [2]. Binz and Kägi [3] have proposed a new classification of MTs taking into account phylogenetic features as an additional classification criterion. Class I and Class II MTs known so far are classified into 15 families, with molluscan MTs belonging to family 2, and crustacean MTs to family 3.

In recent years, new insights into non-mammalian MTs have been gained, and particular interest has been focused on MTs from aquatic organisms, both vertebrates and invertebrates, since these MTs often display atypical features. Owing to the ability of many aquatic species to accumulate toxic heavy metals deriving from both physiological processes and environmental pollution, considerable interest has arisen in using aquatic animals for monitoring the metal contamination in aquatic ecosystems.

This review summarizes the knowledge about MTs from aquatic animals; these MTs often show important differences in amino acid composition and protein structure with respect to mammalian MTs, thus representing an interesting example of evolutionary molecular adaptation. For many MTs from aquatic species the amino acid sequences are now available, but for just four of them the complete three-dimensional (3D) structure has been solved: the MT-1 from the crab *Callinectes sapidus* [4]; the MT-1 from the lobster *Homarus americanus* [5]; the MT-A from the sea urchin *Strongylocentrotus purpuratus* [6]; the MT-A from the Antarctic fish *Notothenia coriiceps* [7].

Fish MTs display the canonical αβ-domain organization found in mammals, with two asymmetric globular domains binding 4 and 3 divalent metal ions [7]. Also echinoderm MTs display the same type of metal-thiolate cluster organization, but the domains are inverted inside the chain depending on the different gene organization [6,8]. In crustacean MTs the two domains are symmetric and bind 3 divalent metal ions each [9,10]. On the other hand, molluscan MTs display a large variety of structures consisting solely of α-domains, solely of β-domains, or of aββ-domains, in addition to the canonical αβ-domain structure [11].

In many aquatic species, MT expression is induced by metals as well as by several organic and inorganic compounds. The sea mussel *Mytilus edulis*, for example, shows marked MT induction upon exposure to cadmium and other heavy metals [12]. Also in crustaceans, MT is induced by metals and represents a biomarker in environmental toxicology, even though in these invertebrates MT may also function as donor of copper to the respiratory protein hemocyanin. MT induction has been widely studied also in fish, in particular in salmonids and nothotenoids, upon exposure to metals and oxidants [13–15].

2. NON-MAMMALIAN METALLOTHIONEINS

The first invertebrate MTs were identified in two aquatic molluscs, the American oyster *Crassostrea virginica* [16] and the limpet *Patella vulgata*

[17]. In 1979, a cadmium-binding metallothionein was isolated from the sea mussel *Mytilus edulis* [12]. Later, new MTs were identified in other molluscs, crustaceans, and echinoderms. Since many of the proteins isolated from invertebrates are rather different from the typical horse kidney MT, firstly they were referred to as 'MT-like proteins'. However, during the Second International Meeting in 1985 it was decided that all proteins resembling the equine renal MT should be named 'metallothioneins'.

Most higher organisms possess two major cadmium/zinc-thionein isoforms whose expression is regulated by a zinc-activated transcription factor (MTF-1). On the other hand, in some unicellular eukaryotes, such as yeasts, copper-thioneins were identified whose synthesis is selectively induced by copper through a copper-activated transcription factor. Recently, the expression of MTs with different metal binding preferences (cadmium/zinc or copper) has been reported also in metazoans, such as crustaceans and molluscs [18,19]. This fact seems to be related to the physiological role exerted by MTs in these invertebrates, where zinc- and copper-thioneins may act as storage forms and metal donors for different metalloproteins, such as carbonic anhydrase (zinc) and hemocyanin (copper). However, in other crustaceans, as the shore crab *Carcinus maenas*, only one MT isoform has been found, and this isoform is induced by zinc, cadmium, and copper [20].

The existence of at least two interactive pools of MTs has been described in the cytosol of almost all organisms: the inducible and the constitutive MT isoforms. These two pools have been observed also in MTs from aquatic species, and they may play different roles. One pool may concern physiological metal regulatory processes, and the other transient changes in free concentration of metals depending on environmental fluctuations.

3. ASPECTS OF METALLOTHIONEIN FUNCTION IN AQUATIC ORGANISMS

3.1. Heavy Metals in Seawater and Freshwater

The so-called heavy metals include zinc, copper, mercury, cadmium, silver, nickel, lead, manganese, chromium, etc. All heavy metals are minor yet ubiquitous components of the biosphere; some of the heavy metals for which no biological function has been identified are considered as xenobiotic (like Cd, Ag, Hg, Pb, and Cr), others are essential for growth and development (like Cu, Zn, Mn) [21].

Heavy metals are usually present in tissues as divalent cations that can be free or complexed to biological ligands. Among the heavy metals, copper shows the interesting characteristic that it may change its oxidation state from 2 to 1. Although heavy metals can bind many chemical groups, several of them have a pronounced preference for sulfhydryl or thiolate groups of polypeptides [22].

Heavy metals are present in seawater and freshwater in different chemical forms resulting from the equilibrium between metal ions and inorganic or organic complexes. There is increasing evidence that metal toxicity depends mainly on the activity of the free metal cations. Metals can enter the organism from the surrounding medium or from food, and different metabolic pathways allow the organisms to utilize, sequester or eliminate them. Metal contamination of aquatic environments constitutes an ever-increasing problem in industrialized countries and encourages research activities aimed at evaluating the effects on the organisms as well as at identifying the molecular mechanisms supporting the protective response.

3.2. Metal Binding and Metal Cluster Organization in Non-mammalian Metallothioneins

MTs can bind monovalent, divalent or trivalent metal ions, such as Cu^+, Ag^+, Zn^{2+}, Cd^{2+}, Hg^{2+}, Sb^{3+}, and Bi^{3+}. Typically, the sulfhydryl groups of MTs bind divalent ions with tetrahedral geometry, and monovalent ions with trigonal coordination [1].

In vitro experiments demonstrated that the affinity of MTs for the most common heavy metals is typically $Hg^{2+} > Cu^+$, Ag^+, $Bi^{3+} \gg Cd^{2+} > Pb^{2+} > Zn^{2+} > Co^{2+}$ [23]. Experiments performed on invertebrates confirmed the above rank of affinity also in mussels, and demonstrated that *in vivo* Cd^{2+}, Cu^+ or Hg^{2+} may displace Zn^{2+} from MTs [24]. However, in mussel digestive gland, Cu^+ is able to displace Zn^{2+}, but not Cd^{2+}, from cadmium- or zinc-thioneins thus indicating that the rank of affinity between metal and sulfhydryl groups is not fully respected *in vivo* [25].

MTs purified from the digestive gland of the crab *C. pagurus* after cadmium injection are mostly complexed with Cd^{2+} and contain only small amounts of Zn^{2+} [26]. Reports on copper release are less clear; the hypothesis that MTs play a role in copper insertion in metalloproteins has been verified *in vitro* on the lobster *H. americanus* [27].

Hemocyanin, the blood oxygen carrier of many molluscs and arthropods, contains two Cu^+ ions in its active site and represents a potential acceptor of copper released from MTs. Among the three different MT isoforms isolated from lobster, only the MT-3 isoform is able to transfer Cu^+ to the active site of apo-hemocyanin [28]. This MT-3 is chemically distinct from the other two; Valls and coworkers [29] proposed that two ancestral MT lineages (copper-thionein *versus* zinc-thioneins) were already present in the early stages of eukaryotic evolution. Moreover, they suggested that the *in vivo* metal preferences of a MT protein could be inferred from the primary sequence patterns. Following this approach, the crustacean zinc-thioneins cluster together and separate from the crustacean copper-thioneins, which instead gather with yeast (*S. cerevisiae* MT CUP1) and arthropoda (*Drosophila* MTN) copper-thioneins [29].

All mammalian MTs have a monomeric dumbbell shape composed of two metal clusters connected by a flexible linker consisting of a Lys-Lys segment. The N-terminal β-domain binds three divalent metal ions; the C-terminal α-domain binds four divalent ions [9]. The β-domain mainly regulates zinc and copper homeostasis, whereas the α-domain may play a central role in heavy metal detoxification [30].

A number of structural studies performed by NMR spectroscopy [6,31,32] and X-ray crystallography [33] showed a constant presence of diamond-shaped cluster structures also in almost all MT proteins from aquatic species. MTs of fish and of some invertebrates display a molecular conformation very similar to that observed in mammalian MTs. On the other hand, in some invertebrates, the α-domain is N-terminal encoded, in contrast to vertebrates where it is C-terminally encoded.

4. METALLOTHIONEINS FROM FISH

4.1. Fish Metallothionein Genes

The structure of MT genes is evolutionary conserved in eukaryotes, even though differences in the number and complexity of genes are observed in different species. In different taxonomic groups, the MT genes may display distinctive features, including a variable number of amino acid substitutions at the positions not occupied by cysteines, different 3′- and 5′-UTR sequences, and a variable length [3].

Fish MTs have been less extensively studied than those of other taxonomic groups. Only recently, studies have been carried out on cDNA or genomic DNA genes from many bony fish species including carp [34,35], goldfish [36], rainbow trout [37], and plaice [38]. By contrast, few reports on MT genes in cartilaginous fish are available, and, only recently, a MT gene from the tiger shark *Scyliorhinus torazame* has been cloned [39].

In mammals, a characteristic of the MT promoters is the presence of multiple metal-responsive elements (MREs) mediating the response to metal ions. MREs have been found in the MT promoters of several teleosts, including the rainbow trout *Oncorhynchus mykiss* [40–43], the pike *Esox lucius* [44], the common carp *Cyrpinus carpio* [36], the zebrafish *Danio rerio* [45], the icefish *Chionodraco hamatus* [46], and the tilapia *Oreochromis mossambicus* [47]. There is a strong resemblance between fish and mammalian MREs, even though the number of MREs in fish MT promoters is rather variable, and almost all fish MT promoters have a dual cluster arrangement of MREs [47].

In fish, the MT system is less complex than in human and other mammals, and consists of one or two genes [40]. Salmoniforms contain two distinct MT genes [37], which are differentially regulated depending on the heavy metal and the developmental stage [14,15]. Among perciforms, only one MT gene has been identified in tilapia [48], whereas two distinct loci *MT-1* and *MT-2* have been identified in notothenioid fish endemic to Antarctic waters [49,50]. In the icefish *C. hamatus*, a member of the major taxonomic group of Antarctic fish, the MT-2 transcript is expressed constitutively, while the MT-1 transcript accumulates in response to cadmium [49]. These isoforms differ in a single amino acid substitution at position 26 (Lys-Thr), which confers a difference in net protein charge. The two MT genes of *C. hamatus* show just a 57% identity. The *MT-1* promoter contains four MREs located in a single proximal cluster within the first 300 bps from the ATG codon. The *MT-2* promoter contains, instead, seven MREs organized into two clusters: a proximal cluster containing three MREs close to the ATG codon, and a distal cluster containing four MREs laying 500-900 bps upstream the ATG. The seven MREs confer to *MT-2* gene the high susceptibility to zinc and cadmium [43,51]. Also in the crucian carp *C. cuvieri* the MT-2 promoter contains seven MREs located in two clusters [51]. Differently from icefish, in *C. cuvieri* also the *MT-1* promoter contains a distal and a proximal MRE cluster, with the last one being the main responsible of the metal induction [52].

Moreover, some interesting differences can be identified between mammalian and fish MT promoters. First, mammalian MT promoters do not contain distal MRE sequences that are instead present in fish MT promoters. Second, mammalian MT promoters are about 75% GC-rich, whereas those of fish are usually AT-rich [40].

Since the couples of MT genes in notothenioids and salmonids display different mechanisms of transcriptional regulation, it has been hypothesized that the two MT genes of notothenioids are the result of a duplication event that occurred in the ancestral lineage of these fish, independently of that which gave origin to the MT isoforms in salmonids [50].

4.2. Fish Metallothionein Proteins: Sequence, Reactivity, and Chemical-Physical Features

As in mammals, fish MTs contain twenty Cys residues and lack aromatic residues. However, fish MTs show a number of distinctive features in the primary structure with respect to their mammalian counterparts (Figure 1). Although most of the amino acid substitutions occurring in fish MTs are identified in the N-terminal region that exhibits species-specific features [53],

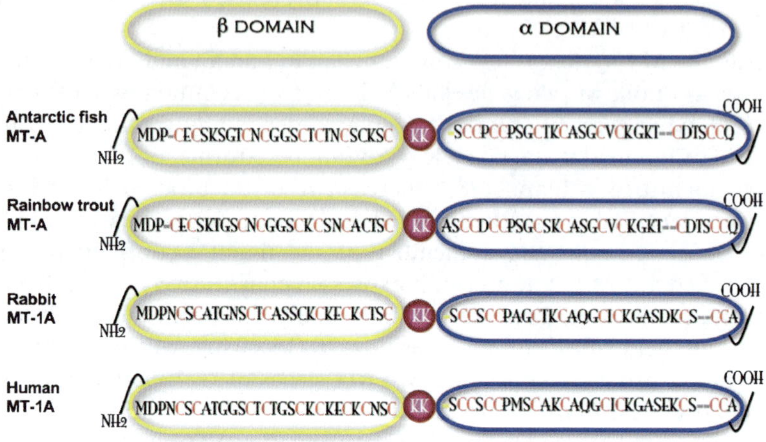

Figure 1. Primary structure comparison of fish and mammalian MTs. Sequence alignment was performed using the program CINEMA (Colour INteractive Editor for Multiple Alignments-Version 2.0), a tool for visualization and manipulation of DNA sequences in the Utopia software package (available at http://utopia.cs.man.ac.uk). As representative of fish and mammalian MTs two fish MT-A isoforms (one from the rainbow trout *O. mykiss* and one from the Antarctic fish *N. coriiceps*) and two mammalian MT-1A isoforms (one from rabbit and one from human) were chosen. In the sequence, Cys residues are labelled in red. The back shift of the ninth Cys located in the C-terminal domain of the fish MTs is shown.

a few but significant variations are found in the remaining portion of the protein. Paradigmatic is the displacement of the ninth cysteine located in the C-terminal domain (Figure 1), together with a lower number of lysines juxtaposed to cysteines, described in both the Antarctic fish *Notothenia coriiceps* [54] and the rainbow trout *Oncorhynchus mykiss* [55]. The main consequence of the cysteine displacement is a different orientation of the K50-G51-K52-T53 loop in the fish MT with respect to the corresponding K50-G51-A52-A53 motif in mouse MT [54].

In line with the high sequence homology, circular dichroism (CD) spectra of fish MTs are rather similar to those of mammalian MTs. MT protein is known to lack substantial secondary structure in the absence of metals and to assume a stable folding only after binding of the correct number of metal equivalents [56]. After cadmium complexation, fish MT displays a positive band around 260 nm, followed by a negative band at 245 nm and a positive peak at 228 nm (Figure 2). As in mammalian MTs, the CD spectrum of fish MT reveals no well-defined secondary structure elements, such as canonical α-helices and β-sheets [57,58].

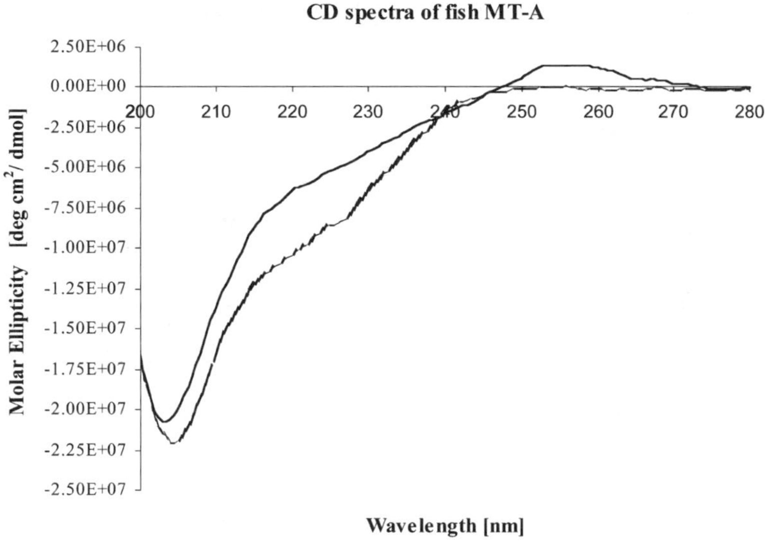

Figure 2. Circular dichroism analysis of *O. mykiss* MT-A. CD spectra were acquired in the near UV region (from 190 to 290 nm) for the recombinant Cd_7-MT-A form (solid line) and for the corresponding apo-thionein form (broken line) obtained by acidification with HCl as described in [58].

The few amino acid substitutions occurring in fish MTs with respect to their mammalian counterparts are able to modify thiol reactivity, thermal stability and metal exchangeability of the protein. In the icefish MT a lower number of accessible/reactive Cys residues was measured with respect to mouse MT; the increased reactivity of thiol groups may depend on the localization of the negative charges on the protein surface, that may hamper the interaction with the anionic reagent DTNB used for the assay [59].

Moreover, fish MTs are less thermostable than mouse MT, as revealed by absorption and CD spectroscopy studies. In fact, when the temperature was increased up to 90°C, a decrease in the absorbance signal characteristic of the cadmium-thiolate clusters was recorded (Figure 3A), and the decrease was more pronounced for fish than for mouse MT, thus suggesting that fish MTs are more sensitive to temperature than their mammalian counterparts [57,58].

Despite the high thermodynamic stability, the metal clusters of MT are usually characterized by a remarkable kinetic reactivity. This fact indicates a dynamic active state leading to a continuous redistribution of the metal ions inside and between the clusters, in spite of the structural rigidity of the molecule. With respect to mammalian MTs, fish MTs display a more

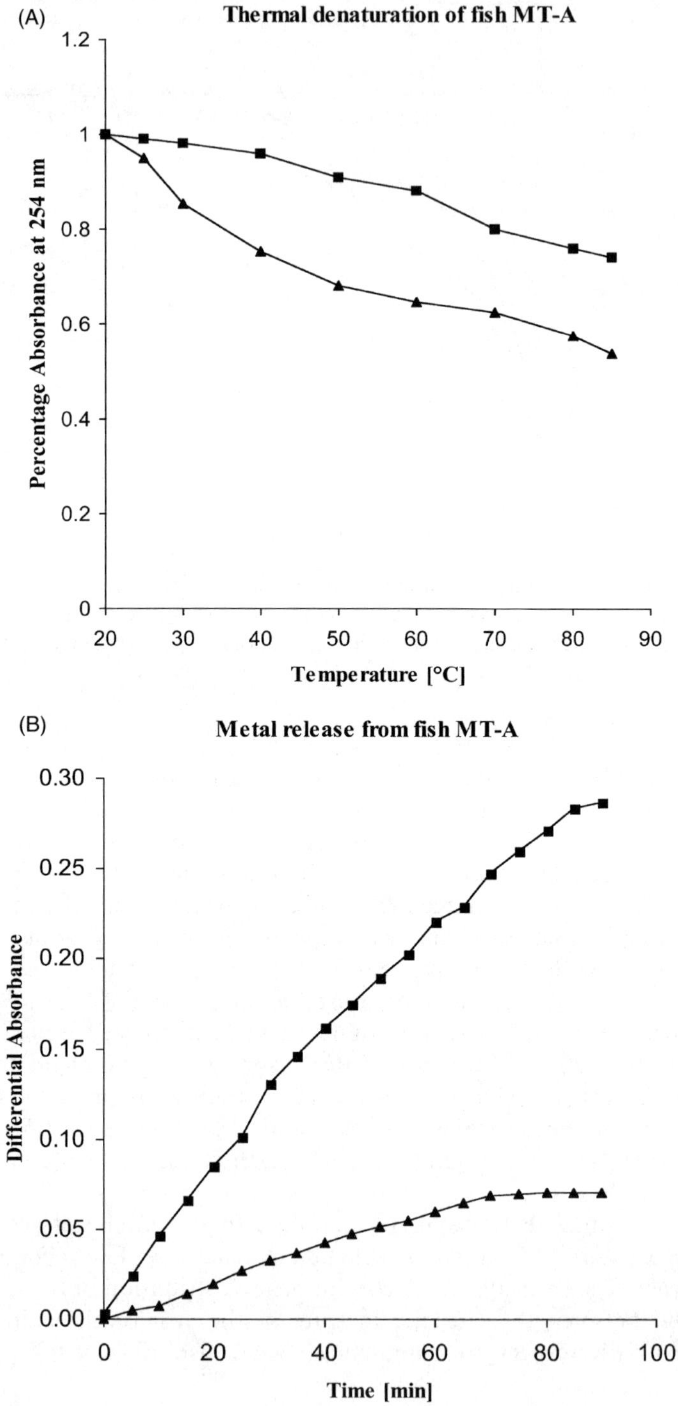
(A)
Thermal denaturation of fish MT-A
Percentage Absorbance at 254 nm
1.2
1
0.8
0.6
0.4
0.2
0
20
30
40
50
60
70
80
90
Temperature [°C]
(B)
Metal release from fish MT-A
Differential Absorbance
0.30
0.25
0.20
0.15
0.10
0.05
0.00
0
20
40
60
80
100
Time [min]

marked release of zinc and cadmium upon cysteine oxidation (Figure 3B). This result demonstrates a greater reactivity of the fish MT against oxidant compounds with respect to the mouse counterpart, as well as a higher metal mobility [58].

Also chemical-physical parameters of fish MTs reveal important differences with respect to those of mammalian MTs. Fish MTs have a lower hydropathic index (a parameter inversely correlated to the flexibility of the polypeptidic chain) with respect to mammalian MTs [7,58]. This feature may be important in functional terms because a higher flexibility can guarantee the conformational changes necessary for the protein for maintaining a good functionality at the low temperatures that fish may experience in aquatic environments [60]. The values of the hydropathic index calculated for different fish MTs are reported in Table 1. In all fish MTs, the hydropathic index is negative, whereas in mammalian MTs this index is usually positive assuming an average value of 0.098. This fact indicates that fish MTs are less hydrophobic and more flexible than their mammalian counterparts [61].

The higher flexibility of MT from the icefish *N. coriiceps* in comparison with mammalian MTs was experimentally confirmed by time-resolved dynamic fluorescence, infrared spectroscopy [59], and NMR spectroscopy [7] studies. Fluorescence measurements showed that fish MT binds a fluorescent probe selective for hydrophobic sites less strongly than mouse MT; NMR analyses unravelled a selective broadening of the heteronuclear spectrum in the case of fish MT with respect to murine MT. Moreover, infrared spectroscopy studies demonstrated for the fish MT a larger accessibility to the solvent [59].

It can be postulated that the greater flexibility of the fish MTs with respect to their mammalian counterparts may respond to a functional request, since it is generally accepted that flexible proteins are more adapted to cope with the effects of low temperatures, whereas thermophilic proteins possess a more rigid structure. For this reason, proteins from cold-adapted species

Figure 3. Functional properties of *O. mykiss* MT-A. (**A**) Effect of increasing temperature: Absorption UV spectra were acquired in the range 220–300 nm for the recombinant fish Cd_7-MT-A (▲) exposed to a thermal gradient (temperature increase from 20 to 90 °C) as described in [58]. The absorbance decrease at 254 nm of the cadmium-thiolate chromophore is plotted as a fraction of the standard absorbance (absorbance at room temperature). For comparison the thermal denaturation profile of rabbit MT-1 (■) is also reported. (**B**) Kinetics of zinc and cadmium release: The kinetics of zinc (▲) and that of cadmium (■) release were acquired for the recombinant fish Cd_7-MT-A following the formation of the metal-$(PAR)_2$ complex at 500 nm, as described in [58]. Each experimental point represents the difference between the absorbance measured in the presence and absence of the proper redox couple: GSH/GSSG for zinc and H_2O/H_2O_2 for cadmium.

Table 1. The values of the hydropathic index for different fish MTs have been calculated by summing the hydropathic values of all the amino acids and dividing by the number of residues in the proteins. The hydropathic values have been obtained by changing the sign of the hydrophilicity index determined with the program MacVector (Oxford Molecular Group) as described in [61].

	Hydropathic Index
Danio rerio	−0.108
Carassius auratus	−0.109
Lex lucius	−0.167
Oncorhynchus mykiss	−0.204
Zoarces viviparus	−0.243
Chionodraco hamatus	−0.246
Chionodraco rastrospinosus	−0.246
Notothenia coriiceps	−0.246
Trematomus bernacchi	−0.246
Gymnodraco acuticeps	−0.246
Oreochromis mossambicus	−0.249
Pleuronectes platessa	−0.251
Gadus morhua	−0.329

often display a lower thermostability compared to that of their mesophilic counterparts.

4.3. Three-dimensional Structure of Fish Metallothioneins

^{113}Cd-NMR spectroscopy experiments demonstrated that the Cd_7-MT from mouse is an asymmetric molecule with the N-terminal domain containing 3-cadmium/9 cysteines, and the C-terminal domain containing 4-cadmium/11 cysteines [62,63].

Up to this time, the only three-dimensional structure of a fish MT available on the PDB is that of Cd_7-MT-A of *N. coriiceps* (Figure 4). The structure has been solved by homonuclear [^{1}H,^{1}H] and heteronuclear [^{1}H,^{113}Cd] NMR correlation experiments performed on the recombinant protein complexed with cadmium [7]. In the icefish MT-A, the architecture of the two domains resembles closely that of mouse Cd_7-MT. A peculiarity of fish MT, with no counterpart in mouse MT, is given by the presence of a α-helical region spanning residues Thr41-Ala44, of a 3_{10} helix from Val48 to Lys50 in the α-domain, and by the presence of an α-helical stretch spanning Glu5-Ser9 in the β-domain. The back shifting by two positions of Cys54 in fish MT, with respect to Cys56 of mammalian MT, displaces the loop in the

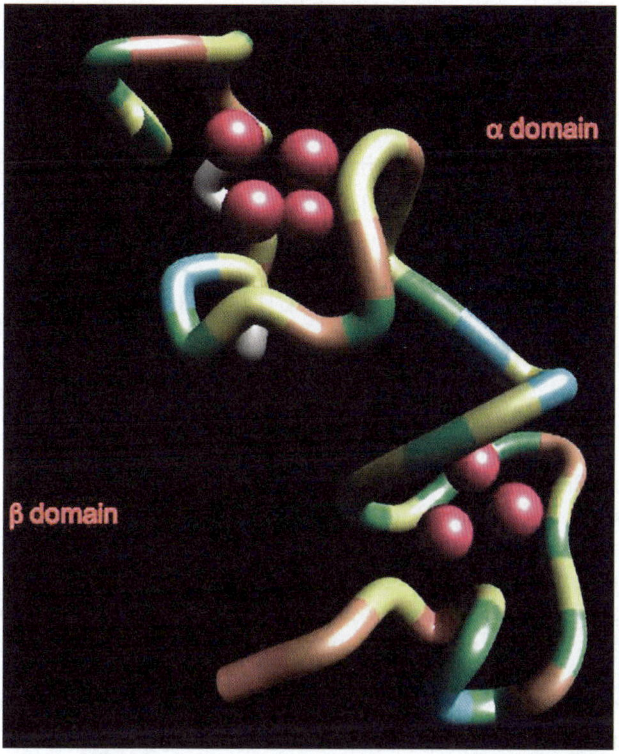

Figure 4. The 3D solution structure of *N. coriiceps* MT-A. The three-dimensional NMR structure of fish Cd_7-MT has been solved by homonuclear [^1H,^1H] and heteronuclear [^1H,^{113}Cd] NMR correlation experiments [7]. The N-terminal β-domain comprises residues 1–28 and a three-metal cluster. The C-terminal α-domain comprises residues 31–60 and four-metal cluster. The Lys29-Lys30 representing the linker segment has been positioned to connect the two domains, despite the lack of NMR information in this regard. Cadmium ions are shown as spheres. The color scheme is the following: blue for polar positive; red for polar negative; green for polar neutral; white for non-polar aliphatic; brown for non-polar P, G; yellow for C. The figure has been adapted from entries made in the Brookhaven Protein Data Bank (ID 1MOG and 1MOJ). The figure was prepared using Ambrosia Molecular Structure Viewer program, a tool of molecular visualization in the Utopia software package (available at http://utopia.cs.man.ac.uk).

α-domain leading to a different distribution of charges on the surface. Moreover, the β-domain of *N. coriiceps* MT hosts an extra charged residue (Glu5), leading once again to a different surface charge distribution with respect to mammalian MTs, that may be responsible for the above described differences in the kinetics of ion exchange and reactivity between fish and mammalian MTs [7].

Another interesting difference between fish and mammalian MTs is the different dynamical behavior between the two domains. In fact, the $[^1H,^{113}Cd_7]$ correlation NMR spectra acquired for *N. coriiceps* MT reveal for the two domains a different intensity of the $[^1H,^{113}Cd_7]$ correlation signals, and this difference is much more pronounced in fish than in mammalian MTs [7].

4.4. Fish Metallothionein Isoforms and Roles

When compared with the multiplicity of mammalian MTs, teleosts possess only two distinct MT isoforms, MT-1 and MT-2 (commonly referred as MT-A and MT-B, respectively). Multiple alignment for the amino acid sequences of different MT-A and MT-B isoforms from different teleosts available on GenBank and EMBL is reported in Figure 5. All the teleost MT-Bs possess 60 amino acids; also the majority of MT-As contain 60 amino acids, with some exceptions. In fact, three MT-As from the salmonids *Oncorhynchus mykiss*, *Salmo salar*, and *Salvelinus alpinus*, and the MT-A from the ayu *Plecoglossus altivelis* have an additional residue at the boundary between β- and α-domains [39].

In fish, the two MT isoforms are differentially regulated by heavy metals. In rainbow trout, MT-B is expressed constitutively, whereas MT-A preferentially accumulates in response to cadmium exposure [13]. In the icefish, a similar behavior is observed for MT-2 and MT-1, respectively [49].

As mammals, fish are able to synthesize MT in response to endogenous and environmental stimuli, and metals are strong inducers of MT expression in different fish cell lines [13,15]. However, a recent study of Bourdineaud and coworkers [64] suggested that, under low contamination pressure, the physiological role of the transcriptional induction of MTs by metals is to compensate for MT degradation, rather than to increase the level of metal sequestration capacities, as it was generally thought. This conclusion was inferred from the observation that no increase in MT proteins was observed in the fish organs, despite the induction of the MT genes. This result is in line with a previous report [65] showing that metals bound to MTs are transported to the lysosomes, thus increasing proteolysis of MTs.

With regards to the free radical scavenging activity of fish MT, no conclusive evidence has so far been obtained *in vivo*, even though many studies indicated that fish MT has antioxidant properties *in vitro* [58].

A last consideration can be made about Antarctic fish. These fish are characterized by genetic isolation and cold-adaptation, and display unique phenotypes that allow them to maintain normal metabolic functions in the absence of both erythrocytes and hemoglobin [66]. Moreover, icefish

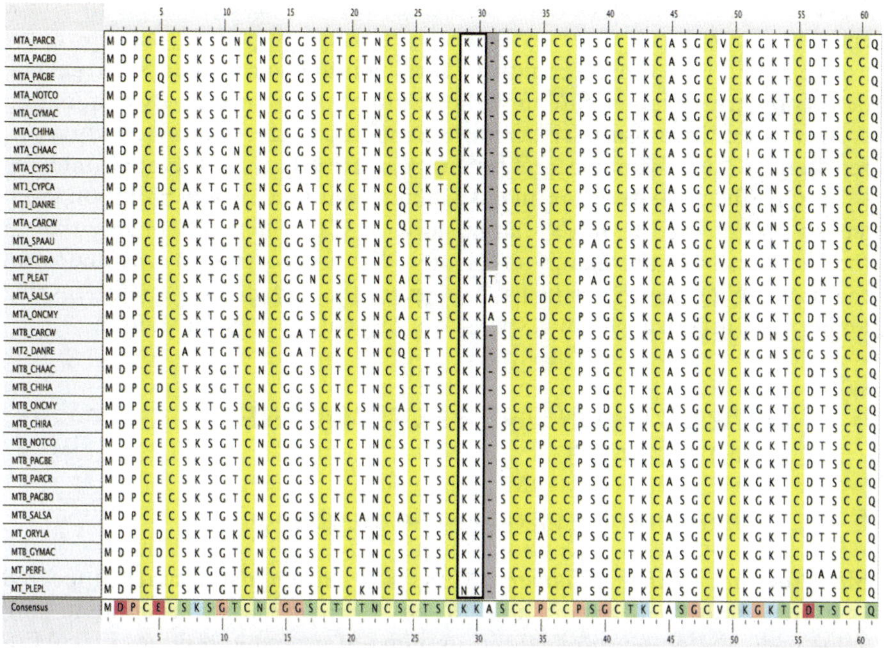

Figure 5. Multiple sequence alignment of different fish MTs. The sequence alignment was performed using the free program CINEMA Version 2.0. From top to bottom the first 16 sequences correspond to the MT-A/MT-1 isoforms of: *P. charcoti* (MTA_PARCR); *P. borchgrevinki* (MTA_PAGBO); *P. bernacchii* (MTA_PAGBE); *N. coriiceps* (MTA_NOTCO); *G. acuticeps* (MTA_GYMAC); *C. hamatus* (MTA_CHIHA); *C. aceratus* (MTA_CHAAC); *Cyprinodon sp* (MTA_CYPS1); *C. carpio* (MT1_CYPCA); *D. rerio* (MT1_DANRE); *C. cuvieri* (MTA_CARCW); *S. aurata* (MTA_SPAAU); *C. rastrospinosus* (MTA_CHIRA); *P. altivelis* (MT_PLEAT); *S. salar* (MTA_SALSA); *O. mykiss* (MTA_ONCMY). The next 13 sequences correspond to the MT-B/MT-2 isoforms of: *C. cuvieri* (MTB_CARCW); *D. rerio* (MT2_DANRE); *C. aceratus* (MTB_CHAAC); *C. hamatus* (MTB_CHIHA); *O. mykiss* (MTB_ONCMY); *C. rastrospinosus* (MTB_CHIRA), *N. coriiceps* (MTB_NOTCO); *P. bernacchii* (MTB_PAGBE); *P. charcoti* (MTB_PARCR); *P. borchgrevinki* (MTB_PAGBO); *S. salar* (MTB_SALSA); *O. latipes* (MT_ORYLA); *G. acuticeps* (MTB_GYMAC). The last 2 sequences correspond to the MTs of *P. fluviatilis* (MT_PERFL) and *P. platessa* (MT_PLEPL). The alignment is colored according to residue type: blue for polar positive; red for polar negative; green for polar neutral; white for non-polar aliphatic; brown for non-polar P, G; yellow for C. The two Lys residues of the linker segment are boxed.

represent an interesting example of 'life without hepatic MT' since they lack endogenous MTs in liver [67]. These unusual organisms have evolved astonishing adaptation mechanisms, turning into an advantage some genetic defects otherwise incompatible with the life of other species. By eliminating erythrocytes, these fish have achieved a marked decrease in blood viscosity, resulting in an advantage at low temperatures of their life environments (–1.9°C). Since Antarctica is an environment with a very low pollution level, icefish do not need to have a large store of MT in liver. It is likely that in these species MTs are mainly involved in the homeostasis of zinc and/or in the scavenging of free radicals. However, the icefish liver maintains the ability to synthesize MT in response to internal or external stimuli.

5. METALLOTHIONEINS FROM MARINE CRUSTACEANS

5.1. Crustacean Metallothionein Sequences

Since the time of their discovery, MTs from marine invertebrates appeared to differ from the vertebrate MTs for the lower cysteine content and, sometimes, the presence of minor amounts of aromatic residues [27]. Multiple alignment for the crustacean MT sequences available on GenBank and EMBL is reported in Figure 6. Most crustacean MTs have only 18 cysteine residues within their sequence, two less compared to the 20 cysteines of mammalian MTs. Exceptions are represented by the 21 cysteines observed in three crustacean MTs (the copper MT of *Callinectes sapidus*, and both MTs of *Portunus pelagicus*), and by the 19 cysteines observed in the MTs of *Panulirus argus, Carcinus maenas, Portunus pelagicus, Necora puber, Cancer pagurus, Eriocheir sinensis.*

Moreover, crustacean MTs typically contain five or six prolines per polypeptide chain, a value definitely higher than the two prolines characteristic of mammalian MTs. Since proline residues can readily accommodate turns and bends in a polypeptide chain and stabilize the secondary structures, α-helix in particular, it is likely that higher proline content may affect the three-dimensional organization of the protein.

The first crustacean MT was isolated from the portunid crab *Scylla serrata* [68]. Another MT isoform was subsequently identified in the same organism [69]. The two isoforms are usually referred to as MT-1 and MT-2; they contain 58 and 57 amino acids, respectively, and show a sequence identity of 83%. In the next decades, one MT isoform from the crab *Carcinus maenas* [20], two isoforms from the blue crab *Callinectes sapidus* [18] and from the tropical crab *Portunus pelagicus* [70] were isolated. All these crab species are marine and show a high degree of sequence homology in their MT isoforms.

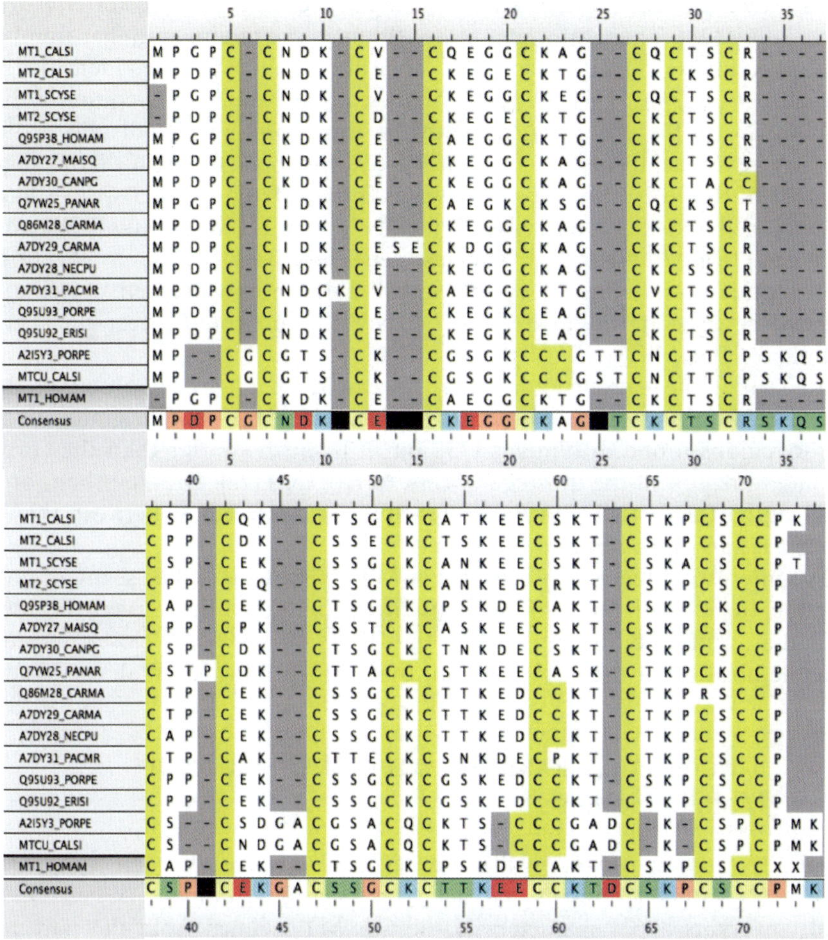

Figure 6. Multiple sequence alignment of different crustacean MTs. The sequence alignment was performed using the free program CINEMA Version 2.0. From top to bottom the sequences correspond to. MTs of the following organisms: *C. sapidus* (MT1_CALSI); *C. sapidus* (MT2_CALSI); *S. serrata* (MT1_SCYSE); *S. serrata* (MT2_SCYSE); *H. americanus* (Q95P38_HOMAM); *H. americanus* (MT1_HOMAM); *M. squinado* (A7DY27_MAISQ); *C. pagurus* (A7DY30_CANPG); *P. argus* (Q7YW25_PANAR); *C. maenas* (Q86M28_CARMA); *C. maenas* (A7DY29_ CARMA); *N. puber* (A7DY28_NECPU); *P. marmoratus* (A7DY31_ PACMR); *P. pelagicus* (Q95U93_PORPE); *E. sinensis* (Q95U92_ERISI); *C. sapidus* (MTCU_CALSI); *P. pelagicus* (A2I5Y3_PORPE). The alignment is colored according to residue type: blue for polar positive; red for polar negative; green for polar neutral; white for non-polar aliphatic; brown for non-polar P, G; yellow for C.

In 1989 a copper-thionein was identified in the American lobster *Homarus americanus* [28]. Then, another copper-thionein containing a higher number of Cys residues (21) with respect to other crustacean MTs was identified in the crab *C. sapidus* [71]. Only 13 of these cysteines share a common distribution with the other crustacean MTs, thus pointing to a distant evolutionary relationship of this MT with the non-copper-thionein of the same organism. Recently, other two MT genes from different lobster species have been cloned and characterized: a MT-cDNA from *H. americanus* [29], and a MT gene from *P. argus* [72]. This last gene putatively encodes a 59 residue polypeptide containing 19 cysteines; the extra non-conserved cysteine is located at the C-terminal domain of the MT and could affect the metal affinity of this MT.

Regarding the gene structure of crustacean MTs, it has been observed that the 5′-UTR region of the lobster *H. americanus* (27 bp) is quite similar to 5′-UTR regions found in humans [73], but it is much shorter than the corresponding region of the lobster *P. argus* MT (67bp) [72]. On the other hand, the length of the 3′-UTR in *P. argus* (451 bp) is intermediate between that of *H. americanus* (595 bp) and humans (200 bp) [72].

An interesting feature of crustacean MTs is that the linker segment connecting the α- with the β-domain often contains two uncharged residues, including a proline, instead of two Lys, that could favor a closer and less flexible position of the domains than that observed in mammalian MTs.

5.2. Three-dimensional Structure of Crustacean Metallothioneins

Typically, mammalian Cd_7-MTs are asymmetric molecules consisting of a smaller N-terminal domain and a larger C-terminal domain. When Otvos and coworkers [9,62] characterized the Cd_7-MT isolated from the portunid crab *S. serrata,* they observed similarities with the mammalian forms, but also an important difference. This crustacean MT, in fact, displays only 18 Cys residues and 6 metal ions bound per molecule, and assumes a symmetric conformation where each domain consists of 9 cysteines binding 3 metal ions.

Further studies performed on the crab *C. pagurus* [10], and on the lobster *H. americanus* [28], confirmed a stoichiometry of 6 metal ions and 18 cysteine residues per thionein molecule, thus pointing at the symmetrical 3-metal clusters (called β_N- and β_C-domain, respectively) as an intrinsic property of the MTs from marine crustaceans. It is likely that the crustacean MTs with the two 3-metal domains could be less stable than the vertebrate MTs with their 3- and 4-metal domains.

The first 3D structure of a crustacean MT solved by heteronuclear [^{1}H,^{113}Cd] and homonuclear [^{1}H,^{1}H] NMR correlation experiments was that of *C. sapidus* MT-1 [4]. A recent study reported the structural characterization of the recombinant MT from the lobster *H. americanus* produced in *E. coli* using zinc- and copper-supplemented media [29]. The M_6-MT stoichiometry was found also for the lobster MT synthesized in zinc-supplemented medium (Zn_6-MT), thus confirming the previous studies performed on the Cd_6-MT prepared *in vitro* from the apo-thionein [74]. Comparison of the CD spectrum of the entire Zn_6-MT with the spectrum obtained by summing the spectrum of the $Zn_3\beta_C$-domain and that of the $Zn_3\beta_N$-domain revealed some slight differences due to a mutual interaction between the domains stronger than that described for mammalian MTs [75].

In recent years, the 3D solution structure of the Cd_6-MT from the lobster *H. americanus* (Figure 7) has been determined by Munoz and coworkers [5]. The 3D structure was solved by NMR experiments for both the native lobster protein, and for the independent C-terminal (β_C) and N-terminal (β_N) domains. Both the crab $Cd_3\beta_N$- and $Cd_3\beta_C$-domain have structures much better defined than the classical rabbit $Cd_3\beta_N$-domain. Moreover, the amide 1D-NMR spectra become progressively broader in the series $Cd_3\beta_C$(crustacean), $Cd_3\beta_N$(crustacean) and $Cd_3\beta_N$ (rabbit), suggesting that the structure of crustacean C-terminal domain is less flexible than the N-terminal domain, and both of them are less flexible than the mammalian N-terminal domain [5]. Overall, these studies demonstrate a higher rigidity for the crustacean MTs in comparison with the mammalian MTs. Similarly to mammalian MTs, also the C-terminal domain ($Cd_3\beta_C$) of lobster MT has a short 3_{10} helix between residues 31–33, and an α-helix segment between residues 42–48 [5].

5.3. Crustacean Metallothionein Isoforms

The presence of two naturally occurring MT isoforms, namely MT-1 and MT-2, was described in several crab species [68–70]. When MTs were purified from both gills and digestive glands of crab, two MTs could be separated by gel filtration: one containing cadmium and zinc, and another containing only copper. Also MTs purified from the digestive glands of lobster were separated by gel chromatography into two fractions: a copper-thionein of around 10 kDa, and a major cadmium-thionein of about 25 kDa [27].

Up to now a high variety of metal-thioneins have been identified in crustaceans, and the number of MT genes varies in the different species. From one to three MT encoding sequences have been identified in *C. pagurus* [76], *C. maenas* [20], *H. americanus* [28,77], and *C. sapidus* [18,78].

In species with only one apparent MT isoform, cadmium is the best MT inducer, and homometalllic Cd_6-MTs are generally identified. On the contrary, for those species with established gene multiplicity, such as the

Figure 7. The 3D solution structure of *H. americanus* MT-1. The three-dimensional NMR structures of both $Cd_3\beta$ domains of lobster MT-1 have been solved by NMR experiments [5]. The β_N-domain comprises residues 1-28; the β_C-domain comprises residues 29–58. The two domains are connected, despite the lack of NMR information to this regard. Cadmium ions are shown as spheres. The color scheme is the following: blue for polar positive; red for polar negative; green for polar neutral; white for non-polar aliphatic; brown for non-polar P, G; yellow for C. The figure has been adapted from entries made in the Brookhaven Protein Data Bank (ID 1J5M and 1J5L). The figure was prepared using Ambrosia Molecular Structure Viewer program.

crab *C. sapidus*, cadmium and copper induce different MT isoforms. In particular, *C. sapidus* has two inducible isoforms (MT-1 and MT-2), and a third isoform (MT-3) induced by copper, but not by cadmium [71]. While *MT-1* and *MT-2* genes encode a 58 and 57 amino acid protein, respectively, each with 18 cysteines, *MT-3* encodes a 63 amino acid protein containing 21 cysteine residues. This last isoform is closely related to molluscan MTs, in particular to the copper-thionein identified in the terrestrial snail *Helix pomatia*, whereas MT-1 and MT-2 cluster with other crustacean MTs. The presence of copper-thioneins in both crustaceans and gastropod molluscs, animals that both depend on hemocyanin for oxygen transport, suggests that copper-thioneins may be involved in the regulation of hemocyanin activity.

6. METALLOTHIONEINS FROM MARINE MOLLUSCS

6.1. Metallothioneins in Molluscan Species

Molluscan MTs belong to the Class I of metallothioneins, together with the MTs of vertebrates and crustaceans. However, molluscan MTs show distinctive differences with respect to vertebrate MTs, in particular they have longer sequences, higher glycine content (about 15%), and higher AT-content in their intron sequences (about 70%). On average, molluscan MTs contain approximately 70 amino acids [79], with the exception of MTs from gastropods that typically contain only 65 amino acids due to an abnormal N-terminal region typical of this class of molluscs.

The first MT identified in molluscs was that from a bivalve, the American oyster *C. virginica* [16]. In the next years, multiple MT isoforms were found in *C. virginica*: at first two isoforms were identified by Engel and Bouwer [80], then four MT isoforms were identified by Jenny and coworkers [11].

Two major groups of MTs have been identified in another family of bivalve molluscs, the MT-10 and MT-20 forms reported in the Mytilidae family [81–83]. Moreover, several reports describe an increasing number of MTs from different molluscs including the MTs from the green mussel *Perna viridis* [84], the Turkish clam *Ruditapes decussates* [85], the keyhole limpet *Megathura crenulata* [86], and the Antarctic clam *Laternula elliptica* [87].

Particular attention has been paid on three bivalve molluscs, the Pacific oyster *Crassostrea gigas*, the sea mussel *Mytilus edulis,* and the freshwater mussel *Dreissena polymorpha.* Since these species filter large amounts of water for respiratory and nutritional purposes, they are usually exposed to, and eventually accumulate, different aquatic pollutants, including heavy metals. These three molluscan species express a different number of MT isoforms. Only one isoform (DpMT) of 73 residues has been described in *D. polymorpha* [88], two isoforms (MT-10 and MT-20) in *M. edulis* [83,89] and three isoforms (CgMT-1, CgMT-2 and CgMT-3) in *C. gigas* [90].

The prototypical structure of most bivalve MTs contains a minimum of 21 cysteines arranged in the canonical $\alpha\beta$-domain structure [82,83,91]. However, additional studies demonstrated in bivalve MTs a diversity of structures in which the canonical organization has been modified, e.g., in the $\alpha\beta\beta$-domain of MT from *C. gigas* [91] and in the multiple α-domain of MTs from *C. virginica* [11].

6.2. Metallothioneins in Bivalve Molluscs

6.2.1. Mytilidae

The MTs of the mussels *D. polymorpha* and *M. edulis* display the canonical αβ-domain structure, typical of the majority of MTs. The freshwater mussel *D. polymorpha* expresses only one isoform [88], while the sea mussel *M. edulis* expresses two MT isoforms which differ in both primary sequences and physiological functions: the constitutive isoform MT-10, and the cadmium-responsive isoform MT-20. The two isoforms differ in the apparent molecular mass: on gel-permeation chromatography MT-10 elutes as monomeric form (10 kDa) and MT-20 as a dimer (20 kDa) [81]. In particular, MT-10 contains 73 amino acids (21 cysteines), whereas MT-20 contains 72 amino acids (23 cysteines) [83,92].

The *Mytilus spp* MTs display an identical gene structure consisting of three exons and two introns [93]. The *MT-10* gene (GenBank accession number AJ577130) is 2413 bp long and contains a 222 bp coding region; the two introns have approximately the same size (1177 versus 1014 bps). However, in *Mytilus spp* two intron-free *MT-10* genes have also been described [93]. The *MT-20* gene (GenBank accession number AJ577131) is 1906 bp long and contains a 222 bp coding region. Contrary to what is observed for *MT-10*, in the *MT-20* gene the two introns have different lengths (719 versus 968 bp). The high similarity in the untranslated region between *MT-10* and *MT-20* genes leads to the hypothesis that they could originate from recent gene-duplication events in the *Mytilus* lineage [82].

The multiple alignment of the amino acid sequences for different mussel MT isoforms available on GenBank and EMBL is reported in Figure 8. It reveals many substitutions occurring in MT-10 isoforms of *Mytilus spp*, mainly in the C-terminal region, thereby suggesting that MT-10 displays a high degree of polymorphism in the same species [93]. A high degree of homology is found instead among the amino acid sequences of the various MT-20 isoforms of *Mytilus spp*, where only four amino acid residues vary.

A recent study [94] compares in structural and functional terms the MT-10 and MT-20 isoforms of *Mytilus spp*. In spite of 68.5% identity and identical Cd^{2+} content, the sequence alignment discloses remarkable differences (Figure 9). Firstly, MT-20 presents two additional Cys and one more Lys with respect to MT-10. Secondly, the higher number of charged residues in MT-20 with respect to MT-10 (12 versus 9) accounts for the higher isoelectric point of the former (7.75 versus 7.22). Thirdly, MT-20 has a higher hydropathic index with respect to MT-10 (0.250 versus 0.199), thus indicating a more rigid structure. As mentioned above, the hydropathic index is usually positive in mammalian MTs, whereas it is negative in fish MTs [54]. Therefore, the positive hydropathic index means that mussel MTs are more

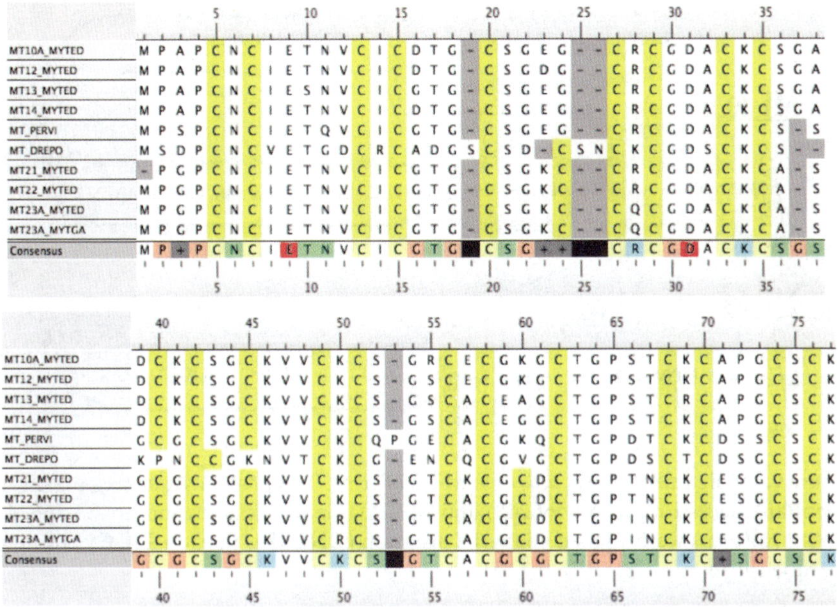

Figure 8. Multiple sequence alignment of different Mytilidae MTs. The sequence alignment was performed using the free program CINEMA-Version 2.0. From top to bottom: four sequences corresponding to different MT-10 isoforms of *Mytilus spp.*; the MT from *P. viridis* (MT_PERVI); the MT from *D. polymorpha* (MT_DREPO); four sequences corresponding to different MT-20 isoforms of *Mytilus spp.* The alignment is colored according to residue type: blue for polar positive; red for polar negative, green for polar neutral; white for non-polar aliphatic; brown for non-polar P, G; yellow for C.

hydrophobic and less flexible than the fish counterparts and resemble mammalian MTs more than fish ones do [94]. An important feature of both mussel isoforms is the lack of the double-lysine segment connecting the α- with the β-domain, which is typical of vertebrate MTs including fish.

Both the MT-10 and MT-20 isoforms were produced as recombinant proteins in *E. coli* and characterized [94]. Circular dichroism analysis of the Cd_7-MTs disclosed marked differences between their secondary structures (Figure 10). Both apo-thioneins display the negative band typical of random coil conformation; after cadmium complexation, a prominent positive band above 252 nm appears, that mainly reflects the chirality of the metal-thiolate bonds. This peak is almost identical in amplitude and shape in the two isoforms, whereas the major differences occur in the high-energy region of

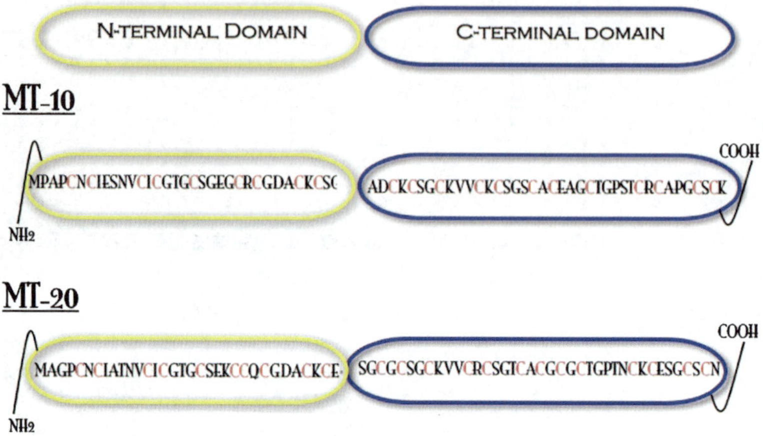

Figure 9. Alignment of *Mytilus spp*. MT isoforms. The protein sequences of MT-10 (GenBank accession no AY566248) and MT-20 (GenBank accession no AY566247) mussel isoforms were aligned using the program CINEMA-Version 2.0. Cys residues are labelled in red.

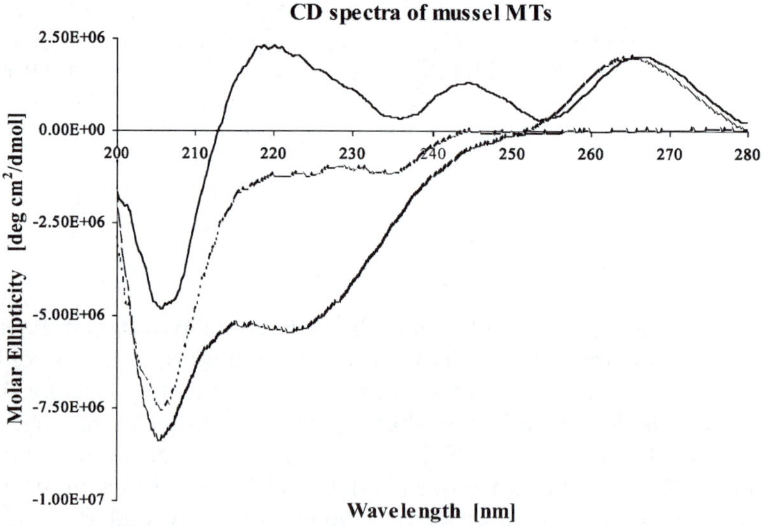

Figure 10. Circular dichroism analysis of *Mytilus spp*. MT isoforms. CD spectra were acquired in the near-UV region (from 190 to 290 nm) for the two recombinant Cd_7-MT-10 (broken line) and Cd_7-MT-20 (solid line) isoforms from the mussel *Mytilus spp* as described in [94]. The metal-free protein (dashed line) were obtained by acidification with HCl.

the spectrum, where the contribution from the polypeptide chain transitions is maximal [95]. Below the crossover at 252 nm, the spectrum of Cd_7-MT-10 (Figure 10) shows a sharp negative band with two shoulders, whereas a positive band with two maxima at 243 and 218 nm, and a shoulder at 235 nm, is found in the CD spectrum of Cd_7-MT-20; below 213 nm, both spectra appear very similar (Figure 10). Even more marked differences are evident when the CD spectra of mussel MTs are compared to the classical CD spectra of vertebrate MTs (Figure 2). These data seem to indicate that the scaffold of the metal-thiolate cluster organization is rather preserved philogenetically from vertebrate to mussel MTs, whereas the folding of polypetide chains seems to be rather variable depending on the kind of MT [94].

The structural differences affect protein stability and function, with MT-10 displaying a higher thermal stability and a better ability to release cadmium ions upon oxidation with respect to MT-20. When the UV spectra are acquired after exposure to a thermal gradient (Figure 11A), the comparison of the denaturation profiles of MT-10 and MT-20 points at a more stable structure for the constitutive isoform MT-10 than for the cadmium-responsive MT-20. Also the kinetics of metal release upon oxidation are markedly different between the two isoforms. MT-20 shows a reduced capability to release Cd^{2+} in response to H_2O_2 exposure (Figure 11B), and this suggests a higher oxidative stability of the cadmium-thiolate bonds in this isoform. Such a result is particular relevant when we consider that MT-20 is the isoform preferentially induced by Cd^{2+}. Therefore, the structural features of the MT-20 seem to perfectly support the physiological role of this isoform of defense against the highly toxic cadmium ions.

In *Mytilus spp*, MT-10 transcripts are more abundant than the MT-20 ones in physiological conditions [96], and the transcription of these two isoforms is differentially regulated by metals. Whereas MT-10 is induced by both essential and toxic metal ions [89], MT-20 is mainly responsive to cadmium [96] and mercury [97]. Contrary to expectations, in freshwater and seawater mussels, exposure to micromolar zinc reveals an increase in the zinc levels in various tissues that was not accompanied by a corresponding induction of MT gene expression [64,89]. From these results the authors inferred that in these molluscs zinc is not necessarily an inducer of MT gene expression. Despite the fact that they are not induced by zinc, the MT isoforms of these molluscs are able to efficiently bind zinc ions *in vitro* [64,98].

The studies performed by Viarengo and coworkers *in vivo* demonstrated only poor MT induction in mussels exposed to strong and mild oxidants, thus suggesting that in mussels MT genes are not particularly responsive to oxidative stress [99]. In line with these results, Khoo and Patel described MREs, but not antioxidant responsive elements (AREs) in *MT* gene promoter of the tropical mussel *Perna viridis* [84].

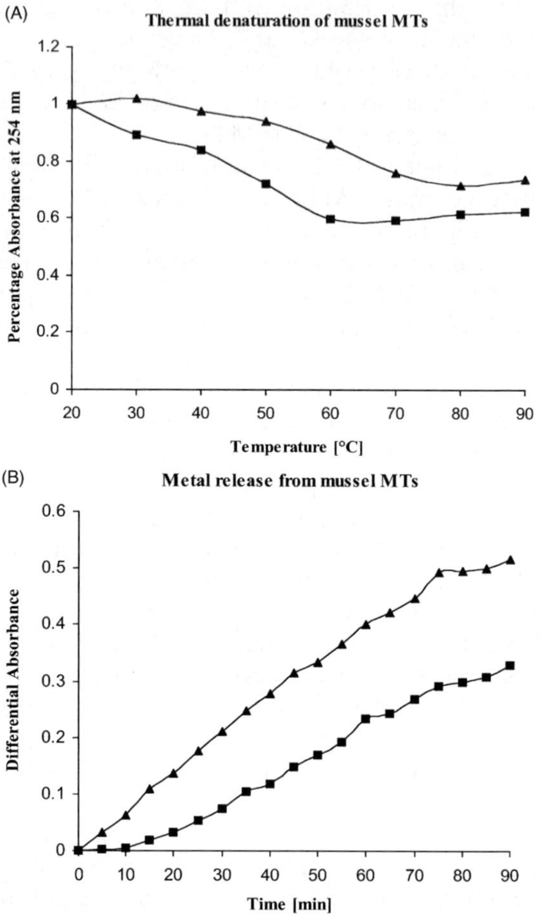

Figure 11. Functional properties of *Mytilus spp.* MT isoforms. (**A**) Effect of increasing temperature: Absorption spectra were acquired in the range 220–300 nm for the recombinant MT-10 (▲) and MT-20 (■) isoforms exposed to a thermal gradient (temperature increase from 20°C to 90°C), as described in [94]. The absorbance decrease at 254 nm of the cadmium-thiolate chromophore is plotted as a fraction of the standard absorbance (absorbance at room temperature). (**B**) Kinetics of cadmium release: Cadmium mobility was measured for MT-10 (▲) and MT-20 (■) isoforms following the formation of the cadmium-$(PAR)_2$ complex at 500 nm induced by 1 mM H_2O_2, as described in [94]. Each point represents the difference between the MT absorbance in the presence and absence of H_2O_2.

6.2.2. *Ostreidae*

In addition to MT isoforms containing the canonical αβ-domain structure, MTs proteins consisting solely of α-domains or of αββ-domains have been described in some molluscs [11]. This is the case of the atypical MT structures identified in different species of oysters. Multiple alignment for the amino acid sequences of different oyster MT isoforms available on GenBank and EMBL is reported in Figure 12.

In the Pacific oyster *C. gigas*, three MT isoforms with different molecular organization have been described. *CgMT-1* and *CgMT-3* encode a typical

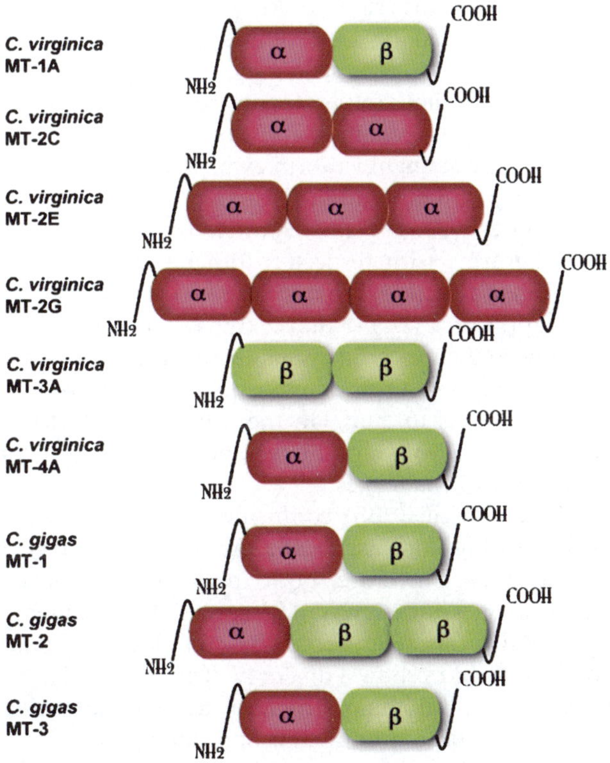

Figure 12. Multiple sequence alignment of different oyster MTs. The sequence alignment was performed using the free program CINEMA-Version 2. From top to bottom: the sequences of six MT isoforms from *C. virginica* (MT-1A, Q53ZD1; MT-2C, Q6VQP5; MT-2E, Q6VQP3; MT-2G, Q6VQP1; MT-3A, Q6VQN9; MT-4A, Q0H9V8) and six MT isoforms from *C. gigas* (MT-1, Q9BIV4; MT-2, Q9NG19; MT-3, Q9U1N5) are shown.

αβ-domain protein [90]. In contrast, *CgMT-2* encodes for a protein with an atypical αββ organization depending on an additional β-domain [91]. This peculiarity of CgMT-2 is unique among the MT proteins described so far.

Also in the American oyster *C. virginica*, multiple MT forms have been described. The canonical αβ-domain structure is observed in CvMT-1. A second subfamily of oyster MTs (designated as CvMT-2) has apparently arisen from a stop mutation that truncates the protein after the α-domain, and from a subsequent series of duplication/recombination events that have led to the development of isoforms containing one to four α-domains without a β-domain [11]. Transcriptomic approaches identified two other oyster MT subfamilies (CvMT-3 and CvMT-4) that do not share significant identity, at the nucleotide level, to CvMT-1 and 2 [100]. While CvMT-4 has the canonical αβ-domain structure, CvMT-3 is unique among molluscan MTs, as it contains a total of 18 cysteines in two highly conserved β-domains. CvMT-4 seems to represent a novel, functionally-derived subfamily of molluscan MTs that are paralogous to the CvMT-1 isoform [100]. On the other hand, CvMT-3 exhibits constitutive expression during larval development and is the dominant isoform expressed in larvae, but it displays low basal levels of expression in adult tissues.

Therefore, the oyster *C. virginica* is characterized by a greater structural diversity of MT isoforms with respect to other organisms, and this may depend on the different functional roles of the various MT isoforms. In fact, a strong up-regulation of CvMT-1 and 2 transcripts after cadmium exposure, but not by zinc and copper, has been documented in adult tissues of *C. virginica* [11], while CvMT-4 displays only a moderate response to cadmium and greater responsiveness to zinc. On the other hand, CvMT-3 seems to play an alternative role during larval development, and to be only moderately responsive to metal challenges.

Overall, the three-domain (αββ) conformation observed in CgMT-2 of *C. gigas* is an interesting contrast to the multiple α-domains observed in CvMT-2 of *C. virginica*. Therefore, both species of oysters seem to have adopted similar strategies for surviving in aquatic environments with wide fluctuations of metal levels [11]. From the complexity of data it can be inferred that molluscan MTs show a diversity of structures that encompasses not only the canonical αβ-domain structure, but also molecular forms in which this structure has been modified in opposite ways.

6.3. Metallothioneins in Marine Gastropods

The first MT identified and characterized in a marine gastropod is the MT from the periwinkle *Littorina littorea*, a model system for studies of natural stress tolerance [101]. The *L. littorea* MT (Ll-MT) contains 100 amino acids and shares 45% and 56% identity with the copper- and cadmium-thioneins, respectively, from another gastropod, the terrestrial snail *Helix pomatia*, and 43–47% identity with marine bivalve MTs. Moreover, Ll-MT displays the [Cys-X-Cys-X(3)-Cys-Thr-Gly-X(3)-Cys-X-Cys-X(3)-Cys-X-Cys-Lys] motif that is typical of molluscan MTs [101]. Multiple alignment for the amino acid sequences of different MT isoforms from marine gastropods available on GenBank and EMBL is reported in Figure 13.

The coding sequence of another molluscan MT has been isolated from the keyhole limpet *Megathura crenulata* [86]. The deduced amino acid sequence of the *M. crenulata* MT (Mc-MT) resembles that of other molluscan MTs. It possesses typically positioned Cys residues, and Gly residues are twice the number of vertebrate MTs, similarly to mussel MTs. Mc-MT contains

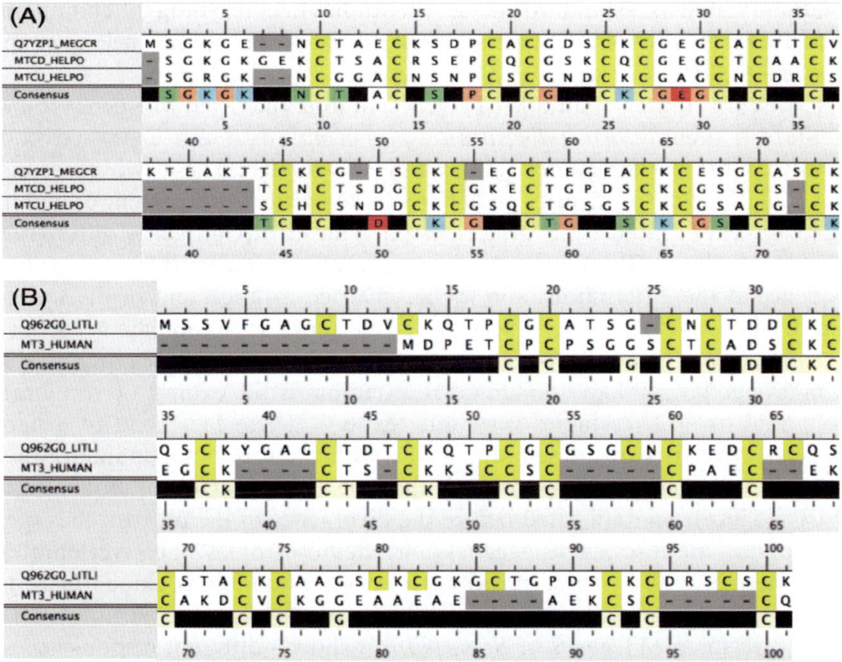

Figure 13. Multiple sequence alignment of different gastropodan MTs. The sequence alignment was performed using the free program CINEMA-Version 2.0. From top to bottom the following sequences are shown: the MT of *M. crenulata* (Q7YZP1_MEGCR); the Cd-MT of *H. pomatia* (MTCD_HELPO); the Cu-MT of *H. pomatia* (MTCU_HELPO); the MT of *L. littorea* (Q962G0_LITLI); the MT-3 of *H. sapiens*. The alignment is colored according to residue type: blue for polar positive; red for polar negative; green for polar neutral; white for non-polar aliphatic; brown for non-polar P, G; yellow for C.

70 amino acids, thus being longer than the MT isoforms identified in the terrestrial gastropod *H. pomatia* [102]. The most interesting characteristic of the Mc-MT is the acidic pentapeptide sequence (Lys-Glu-Gly-Glu-Ala) positioned in the α-domain. This sequence has only been observed in mammalian MT-3, a type of MT expressed mainly in the central nervous system where it inhibits growth and survival of neurons.

7. METALLOTHIONEINS FROM ECHINODERMS

7.1. Echinoderm Metallothionein Genes

Echinoderms are a phylum of marine animals found at all ocean depths; sea urchins are small echinoderms living in oceans all over the world and are commonly used as a standard subject for studies in embryology. Echinoderm MTs are rather different from those of vertebrates, indeed they belong to the class II of MTs.

Sea urchin contains multiple MT genes; among the seven MT genes found in the sea urchin *Strongylocentrotus purpuratus*, two are expressed constitutively and are designated *MT-A* and *MT-B* [103]. These two genes display a high degree of identity and share the tripartite arrangement of coding exons typical of almost all vertebrates [2]. The sea urchin genes differ from the mammalian MT genes for the presence of a large, entirely non-coding sequence at the 3′ terminus, and for an inverted arrangement of the second and third exons [81]. While in vertebrate MT genes, the β-domain is encoded by exons 1 and 2, and the α-domain is encoded by exon 3, in sea urchin MT, exon 3 encodes a β-like domain corresponding to the exon 2 of vertebrate MTs, and exon 2 resembles vertebrate exon 3. These facts lead to a poor sequence similarity of the sea urchin MT with the vertebrate counterparts.

A generally accepted hypothesis is that the central region of an ancestral MT gene has been duplicated before the divergence of the protostomes (i.e., drosophila, molluscs, crustaceans) and deuterostomes (i.e., vertebrates, echinoderms). The switched positions of exons 2 and 3 in echinoderms with respect to vertebrates may have resulted from a divergent evolution [103].

The two main MT genes of *S. purpuratus* show a different responsiveness to zinc and other metals [2]. The responsiveness to zinc is accounted by the presence of multiple MREs in the 5′-upstream region of both *MT-A* and *MT-B* genes [104]. In particular, the *MT-A* gene promoter shows two proximal MREs that are identical to those identified in mammalian MTs [103]. It has been reported that the two MT genes are developmentally regulated, and in embryos their expression differs depending on the tissue fractions. MT-A is mainly expressed in ectodermal tissues, whereas MT-B in both ectodermal and endodermal fractions [105].

7.2. Sea Urchin Metallothionein Proteins: Sequence and Structure

Sea urchin MTs have a rather poor homology with the MTs of mammals and fish [81]. A well characterized sea urchin MT is the MT-A isoform of *S. purpuratus*; it contains 64 residues and 20 cysteines, with essentially the same number of Cys-Cys and Cys-X-Cys motifs (where X stands for any amino acid, other than Cys) found in mammalian MTs [8]. Unlike mammalian MTs, where all three Cys-Cys pairs are located in the C-terminal half of the polypeptide chain, in sea urchin MT-A all four Cys-Cys pairs occur in the N-terminal portion. A further remarkable difference is the presence of a single Phe residue in the α-domain of MT-A, whereas most MTs are devoid of aromatic residues [1]. Moreover, sea urchin MT-A differs from the mammalian proteins by its larger negative charge [8].

Looking at the primary sequences, sea urchin MTs appear to be less similar to each other than MTs of closely related vertebrates. Indeed, many differences in the amino acid sequences of MTs from different sea urchins can be observed (Figure 14). The most remarkable is the insertion of two

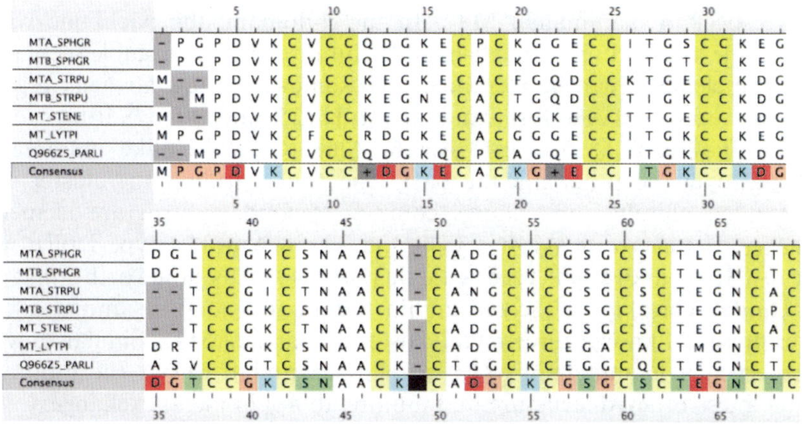

Figure 14. Multiple sequence alignment of different echinoderm MTs. The sequence alignment was performed using the free program CINEMA-Version 2.0. From top to bottom the sequences correspond to: the MT-A (MTA_SPHGR) and the MT-B (MTB_SPHGR) isoforms of *S. granularis*; the MT-A (MTA_STRPU) and the MT-B (MTB_STRPU) isoforms of *S. purpuratus*; the MT of *S. neumayeri* (MT_STENE); the MT of *L. pictus* (MT_LYTPI); the MT of *P. lividus* (Q966Z5_PARLI). The alignment is colored according to residue type: blue for polar positive; red for polar negative; green for polar neutral; white for non-polar aliphatic; brown for non-polar P, G; yellow for C.

residues (Pro and Gly) at the N-terminus of two MT isoforms, the *S. granularis* and *L. pictus* MTs; these additional amino acids have not been found in other sea urchin species. Moroever, the MT from *P. lividus* shows high similarity to the MT from *S. purpuratus*, but the presence of non-conservative amino acid substitutions, together with the insertion of two residues (Ser and Val) in the *P. lividus*, makes the similarity scores of these two sea urchin proteins lower than that of MTs from vertebrates of the same order [106].

For structural studies the *S. purpuratus* MT-A was produced as recombinant protein in *E. coli* and analyzed by homonuclear [^{1}H,^{1}H] [8] and heteronuclear [^{1}H,^{113}Cd] NMR correlation experiments [6]. As in mammals, the *S. purpuratus* MT-A binds seven cadmium ions per molecule through tetrahedral coordination to the Cys sulfurs; all cysteines are bound to the seven metal ions and no free Cys side-chains are observed. The 3D structure (Figure 15) reveals the presence of two globular domains, an N-terminal, four-metal (Cd_4-Cys_{11}) cluster comprising residues 1–36, and a C-terminal, three-metal (Cd_3-Cys_9) cluster comprising the other residues. Thus, sea urchin MT-A, like the mammalian counterparts, is made up of two separate domains connected by a flexible linker, but the domains are inverted inside the chain depending on the different gene organization described above. In fact, in the sea urchin MT-A, the positions of exons 2 and 3 are switched with respect to mammalian MTs. In the α-domain, the backbone winds around the 'metal core' with left-handed chirality and some stable secondary structure elements can be identified: two somewhat irregular helix regions with the residues Cys6-Glu11 and Asp21-Thr25, and three half-turns with the residues Pro2-Lys5, Lys13-Ala16, Ala16-Gly19. Unlike mammalian MTs, also in the β-domain the backbone winds around the cluster with left-handed chirality and shows common secondary structure elements: two type II turns with Ala46-Cys49 and Thr58-Asn61, a type I turn with Asn40-Cys43, and a half turn with Ile37-Asn40. It is noteworthy that the left-handed chirality of the sea urchin β-domain is unique among all the other presently known three-metal cluster domains of MTs. Similar to what is observed in mammalian MTs, also in sea urchin MT-A the β-domain shows a less ordered structure and a more ability in metal exchange.

Therefore, the dominant impression obtained by comparing sea urchin MT-A with mammalian MTs is a similar two-domain structure. However, the sequential order of the α- and β-domains is inverted between the two MTs, and a significantly different connectivity pattern of the Cys-metal coordination bonds and novel local structures of the polypeptide chain are present.

Very interesting is a comparative study performed on two MTs isolated from sea urchin species living in very different natural habitats: the Mediterranean echinoid *S. granularis*, and the Antarctic urchin *S. neumayeri*

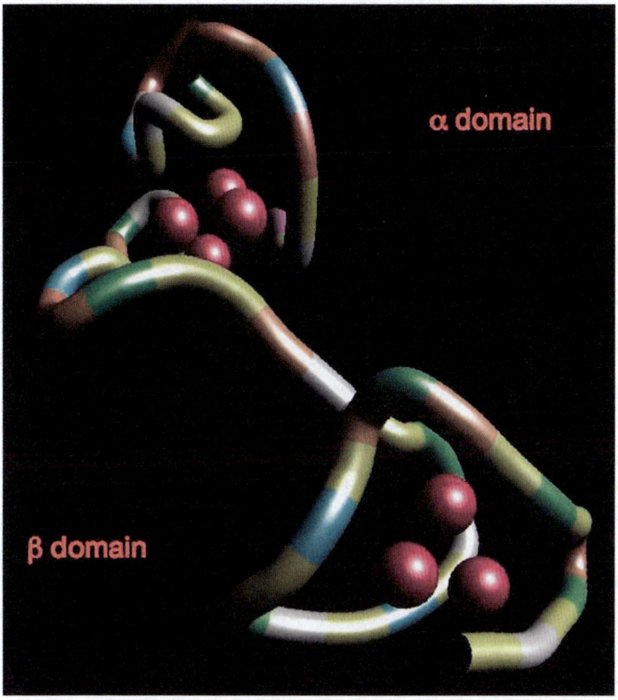

Figure 15. The 3D solution structure of *S. purpuratus* MT-A. The three-dimensional NMR structure of *S. purpuratus* Cd_7-MT-A has been solved by homonuclear [^{1}H,^{1}H] and heteronuclear [^{1}H,^{113}Cd] NMR correlation experiments [6]. The N-terminal α-domain comprises residues 1–36 and a four-metal cluster. The C-terminal β-domain comprises residues 37–66 and a three-metal cluster. The two domains are connected, despite the lack of NMR information to this regard. Cadmium ions are shown as spheres. The color scheme is the following: blue for polar positive; red for polar negative; green for polar neutral; white for non-polar aliphatic; brown for non-polar P, G; yellow for C. The figure has been adapted from entries made to the Brookhaven Protein Data Bank [6] (ID 1QJK and 1QJL). The figure was prepared using Ambrosia Molecular Structure Viewer program, a tool of molecular visualization in the Utopia software package.

[106]. In the embryos of the Mediterranean species a high MT content was detected, whilst the MT level was almost negligible in the Antarctic embryos. It is worth noting that these two species of sea urchin, albeit related phylogenetically, display distinctive developmental patterns. The development of *S. neumayeri* is much slower than that of other sea urchin species, including *S. granularis*. Since MT constitutes the main zinc depot of the developing embryo and a large amount of zinc is required for the metabolic

processes occurring during embryo development, this explains the higher level of MT observed for the fast-growing species [106].

8. CONCLUDING REMARKS

MTs are usually considered as highly conserved proteins during evolution, and their moderate variability across phylogenetically distant organisms may prove the crucial function that they exert in biological systems. Phylogenetic relationships among the various MT families have recently been established by different methods and approaches, leading to the classification of MTs into a variety of phylogenetically related subfamilies [3].

The discovery of atypical MT isoforms in several invertebrates, in aquatic species in particular, has provided new insights into the biological role of this class of proteins. An increasing body of evidence suggests that MTs from invertebrates, owing to some structural distinctive features, may play rather different roles with respect to vertebrate counterparts both in terms of metal specificity (copper *versus* cadmium/zinc) and antioxidant activity.

The canonical αβ-domain organization observed in almost all MTs may be the result of a long evolutionary process, where the three-metal cluster β-domain could be the ancestral domain, with a primary role in the homeostasis of physiologically relevant metals, such as zinc and copper. Duplication of the β-domain could have given rise to a ββ-domain MT, and the subsequent divergence of the two domains then could have produced the αβ-organization [75]. Indeed, selective pressure may have led to the evolution of a two-domain MT with specific functions carried out by the two domains. The β-domain should be mainly devoted to metal homeostasis, and the α-domain more important for metal detoxification. This hypothesis is supported by the presence of single β-domain structures in some copper-thioneins (i.e., in *Drosophila* and yeasts), as well as by the existence of two β-domain structures (i.e., in oysters).

However, this theory experiences some difficulties when invertebrate and vertebrate MT gene structures are compared. In all vertebrates, the β-domain is N-terminally encoded, whereas in some invertebrates the reverse is the case, the β-domain being C-terminally encoded [107]. The homology between the α- and the β-domains of invertebrate and vertebrate MTs would require an inversion between the α- and the β-encoding regions within the MT gene. In line with this hypothesis is the identification of a single-domain peptide (containing 41 residues) selective for cadmium binding in the terrestrial worm *Eisena foetida* [108]. The four-metal cluster organization of this peptide suggests functional analogy to the MT α-domain.

Therefore, further comparative analyses of the functional and structural features of MTs from different organisms are needed to gain a more complete insight of these proteins.

ACKNOWLEDGMENTS

I would like to thank all my colleagues for helpful discussions and critical reading of the manuscript. A special thank to Aldo Viarengo for introducing me into the 'metallothionein world' and Elena Grasselli for her assistance in the manuscript preparation.

ABBREVIATIONS AND DEFINITIONS

3D	three-dimensional
ARE	antioxidant responsive element
CD	circular dichroism
DTNB	5′,5′-dithio-bis(2-nitrobenzoic acid)
EMBL	European Molecular Biology Laboratory
MRE	metal responsive element
MT	metallothionein
MTF-1	metal-regulatory transcription factor 1
NMR	nuclear magnetic resonance
PAR	4-(2-pyridylazo)resorcinol
PDB	Brookhaven Protein Data Bank
UTR	untranslated region

REFERENCES

1. J. H. R. Kägi, in *Metallothionein III. Biological Roles and Implications*, Ed. K. T. Suzuki, N. Imura, M. Kimura, Birkhäuser Verlag, Basel, 1993, pp. 29–55.
2. D. G. Wilkinson and M. Nemer, *Mol. Cell Biol.*, 1987, **7**, 48–58.
3. P. A. Binz and J. H. R. Kägi, *Metallothioneins: Molecular Evolution and Classification*, Birkhäuser Verlag, Basel, 1999, pp. 7–13.
4. S. S. Narula, M. Brouwer, Y. Hua and I. M. Armitage, *Biochemistry*, 1995, **34**, 620–631.
5. A. Munoz, F. H. Forsterling, C. F. Shaw 3rd and D. H. Petering, *J. Biol. Inorg. Chem.*, 2002, **7**, 713–724.
6. R. Riek, B. Precheur, Y. Wang, E. A. Mackay, G. Wider, P. Guntert, A. Liu, J. H. R. Kägi and K. Wüthrich, *J. Mol. Biol.*, 1999, **291**, 417–428.
7. C. Capasso, V. Carginale, O. Crescenzi, D. Di Maro, E. Parisi, R. Spadaccini and P. A. Temussi, *Structure*, 2003, **11**, 435–443.
8. Y. Wang, E. A. Mackay, M. Kurasaki and J. H. R. Kägi, *Eur. J. Biochem.*, 1994, **225**, 449–457.
9. J. D. Otvos, R. W. Olafson and I. M. Armitage, *J. Biol. Chem.*, 1982, **257**, 2427–2431.
10. J. Overnell, M. Good and M. Vašák, *Eur. J. Biochem.*, 1988, **172**, 171–177.

11. M. J. Jenny, A. H. Ringwood, K. Schey, G. W. Warr and R. W. Chapman, *Eur. J. Biochem.*, 2004, **271**, 1702–1712.
12. S. G. George, E. Carpene, T. L. Coombs, J. Overnell and A. Youngson, *Biochim. Biophys. Acta.*, 1979, **580**, 225–233.
13. L. Vergani, C. Lanza, C. Borghi, L. Scarabelli, I. Panfoli, B. Burlando, F. Dondero, A. Viarengo and G. Gallo, *Mol. Cell Endocrinol.*, 2007, **263**, 29–37.
14. P. E. Olsson, S. J. Hyllner, M. Zafarullah, T. Andersson and L. Gedamu, *Biochim. Biophys. Acta.*, 1990, **1049**, 78–82.
15. M. Zafarullah, P. E. Olsson and L. Gedamu, *Biochim. Biophys. Acta.*, 1990, **1049**, 318–323.
16. J. L. Casterline Jr. and G. Yip, *Arch. Environ. Contam. Toxicol.*, 1975, **3**, 319–329.
17. A. G. Howard and G. Nickless, *Chem. Biol. Interact.*, 1977, **16**, 107–114.
18. M. Brouwer, J. Enghild, T. Hoexum-Brouwer, I. Thogersen and A. Truncali, *Biochem. J.*, 1995, **311**, 617–622.
19. R. Dallinger, B. Berger, P. Hunziker and J. H. R. Kägi, *Nature*, 1997, **388**, 237–238.
20. S. N. Pedersen, K. L. Pedersen, P. Hojrup, J. Knudsen and M. H. Depledge, *Comp. Biochem. Physiol. C Pharmacol. Toxicol. Endocrinol.*, 1998, **12**, 251–259.
21. F. A. Cotton and G. Wilkinson, *Advanced Inorganic Chemistry*, New York, 1967.
22. H. C. Freeman, *Metal Complexes of Amino Acids and Peptides*, in *Inorganic Biochemistry*, Ed. G. L. Eichhorn, Elsevier, Amsterdam, 1973, pp. 121–167.
23. M. Vašák, *Methods Enzymol.*, 1991, **205**, 452–458.
24. A. Viarengo, S. Palmero, G. Zanicchi, R. Cappelli, R. Vaissiere and M. Orunesu, *Lam. Marine Environ. Res.*, 1985, **16**, 23–26.
25. A. E. Funk, F. A. Day and F. O. Brady, *Comp. Biochem. Physiol. C*, 1987, **86**, 1–6.
26. J. Overnell, *Environ. Health Perspect.*, 1986, **65**, 101–105.
27. D. W. Engel and M. Brouwer, *Environ. Health. Perspect.*, 1986, **65**, 87–92.
28. M. Brouwer, D. R. Winge and W. R. Gray, *J. Inorg. Biochem.*, 1989, **35**, 289–303.
29. M. Valls, R. Bofill, R. Gonzalez-Duarte, P. Gonzalez-Duarte, M. Capdevila and S. Atrian, *J. Biol. Chem.*, 2001, **276**, 32835–32843.
30. Y. Zhou, L. Li and B. Ru, *Biochim. Biophys. Acta.*, 2000, **1524**, 87–93.
31. K. Zangger, G. Oz, J. D. Otvos and I. M. Armitage, *Protein Sci.*, 1999, **8**, 2630–2638.
32. G. Oz, K. Zangger and I. M. Armitage, *Biochemistry*, 2001, **40**, 11433–11441.
33. A. H. Robbins and C. D. Stout, *Methods Enzymol.*, 1991, **205**, 485–502.
34. H. W. Ren, N. Itoh, M. Kanekiyo, S. Tominaga, J. Kohroki, G. S. Hwang, T. Nakanishi, N. Muto and K. Tanaka, *Biol. Pharm. Bull.*, 2000, **23**, 145–148.
35. E. Hermesz, M. Abraham and J. Nemcsok, *Comp. Biochem. Physiol. C Toxicol. Pharmacol.*, 2001, **128**, 457–465.
36. P. C. Chan, C. K. Shiu, F. W. Wong, J. K. Wong, K. L. Lam and K. M. Chan, *Biochim. Biophys. Acta*, 2004, **1676**, 162–171.
37. K. Bonham, M. Zafarulla and L. Gedamu, *DNA*, 1987, **6**, 519–528.

38. S. George, M. Leaver, N. Freich and D. Burgess, *Mar. Environ. Res.*, 1990, **28**, 173–177.
39. Y. S. Cho, B. N. Choi, E. M. Ha, K. H. Kim, S. K. Kim, D. S. Kim and Y. K. Nam, *Mar. Biotechnol. (NY)*, 2005, **7**, 350–362.
40. M. Zafarullah, K. Bonham and L. Gedamu, *Mol. Cell Biol.*, 1988, **8**, 4469–4476.
41. P. E. Olsson, P. Kling, L. J. Erkell and P. Kille, *Eur. J. Biochem.*, 1995, **230**, 344–349.
42. S. L. Samson and L. Gedamu, *J. Biol. Chem.*, 1995, **270**, 6864–6871.
43. S. L. Samson, W. J. Paramchuk and L. Gedamu, *Biochim. Biophys. Acta*, 2001, **1517**, 202–211.
44. P. Kille, J. Kay and G. E. Sweeney, *Biochim. Biophys. Acta*, 1993, **1216**, 55–64.
45. C. H. Yan and K. M. Chan, *Biochim. Biophys. Acta*, 2004, **1679**, 47–58.
46. R. Scudiero, V. Carginale, C. Capasso, M. Riggio, S. Filosa and E. Parisi, *Gene*, 2001, **274**, 199–208.
47. A. Cheung, Pok. Lap, V. Kwok Lim Lam and K. M. Chan, *Biochim. Biophys. Acta*, 2005, **1731**, 191–201.
48. K. M. Chan, *Biochem. Biophys. Res. Com.*, 1994, **205**, 368–374.
49. V. Carginale, R. Scudiero, C. Capasso, A. Capasso, P. Kille, G. Di Prisco and E. Parisi, *Biochem. J.*, 1998, **332**, 475–481.
50. L. Bargelloni, R. Scudiero, E. Parisi, V. Carginale, C. Capasso and T. Patarnello, *Mol. Biol. Evol.*, 1999, **16**, 885–897.
51. H. Ren, M. Xu, P. He, N. Muto, N. Itoh, K. Tanaka, J. Xing and M. Chu, *Biochem. Biophys. Res. Commun.*, 2006, **342**, 1297–1304.
52. P. He, M. Xu and H. Ren, *Int. J. Biochem. Cell Biol.*, 2007, **39**, 832–841.
53. J. Kay, A. Cryer, B. M. Darke, P. Kille, W. E. Lees, C. G. Norey and J. M. Stark, *Int. J. Biochem.*, 1991, **23**, 1–5.
54. R. Scudiero, P. A. Temussi and E. Parisi, *Gene*, 2005, **345**, 21–26.
55. L. Vergani, M. Grattarola, F. Dondero and A. Viarengo, *Protein Expr. Purif.*, 2003, **27**, 338–345.
56. H. Willner, M. Vašák and J. H. Kägi, *Biochemistry*, 1987, **26**, 6287–6292.
57. S. D'auria, V. Carginale, R. Scudiero, O. Crescenzi, D. Di Maro, P. A. Temussi, E. Parisi and C. Capasso, *Biochem. J.*, 2001, **354**, 291–299.
58. L. Vergani, M. Grattarola, C. Borghi, F. Dondero and A. Viarengo, *FEBS J.*, 2005, **272**, 6014–6023.
59. C. Capasso, O. Abugo, F. Tanfani, A. Scire, V. Carginale, R. Scudiero, E. Parisi and S. D'auria, *Proteins*, 2002, **46**, 259–267.
60. C. J. Marshall, *Trends Biotechnol.*, 1997, **15**, 359–364.
61. C. Capasso, V. Carginale, R. Scudiero, O. Crescenzi, R. Spadaccini, P. A. Temussi and E. Parisi, *J. Mol. Evol.*, 2003, **57**(Suppl 1), S250–S257.
62. J. D. Otvos, H. R. Engeseth and S. Wehrli, *Biochemistry*, 1985, **24**, 6735–6740.
63. M. Vašák, G. E. Hawkes, J. K. Nicholso and P. J. Sadler, *Biochemistry*, 1985, **24**, 740–747.
64. J. P. Bourdineaud, M. Baudrimont, P. Gonzalez and J. L. Moreau, *Biochimie*, 2006, **88**, 1787–1792.
65. A. Viarengo and J. A. Nott, *Comp. Biochem. Physiol. C*, 1993, **104**, 355–372.
66. J. T. Ruud, *Nature*, 1954, **173**, 848–850.

67. R. Scudiero, V. Carginale, M. Riggio, C. Capasso, A. Capasso, P. Kille, G. Di Prisco and E. Parisi, *Biochem. J.*, 1997, **322**, 207–211.
68. R. W. Olafson, R. G. Sim and A. Kearns, *Experientia Suppl.*, 1979, **34**, 197–204.
69. K. Lerch, D. Ammer and R. W. Olafson, *J. Biol. Chem.*, 1982, **257**, 2420–2426.
70. S. G. Ang and P. S. Chong, *Talanta*, 1998, **45**, 693–701.
71. R. A. Syring, T. Hoexum Brouwer and M. Brouwer, *Comp. Biochem. Physiol. C Toxicol. Pharmacol.*, 2000, **125**, 325–332.
72. E. Molto, E. Bonzon-Kulichenko, A. Del Arco, D. M. Lopez-Alanon, O. Carrillo, N. Gallard and A. Andres, *Gene*, 2005, **361**, 140–148.
73. A. K. West, R. Stallings, C. E. Hildebrand, R. Chiu, M. Karin and R. I. Richards, *Genomics*, 1990, **8**, 513–518.
74. Z. Zhu, E. F. Derose, G. P. Mullen, D. H. Petering and C. F. Shaw 3rd, *Biochemistry*, 1994, **33**, 8858–8865.
75. N. Cols, N. Romero-Isart, R. Bofill, M. Capdevila, P. Gonzalez-Duarte, R. Gonzalez-Duarte and S. Atrian, *Protein Eng.*, 1999, **12**, 265–269.
76. J. Overnell, *Comp. Biochem. Physiol. C*, 1984, **77**, 237–243.
77. C. L. Chou, R. D. Guy and J. F. Uthe, *Sci. Total Environ.*, 1991, **105**, 61–71.
78. M. Brouwer, D. Schlenk, A. H. Ringwood and T. Brouwer-Hoexum, *Arch. Biochem. Biophys.*, 1992, **294**, 461–468.
79. D. H. Hamer, *Annu. Rev. Biochem.*, 1986, **55**, 913–951.
80. D. W. Engel and M. Brouwer, *Detoxification of accumulated trace metals by the american oyster C. virginica*, 1982, pp. 89–107.
81. J. M. Frazier, S. S. George, J. Overnell, T. L. Coombs and J. H. R. Kägi, *Comp. Biochem. Physiol. C*, 1985, **80**, 257–262.
82. E. A. Mackay, J. Overnell, B. Dunbar, I. Davidson, P. E. Hunziker, J. H. R. Kägi and J. E. Fothergill, *Eur. J. Biochem.*, 1993, **218**, 183–194.
83. D. Barsyte, K. N. White and D. A. Lovejoy, *Comp. Biochem. Physiol. C Pharmacol. Toxicol. Endocrinol.*, 1999, **122**, 287–296.
84. H. W. Khoo and K. H. Patel, *J. Exp. Zool.*, 1999, **284**, 445–453.
85. D. C. Simes, M. J. Bebianno and J. J. G. Moura, *Aquat. Toxicol.*, 2002, 1–12.
86. B. Lieb, *Comp. Biochem. Physiol. C Toxicol. Pharmacol.*, 2003, **134**, 131–137.
87. H. Park, I. Y. Ahn, H. J. Choi, S. H. Pyo and H. E. Lee, *Protein Expr. Purif.*, 2007, **52**, 82–88.
88. J. Engelken and A. Hildebrandt, *Biochem. Cell Biol.*, 1999, **77**, 237–241.
89. S. Lemoine, Y. Bigot, D. Sellos, R. P. Cosson and M. Laulier, *Mar. Biotechnol. (NY)*, 2000, **2**, 195–203.
90. A. Tanguy, C. Mura and D. Moraga, *Aquat. Toxicol.*, 2001, **55**, 35–47.
91. A. Tanguy and D. Moraga, *Gene*, 2001, **273**, 123–130.
92. F. Geret and R. P. Cosson, *Arch. Environ. Contam. Toxicol.*, 2002, **42**, 36–42.
93. V. Leignel and M. Laulier, *Comp. Biochem. Physiol. C Toxicol. Pharmacol.*, 2006, **142**, 12–8.
94. L. Vergani, M. Grattarola, E. Grasselli, F. Dondero and A. Viarengo, *Arch. Biochem. Biophys.*, 2007, **465**, 247–253.
95. R. H. Buhler and J. H. R. Kägi, *Experientia Suppl.*, 1979, **34**, 211–220.
96. B. Raspor, Z. Dragun, M. Erk, D. Ivankovic and J. Pavicic, *Sci. Total Environ.*, 2004, **333**, 99–108.

97. F. Dondero, L. Piacentini, M. Banni, M. Rebelo, B. Burlando and A. Viarengo, *Gene*, 2005, **345**, 259–270.
98. M. Grattarola, M. Carloni, F. Dondero, A. Viarengo and L. Vergani, *Mol. Biol. Rep.*, 2006, **33**, 265–272.
99. A. Viarengo, B. Burlando, M. Cavaletto, B. Marchi, E. Ponzano and J. Blasco, *Am. J. Physiol.*, 1999, **277**, R1612–R1619.
100. M. J. Jenny, G. W. Warr, A. H. Ringwood, D. A. Baltzegar and R. W. Chapman, *Gene*, 2006, **379**, 156–165.
101. T. E. English and K. B. Storey, *J. Exp. Biol.*, 2003, **206**, 2517–2524.
102. B. Berger, R. Dallinger, P. Gehrig and P. E. Hunziker, *Biochem. J.*, 1997, **328**, 219–224.
103. P. Harlow, E. Watkins, R. D. Thornton and M. Nemer, *Mol. Cell Biol.*, 1989, **9**, 5445–5455.
104. G. Bai, E. W. Stuebing, H. R. Parker, P. Harlow and M. Nemer, *Mol. Cell Biol.*, 1993, **13**, 993–1001.
105. M. Nemer, R. D. Thornton, E. W. Stuebing and P. Harlow, *J. Biol. Chem.*, 1991, **266**, 6586–6593.
106. R. Scudiero, C. Capasso, V. Carginale, M. Riggio, A. Capasso, M. Ciaramella, S. Filosa and E. Parisi, *Cell. Mol. Life Sci.*, 1997, **53**, 472–477.
107. M. Nemer, D. G. Wilkinson, E. C. Travaglini, E. J. Sternberg and T. R. Butt, *Proc. Natl. Acad. Sci. USA*, 1985, **82**, 4992–4994.
108. C. Gruber, S. Stürzenbaum, P. Gehrig, R. Sack, P. Hunziker, B. Berger and R. Dallinger, *Eur. J. Biochem.*, 2000, **267**, 573–582.

Met. Ions Life Sci. **2009**, *5*, 239–277

9

Metal Detoxification in Freshwater Animals. Roles of Metallothioneins

Peter G. C. Campbell and Landis Hare

Université du Québec, Institut National de la Recherche Scientifique, INRS-T, Québec, QC, Canada G1K 9A9

<peter.campbell@ete.inrs.ca>

<landis.hare@ete.inrs.ca>

Metal Ions in Life Sciences, Volume 5 Edited by Astrid Sigel, Helmut Sigel, and Roland K. O. Sigel

Published by the Royal Society of Chemistry, www.rsc.org DOI: 10.1039/9781847558992-00239

ABSTRACT: In laboratory experiments with a variety of aquatic animals, the toxicity of non-essential metals normally exhibits a threshold response. At low exposure concentrations (low internal doses), the organism can detoxify the incoming metal and thus tolerate the exposure, whereas at higher concentrations, i.e., above some threshold, the detoxification mechanism is no longer able to protect the organism completely, the incoming metal binds to metal-sensitive sites within the cell, and deleterious effects begin to occur. Several metal-detoxification strategies have been identified in laboratory experiments, including metal sequestration in insoluble granules and metal complexation by metallothionein or metallothionein-like peptides (MTLP), but the ability of these mechanisms to prevent metals from binding to metal-sensitive sites in the intracellular environment has not been rigorously tested in field situations.

In this chapter we briefly summarize the laboratory evidence supporting the threshold model for metal toxicity and then present the results of our field studies on chronically exposed freshwater animals (*Pyganodon grandis*, a unionid bivalve mollusc; *Perca flavescens*, a percid fish), studies that were designed to test the threshold model under field conditions. The two biomonitor species were collected from lakes located along a metal concentration gradient, as defined by ambient dissolved metal concentrations. Subcellular metal partitioning was determined in target organs (bivalve gills and digestive gland; perch liver) by differential centrifugation, and MTLP were measured independently by a mercury saturation assay. From these studies we extract evidence for the inadequacy of the threshold model in the case of chronic/life-long metal exposures (i.e., evidence for incomplete detoxification even at low metal exposures in the field). We speculate that under chronic exposure conditions the animals establish a trade-off between the "cost" of detoxifying non-essential metals and the "cost" of allowing some of these metals to spill over onto metal-sensitive sites. Finally, we discuss the implications of the apparent absence of an exposure threshold below which metal detoxification is complete.

KEYWORDS: bivalve · cadmium · chronic lifetime exposure · detoxification · fish · granules · metallothionein · nickel

1. INTRODUCTION

Aquatic toxicology, the study of the interaction of toxic contaminants with aquatic organisms, is traditionally based on the threshold theory of toxicity, whereby dose-response or concentration-response relations are assumed to exhibit a "hockey-stick" or threshold response (Figure 1A). At low exposure concentrations (or low internal doses), it is assumed that organisms can detoxify incoming contaminants and thereby tolerate exposure to them, whereas at higher concentrations, i.e., above some threshold, the detoxification mechanism is no longer able to protect the organisms completely and deleterious effects begin to occur.

In the case of non-carcinogenic organic contaminants or xenobiotics, the threshold approach seems to work reasonably well [1], but for metals the applicability of a threshold model for predicting toxic effects does not appear to have been tested rigorously, particularly for chronically exposed organisms. Several metal-detoxification strategies have been identified, including metal sequestration in insoluble granules [2] and metal complexation by metallothionein (MT) or metallothionein-like (MTLP) peptides [3,4], but the ability of these mechanisms to prevent metals from binding to metal-sensitive sites in the intracellular environment has not been tested in field situations.

In this chapter we first present a brief summary of the laboratory evidence supporting the threshold model for metals, drawing on the excellent review of Mason and Jenkins [5]. We then present the results of our field studies on chronically exposed freshwater animals (a bivalve mollusc and a percid fish),

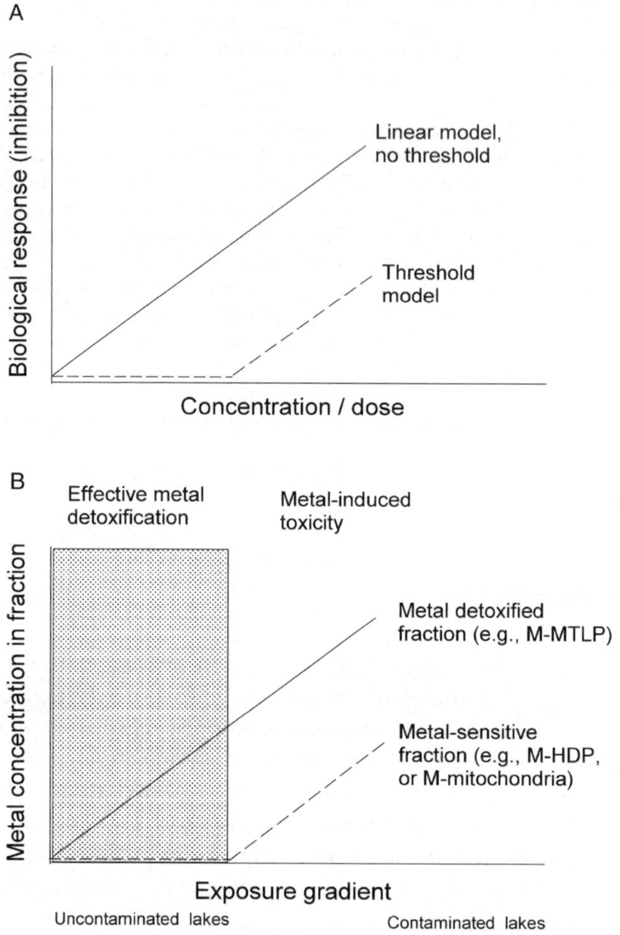

Figure 1. **(A)** Simplified representation of the inhibition response of an aquatic animal exposed to a non-essential metal – linear model, no threshold; threshold model. **(B)** Hypothetical relationship between chronic exposure to a non-essential metal and the sub-cellular metal partitioning of this metal, showing initial protection of the metal-sensitive compartments up to a threshold metal exposure. Panel (B) is reproduced from [49] with permission of Elsevier, copyright (2005).

studies that were designed to test the threshold model under field conditions. From these latter studies we extract evidence for the inadequacy of the threshold model in the case of chronic/life-long metal exposures (i.e., evidence for incomplete detoxification even at low metal exposures in the field). Finally, we discuss the implications of this apparent absence of an exposure threshold below which metal detoxification is complete.

2. BASIC CONCEPTS

2.1. Metal Speciation in the External Environment and Bioavailability

Relationships between metal speciation and metal bioavailability in the aquatic environment have been explored extensively over the past 30 years and for cationic metals there is now general acceptance of the idea that the free metal ion concentration is normally the best predictor of a metal's bioavailability [6,7]. This dependency on the external free ion concentration is readily explained in terms of the formation of a metal complex at the biological surface (Figure 2):

$$M^{z+} + \text{L-cell membrane} \overset{K_M}{\longleftrightarrow} \text{ML-cell membrane} \quad (1)$$

$$K_M = \frac{[\text{ML-cell membrane}]}{[M^{z+}][\text{L-cell membrane}]} \quad (2)$$

The biological response is assumed to be proportional to the concentration of the species ≪ML-cell membrane≫ at the biological surface (where

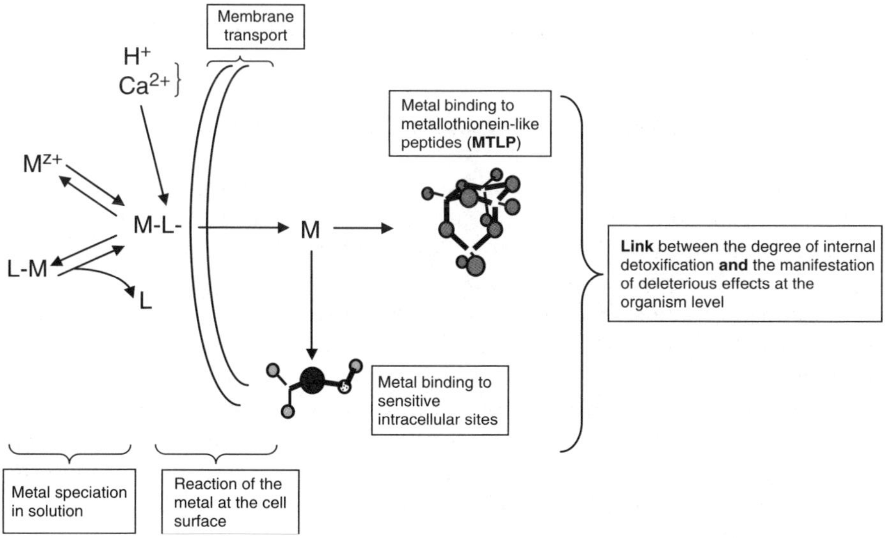

Figure 2. Surface complexation, transmembrane transport and subcellular partitioning of a non-essential metal ≪M≫. Adapted from [74] with permission from the Presses de l'Université du Québec, copyright (2004).

K_M is an equilibrium constant); under conditions of constant water hardness and pH, [ML-cell membrane] will vary as a function of the free metal ion concentration in the exposure solution, $[M^{z+}]$. The molecular identity of «L-cell membrane», the surface ligand to which the metal binds, has rarely been determined, but it is often assumed that «L-cell membrane» is a metal transport site, through which the metal M can enter the cell, thereby gaining access to potentially metal-sensitive intracellular sites.

Several refinements of this basic model have been proposed, most importantly to account for the protection against metal toxicity afforded by water hardness (i.e., competition of the hardness cations, Ca^{2+} and Mg^{2+}, with M^{z+} for binding at the biotic ligand; equations 2a,b) and by acidity (i.e., analogous competition between the proton, H^+, and M^{z+}; equation 3) (Figure 2).

$$Ca^{2+} + \text{L-cell membrane} \overset{K_{Ca}}{\longleftrightarrow} \text{Ca-L-cell membrane} \tag{2a}$$

$$Mg^{2+} + \text{L-cell membrane} \overset{K_{Mg}}{\longleftrightarrow} \text{Mg-L-cell membrane} \tag{2b}$$

$$H^{+} + \text{L-cell membrane} \overset{K_{H}}{\longleftrightarrow} \text{H-L-cell membrane} \tag{3}$$

For a more complete discussion of the strengths and limitations of this "biotic ligand model", the reader may wish to consult several recent review articles [8–10].

2.2. Metal Uptake, Homeostasis, and Detoxification

Cationic metals such as Cd^{2+} and Ni^{2+} can enter biological cells via three generic mechanisms [6]: (i) facilitated cation transport; (ii) facilitated transport of an anionic metal-ligand complex; and (iii) simple diffusive uptake of lipophilic metal-ligand complexes. At epithelial surfaces, i.e., those exposed to waterborne metals, facilitated cation transport is by far the most common uptake mechanism, both for essential metals (for which there exist specific transport systems designed to ensure an adequate supply of these essential oligoelements) and for non-essential metals such as Cd (which succeed in "fooling" the transport systems designed for essential metals). When the facilitated cation transport mechanism dominates, metal-organism interactions tend to follow the predictions of the biotic ligand model described above.

Once inside the cell, the metal encounters an environment rather different from that outside the cell. The intracellular pH is close to 7 and remains virtually constant (except in digestive vacuoles), whereas in the external environment the pH may vary over 4 or more pH units. Likewise, intracellular ligand:metal ratios are much higher than in a typical aquatic system, favoring metal complexation, and the nature of the ligands is also very different (e.g., complexation by thiol groups is far more important in the intracellular environment than outside the cell). One of the consequences of this ligand-rich environment is that the free metal ion concentration within the cell is calculated to be vanishingly small (e.g., $< 10^{-15}$ M; see [11] for a discussion of this point). It follows that metal exchange reactions that proceed by dissociation of the original metal complex followed by binding of the free metal to the new ligand will be so slow as to be ineffective in the intracellular environment. Instead, it has been suggested that incoming (essential) metals are bound to metal chaperone ligands, which then bind to the appropriate receptor ligand and relinquish their metal [11,12]. Metallothionein, or at least certain metallothionein isoforms, may play a role in maintaining a labile pool of essential metals in the intracellular environment [13].

Given that even essential metals can be toxic if present in excess within the cell, one can appreciate that living cells have developed a variety of complementary mechanisms for controlling metal concentrations in the intracellular environment [14]: (i) up- or down-regulation of metal uptake; (ii) up- or down-regulation of metal efflux; (iii) metal sequestration in granules; and (iv) metal complexation by metallothionein. Of these various homeostatic mechanisms, metal sequestration and complexation (i.e., (iii) and (iv)) are of particular interest in the present context, since in principle they lead to changes in the subcellular partitioning of the metal and thus should be detectable analytically in field-exposed organisms (see Section 2.3).

2.3. Determination of Subcellular Metal Distribution

To study changes in the subcellular distribution of metals in living organisms, one must first isolate the different metal-binding ligands and then determine the amount of metal associated with each ligand pool. Methods currently used to perform the isolation step tend to fall into one of two classes, differential centrifugation or high performance liquid chromatography (HPLC) or some combination of the two (Table 1). In all cases the biological tissue of interest must first be homogenized, normally in a cold, isotonic medium and as gently as possible so as to avoid disrupting membrane-bound cellular organelles (e.g., nuclei, mitochondria, microsomes, lysosomes).

Once the sample has been homogenized, differential centrifugation can be used to separate the various organelles [14–17] as a function of their settling velocities (Figure 3). Centrifugation does not yield perfectly clean separations among fractions [18], and various potential artefacts can complicate the interpretation of centrifugal fractionation results (Table 1), such as, for example, breakage or clumping of particles and leakage of soluble

Table 1. Summary of the steps used to determine the subcellular partitioning of metals – advantages and limitations [16,17,19,20,24].

Operation	Consideration	Limitation
Tissue homogenization	• essential step to gain access to the intracellular environment • choice of homogenization buffer (minimize organelle breakage; maintain compatibility with HPLC separation)	• possible leakage from or inadvertent breakage of organelles, leading to redistribution of metals
Differential centrifugation	• inherently gentle • metal analyses necessarily carried out « off-line » • important to perform metal mass balances to evaluate metal losses or contamination during the separation steps	• inevitable overlap between some fractions, notably microsomes/ lysosomes • potential for metal redistribution following heat denaturation of the cytosol
Liquid chromatography	• choice of eluant (sufficiently high ionic strength to minimize loss of ligands by adsorption to chromatographic support) • better distinction among cytosolic ligands than with simple heat denaturation protocol • metal analyses can be carried out « on-line » • important to perform metal mass balances to evaluate metal losses or contamination during the separation steps	• potential adsorptive losses of cytosolic ligands during chromatography • potential for metal dissociation from ML complexes during the chromatography (especially for weaker complexes, e.g. with Zn) • potential for metal contamination from the chromatographic column and support during the separation step
Metal analyses	• if carried out « off-line », many different techniques can be used • if carried out « on-line », ICP-MS offers major advantages (high sensitivity, small sample volumes, rapid response times)	• to ensure low background metal concentrations, high purity reagents needed • high ionic strength eluants tend to favor corrosion of HPLC columns, pumps, fittings

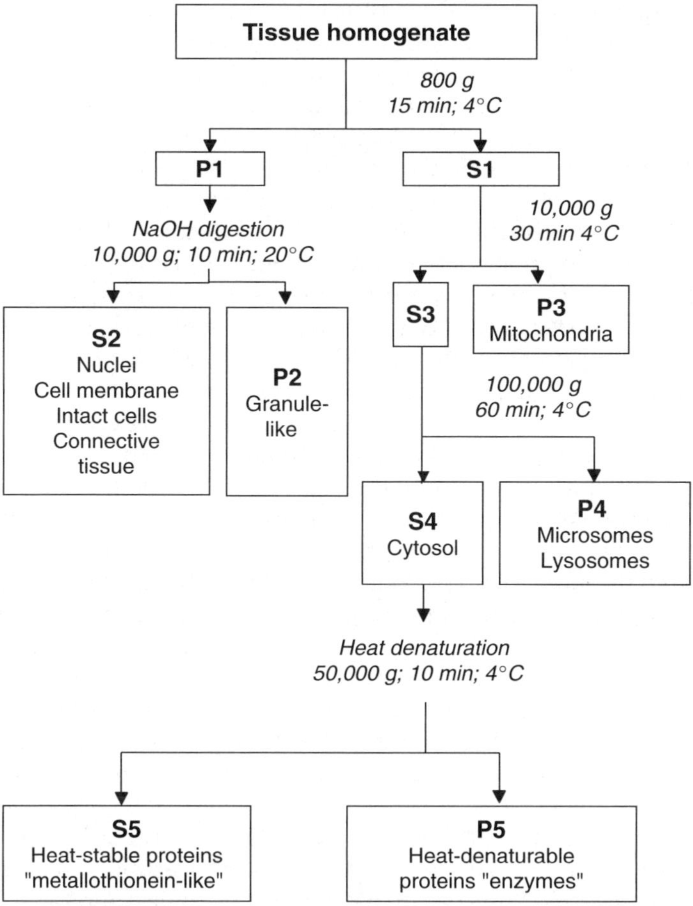

Figure 3. Differential centrifugation method used to separate the various subcellular fractions. P = pellet; S = supernatant. Adapted from [17] with permission from Elsevier, copyright (2006).

constituents from organelles [16,17]. Marker enzymes can be used to evaluate the prevalence of these artefacts and to optimize the centrifugation conditions [18], but nevertheless the fractions referred to in Figure 3 are operational in nature, and their designation as granule-like, mitochondria, lysosomes + microsomes, etc. should be considered with caution (see also Section 3.4.2). Similarly, the convenient grouping of the fractions into "potentially metal-sensitive" (heat-denaturable proteins or HDP, mitochondria, microsomes) and "detoxified" (granules and metallothionein-like or HSP) categories [16] is likely an oversimplification.

To complete the differential centrifugation approach and to determine the speciation of the dissolved cytosolic metals, the supernatant remaining after the final ultracentrifugation (Figure 3: supernatant S4) can be heated, cooled, and subjected to a second ultracentrifugation step. The heat-denaturable proteins will be found in the pellet, whereas heat-stable proteins such as metallothionein will remain in the supernatant. Alternatively, the heating and cooling step can be skipped and liquid chromatography used to separate the different cytosolic molecules involved in metal binding [19]. A variety of chromatographic approaches have been adopted to perform this separation. Size-exclusion chromatography is often used as the first step [20], followed in some cases by ion-exchange chromatography [21]. Lobinski and Szpunar [22] provide a useful summary of the various "hyphenated" techniques that can be used to separate and quantify metal complexes with biomacromolecules.

Chromatographic separations of metal-ligand complexes (Table 1) are deceptively simple to perform, but it must be admitted that conditions during the chromatographic separation differ greatly from those that existed in the original cytosol. Metal-ligand (ML) complexes may dissociate during the chromatographic step, or they may undergo exchange reactions with metals derived from the mobile phase, column or packing material. One obvious way to test for such phenomena is to calculate a mass balance for each metal of interest, i.e., to determine the percent recovery for each metal. Recoveries of strongly bound metals such as Cd or Cu are normally satisfactory, but there are several reports of major losses of Zn during the chromatographic separation step [20,23,24]. The potential influence of these reactions on the results of HPLC separations does not seem to be widely appreciated.

The various fractions isolated in the separation steps described above must obviously be analyzed for their metal content (Table 1). For the differential centrifugation part of the protocol, the fractions are necessarily digested and analyzed "off-line" in batch mode. A similar approach can also be used for the chromatographic separation step, i.e., by collecting and analyzing individual fractions of the column eluate. However, the successful coupling of high performance liquid chromatography to inductively coupled plasma mass spectroscopy (HPLC/ICP-MS) now allows on-line analysis of the column eluate. The virtues and limitations of ICP-MS as a multi-element HPLC detector for the analysis of metal-bearing peptides and proteins have been discussed in detail by Mason et al. [20,25].

2.4. Consequences of Metal Binding to Various Subcellular Components

Mason and Jenkins [5] have compiled a very useful review of metal-organism interactions, using "classical" metal toxicology concepts as the framework for their analysis. Using data from exposures of test organisms to metals under controlled conditions in the laboratory, they considered the traditional paradigm whereby (i) metals cause toxicity when they bind to "inappropriate", metal-sensitive intracellular sites, and (ii) there is a threshold concentration below which exposed organisms manage to prevent metals from binding to these metal-sensitive sites.

These are in effect the basic tenets of metal toxicology. When present in excess in living cells, trace metals may bind to physiologically sensitive target molecules (small peptides such as glutathione; metalloenzymes; DNA and RNA) or organelles (mitochondria, endoplasmic reticulum and nuclei), leading to deleterious effects. This binding of an "inappropriate" metal to a metal-sensitive site has been called "spillover" [26,27]. It follows that metal tolerance/resistance can be taken as evidence of the organism's ability to prevent "inappropriate" metals from binding to such sites.

In their analysis of metal detoxification [5], Mason and Jenkins distinguished among three classes of intracellular ligands on the basis of the consequences of metal binding. For ligands in the first class, metal binding leads to beneficial results (e.g., an apo-metalloenzyme that requires a specific metal in order to function normally). The opposite reasoning applies to ligands in the second class, where binding of the metal results in a negative effect (e.g., molecules that are inactivated when bound to an inappropriate metal). Detoxification thus corresponds to the efforts of the organism to limit the formation of species of this type. Finally, ligands in the third class are thought to be metal-insensitive, demonstrating neither beneficial nor deleterious effects when the metal binds to them. Complexes of this third type would represent metal sequestered in inactive intracellular pools. Various subcellular systems have evolved to control the distribution of metals among these three types of ligand, and to permit the accumulation, regulation, and immobilization of trace metals; these include metal binding proteins such as metallothionein (MT), lysosomes, granules, and membrane-bound vesicles. Regulation of metals may also involve control of the uptake and loss of metals [14], but in the present discussion we will concentrate on metal handling within the cell.

A limited number of laboratory exposures have been used to test this model of metal toxicity – representative examples are described in the following section.

2.5. Examples of Laboratory Experiments Linking Changes in Metal Exposure to Changes in Metal Subcellular Distribution and the Onset of Deleterious Effects

In this section we briefly review some of the laboratory evidence for the "spillover" hypothesis. In these experiments, metal-naïve aquatic organisms were exposed to a range of free-metal concentrations under controlled conditions in the laboratory and organism responses were followed at several different levels: total metal concentrations in the subcellular environment, changes in metal subcellular partitioning, and the manifestation of metal-induced deleterious effects. The metals studied include both essential and non-essential elements. The goal of the experiments was to determine if the onset of the deleterious effects coincided with detectable changes in the subcellular partitioning of a given metal. We have chosen two examples: the first conforms to the threshold model, with complete detoxification suddenly breaking down (Cu in crab larvae), whereas the second illustrates a variant of this model with partial detoxification followed by the onset of overt toxicity (Cd in a polychaete).

In the first case, Sanders and Jenkins [28] exposed a crab (*Rhithropanopeus harrisii*) throughout its period of larval development to a range of Cu^{2+} concentrations in sea water buffered with nitrilotriacetate (NTA) ($10^{-12.4}$ M to $10^{-7.9}$ M Cu^{2+}; concentrations of potentially competitive metals were kept constant). Crab cytosol was isolated by ultracentrifugation and cytosolic Cu was then separated by HPLC into three fractions: high molecular weight or HMW (>20 kDa; Cu-metalloenzymes + hemocyanin); metallothionein-like (10–12 kDa; resolved into two metal-binding ligands when separated by ion exchange chromatography); and low molecular weight LMW (<5 kDa). The authors examined relationships among [Cu^{2+}], cytosolic [Cu], copper-thionein concentrations and growth.

At low copper concentrations, i.e., from $10^{-12.4}$ to $10^{-10.6}$ M Cu^{2+}, cytosolic Cu was associated with metallothionein-like ligands and HMW ligands (Figure 4A), and was independent of external [Cu^{2+}]. At higher [Cu^{2+}], copper appeared in a LMW fraction (<5 kDa) and accumulated in this pool (Figure 4A), as well as in the MT pool, as [Cu^{2+}] increased. Copper concentrations in the HMW fraction initially declined, presumably in response to the induction of metallothionein, and then increased to their original levels. Neither crab survival nor the duration of its larval stage was affected by Cu^{2+} over the studied range, but reductions in larval growth (Figure 4B) coincided with Cu accumulation in the MT and LMW pools. Crab growth rate slowed at [Cu^{2+}] > 10^{-11} M, a concentration just above the estimated ambient range. As shown clearly in Figure 4, above this Cu^{2+} concentration the induction of MT-like proteins was insufficient to keep Cu out of the LMW fraction even though MT concentrations increased. The authors suggested that of the three cytosolic pools, it was accumulation of Cu in the LMW pool that was most indicative of toxic effects on the organism.

In the second example, Jenkins and Mason [29] examined the effects of a non-essential metal, Cd, on the marine polychaete *Neanthes arenaceodentata*. Organisms were exposed to a range of Cd^{2+} concentrations (10^{-12} to 10^{-8} M) in NTA-buffered seawater for 11 weeks (and allowed to mate after 5 weeks). Regardless of the Cd exposure concentration, total Cd

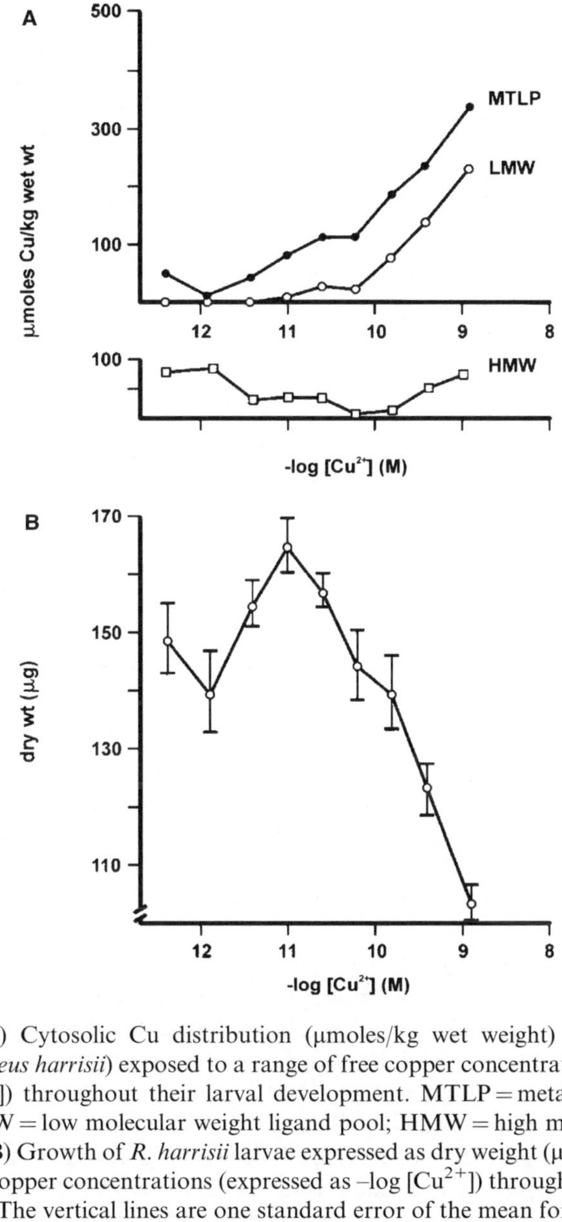

Figure 4. (**A**) Cytosolic Cu distribution (µmoles/kg wet weight) in crab larvae (*Rhithropanopeus harrisii*) exposed to a range of free copper concentrations (expressed as –log [Cu^{2+}]) throughout their larval development. MTLP = metallothionein-like peptides; LMW = low molecular weight ligand pool; HMW = high molecular weight ligand pool. (**B**) Growth of *R. harrisii* larvae expressed as dry weight (µg), exposed to a range of free copper concentrations (expressed as –log [Cu^{2+}]) throughout their larval development. The vertical lines are one standard error of the mean for five replicates. Adapted from [28] with permission from *The Biological Bulletin*, Woods Hole, MA, USA, copyright (1984).

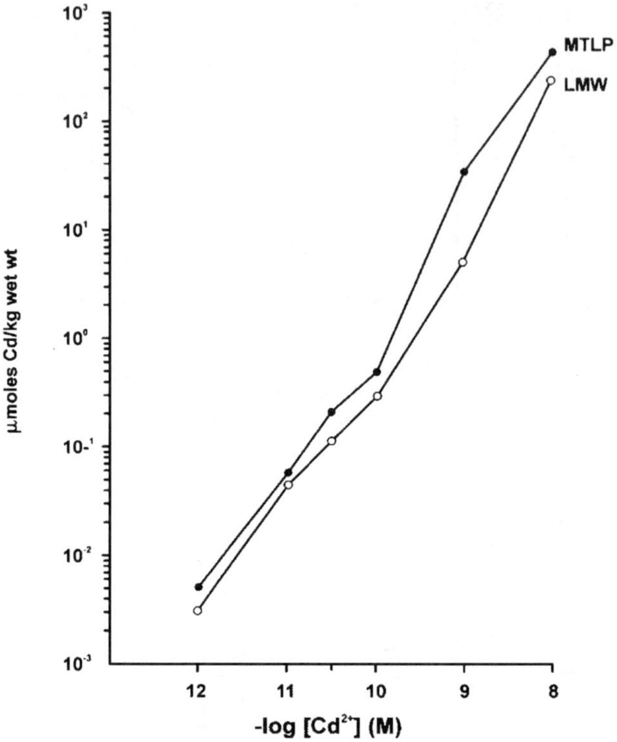

Figure 5. Cytosolic Cd distribution (µmoles/kg wet weight) in the marine polychaete *Neanthes arenaceodentata* exposed to a range of free cadmium concentrations (expressed as –log [Cd^{2+}]) for 11 weeks. Cytosol = total cytosolic metal; MTLP = metallothionein-like peptides; LMW = low molecular weight ligand pool. Adapted with from [29] with permission from Elsevier, copyright (1988).

accumulation reached steady-state within the first 7 days. The subcellular distribution of Cd was determined by a combination of differential centrifugation (cellular debris 200 *g*; nuclei-mitochondria 10,000 *g*; microsomes 100,000 *g*; cytosol) and HPLC. Cadmium concentrations decreased in the sequence cytosol > debris > mitochondria > microsomes.

Unlike the previous example, here the accumulation of Cd in the whole organism, *and* in all the subcellular fractions (Figure 5), was more or less proportional to [Cd^{2+}] over the entire concentration range. In the cytosol, which represented the major intracellular pool for Cd (40–65% of the accumulated metal), Cd accumulated in both the MTLP and LMW fractions; there was no significant accumulation of Cd in the HMW fraction. The observation of Cd in the LMW pool, even at low Cd exposures, is contrary to the threshold model. There was no hint of a step function or concentration threshold below which the polychaete was able to prevent Cd accumulation in potentially metal-sensitive fractions. Perturbations in reproduction were observed at an exposure concentration of 10^{-8} M Cd^{2+}, which coincided with an increased accumulation of Cd-LMW (change of slope in Figure 5). The authors concluded that the polychaete was incapable of controlling the subcellular distribution of Cd (i.e., ineffective detoxification).

These two laboratory examples were chosen to illustrate the range of metal detoxification responses that might be expected in field-exposed aquatic organisms. In the first case, which conformed closely to the threshold model, the crab larvae were able to prevent Cu binding to inappropriate subcellular ligands at low copper concentrations, and "spillover" only occurred at exposure concentrations above $\sim 10^{-11}$ M Cu^{2+}. In contrast, in the second case the polychaete was unable to prevent Cd from binding to inappropriate subcellular ligands even at the lowest exposures; the results suggest only partial detoxification at low Cd exposure followed by a breakdown in this strategy at higher metal concentrations (giving rise to the positive deviations from proportionality). In the following section, we discuss how the threshold model of toxicity might apply in the field and we present the results of studies that were designed to test the applicability of the model with native aquatic animals living along a metal contamination gradient.

3. REVIEW OF FIELD OBSERVATIONS LINKING CHANGES IN METAL EXPOSURE TO CHANGES IN SUBCELLULAR DISTRIBUTION AND THE ONSET OF DELETERIOUS EFFECTS

3.1. Conceptual Link between Laboratory Experiments and Field Observations

As discussed in the previous section, in laboratory experiments with metal-naïve organisms there is some evidence linking the onset of metal toxicity to the appearance of metals in putative metal-sensitive pools. However, there are very few, if any, tests of this model in the field for chronically exposed field populations.

In principle, one might compare the collection of native animals along metal concentration gradients in the field to the experimental manipulation of metal exposure concentrations in the laboratory. If a spatial study along a metal concentration gradient is analogous to experimental exposures at different metal concentrations, and if the native organisms behave according to the threshold model, then up to a certain exposure threshold they should be able to detoxify the metal effectively, but beyond this threshold we should observe metal binding to inappropriate subcellular ligands and the appearance of metal-induced effects (i.e., we might anticipate a "hockey-stick" response on proceeding from lake to lake along the metal concentration gradient; Figure 1B). On the other hand, if the partial detoxification model applies, then we would expect to see metal binding to inappropriate ligands even at the low end of the metal exposure gradient.

Our objective was thus to test the threshold hypothesis in the field with two different types of aquatic animals, yellow perch (*Perca flavescens*) and the floater mollusc (*Pyganodon grandis*). We worked with native animals, collected in lakes representing a metal concentration gradient, and compared different *metals* in the same organ, different *organs* in the same animal, and similar organs in different *animals*. In all the organs we determined both the total accumulation of each metal and its subcellular partitioning.

In the following sections we briefly describe the study areas, the sampling protocols and the methods used to determine subcellular metal partitioning. We then present the results for the spatial studies and compare these observations with the predictions of the threshold model. Interested readers will find more detailed descriptions of the materials and methods in the original journal papers, which are cited throughout the text.

3.2. Description of the Experimental Approach

3.2.1. Native Organisms

We chose juvenile yellow perch (*Perca flavescens*) and the adult floater mollusc (*Pyganodon grandis*) as the freshwater sentinel organisms for our studies, on the basis of their ubiquity, abundance, relative immobility, metal tolerance and known ability to synthesize metallothionein-like proteins in response to metal exposure. Yellow perch are abundant and widely distributed across North America; this species does not travel long distances and therefore its metal body concentrations tend to represent local sources [30,31]. The floater mollusc, a sedentary unionid bivalve, lives half-buried in sediments of the littoral zone of lakes and reflects the contamination of its immediate surroundings [32,33].

3.2.2. Metals Studied

Data were collected for four metals (Cd, Cu, Ni, and Zn), but in the present paper we will only consider those metals that are non-essential to fish and invertebrates (Cd, Ni) so as to be able to detect "inappropriate" metal binding without ambiguity. At the biochemical level, Cd is recognized to be an excellent inducer of metallothionein biosynthesis (in the laboratory) and this metal binds to MT in the cytosol. Exposure to Ni has also been shown to give rise to increased MT concentrations in fish [34], but it is not nearly as potent an inducer of metallothionein synthesis as Cd, at least in part because it is a "harder" cation than Cd in the hard-soft acid-base classification of metals and thus has very different ligand binding affinities [35].

3.2.3. Study Area

Bivalves were collected from lakes situated near Rouyn-Noranda in northwestern Québec, Canada, where a large metal smelter has been operating since 1927. Although atmospheric emissions are currently controlled, prior to 1985 the smelter was an important source of metals (Cd, Cu, Zn) and atmospheric deposition contaminated many of the surrounding lakes [36]. Sampling for yellow perch included lakes located both in the vicinity of Rouyn-Noranda and near Sudbury in north-central Ontario, Canada. Sudbury is also a mining and smelting area, but the surrounding lakes are contaminated with a somewhat different suite of metals (Cd, Cu, Ni) than the Rouyn-Noranda lakes (Cd, Cu, Zn). Within these study areas we chose lakes that were known from our earlier studies to have low, moderate or high metal concentrations in water and animals [33,37,38].

3.2.4. Spatial Studies. Sampling and Analysis

For each lake, we studied the response of the indigenous molluscs and fish to the metal gradient at the cellular/biochemical level (metal concentrations in target organs, subcellular metal partitioning, metallothionein induction), and this response was linked to effects at the physiological level (endocrine and metabolic status; indicators of oxidative stress).

3.2.4.1. Water. Water samples were collected in the epilimnion of each of the studied lakes, using in situ diffusion samplers, to estimate metal exposure. After equilibration with the ambient water ($\sim$2 weeks), the samplers were retrieved and sub-samples for trace metal, major cation, major anion, and organic carbon determinations were removed. The ambient pH was also measured in water collected at a depth of 0.5 m, near the diffusion samplers. Sample collection from the diffusion samplers is described in detail in Giguère et al. [39], as are the analytical methods used to determine the concentrations of the various analytes. Metal speciation at equilibrium was calculated for each lake with the Windermere Humic Aqueous Model (WHAM 6.0.1) [40].

3.2.4.2. Fish. Juvenile yellow perch were collected from each lake using a seine net and kept alive in aerated coolers for <3 h. In the laboratory, the fish were killed with a blow to the head and their liver was removed for study. The liver was chosen as the target organ on the basis of a previous study on wild perch in which this organ was shown to have high metal concentrations [39,41]. To minimize possible variability related to fish age, we collected juvenile fish of the same size class (7–10 g; 1–2 y) for the spatial study.

3.2.4.3. Molluscs. Adult molluscs (7–9 cm shell length; 5-6 y) were collected by divers in each lake, at a depth of 3–4 m, over a sediment surface area of $\leq$1,200 m^2; sampling was carried out in early summer just before bivalve reproduction. The bivalves collected from each lake were returned live to the field laboratory where gills and the digestive gland were dissected within 4 h of collection [42].

3.2.4.4. Total Metal Analyses. Biological samples were lyophilized, weighed, digested in an autoclave with nitric acid, and analyzed for Cd and Ni by inductively coupled plasma atomic emission spectrometry (ICP-AES). Analytical procedures and details of the quality assurance/ quality control measurements can be found in Bonneris et al. [42] and in Giguère et al. [17,39,43].

3.2.4.5. Metal Subcellular Partitioning. The biological samples were partially thawed, gently homogenized and kept on ice. Details of the homogenization procedures, which differed for the two species, can be found in Bonneris et al. [42] (mollusc gills and digestive glands) and in Giguère et al. [17] (fish livers). A differential centrifugation procedure, adapted from that of Wallace et al. [16] and shown in Figure 3, was used to separate various subcellular compartments: (1) a fraction comprising nuclei, cell membranes, intact cells and connective tissue, termed «nuclei/ debris» hereafter; (2) a granule-like or resistant fraction; (3) mitochondria; (4) a fraction combining microsomes and lysosomes; (5) cytosolic heat-stable proteins, including metallothioneins, designated ≪HSP≫; and (6) cytosolic heat-denaturable proteins, termed ≪HDP≫ (Figure 3). Four of the subcellular compartments are "particulate" (fractions 1 to 4), whereas two others (HSP and HDP) are dissolved "cytosolic" fractions. Details of the differential centrifugation approach, a critical evaluation of its strengths and weaknesses, and metal mass balances can be found in Giguère et al. [17].

Aliquots taken from each whole homogenate and all centrifugation pellets were digested with nitric acid; concentrations of Cd and Ni in the digested samples and in the supernatants were measured by ICP-AES. For both the supernatants and the pellets, metal concentrations in a given fraction are expressed on a total organ dry weight basis, e.g., total Cd or Ni burden (nmol) measured in a particular fraction:weight of the organ (g dry weight). The digestion procedures are described in detail in Bonneris et al. [42] and Giguère et al. [17].

3.2.4.6. Metallothionein Determinations. Biological samples were partially thawed, homogenized under a nitrogen atmosphere in TRIS buffer (25 mM; pH 7.2) and kept on ice. A sub-sample of each homogenate was dried at 65 °C for 48 h to determine the dry-weight to wet-weight ratio. Another sub-sample of the homogenate was centrifuged at 30,000 *g* for 30 min at 4 °C, and the supernatant was analyzed for metallothionein-like proteins with a Hg saturation assay [44,45]. As a quality control, recovery of a MT standard (MT from rabbit liver, Sigma Chemical Co.) was determined with every assay; mean recoveries ranged from 95 to 101%. Metallothionein concentrations in each organ are expressed in nanomoles of Hg-binding sites per gram dry weight (dw).

3.3. Metal Detoxification Strategies in Molluscs and Fish Living at Sites along a Metal Concentration Gradient

3.3.1. Exposure Gradient and Bioaccumulation

The choice of lakes for the perch sampling was not identical to that for the bivalves (only lakes Opasatica and Vaudray were common to the two sets; Tables 2 and 3). This separation between the two groups of lakes resulted in part because perch are more tolerant of acidic conditions than are the bivalves, and in some lakes perch were present but the bivalve was not. Nevertheless, for both the perch and mollusc spatial studies, the selected lakes clearly represent a marked Cd and Ni concentration gradient, as reflected in the ambient free metal ion concentrations calculated with the WHAM chemical speciation model (Tables 2 and 3). For the nine lakes sampled for bivalves, the highest free Cd^{2+} concentrations were $\sim$14-fold higher than those at the least contaminated sampling stations (Table 2). The eight lakes retained for the perch study covered a wider geographical area, and the variation in calculated free Cd^{2+} concentrations ($\sim$115-fold) and free Ni^{2+} concentrations ($>$700-fold) was greater (Table 3) than for the Rouyn-Noranda suite of bivalve lakes.

Table 2. Aqueous free Cd^{2+} concentrations calculated for the bivalve suite of lakes and total organ Cd concentrations (± SD) in the bivalves collected from these lakes.

Lake (code)	Aqueous $[Cd^{2+}]$ (pM)	Gill [Cd] (nmol/g dw)	Dig. Gland [Cd] (nmol/g dw)
Évain (EV)	7.3	160 ± 42	150 ± 36
Opasatica (OP)	9.4	95 ± 11	123 ± 15
Ollier (OL)	16	53 ± 3	55 ± 8
Dufay (DF)	20	416 ± 153	300 ± 112
Joannès (JO)	42	1000 ± 191	325 ± 28
Héva (HE)	39	641 ± 78	388 ± 23
Caron (CA)	99	243 ± 34	245 ± 31
Bousquet (BO)	89	1472 ± 129	452 ± 63
Vaudray (VA)	62	979 ± 152	969 ± 108
max/min	**14**	**28**	**18**

Table 3. Aqueous free Cd^{2+} and Ni^{2+} concentrations calculated for the fish suite of lakes and total liver Cd and Ni concentrations (± SD) in perch collected from these lakes.

Lake (code)	Aqueous $[Cd^{2+}]$ (pM)	Aqueous $[Ni^{2+}]$ (nM)	Liver [Cd] (nmol/g dw)	Liver [Ni] (nmol/g dw)
Rouyn-Noranda				
Opasatica (OP)	20	3.1	36 ± 1	2.3 ± 0.9
Vaudray (VA)	30	2.1	320 ± 62	4.1 ± 1.4
Osisko (OS)	250	2.9	282 ± 19	6.8 ± 3.2
Dufault (DU)	2300	4.9	421 ± 34	5.6 ± 4.4
Sudbury				
Laurentian (LA)	260	520	42 ± 9	16.1 ± 3.3
Wavy (WA)	1200	750	118 ± 6	20.9 ± 2.4
Raft (RA)	880	1280	509 ± 82	82.8 ± 11.7
Hannah (HA)	1000	1600	92 ± 16	26.6 ± 4.1
max/min	**115**	**760**	**14**	**36**

In that our biomonitor organisms have lived their entire lives in the lakes from which they were collected, we can safely assume that their bioaccumulated metal concentrations represent a steady state between uptake, storage, and elimination [46]. For a given type of animal, metal concentrations varied among its organs; for example, the gills and digestive gland were important sites for metal accumulation in the bivalve [33,42], as was the liver in yellow perch [39,41].

The metal exposure gradient was reflected in metal concentrations in these target organs. Cadmium concentrations in the gills and digestive gland of *P. grandis* varied from lake to lake (Table 2) by a factor of 28 for the gills and 18 for the digestive gland. Total hepatic Cd and Ni in *P. flavescens* also varied from lake to lake (Table 3) with hepatic concentrations increasing 14-fold for Cd and 36-fold for Ni along the contamination gradient. This higher metal bioaccumulation along the metal exposure gradient was associated with higher steady-state concentrations of metallothionein-like peptides, as determined by the mercury-saturation assay. Significant relationships were obtained between MTLP concentrations and bioaccumulated Cd for the bivalve (Figure 6), an observation that is consistent with the results of our earlier work in this region on *P. grandis* [33] and *P. flavescens* [47].

In summary, we have the desired metal concentration gradient in the studied lakes and along this exposure gradient metal bioaccumulation (bivalve gills and digestive gland; fish liver) increases, as does the steady state concentration of metallothionein. The conditions are thus appropriate to answer the question: Is the induction of metallothionein along the gradient sufficient to prevent Cd and Ni from binding to inappropriate subcellular ligands, at least in the lakes at the lower end of the contamination gradient?

3.3.2. Subcellular Metal Partitioning. General Observations

Application of the differential centrifugation protocol (Figure 3) yielded six operationally-defined fractions, each of which was analyzed for Cd and Ni. Since the toxicological significance of metal accumulation in the ≪nuclei + cellular debris≫ fraction is unclear (see discussion in Giguère et al. [17] and Cain et al. [48]), we will focus our attention on the remaining subcellular fractions in the following discussion. Metal partitioning patterns differed between organs (for the bivalve), between metals (for a given organ), and to a lesser extent from lake to lake (see below).

Comparison of the bivalve gills and digestive gland reveals marked differences between the two organs. Metal partitioning in the gills was dominated by calcium granules, which represented $51 \pm 7\%$ (mean $\pm$ SD) of the gill dry weight and accounted for the majority of the total gill Cd burden ($58 \pm 13\%$) [42]. In the digestive gland the granule fraction was generally much less important as a Cd sink ($10 \pm 10\%$), a result that reflects the much lower abundance of granules in this organ ($5 \pm 3\%$ of the gill dry wt). Despite this difference, the partitioning of cytosolic Cd between the HDP and HSP protein fractions was similar in the two bivalve organs, with the higher percentage of Cd being found in the HSP fraction.

Comparison of the bivalve digestive gland and the perch liver, two organs with a similar metabolic function, showed similar Cd partitioning patterns, the HSP fraction being the single most important subcellular fraction

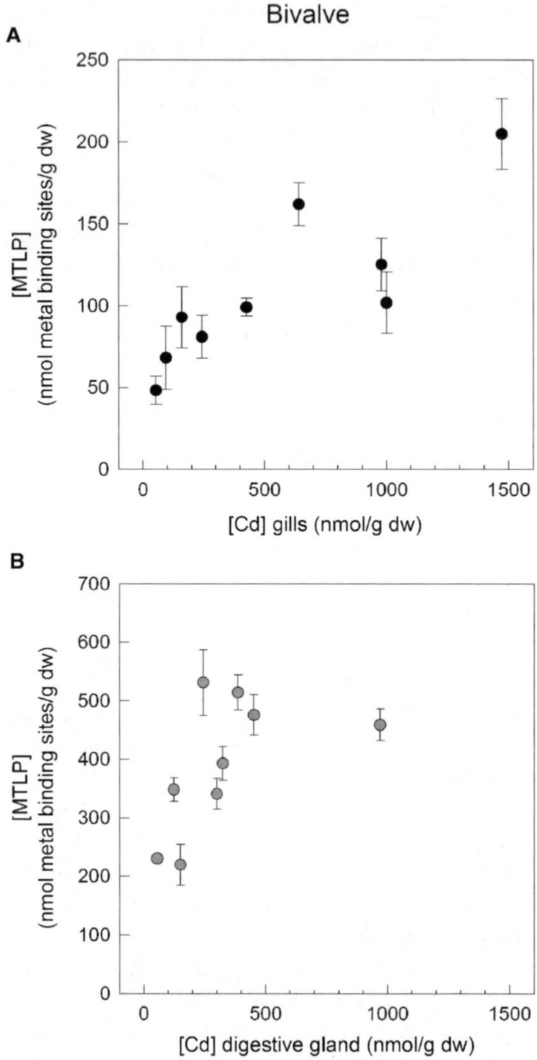

Figure 6. Relationships between the steady-state concentrations of accumulated Cd and metallothionein-like peptides in organs of the bivalve, *Pyganodon grandis*. (**A**) bivalve gills; (**B**) bivalve digestive gland. Each point represents data for bivalves collected in a single lake (mean $\pm$ SD; N = 3 composite samples from 4 different individuals). Data are replotted from [49] with permission from Elsevier, copyright (2005).

(mean ± SD: 39 ± 12% in bivalve digestive gland; 47 ± 4% in perch liver). Within the particulate fraction (organelles, granules), the contributions of the "mitochondria" and "lysosomes + microsomes" fractions were consistently higher in the bivalve digestive gland than in the bivalve gills, and were even more important in the perch liver.

Lastly, within the perch liver, the partitioning of Ni was very different from that of Cd. The HDP fraction was the single most important contributor to subcellular Ni (26 ± 4%), with a smaller proportion present in the HSP fraction (18 ± 4%). Somewhat lower contributions were afforded by the mitochondrial (13 ± 1%) and lysosomes + microsomes (16 ± 4%) fractions. Nickel contamination of the resistant and nuclei + debris fractions (from the NaOH reagent used to digest the tissue) prevented us from evaluating Ni detoxification in granules, but the low levels of this contamination set an upper limit for the contribution of these fractions and allow us to conclude that only a small proportion of Ni could be accumulated in these two fractions.

3.3.3. *Metal Subcellular Partitioning. Spatial Trends*

To evaluate changes in metal partitioning along the exposure gradient, we have plotted the Cd or Ni concentration in each subcellular fraction against the total organ metal concentration (Figures 7 and 8), where the total organ metal concentration (x-axis) is a reflection of the past metal exposure experienced by the biomonitor species. These figures are analogous to the plots shown earlier for the laboratory studies (Figures 4 and 5). Note that the metal concentrations in a given fraction are expressed on a total organ dry weight basis, i.e., total metal burden (nmol) measured in a particular fraction:weight of the organ (g dw), which allows us to compare the bivalve and the perch on a common basis [49].

3.3.3.1. Bivalve (Cd). For bivalves collected along the metal exposure gradient, Cd concentrations increased in all but one of the subcellular compartments, in both the gills and the digestive gland (Figure 7). The HDP fraction constituted a notable exception to this trend; Cd concentrations in this fraction remained quite low even in the most contaminated lakes (Figure 7a,b), especially in the gills, and showed no tendency to increase along the metal concentration gradient. As expected, Cd concentrations in the HSP fraction (Figure 7e,f) were positively related to variations in MT concentrations along the exposure gradient ($P < 0.001$), both in the gills and in the digestive gland [42].

3.3.3.2. Perch (Cd and Ni). Cadmium concentrations in all subcellular compartments, with the exception of the granule fraction, were significantly

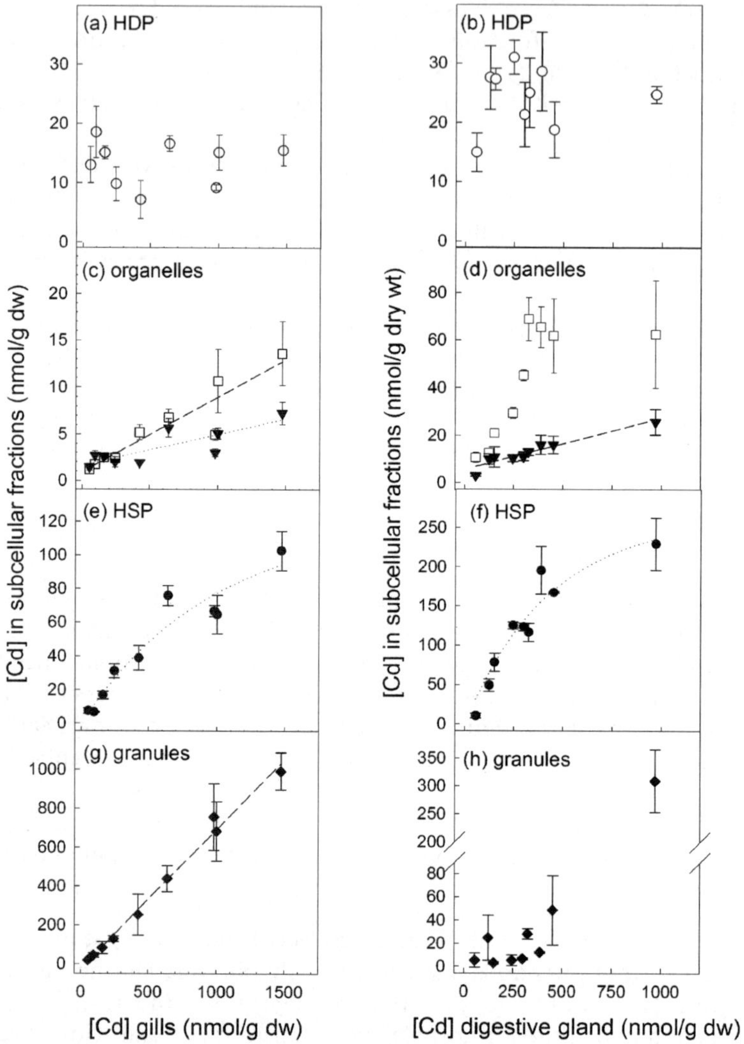

Figure 7. Cadmium sub-cellular partitioning in the gills and digestive gland of the bivalve, *Pyganodon grandis*, collected from nine lakes along a metal contamination gradient (see Table 2). Cadmium concentrations in the various fractions are compared to Cd concentrations in the whole organ: (a,b) = heat-denaturable proteins (HDP); (c,d) = organelles (open squares = mitochondria; inverted triangles = lysosomes + microsomes); (e,f) = heat-stable proteins (HSP); (g,h) = granules. Each point represents data for bivalves collected in a single lake (mean ± SD; N = 3 composite samples from 4 different individuals). Metal concentrations for the metal-sensitive fractions are plotted in panels a-d; metals in the metal-detoxified fractions are plotted in panels e-h. Curves are lines of best fit. Data are replotted [42] with permission from Elsevier, copyright (2005).

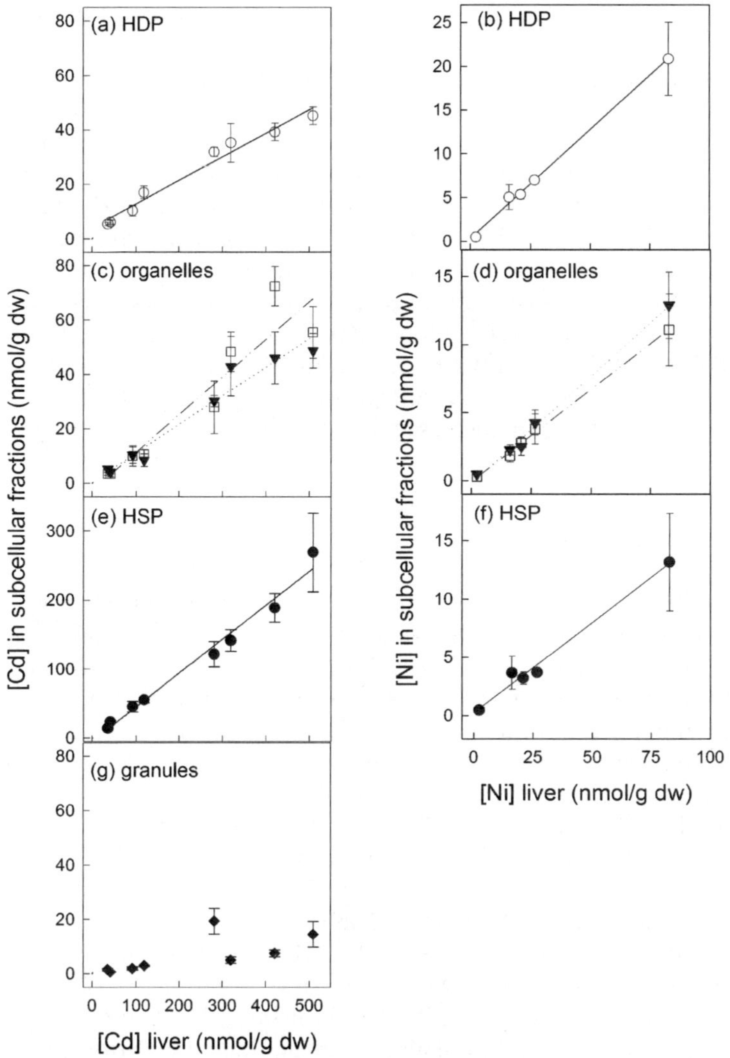

Figure 8. Cadmium and nickel partitioning in the liver of yellow perch, *Perca flavescens*, collected from eight (Cd) or five (Ni) lakes along a metal contamination gradient (see Table 3). Cadmium and Ni concentrations in the various fractions are compared to Cd and Ni concentrations in the whole organ: (a,b) = heat-denaturable proteins (HDP); (c,d) = organelles (open squares = mitochondria; inverted triangles = lysosomes + microsomes); (e,f) = heat-stable proteins (HSP); (g) = granules. Each point represents data for juvenile perch collected in a single lake (mean ± SD; N = 3 composite samples). Results for Ni in the granules fraction are unavailable (see text for details). Metal concentrations for the metal-sensitive fractions are plotted in panels a-d; metals in the metal-detoxified fractions are plotted in panels e-g. Curves are lines of best fit. Data are replotted from [17] with permission from Elsevier, copyright (2006).

($P < 0.001$) linearly related to total hepatic [Cd] (Figure 8a,c,e). Most of the hepatic Cd burden was in the cytosolic portion of the cells (average for all lakes of $59 \pm 7\%$). Similarly, plots of Ni concentrations in the HDP, organelle, and HSP fractions as a function of total hepatic Ni showed linear trends with no hint of a threshold response (Figure 8b,d,f). Approximately half of the total hepatic Ni burden was found in the cytosolic portion of liver cells (average for all lakes of $44 \pm 7\%$). Marked Ni bioaccumulation occurred in the HDP fraction (43-fold difference between the most and least contaminated lakes, followed by the mitochondria (38-fold), HSP (28-fold) and microsomes/lysosomes (27-fold) fractions. However, the contribution of each fraction to the total Ni burden remained constant as hepatic [Ni] increased (as reflected in the similar slopes in Figure 8b,d,f).

3.3.4. Summary of Field Observations

3.3.4.1. Bivalve. Gill tissues contained abundant calcium concretions that accounted for over 60% of the total gill Cd burden. Cadmium concentrations in this granule fraction reflected ambient metal concentrations. Metal concentrations in the digestive gland also responded to the metal contamination gradient, but to a lesser extent than the gills, reflecting the lower abundance of granules in the digestive gland. In bivalve gills, as total gill Cd increased along the metal exposure gradient, Cd concentrations increased markedly in the granule and HSP fractions, but the HDP fraction did not respond and the increase in the ≪mitochondria≫ fraction was modest (Figure 7a,c,e). Trends for the digestive gland were somewhat similar: Cd concentrations in the HSP fraction again increased along the contamination gradient, and Cd concentrations in the HDP fraction, though somewhat higher than in the gill tissue, did not increase along the exposure gradient (Figure 7b,d,f). However, unlike the case for the gills, Cd concentrations in the granule fraction varied little along the contamination gradient (with the exception of a single lake, Lake Bousquet), and Cd concentrations in the ≪mitochondria≫ fraction increased more markedly than in the gills, with no hint of protection in the less contaminated lakes.

3.3.4.2. Perch. In juvenile yellow perch, hepatic Cd and Ni also accumulated to higher levels in lakes at the high end of the exposure gradient than in the control lakes. Accumulation of Cd was accompanied by the induction of metallothionein-like peptides, but even in the moderately contaminated lakes, some of the Cd was present in putative metal-sensitive fractions (HDP; mitochondria). Unlike the observations with the bivalve, Cd concentrations in the HDP fraction increased as the total Cd concentration in the organ increased. The relative contribution of this fraction to the total Cd burden did however decrease significantly (16 to 9%; $P = 0.008$) [17], suggesting that partial protection is afforded to the proteins present in this fraction by the initiation of some detoxification mechanism (e.g., metallothionein induction). There was no indication of a threshold level below which all the Cd was completely bound by the metallothionein (Figure 8a,c,e). Similarly, along the metal exposure gradient, Ni concentrations increased in the organelles fraction and in the "enzyme" or heat-denaturable proteins fraction. Again, there was no threshold exposure concentration below which Ni did not appear in these potentially metal-sensitive fractions (Figure 8b,d).

3.4. Critical Analysis of the Threshold Toxicity Model as Applied in a Field Setting and the Role of Metallothionein

3.4.1. Apparent Failure of the Threshold Model

The goal of the studies described in Section 3.3 was to determine whether the threshold model applies to aquatic animals that are chronically exposed to sublethal but elevated metal concentrations in their natural habitat. To compare metal-handling strategies in our two biomonitor species, we plotted the metal concentration in each sub-cellular fraction against the total organ metal concentration (Figures 7 and 8). If the threshold model applied to chronically exposed indigenous aquatic organisms, the concentration of "inappropriate" metals such as Cd or Ni in potentially metal-sensitive fractions should remain low and constant in specimens collected in lakes resulting in low tissue concentrations, whereas above a certain threshold exposure, towards the high end of the exposure gradient, one should begin to see Cd and Ni appearing in the metal-sensitive fractions. Such a trend gives the "hockey-stick" function shown in Figure 1.

From an examination of Figures 7 and 8, it is clear that the threshold model does not apply to field-collected animals. Although metallothionein induction increases as a function of metal exposure (as indicated by the increase in the metals bound to the HSP fraction), the resulting steady-state concentrations of metallothionein are insufficient to prevent Cd and Ni from binding to potentially metal-sensitive sites within the subcellular environment (mitochondria and heat-denaturable proteins). Indeed, for virtually all of the subcellular compartments examined, the amount of Cd or Ni found in a given subcellular fraction is directly proportional to the total amount of the metal in the organ. This proportionality is analogous to what Jenkins and Mason [29] reported for their experimental long-term exposures of the marine polychaete, *N. arenaceodentata* (compare Figure 5 with Figures 7c–d; 8a–d), and conflicts with the predictions of the threshold model. Only in the case of the bivalve, for Cd binding to the heat-denaturable protein fraction (Cd–HDP), was this proportionality not observed (Figures 7a,b). Cadmium concentrations in this fraction remained relatively low and did not respond to the increase in ambient $[Cd^{2+}]$. For example, in the most contaminated lakes, the gill Cd-HSP fraction increased 15-fold (up to 103 nmol Cd/g dry weight), but the gill HDP fraction remained unchanged ($\sim$15 nmol/g dry weight).

3.4.2. *Possible Pitfalls in the Interpretation of the Subcellular Partitioning Results*

As mentioned earlier (Section 3.2.4), the fractions obtained with our subcellular partitioning protocol (Figure 3) are clearly operational in nature. For example, metal accumulation in fraction P4 (lysosomes + microsomes, see Figures 7c,d; 8c,d) is inherently difficult to interpret, given the fact that its components can play a variety of biochemical roles [17]. On the one hand, if metals detected in this faction were mostly attributable to an accumulation in the lysosomal fraction, then their presence would be indicative of metal storage for eventual elimination. On the other hand, if metals detected in this fraction were mostly in microsomes or in fragmented endoplasmic reticulum, then their presence could be interpreted as indicative of metal binding to a "metal-sensitive site", since these organelles are responsible for protein synthesis and transport and for glycogen storage [50].

In the cytosolic fraction (S4), the heat-denaturable protein component is assumed to correspond largely to (metallo)-enzymes and other non-enzymatic proteins (e.g., albumin, respiratory pigments such as hemoglobin). The presence of Cd and Ni in this fraction clearly reflects "inappropriate metals bound to a potentially metal-sensitive site", i.e., a potential harbinger of metal toxicity (see Section 3.4.3). We considered the possibility that the proteins contributing to the HDP fraction might lose their metals on denaturation, leading to an underestimate of the metal originally bound to these ligands [19]. However, in control experiments with perch from the reference lake (Opasatica), we isolated the S4 fraction (operationally defined cytosol), equilibrated it with a ^{109}Cd spike, and then proceeded with the heat denaturation step. The HDP fraction retained a ^{109}Cd label, demonstrating that the heat-denaturable cytosolic proteins could still complex the Cd after denaturation. This spiking experiment does not rule out the possibility that some "leakage" of metal from the HDP fraction to the HSP supernatant occurs during the denaturation step. However, if such leakage did occur, it would lead to an *under*estimation of the amount of Cd bound to the HDP fraction and thus an underestimate of the binding of Cd to "inappropriate" sites. The fact that we can detect Cd in the HDP fraction despite this possible leakage actually strengthens our conclusion that there is no threshold below which Cd does not bind to heat-denaturable proteins. Note too that if this leakage were quantitatively important, the relationship between Σ(Cd + Cu + Zn) in the HSP fraction and total metallothionein concentrations would break down; this is not the case (see below).

In the speculative realm, it is at least possible that the HDP fraction might include specific ligands that have been synthesized to cope with the influx of Cd and Ni. Whereas some limited evidence does exist for the synthesis of metal-binding ligands other than MT or MTLP in response to metal stress, notably in some marine invertebrates [19,51,52], in all reported cases these ligands have been heat-stable (not heat-denaturable).

The other cytosolic fraction, heat-stable proteins and peptides, obviously includes metallothionein and metallothionein-like proteins. However, various amino acids and the low molecular weight peptides (e.g., GSH) would also report to this fraction. All these ligands are heat-stable and could complex soft, polarizable metals such as Cd. However, independent determinations of MT (Hg saturation assay), in bivalve gill and digestive gland (Figure 9a,b) and in yellow perch liver (Figure 9c), were closely related to the sum of the metals (Cd + Cu + Zn) accumulated in the HSP fraction, indicating that MTLP are likely the major heat-stable ligands in this fraction. In addition, in the perch study, hepatic GSH concentrations were determined independently [43] and showed a significant *decrease* along our exposure gradient, as [Cd-HSP] increased. Based on these complementary observations, we conclude that the metals found in the HSP fraction are largely bound to metallothionein-like proteins. We did not include Ni in these regressions, as Ni concentrations in the HSP fraction were unavailable for these animals (all of which came from the Rouyn-Noranda lakes, where Ni ambient concentrations are low).

3.4.3. *Evidence for Metal-Induced Adverse Effects*

In the foregoing discussion we have emphasized metal subcellular partitioning, and have implied that binding of non-essential metals such as Cd or Ni to the mitochondria or to the heat-denaturable protein fraction can be construed as an indication of sublethal metal toxicity. Indeed, for both the bivalve and the perch there is good evidence that the animals are adversely affected by their metal exposure, in a dose-dependent manner.

At the biochemical level, malondialdehyde measurements showed some evidence of oxidative stress in the bivalve gills but not in their digestive gland [38,42]. In a separate study [53] we measured progressive population-level effects for bivalves living in lakes along the Cd exposure gradient (littoral density, live weight, dry viscera biomass, production, and cumulative fecundity) and demonstrated that all these population-level parameters

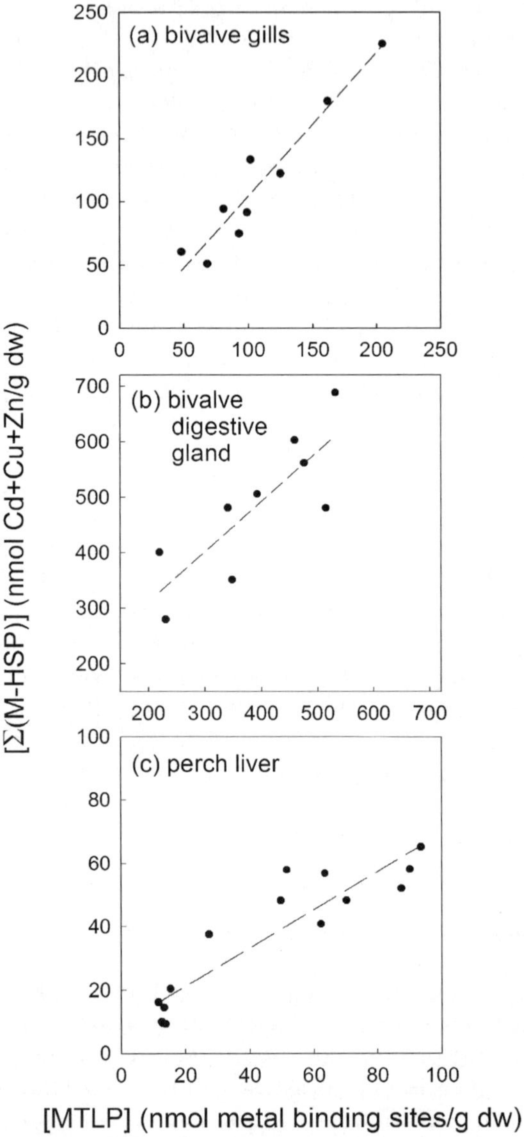

Figure 9. Relationships between the sum of Cd, Cu, and Zn concentrations in the heat-stable proteins (HSP) fraction, obtained after differential centrifugation and heat-denaturation of organ homogenates, and the concentration of metallothionein-like peptides (MTLP) as determined by the Hg saturation assay in the same organs. (a) *Pyganodon grandis* gills; (b) *P. grandis* digestive gland; (c) *Perca flavescens* liver. Data in panel (a) are replotted from [75] with permission from Elsevier, copyright (2002).

decreased gradually with increasing concentrations of free Cd^{2+} at the sediment-water interface, without any apparent threshold.

Measures of oxidative stress in juvenile yellow perch were inconclusive [43], but endocrine and metabolic perturbations have been noted in perch from the more contaminated lakes. These perturbations were characterized by an attenuated cortisol stress response *in vivo* and a lower secretory capacity in response to ACTH *in vitro* [47,54,55]. In addition to the endocrine responses, anomalies in intermediary metabolism and use of energy reserves were observed, suggesting an impaired capacity to mobilize liver glycogen reserves [54,56]. Metabolic impairment was also reported by Couture and Rajotte [57], who demonstrated that chronic metal exposure of yellow perch in Sudbury lakes was associated with a reduction of their aerobic capacities.

At the whole organism level, Sherwood et al. [58] studied the bioenergetics of yellow perch in the Rouyn-Noranda lakes. Using field estimates of fish growth coupled to *in situ* estimates of food consumption, they calculated that growth efficiency (i.e., the annual growth increment relative to the total energy consumed) was three times lower in cortisol-impaired yellow perch from the metal-contaminated lakes relative to fish from reference lakes. The authors concluded that metal-exposed perch experience greater total energetic costs, including both the "cost" of detoxification and the "cost" of hunting for smaller and scarcer prey items in an impoverished benthic community [59]. Somewhat different trends were reported by Couture, Pyle and co-workers on the basis of their studies of both the Sudbury and Rouyn-Noranda lakes [60,61]; yellow perch grew faster in contaminated than clean lakes but the longevity of contaminated fish was greatly reduced.

As a cautionary note, it should be pointed out that animals living in the Rouyn-Noranda or Sudbury lakes are exposed to metal mixtures, and that the various impairments described above cannot at this stage be unequivocally attributed to a single metal. Nevertheless, with particular reference to the Sudbury lakes, the low affinity of Ni for thionein cytosolic proteins relative to other metals, its consequent accumulation in the HDP fraction and the apparent lack of detoxification of this metal at the cellular level all point to Ni being the principal stressor in these lakes. On the other hand, in the Rouyn-Noranda area Cd consistently shows the strongest statistical link to metallothionein induction and to the degree of metabolic impairment. Borgmann et al. [62,63] have also linked Cd and Ni to deleterious effects in, respectively, Rouyn-Noranda and Sudbury area lakes.

3.4.4. Literature Precedents

Although the results of our field measurements appear to contradict the threshold model of metal toxicity, the detection of non-essential metals in potentially metal-sensitive subcellular pools even at low metal exposures is not without precedent. In the examples given below, for invertebrates and fish, we first present some laboratory evidence for this phenomenon and then consider field observations.

3.4.4.1. Laboratory. No results were found in the literature for freshwater molluscs and realistic Cd or Ni concentrations, but several studies do report binding of non-essential metals to non-thionein ligands in marine species. Ng and Wang [64] exposed the marine green mussel (*Perna viridis*) to waterborne ^{110m}Ag (0.6 nM) and ^{109}Cd (18 nM) for 24 h, and followed the dynamics of the subcellular distribution of Ag and Cd using a differential centrifugation scheme similar to that employed in our work. Over the first 24 h most of the accumulated Ag and Cd was associated with the insoluble pellet (50–60%). The remaining cytosolic metals were partially bound to a metallothionein-like fraction (HSP), but an equivalent amount was found in the HDP fraction. Similar results were obtained for a 96-h pulse-chase experiment with radio-labelled phytoplankton; although metal distributions among the three fractions (insoluble; MT; HDP) differed slightly from those determined for the waterborne exposures, in both cases an appreciable amount of Cd (20–30%) was associated with the HDP fraction, and interestingly this proportion remained virtually unchanged during a 28-d depuration period (i.e., no metallothionein-mediated rescue function; cf. [65]).

In 24-h experiments on zebrafish (*Brachydanio rerio*), at quite low Cd concentrations (0.1 to 15 nM), Wicklund-Glynn [66] observed binding of Cd to both metallothionein and to non-thionein HMW subcellular ligands. Even at the lowest exposure concentration (0.1 nM), Cd was associated with the organelle fraction and the HMW cytosolic ligands after these short-term exposures. Hamilton et al. [67] carried out a long-term laboratory exposure of brook trout (*Salvelinus fontinalis*) to Cd and noted a marked increase in metallothionein levels during the course of the exposure. However, despite this induction of metallothionein they also found appreciable Cd in their "non-thionein" fraction, even at the lowest exposure concentration (it should be noted however that this "low" concentration (32 nM) greatly exceeded the highest Cd concentrations observed in our most contaminated lakes (< 3 nM)).

Brown et al. [68] reported analogous results for scorpionfish (*Scorpaena guttata*), albeit once again for exposure to environmentally unrealistic Cd concentrations (900 to 180,000 nM). They analyzed both 4-day acute and 4-8 week chronic exposures, and noted different patterns in the subcellular distribution of Cd. Acute exposures seemed to indicate saturation of the available metallothionein and a saturation-dependent "spillover" of Cd into the heat-denaturable protein pool, whereas chronic exposures appeared to involve an equilibrium-dependent distribution of Cd between the HMW and MT pools. Although metallothionein sequestered a large portion of Cd in exposed scorpionfish, in most instances a portion escaped detoxification by metallothionein and was associated with the HMW pool. Interestingly, the authors concluded that "in fact, a threshold of MT-Cd increase without a corresponding HMW-Cd increase may not exist" and suggest that the concept of saturation of MT and spillover into the HMW pool may be applicable for acute exposures but not chronic exposures (note that the lead author was in fact one of the first researchers to formulate the "spillover" hypothesis). Our results indicate that binding of metals such as Cd to the HDP pool also occurs for chronic exposures at environmentally realistic metal concentrations.

3.4.4.2. Field. In one of the first field tests of the threshold model, in a marine setting, Jenkins et al. [69] collected white croakers (*Genyonemus lineatus*) from two stations in the southern California bight (one more contaminated than the other) and used liquid chromatography to determine the cytosolic metal distributions in liver, muscle, kidney and gastrointestinal tract tissues. Although most of the hepatic Cd was found under the metallothionein peak (95–97%), appreciable binding of Cd to the HMW peak was observed in the gastrointestinal tract (33–40%), in the muscle (71–93%), and in the kidney tissue of fish from the less contaminated site (55%). In all cases the contribution of the LMW peak was low to negligible. These results provide clear evidence that binding of non-essential metals such as Cd to potentially metal-sensitive sites does occur in wild fish, but further interpretation of these results is limited by the fact that only two sites were sampled (i.e., no progressive contamination gradient).

In a second marine example, Mouneyrac et al. [70] used a combination of centrifugation, heat-denaturation, and liquid chromatography to determine the subcellular distribution of Cd in the soft tissues of oysters (*Crassostrea gigas*) collected from a metal-contaminated site and from a clean control site. At the control site, total Cd concentrations in the soft tissues were low and 95% of the accumulated metal was associated with the HMW heat-denatured fraction. Concentrations of metallothionein-like proteins, as determined by differential pulse polarography, were much higher at the contaminated site, and the cytosolic Cd distribution reflected their presence; 67% of the accumulated Cd was found under the MTLP peak, but 33% nevertheless remained associated with the HMW fraction.

In one of the rare field examples from the freshwater environment, Cain et al. [71] collected benthic invertebrates along a metal concentration gradient in the Clark Fork River, Montana, USA, and determined the subcellular partitioning of cytosolic metals. Just as in our study, Cd concentrations increased progressively along the metal exposure gradient, in both the "metal-detoxified" and the "metal-sensitive" fractions, without any suggestion of a lower threshold below which metal-sensitive fractions were protected.

3.4.5. Ecotoxicological Implications

Although the results of our field studies represent compelling evidence for the incomplete detoxification of Cd and Ni, even in animals living in relatively clean lakes, it is also apparent that metallothionein is induced in these organisms in a dose-dependent manner (Figures 7e–f; 8 e–f). In addition, there is no clear evidence for saturation of the capacity of the perch or the bivalve to synthesize metallothionein (cf. [72]). In other words, bivalves and perch living in the moderately contaminated lakes possess the metabolic capacity to increase their steady-state metallothionein concentrations beyond those observed. For example, if one compares the animals collected from lakes in the middle of the contamination gradient with those at the high end, it is clear that those in the middle have not exhausted their ability to synthesize metallothionein – they could indeed have synthesized additional MT, but seem to have "chosen" not to do so. An obvious question arises: Why don't the perch and the bivalve synthesize enough MT to keep Cd and Ni off the metal-sensitive sites and avoid the associated endocrine and metabolic perturbations?

We speculate that under chronic exposure conditions the animal establishes a trade-off between the "cost" of detoxifying the nonessential metals and the "cost" of allowing some of these metals to spill over onto the metal-sensitive sites. This ability or capacity to trade-off may be an acquired trait, found in adapted or acclimated animals. In our field situation, metal smelting activities in the Rouyn-Noranda and Sudbury regions have been carried out for more than 75 years, a period long enough for natural selection to have exerted an effect on the local populations [73].

4. CONCLUDING REMARKS AND FUTURE DIRECTIONS

Based on our observations on two different wild freshwater organisms, a bivalve mollusc and a percid fish, we suggest that the classical threshold model does not apply in the case of chronic/life-long metal exposures. We have presented evidence that Cd and Ni detoxification is incomplete in these animals, even in our least contaminated lakes – there is no evidence for a threshold exposure concentration below which the incoming Cd or Ni is completely detoxified.

Consistent with this diagnosis of incomplete metal detoxification, a variety of progressive responses (biochemical, physiological, morphological, demographic, etc.) can be detected in animals collected along the metal-concentration gradient. However, there is also no evidence from our field study of an *upper* concentration threshold beyond which metal detoxification completely ceases to function. Steady-state metallothionein concentrations continue to increase in a dose-dependent manner along the exposure gradient, in both the bivalve and the perch. In other words, within the range of exposure conditions represented by the Rouyn-Noranda and Sudbury lakes, our biomonitor species are metal-stressed but are nevertheless able to survive, mature and reproduce successfully.

Given the long history of metal contamination of the lakes in the Rouyn-Noranda and Sudbury areas, we suggest that adaptation to metals may have favored organisms that can tolerate some metal "spillover" without unduly compromising their eventual growth, survival, and reproduction, presumably at lower overall cost than turning on the detoxification apparatus at full capacity to completely suppress metal binding to metal-sensitive sites (as might occur in a short-term exposure of a non-adapted organism). For example, if the action of a particular enzyme in the HDP fraction is inhibited by Cd or Ni binding, the organism could presumably compensate successfully by increasing the steady-state enzyme concentration. Such a response would in principle be detectable analytically, e.g., at the gene expression level or by proteomic analysis. In other words, we suggest that tolerance to non-essential metals like Cd and Ni must involve both (incomplete) detoxification *and* some metabolic compensation for any deleterious effects caused by the binding of the metals to inappropriate cellular sites.

ACKNOWLEDGEMENTS

The authors would like to acknowledge the key contributions of the many graduate students and postdoctoral fellows who carried out the original research work that is summarized in this chapter, and who participated fully in the original interpretation of the results. The research was principally supported by the Metals in the Environment Research Network (MITE-RN). This network received financial contributions from the Natural Sciences and Engineering Research Council of Canada (NSERC), the Mining Association of Canada, Ontario Power Generation Inc., the International Copper Association, the International Lead Zinc Research Organization, and the Nickel Producers Environmental Research Association. P.G.C. Campbell is supported by the Canada Research Chair program.

ABBREVIATIONS AND DEFINITIONS

ACTH	adrenocorticotropic hormone
BLM	biotic ligand model
dw	dry weight
GSH	glutathione
HDP	heat denaturable proteins
HMW	high molecular weight
HPLC	high performance liquid chromatography
HSP	heat stable proteins
ICP-AES	inductively coupled plasma-atomic emission spectroscopy
ICP-MS	inductively coupled plasma-mass spectrometry
K_M	equilibrium constant for the minding of metal "M" to a ligand, either in solution or at the cell surface
kDa	kiloDalton = 1000 atomic mass units
LMW	low molecular weight
L-cell membrane	metal binding ligand present at the cell surface (the "biotic ligand" postulated in the biotic ligand model)
M^{z+}	free metal ion (hydrated; water molecules not shown)
ML	metal-ligand complex
MT	metallothionein
MTLP	metallothionein-like peptide
NTA	nitrilotriacetate
SD	standard deviation
SE	standard error
TRIS	tris(hydroxymethyl)aminomethane
WHAM	Windermere Humic Aqueous Model

REFERENCES

1. D. A. Wright and P. M. Welbourn, in *Environmental Toxicology*, Cambridge University Press, Cambridge, UK, 2002, pp. 21–69.
2. J. A. Nott and W. J. Langston, *Aquat. Toxicol.*, 1993, **25**, 43–54.
3. J.-C. Amiard, C. Amiard-Triquet, S. Barka, J. Pellerin and P. S. Rainbow, *Aquat. Toxicol.*, 2006, **76**, 160–202.
4. G. Roesijadi, in *Metallothionein III*, Ed. N. Imura, M. Kimura and K. T. Suzuki, Birkhäuser Verlag, Basel, CH, 1993, pp. 141–158.
5. A. Z. Mason and K. D. Jenkins, in *Metal Speciation and Bioavailability in Aquatic Systems*, Ed. A. Tessier and D. Turner, John Wiley & Sons, Chichester, UK, 1995, pp. 479–608.

6. P. G. C. Campbell, in *Metal Speciation and Bioavailability in Aquatic Systems*, Ed. A. Tessier and D. Turner, John Wiley & Sons, Chichester, UK, 1995, pp. 45–102.
7. K. J. Wilkinson and J. Buffle, in *Physicochemical Kinetics and Transport at Biointerfaces*, Ed. H. P. van Leeuwen and W. Köster, John Wiley & Sons Ltd., Chichester, UK, 2004, pp. 445–533.
8. P. G. C. Campbell, O. Errécalde, C. Fortin, V. P. Hiriart-Baer and B. Vigneault, *Comp. Biochem. Physiol. Part C*, 2002, **133**, 185–202.
9. S. Niyogi and C. M. Wood, *Environ. Sci. Technol.*, 2004, **38**, 6177–6192.
10. V. I. Slaveykova and K. J. Wilkinson, *Environ. Chem.*, 2005, **2**, 9–24.
11. L. A. Finney and T. V. O'Halloran, *Science*, 2003, **300**, 931–936.
12. T. V. O'Halloran and V. C. Culotta, *J. Biol. Chem.*, 2000, **275**, 25057–25060.
13. J. D. Otvos, H. R. Engeseth, D. G. Nettesheim and C. R. Hilt, *Experientia Suppl.*, 1987, **52**, 171–178.
14. P. S. Rainbow, *Environ. Pollut.*, 2002, **120**, 497–507.
15. W. W. Nash, B. W. Poor and K. D. Jenkins, *Comp. Biochem. Physiol.*, 1981, **69C**, 205–211.
16. W. G. Wallace, B.-G. Lee and S. N. Luoma, *Mar. Ecol. Progr. Ser.*, 2003, **249**, 183–197.
17. A. Giguère, P. G. C. Campbell, L. Hare and P. Couture, *Aquat. Toxicol.*, 2006, **77**, 178–189.
18. C. De Duve, *Science*, 1975, **189**, 186–194.
19. V. Bragigand and B. Berthet, *Comp. Biochem. Physiol. Part A: Mol. Integrat. Physiol.*, 2003, **134**, 55–61.
20. A. Z. Mason, S. D. Storms and K. D. Jenkins, *Anal. Biochem.*, 1990, **186**, 187–201.
21. A. Z. Mason and R. F. Meraz, *Mar. Environ. Res.*, 1998, **46**, 573–577.
22. R. Lobinski and J. Szpunar, *Anal. Chim. Acta*, 1999, **400**, 321–332.
23. D. R. Winge and M. Brouwer, *Environ. Health Perspect.*, 1986, **65**, 211–214.
24. S. Micallef, Y. Couillard, P. G. C. Campbell and A. Tessier, *Talanta*, 1992, **39**, 1073–1079.
25. A. Z. Mason and S. D. Storms, *Mar. Environ. Res.*, 1993, **35**, 19–23.
26. D. Winge, J. Kanso and A. V. Colucci, in *Trace Element Metabolism in Animals*, Ed. W. G. Hoekstra, J. W. Suttie, H. E. Ganther and W. Mertz, University Park Press, University of Wisconsin, Madison, WI, USA, 1973, pp. 500–502.
27. D. A. Brown, C. A. Bawden, K. W. Chatel and T. R. Parsons, *Environ. Conserv.*, 1977, **4**, 213–216.
28. B. M. Sanders and K. D. Jenkins, *Biol. Bull.*, 1984, **167**, 704–711.
29. K. D. Jenkins and A. Z. Mason, *Aquat. Toxicol.*, 1988, **12**, 229–244.
30. S. K. Aalto and G. E. Newsome, *Can. J. Fish. Aquat. Sci.*, 1990, **47**, 1959–1962.
31. A. Hontela, P. Dumont, D. Duclos and R. Fortin, *Environ. Toxicol. Chem.*, 1995, **14**, 725–731.
32. A. Tessier, Y. Couillard, P. G. C. Campbell and J.-C. Auclair, *Limnol. Oceanogr.*, 1993, **38**, 1–17.
33. Y. Couillard, P. G. C. Campbell and A. Tessier, *Limnol. Oceanogr.*, 1993, **38**, 299–313.

34. M. D. Ptashynski, R. M. Pedlar, R. E. Evans, C. L. Baron and J. F. Klaverkamp, *Aquat. Toxicol.*, 2002, **58**, 229–247.
35. J. J. R. Frausto da Silva and R. J. P. Williams, *The Biological Chemistry of the Elements: The Inorganic Chemistry of Life*, Oxford University Press, Oxford, UK, 2001.
36. Y. Couillard, M. Courcelles, A. Cattaneo and S. Wunsam, *J. Paleolimnol.*, 2004, **32**, 149–162.
37. O. Perceval, B. Pinel-Alloul, G. Méthot, Y. Couillard, A. Giguère, P. G. C. Campbell and L. Hare, *Environ. Pollut.*, 2002, **118**, 5–17.
38. A. Giguère, Y. Couillard, P. G. C. Campbell, O. Perceval, L. Hare, B. Pinel-Alloul and J. Pellerin, *Aquat. Toxicol.*, 2003, **64**, 185–200.
39. A. Giguère, P. G. C. Campbell, L. Hare, D. G. McDonald and J. B. Rasmussen, *Can. J. Fish. Aquat. Sci.*, 2004, **61**, 1702–1716.
40. Centre for Hydrology and Ecology, 2001. Natural Environmental Research Council, NERC, Windermere, UK.
41. S. Eastwood and P. Couture, *Aquat. Toxicol.*, 2002, **58**, 43–56.
42. E. Bonneris, O. Perceval, S. Masson, L. Hare and P. G. C. Campbell, *Environ. Pollut.*, 2005, **135**, 195–208.
43. A. Giguère, P. G. C. Campbell, L. Hare and C. Cossu-Leguille, *Can. J. Fish. Aquat. Sci.*, 2005, **62**, 563–577.
44. M. D. Dutton, M. Stephenson and J. F. Klaverkamp, *Environ. Toxicol. Chem.*, 1993, **12**, 1193–202.
45. J. F. Klaverkamp, K. Wautier and C. L. Baron, *Aquat. Toxicol.*, 2000, **50**, 13–25.
46. S. N. Luoma and P. S. Rainbow, *Environ. Sci. Technol.*, 2005, **39**, 1921–1931.
47. J.-S. Laflamme, Y. Couillard, P. G. C. Campbell and A. Hontela, *Can. J. Fish. Aquat. Sci.*, 2000, **57**, 1692–1700.
48. D. J. Cain, S. N. Luoma and W. G. Wallace, *Environ. Toxicol. Chem.*, 2004, **23**, 1463–1473.
49. P. G. C. Campbell, A. Giguère, E. Bonneris and L. Hare, *Aquat. Toxicol.*, 2005, **72**, 83–97.
50. B. A. Fowler, G. W. Lucier and A. W. Hayes, in *Principles and Methods in Toxicology*, Ed. A. W. Hayes, Raven Press Ltd., New York, NY, USA, 1989, pp. 815–833.
51. W. J. Langston, M. J. Bebianno and Z. Mingjiang, *Mar. Environ. Res.*, 1989, **28**, 195–200.
52. W. J. Langston, M. J. Bebianno and G. R. Burt, in *Metal Metabolism in Aquatic Environments*, Ed. W. J. Langston and M. J. Bebianno, Chapman & Hall, London, 1998, pp. 219–283.
53. O. Perceval, Y. Couillard, B. Pinel–Alloul, A. Giguère and P. G. C. Campbell, *Aquat. Toxicol.*, 2004, **69**, 327–345.
54. H. M. Levesque, T. W. Moon, P. G. C. Campbell and A. Hontela, *Aquat. Toxicol.*, 2002, **60**, 257–267.
55. A. Gravel, P. G. C. Campbell and A. Hontela, *Can. J. Fish. Aquat. Sci.*, 2005, **62**, 982–990.
56. H. Levesque, J. Dorval, G. J. Van Der Kraak, P. G. C. Campbell and A. Hontela, *J. Toxicol. Environ. Health*, 2003, **66**, 657–676.

57. P. Couture and J. W. Rajotte, *J. Environ. Monit.*, 2003, **5**, 216–221.
58. G. D. Sherwood, J. B. Rasmussen, D. J. Rowan, J. Brodeur and A. Hontela, *Can. J. Fish. Aquat. Sci.*, 2000, **57**, 441–450.
59. J. B. Rasmussen, J. M. Gunn, A. Iles, A. Gagnon, A. Lacroix, P. G. C. Campbell and A. Hontela, *Human Ecol. Risk Assess.*, 2008, **14**, 317–350.
60. G. Pyle, P. Busby, C. Gauthier, J. Rajotte and P. Couture, *Hum. Ecol. Risk Assess.*, 2008, **14**, 126–145.
61. P. Couture and G. Pyle, *Hum. Ecol. Risk Assess.*, 2008, **14**, 73–96.
62. U. Borgmann, W. P. Norwood, T. B. Reynoldson and F. Rosa, *Can. J. Fish. Aquat. Sci.*, 2001, **58**, 950–960.
63. U. Borgmann, M. Nowierski, L. C. Grapentine and D. G. Dixon, *Environ. Pollut.*, 2004, **129**, 39–48.
64. T. Y. T. Ng and W. X. Wang, *Environ. Toxicol. Chem.*, 2005, **24**, 2365–2372.
65. G. Roesijadi and W. E. Robinson, in *Aquatic Toxicology Molecular Biochemical and Cellular Perspectives*, Ed. D. C. Malins and G. K. Ostrander, Lewis, Boca Raton, FL, USA, 1994, pp. 387–420.
66. A. Wicklund Glynn, *Aquat. Toxicol.*, 1996, **35**, 47–58.
67. S. J. Hamilton, P. M. Mehrle and J. R. Jones, *Trans. Amer. Fish Soc.*, 1987, **116**, 551–560.
68. D. A. Brown, S. M. Bay and G. P. Hershelman, *Aquat. Toxicol.*, 1990, **16**, 295–310.
69. K. D. Jenkins, D. A. Brown, G. P. Hershelman and W. C. Meyer, in *Physiological Mechanisms of Marine Pollutant Toxicity*, Academic Press Inc, New York, NY, USA, 1982, pp. 177–196.
70. C. Mouneyrac, B. Berthet and J.-C. Amiard, *Wat. Air Soil Pollut.*, 1999, **112**, 187–196.
71. D. J. Cain, S. N. Luoma and W. G. Wallace, *Environ. Toxicol. Chem.*, 2004, **23**, 1463–1473.
72. M. Baudrimont, S. Andres, J. Metivaud, Y. Lapaquellerie, F. Ribeyre, N. Maillet, C. Latouche and A. Boudou, *Environ. Toxicol. Chem.*, 1999, **18**, 2472–2477.
73. V. Bourret, P. Couture, P. G. C. Campbell and L. Bernatchez, *Aquat. Toxicol.*, 2008, **86**, 76–90.
74. P. G. C. Campbell and Y. Couillard, in *Écotoxicologie Moléculaire-Principes Fondamentaux et Perspectives de Développement*, Ed. E. Pelletier, P. G. C. Campbell and F. Denizeau, Les Presses de l'Université du Québec, Ste-Foy, Quebec, Canada, 2004, pp. 9–61.
75. E. Bonneris, A. Giguère, O. Perceval, T. Buronfosse, S. Masson, L. Hare and P. G. C. Campbell, *Aquat. Toxicol.*, 2002, **71**, 319–334.

Met. Ions Life Sci. **2009**, *5*, 279–317

10

Structure and Function of Vertebrate Metallothioneins

***Juan Hidalgo,*[1] *Roger Chung,*[2] *Milena Penkowa,*[3] *and Milan Vašák*[4]**

[1]Unidad de Fisiología Animal, Facultad de Ciencias, Universidad Autonoma de Barcelona, Bellaterra, E-08193 Barcelona, Spain
<juan.hidalgo@uab.es>

[2]NeuroRepair Group, Menzies Research Institute, University of Tasmania, Tasmania, Australia
<rschung@utas.edu.au>

[3]Panum Institute, Faculty of Health Sciences, University of Copenhagen, Blegdamsvej 3, DK-2200 Copenhagen, Denmark
<m.penkowa@mai.ku.dk>

[4]Department of Biochemistry, University of Zürich, Winterthurerstrasse 190, CH-8057 Zürich, Switzerland
<mvasak@bioc.uzh.ch>

Metal Ions in Life Sciences, Volume 5 Edited by Astrid Sigel, Helmut Sigel, and Roland K. O. Sigel

Published by the Royal Society of Chemistry, www.rsc.org DOI: 10.1039/9781847558992-00279

ABSTRACT: In 1957, Margoshes and Vallee reported on the isolation of a protein from horse kidney, which showed a high affinity for cadmium, and soon thereafter the protein was named metallothionein (MT) by the leading scientists Kägi and Vallee. Fifty years of intense research has dissected out many of the biochemical, molecular, and genetic aspects of these proteins, yet not that much is understood on its physiological putative functions. Being a highly conserved family of proteins would suggest essential biological functions, but these may be dispensable and/or assumed by other proteins as demonstrated by the phenotype of knock-out mice in normal conditions. Nevertheless, under challenging conditions (such as tissue injury) a strong phenotype appears that is suggestive of important physiological functions. This has been particularly well shown in the brain, where antiinflammatory, antioxidant and antiapoptotic effects of MT have been demonstrated. To date, the results gathered strongly support a therapeutic value of these proteins that deserve attention in clinical studies.

KEYWORDS: apoptosis · brain metallothioneins · inflammation · metal binding · metallothionein regulation · metallothionein structure and reactivity · oxidative stress · vertebrate metallothioneins

1. INTRODUCTION

In 1957, Margoshes and Vallee reported on the isolation of a protein from horse kidney, which showed a high affinity for cadmium [1]. This protein was soon biochemically characterized, showing an unusually high content of metals and cysteine residues, and was named metallothionein (MT) [2,3]. In these early studies, putative physiological functions such as transport and storage of essential heavy metals (zinc, copper) and detoxification of non-essential ones (cadmium, mercury) were proposed, but currently it is envisaged that these proteins could also fulfil additional functions (see below).

In the last 50 years an overwhelming flow of information on these proteins has been published. It is clear that they constitute a superfamily of proteins, widely distributed in the animal kingdom as well as other phylogenetic groups. Attempts to subdivide it into several families on the basis of structural relationships have been carried out, with vertebrate MTs constituting Family 1 [4]. Mammalian MTs are comprised of several subfamilies, m1 to m4. In the mouse, the m1 to m4 subfamilies are composed of a single member (MT-1 in the m1 subfamily, MT-2 in the m2, and so on), in contrast to ungulates and primates where an increased genetic polymorphism exists. Non-mammalian MTs receive different nomenclatures with this classification: a1 and a2 for avian MTs, b for batracian MT, and t for teleost MT, while those proteins with uncertain structural information as for full assignment would receive a n.d. label (n.d. mammalian MT, n.d. avian MT, and so on). It should be recognized that this classification does not provide information on the putative function of MTs; an alternative classification of MTs based in function rather than structure has been proposed [5].

2. MAMMALIAN METALLOTHIONEIN GENE AND PROTEIN STRUCTURE

2.1. Structure of Mammalian Gene Loci

In humans, the MT genes are tightly clustered in the q13 region of chromosome 16 [6–9], consisting of 7 functional MT-1 genes (*MT-1A, -B, -E, -F, -G, -H,* and *-X*) and a single gene encoding each of the other MT isoforms, namely MT-2 (the *MT-2A* gene), MT-3 and MT-4. A number of non-functional MT pseudogenes exist within the human genome, which show significant homology to functional MT genes, but either lack promoter sequences or contain missense mutations that prevent correct transcription of a functional protein. These include *MT-1C, -D, -I, -J, -K, -L,* and *MT-2B.* In humans, MT-2A is the most highly expressed gene, accounting for approximately 50% of total MT expression. Human MT-1/-2 genes exhibit a large number of genetic polymorphisms, although to date none of these mutations have been linked to specific human diseases.

Mice have a much simpler MT gene structure, with only one functional gene for each isoform MT-1 to MT-4, all located on chromosome 8 [8–11].

All human and murine MT genes contain 3 exons. The first encodes the N-terminal region of the α-domain, exon 2 encodes the rest of the α-domain, while exon 3 encodes the entire β-domain (reviewed in [12]).

2.2. Protein Structure of Mammalian Metallothioneins

Metallothionein was first identified as a cadmium-binding protein in equine kidney [1]. Subsequent purification and characterization determined MT as a low molecular weight (6–7 kDa) metal binding protein, with a characteristic amino acid composition [2,3]. They are characterized by a high number of cysteine residues and the absence of disulfides, aromatic amino acids, and histidine. This and the presence of a metal/cysteine ratio of about 1 to 3 have been taken to indicate a highly unusual structure of this protein [13].

Mammalian MT-1 to MT-4 consist of a single polypeptide chain of 61 to 68 amino acids, with an N-terminal acetylmethionine and often alanine at the carboxyterminal end (Figure 1).

One of the most striking features of the mammalian MT family of proteins is their high cysteine content, being comprised of 20 highly conserved cysteine residues, present in Cys-Xaa-Cys, Cys-Xaa-Yaa-Cys, and Cys-Cys motifs, where Xaa and Yaa stand for non-cysteine amino acids [14]. The best characterized mammalian MT-1 and MT-2 isoforms are involved in the preferential binding of 7 divalent (Zn^{II}) and up to 12 monovalent (Cu^{I}) metal ions *in vivo* through metal-thiolate coordinative bonds. However, *in vitro* MT-1/-2 also show binding to other metals, such as Cd^{II}, Bi^{III}, Pt^{II}, Ag^{I}, and Hg^{II}, and can readily bind these metals when they are administered to animals [15]. The binding affinity of these metals to MT follows the order of thiolate model complexes, i.e., $Zn^{II} < Pb^{II} < Cd^{II} < Cu^{I} < Ag^{I}$, Hg^{II}, Bi^{III}, allowing that lower affinity metals can be displaced by higher affinity ones [16]. This contributes to the likelihood that, besides Zn^{II}, other pure metal-MT complexes do not occur naturally *in vivo*, but rather as mixed metal proteins [17,18]. Furthermore, differences in metal exposure are likely to explain interindividual and tissue variations in the metal composition of MT-1/-2 [14].

The metal-free protein, also named the apoprotein or thionein, possesses a predominantly disordered structure, which renders it vulnerable to

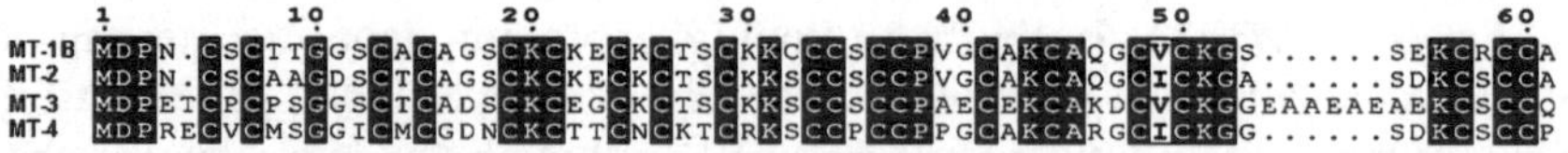

Figure 1. ClustalW amino acid sequence alignments of the four human metallothionein isoforms with conserved residues highlighted. The figure was generated with the program ESPript version 2.2.

proteolysis [19]. However, a well-defined structure develops upon metal binding. Mammalian MTs constitute a class of proteins where the structural features were forthcoming almost exclusively from spectroscopic studies. The first and most direct evidence for the existence of metal-thiolate clusters in MTs came from the ^{113}Cd NMR studies of the reconstituted $^{113}Cd_7$MT-2 [20]. These studies revealed that 20 cysteine residues and 7 divalent metal ions are partitioned between two metal-thiolate clusters, a cyclohexane-like 3-metal cluster ($M_3^{II}(Cys)_9$) in the N-terminal β-domain (residues 1–30) and an adamantane-related 4-metal cluster ($M_4^{II}(Cys)_{11}$) in the C-terminal α-domain (residues 31–61). In these clusters the metal ions are coordinated by both terminal and μ_2-bridging thiolate ligands in a tetrahedral-type of symmetry [21]. A great deal of knowledge on the chemical features of MT-1/-2 pertinent to their function has arisen from the determination of their three-dimensional (3D) structures. The 3D structures of MT-1/-2 from various species were obtained mainly by NMR spectroscopy [22–25], but also by X-ray crystallography [26,27] (Figure 2). The reported 3D structures reveal a similar monomeric two-domain protein of a dumbbell-like shape featuring the already described cluster topology (Figure 3) and identical polypeptide folding. The two protein domains are connected by a flexible hinge region of a conserved Lys-Lys sequence in the middle of the polypeptide chain. The discontinuity in the NMR structure presented in Figure 2, which is also seen in other NMR 3D structures of MT-1/-2, is due to the lack of NOE-based information as to the mutual orientation of both domains.

The flexibility of the protein backbone structure enfolding the metal core in MT-1/-2 is well documented. Both the calculated root-mean-square deviation (rmsd) values from NMR data and the crystallographic B-factors indicate that a considerable degree of dynamic structural disorder exists [23,26]. More direct evidence for the nonrigid MT structure came from the ^{1}H NMR studies of ^{1}H-^{2}H amide exchange in Cd_7MT-1/-2 [24,25]. In these studies, the enhanced flexibility of the less structurally constrained β-domain compared to the α-domain in both isoforms has been demonstrated. Molecular dynamics simulations of the β-domain of rat liver MT-2 in aqueous solution also show that the polypeptide loops between cysteine ligands exhibit an extraordinary flexibility without disrupting the geometry of the 3-metal cluster [28].

Because of the flexibility of the polypeptide loops between cysteine ligands it was generally assumed that the protein can accommodate a wide range of divalent metal ions of different sizes without any selectivity. However, in polynuclear adamantane-like cages both the metal size and the varying length of the metal-thiolate bonds give rise to widely differing cluster sizes and thus cluster volumes [31]. As a consequence of this effect an expansion of the cluster core in MT should influence the energetic of the folding

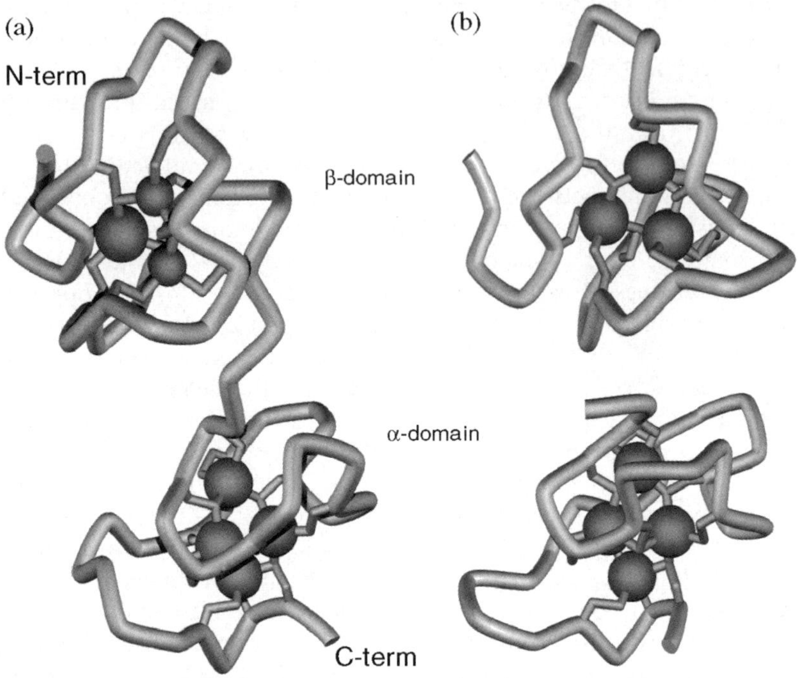

Figure 2. Three-dimensional structure of rat MT-2 as determined (**a**) by X-ray crystallography [26], Cd_5,Zn_2MT-2 and (**b**) by NMR in aqueous solution [23], $^{113}Cd_7MT$-2. Metals are shown as shaded spheres connected to the protein backbone by cysteine thiolate ligands. The models were generated with the program DS Viewer version 5.0 (Accelrys) using the PDB coordinates 4mt2, 1mrt, and 2mrt. Reproduced from [274] with permission from John Wiley and Sons, Ltd., Chichester, UK (copyright 2005).

process. The metal selectivity of MT-1/-2 structure has been studied by offering seven equivalents of two divalent metal ions, i.e., Co^{II}/Cd^{II}, Zn^{II}/Cd^{II}, Co^{II}/Zn^{II}, and Fe^{II}/Cd^{II} in a 3 to 4 ratio to the apoprotein followed by the determination of the respective metal distributions within and between the clusters by electronic absorption, MCD, ^{113}Cd NMR, ^{57}Fe Mössbauer, and EPR spectroscopy, as required [29,30]. As a result, offering Co^{II}/Cd^{II} ions resulted in $Co^{II}_3Cd_4MT$-2 in which two homometallic clusters, i.e., the Co^{II}_3-thiolate cluster in the β-domain and the Cd_4-thiolate cluster in the α-domain, were present. In the case of Fe^{II}/Cd^{II}, the generated homometallic species Fe^{II}_7MT-1 and Cd_7MT-2 in the ratio of added metal ions further signify a substantial selectivity of the protein. Although Co^{II}/Zn^{II} and Zn^{II}/Cd^{II} gave rise to heterometallic clusters in $Co^{II}_3Zn_4MT$-1 and Zn_3Cd_4MT-2,

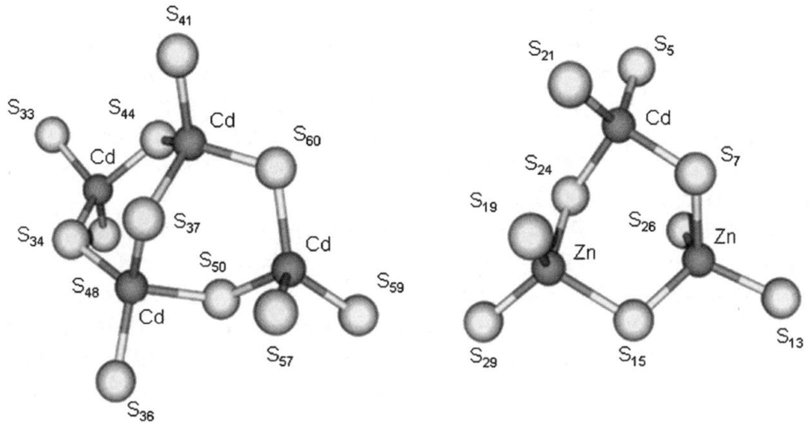

Figure 3. Schematic drawing of the two metal-thiolate clusters in rat Zn_2,Cd_5MT-2: A Cd_4Cys_{11} cluster in the C-terminal α-domain and a Zn_2CdCys_9 cluster in the N-terminal β-domain. Sγ atoms are labeled by Cys residue number. The figure was generated with the program DS Viewer version 5.0 (Accelrys) using the PDB coordinates 4 mt2. Reproduced from [274] with permission from John Wiley and Sons, Ltd., Chichester, UK (copyright 2005).

they showed a distinct metal distribution. Thus, an appreciable selectivity in metal partitioning among and within the clusters is indeed observed in this protein. These results are opposite to those obtained in the studies of inorganic adamantane-like cages with monodentate thiolate ligands of the general formula $[M_4(SPh)_{10}]^{2-}$ ($M = Cd^{II}$, Zn^{II}, Co^{II}, Fe^{II}) where, despite the differences in the metal-thiolate affinities and in the homometallic cluster volumes, a simple mixing of two homometallic cages always produced heterometallic cage complexes with an almost statistical distribution of both metal ions [31]. Since the same metal ions were employed in the studies of MT-1/-2, properties of the MT structure are responsible for this effect. It has been concluded that the interplay between both the chemistry of metal ions and the steric requirements of the protein structure would determine the diversity of metal-thiolate cluster structures in MT. In this context it should be noted that small structural differences between Zn_7- and Cd_7MT-2, due to differences in the cluster volumes (approx. 20%), have been observed [27].

The coordination of the metal ions is the major determinant in the folding of the polypeptide backbone around the two clusters of MT-1/-2. The pathway of cluster formation has been determined for Cd_7MT-2 and Co_7MT-2 by ^{113}Cd NMR and EPR/paramagnetic ^{1}H NMR, respectively. Because of the similar size and chemistry of Co^{II} and Zn^{II}, the isostructural substitution of Zn^{II} by Co^{II} is a widely used method to probe the spectroscopically silent zinc sites in proteins. The studies revealed that the formation of both clusters in Cd_7MT-2 are cooperative and sequential with the 4-metal cluster in the α-domain being formed first [32]. However, the formation of Co^{II} clusters, and by inference also Zn^{II} clusters, proceeds by a different pathway. In this case, prior to the cluster formation in the α-domain, the first four Co^{II} are bound to independent Co^{II} sites in tetrathiolate coordination [33,34].

At neutral pH, closely similar average apparent stability constants have been determined for MT-1/-2 by different methods, with values of the order of 10^{11} M^{-1} and 10^{14} M^{-1} for Zn^{II} and Cd^{II}, respectively (reviewed in [35]). The recent development of fluorescent metal chelators with different metal affinities allowed the dissection of average apparent stability constants. Although each of the seven Zn^{II} ions in Zn_7MT-2 is bound in a tetrathiolate coordination environment, analysis of Zn^{II} binding to thionein revealed at least three classes of sites with affinities that differ by 4 orders of magnitude [36]. One Zn^{II} ion is relatively weakly bound ($\log K = 7.7$), making MT a zinc donor. Moreover, it has been suggested that physiological ligands including thionein would further modulate the metal binding and the redox reactivity of thiols in cellular environment. These chemical characteristics suggest how the molecular structures and redox chemistries of fully and partially metallated MT and thionein may contribute to the variety of different functions that MT may serve in the cell.

Copper binds readily to MT *in vivo* and *in vitro* [16]. Both Wilson's and Menkes' diseases in humans are inborn errors of copper metabolism in which excess copper accumulates intracellularly in MT. In all instances examined to date, copper is bound to MT as Cu^{I}. The only available 3D crystal structure of Cu^{I}-thiolate clusters in MT is that of yeast Cu_8MT. The structure shows the largest known oligonuclear Cu^{I}-thiolate cluster in biomolecules, consisting of six trigonally and two diagonally coordinated Cu^{I} ions [37]. However, the 3D structure of Cu^{I} containing mammalian MT-1/-2 is so far unknown. The structural features of Cu^{I}-thiolate clusters in mammalian MTs have been studied by various spectroscopic techniques [38]. The current knowledge is limited to the fact that the fully Cu^{I}-loaded MT form binds 12 Cu^{I} ions into two metal-thiolate clusters, where, in contrast to divalent metal ions, the monovalent copper ions are coordinated by two or three cysteine ligands forming two independent Cu^{I}_{6}-thiolate clusters [38,39]. At lower Cu^{I}/protein stoichiometries, the presence of two distinct Cu^{I}_{4}-thiolate clusters in both protein domains has also been reported [40]. There is evidence to suggest that Cu^{I} binds preferably to the less structurally constrained β-domain [41], whereas Cd^{II} and Zn^{II} bind to the α-domain [42]. As isolated, MT-3 contains both Cu^{I} and Zn^{II} ions contained in Cu_4Zn_{3-4} MT-3. The extended X-ray absorption fine structure (EXAFS) studies on this species revealed the presence of two homometallic clusters, a

Zn_{3-4}-thiolate cluster and a Cu^{I}_{4}-thiolate cluster with Zn^{II} and Cu^{I} ions tetrahedraly and trigonally coordinated, respectively [43,44].

A striking feature of the Cu^{I}_{4}-thiolate cluster, which by immunochemical methods was found to be localized in the β-domain [45], is its remarkable stability against air oxidation. By contrast, the studies on a well defined Cu_4,Zn_4MT-3 form prepared by selective metal filling of both domains *in vitro* revealed the presence of a redox labile zinc site in the Zn_4-thiolate cluster located in the α-domain. While under anaerobic or reducing conditions the Zn_4-thiolate cluster is stable, a partial oxidation of specific thiolate ligands in air results in Zn^{II} release giving rise to a Zn_3-thiolate cluster in the α-domain [46]. It may be noted that the neuroinhibitory activity of MT-3 in cell cultures was found for both Cu_4Zn_{3-4}MT-3 and Zn_7MT-3 metalloforms [47,48].

While MT-1/-2 and MT-3 share similar primary structures and the same number of cysteine residues, MT-3 has two regions that distinguish it from all MT-1/-2 isoforms. These are a T_5 insert followed by a unique C_6-P-C-P_9 motif within the β-domain, and an acidic hexapeptide insert within the α-domain (Figure 1). Interestingly, ^{113}Cd NMR studies have identified unprecedented dynamic processes including the temporary breaking and reforming of coordinative bonds within the 3-metal cluster in the β-domain of Cd_7MT-3, which have not been observed in Cd_7MT-1/-2 [49]. The mutation of conserved proline residues in the T_5-C-P-C-P_9 motif to Ala and Ser, amino acids present in MT-2, abolished the neuroinhibitory activity and cluster dynamics [50]. Thus, changes in the primary structure of MT-3, when compared with MT-1/-2, may account for variations in the biological activity of MT-3 in comparison to MT-1/-2.

Because of dynamic processes in the β-domain, only the NMR structure of the α-domain of mouse and human Cd_7MT-3 could be determined [51,52]. The structure of this domain reveals a peptide fold and a cluster organization very similar to those found in the α-domain of Cd_7MT-1/-2, with the exception of an extended flexible loop encompassing the acidic hexapeptide insert. The presence of a 3-metal cluster in the β-domain of Cd_7MT-3 was inferred from the studies of separate α- and β-domains, which in the whole protein interact with each other [49]. The conserved T_5-C-P-C-P_9 motif within the β-domain, which has been found essential for bioactivity and dynamics of the 3-metal cluster [53], and the acidic loop within the α-domain [51] were proposed as a possible interaction site with other proteins. Furthermore, the metal-binding affinity of MT-3 is observed to be weaker than MT-1/-2 and the metal binding to MT-3 is non-cooperative [54,55]. A more detailed account of the structural and functional properties of MT-3 is given in Chapter 11 of this volume.

The last identified member of the mammalian MT family is MT-4. This isoform consists of 62 amino acids, showing an insert of Glu in position 5 relative to MT-1/-2 proteins (Figure 1), and appears to be present exclusively in cornified and stratified squamous epithelia. Much of the information on MT-4 deals with molecular biology and expression profiles in mammalian maternal deciduum [56] and during epithelia development and physiology [9,57]. All of these studies revealed that the MT-4 gene is subject to a strict developmental regulation. However, the question of whether MT-4 is involved in copper or zinc metabolism in epithelia is still debated. Insights into the function of a protein can be inferred from its structural properties. From the studies on the metal binding abilities of MT-4 using heterologously expressed MT-4 with zinc, cadmium, and copper in combination with the *in silico* protein sequence analyses, the copper binding nature has been suggested [58]. However, from biological studies a special role of MT-4 in the regulation of zinc-dependent processes in keratinocytes has also been suggested [9]. This function is supported by the structural studies on Cd_7MT-4 in which similar average apparent metal-binding affinities of two metal-thiolate clusters in Cd_7MT-4 and Cd_7MT-1/-2, overall similarities in cluster topologies to that determined for Cd_7MT-1/-2 (Figure 3), and the pathway of cluster assembly have been found [59]. Thus, MT-4 may be involved in both zinc and copper metabolism in keratinocytes.

2.3. Structure and Reactivity

In spite of the high thermodynamic stability of the metal-thiolate clusters in MT, and as a consequence of the dynamic properties of the MT structure, they are kinetically very labile, i.e., the thiolate ligands allow a rapid metallation and demetallation. The rate of metal exchange in MT-1/-2 follows the order $Hg^{II} > Cd^{II} > Zn^{II}$. A comparison with similar studies on inorganic complexes affords the most plausible explanation for this effect [60]. In these studies, a correlation between the level of ligand preorganization and complex lability revealed that a decreasing rigidity of the ligand structure results in an increasing lability of the metal complexes. Based on the already discussed properties of metal-free and metal-containing MT forms, the apoprotein with multidentate cysteine thiolate ligands resembles chelating inorganic ligands with long bridges, for which a low level of ligand preorganization and hence a high kinetic lability has been shown.

Evidence for the kinetic lability of metals in Cd_7MT was provided by ^{113}Cd NMR saturation transfer experiments, which established the presence of inter- and/or intramolecular metal exchange within the 3-metal cluster of the β-domain with a half-life of about 0.5 s. The occurrence of similar processes taking place within the 4-metal cluster, but with a half-life of about 16 minutes, was afforded by metal exchange studies using the radioactive ^{109}Cd isotope [61]. The importance of protein dynamics for metal exchange has been recognized in the NMR studies of Cd_7MT-1 in which the enhanced backbone flexibility, when compared to Cd_7MT-2, resulted in a much faster intersite cadmium exchange in the 3-metal cluster [25]. In this context it may be noted that in zinc enzymes such as alkaline phosphatase and carboxypeptidase, where a well-defined catalytic site is present in a rather rigid protein structure, the exchange half-life of zinc is in the order of hours and days. Thus, the actual exchange rates of metal ions in proteins are not an intrinsic attribute of their binding properties, but rather they are determined by the energetic and kinetics of protein folding.

Intermolecular zinc transfer between zinc proteins and Zn_7MT *in vitro* has also been studied, leading to apoenzyme activation [62,63] and zinc-dependent transcription factors modulation [64,65]. Moreover, Zn_7MT has been shown to be able to transfer one zinc ion to apoenzymes possessing a much lower affinity compared to the apparent average binding constant for zinc in Zn_7MT. However, the presence of one weakly bound metal ion in the structure of Zn_7MT has recently been demonstrated [36].

Although the thiol groups in MT are masked through their interaction with metal ions, they retain a substantial degree of the nucleophilicity seen with the metal-free protein. This property is reflected by the extremely high reactivity of the coordinated cysteine ligands with alkylating and oxidizing agents such as iodoacetamide or 5,5′-dithiobis-(2-nitrobenzoic acid) (DTNB), respectively [66]. Differential sulfur reactivities in both domains have also been reported. Kinetic, mass spectrometric, and NMR studies have shown that the kinetically preferred reaction of mammalian MT with electrophiles may be localized in either the α- or the β-domain depending on the specific attacking reagent. For instance, whereas iodoacetamide and *p*-(hydroxymercuri)benzoate have been found to react preferentially with the β-domain, DTNB, aurothiomalate, melphalan, and chlorambucil have been shown to react preferentially with the α-domain [67–69].

Cysteine residues of the zinc-thiolate clusters in Zn_7MT-1/-2 can also be oxidized by mild cellular oxidants like oxidized glutathione (GSSG) [63] releasing bound metal ions in this process. Experiments in the presence of the GSH/GSSG redox pair provided evidence for an oxidoreductive mechanism modulating the zinc affinity of the cysteine thiolate ligands in Zn_7MT *in vitro* [70,71]. The importance of redox-active cysteine ligands in zinc proteins, especially in MTs, in changing redox signals to zinc signals in a cellular environment has also been discussed [72]. In metal-linked neurodegenerative disorders like Alzheimer's disease (AD), a dysregulated copper homeostasis and related neurotoxicity are linked to the ROS production. A protective role of extracellularly occurring MT-3 from copper-mediated toxicity has been suggested based on *in vitro* studies. In these studies, Zn_7MT-3 through Cu^{II} reduction to Cu^{I} by thiolate ligands and binding to the protein, forming an air stable $Cu(I)_4Zn_4$MT-3 species, efficiently scavenges and redox silences the toxic free Cu^{II} ions [73].

Another interesting aspect of MT reactivity is its ability to react with radical species. Thus, it has been shown that mammalian M_7^{II}-MT ($M = Zn^{II}$ and/or Cd^{II}) are efficient scavengers of free radicals such as hydroxyl ($OH^{\bullet}$), superoxide ($O_2^{-\bullet}$) or nitric oxide ($NO^{\bullet}$) [74,75]. In all instances, the attack of free radicals occurs at the metal-bound thiolates, leading to the protein oxidation and/or modification and subsequent metal release. Interestingly, in many instances these effects could be reversed under reductive conditions and in the presence of the appropriate metal ion. The protective effect of individual MT isoforms from free radical damage in biological systems is discussed below (see also Chapter 12).

Studies on ligand substitution reactivity have revealed that small multidentate ligands, e.g., polyamines, polyaminocarboxylates, and bis(thiosemicarbazones), are effective competitors for zinc bound to MT in reactions which are much faster than the dissociation rate constant for zinc in ZnMT-1/-2, implying a direct competition mechanism [76]. Biphasic kinetics and differential reactivity of the two metal clusters, α- and β-domain, in mammalian M_7^{II}MT have been measured with ethylenediamine-N,N,N′,N′-tetraacetic acid (EDTA) $\beta > \alpha$ [77] and, in opposite order $\alpha > \beta$, with nitrilotriacetate (NTA) [78]. On the other hand, ligands such as ethylenediamine diacetate and triethylenetetramine were found to be ineffective even at thermodynamically competent concentrations. From these studies it has been suggested that a tripod configuration of chelating ligands is required in thiolate substitution and that only specific regions of the protein domains may provide easy access to the metal clusters. In this context it may be noted that in the crystal structure of Zn_2Cd_5MT-2 each domain contains a solvent-exposed cleft containing three accessible cysteine sulfurs. In summary, the reactivity of the MT structure, which is pertinent to its function, is dominated by the chemistry of the nucleophilic thiolate groups.

2.4. Metallothionein Regulation

It is generally accepted that MT-1/-2 expression is highly inducible by a range of stimuli, including metals, hormones, cytokines, inflammation, and stress [79]. MT-1 and -2 expression is coordinately regulated in mice by metals and glucocorticoids [11,80], however interestingly in humans, while metals induce expression of all MT-1/-2 isoforms, only expression of MT-1e and MT-2a are induced by glucocorticoids [81]. MT-3 is regulated in a different manner (see Chapter 11).

A complex system of transcriptional regulation exists governing induction of MT expression. Metal-induced synthesis is mediated through the action of short *cis*-acting DNA sequences known as metal responsive elements (MREs), which are present in the promotor region of all mammalian MT genes [82,83]. Similarly, glucocorticoid responsive elements (GREs) are responsible for MT expression in response to glucocorticoids, although in this case they are present in the promotor region of some but not all MT genes [82,84]. MT expression is also regulated by antioxidant response elements (AREs), although some MREs also respond to oxidants [85].

Metal-induced expression of MT has been extensively characterized, and it is now understood that this involves a complex mechanism of regulation, with numerous factors of different sizes and properties able to bind MREs and regulate MT expression either positively or negatively [86,87]. These include MREBP (MRE binding protein) which specifically binds MREs of the human MT-2a gene (the major expressed human MT-1/-2 isoform) and MTF-1 (MRE binding transcription factor), a zinc-sensitive transcription factor. It is important to note that these proteins act differently, with MREBP inhibiting transcription [88] while MTF-1 has an important role in the control of MT gene expression [83,89,90]. A number of other MRE binding proteins have been identified, including MRE binding factor-1 (MBF-1), zinc activated protein (ZAP) and zinc regulated factor (ZiRF1) and p108, although it is not yet clear how these proteins affect transcription of MT genes (see also Chapter 2).

MT expression also appears to be regulated at the post-transcriptional level, however, protein levels of MT-1/-2 do not always reflect mRNA levels. As an example of this, it has been reported that copper treatment of adult rats reduces renal levels of MT-1/-2 mRNA, but increases renal protein levels of MT-1/-2 [91]. In a previous study by the same group, it was found that copper treatment induced MT-2 mRNA levels dramatically (over 30-fold) while MT-1/-2 protein levels were only increased 7-fold [92]. Discrepancies between gene and protein expression have also been observed in other organisms, such as molluscs [93], further supporting the notion that MT expression can also be regulated post-transcriptionally.

3. NON-MAMMALIAN VERTEBRATE METALLOTHIONEINS

3.1. Avian Metallothioneins

Avian MTs are the closest vertebrate descendent of all non-mammalian vertebrate MTs (for an interesting insight into MT evolution see the work of Pierre-Alain Binz; http://www.bioc.unizh.ch/mtpage/poster/posterevol.html), and share a highly homologous structure with mammalian MTs. Chicken (*Gallus gallus*) MT (ckMT) has been the model for the study of avian MTs [94], although it has been determined that there is an extremely high level of conservation (no amino acid substitutions and 97% homology at the cDNA level) between avian MTs across a range of species including turkey (*Meleagris gallopavo*), pheasant (*Phasianus colchicus*), ducks (*Cairina moschata* and *Anas platyrhyncos*), and quail (*Coturnix coturnix*) [95–97]. Until recently it was considered that a single gene encodes for MT in birds [98,99], except for the unique report of two MT isoforms in pigeons (*Columba livia*), neither of which is similar to other avian MTs [100]. However, strong evidence for the presence of a second MT gene in chicken, ckMT2, has recently been reported, further supporting the hypothesis that there are at least two avian MT genes [101].

CkMT protein was first isolated and characterized in 1973 by Weser and colleagues [102] as a 63 amino acid polypeptide, and later was found to share a 68% sequence homology to mouse MT-1 [98]. It was found that there is one insertion (and two substitutions) within the amino-terminal region, a region considered invariant among mammalian MTs. Furthermore, while 18 of the 20 cysteines in ckMT were aligned with cysteines from other mammalian sequences, two cysteines near the carboxyl terminus were shifted by one residue due to the insertion of an additional proline in that region. Interestingly, the ckMT gene shares the same exon/intron structure of mammalian MTs, and ckMT expression is also regulated by similar stimuli, including metal overdose, oxidative stress, and glucocorticoids [94,103]. Detailed *in vivo* and *in vitro* metal binding analyses have revealed that the Zn^{II}, Cd^{II} and Cu^{I} binding properties of ckMT lie between those of mammalian MT-1 and MT-4, with zinc and cadmium binding properties closer to MT-1 and more closely to MT-4 for copper binding. To date there is no crystal structure data regarding ckMT, however, theoretical analysis of the primary structure predicts that ckMT does not markedly differ from mammalian MTs [98].

3.2. Fish Metallothioneins

As in other vertebrates, fish MTs retain the dumbbell shape two domain structure that characterizes vertebrate MTs (see also Chapter 8). Interestingly however, while the overall 3D NMR structure of fish MTs remains similar to other vertebrates, the misalignment of the ninth cysteine residue within the α-domain of fish MT has been reported, as well as differences in the dynamic behavior of the α- and β-domains as determined from spectroscopic studies [104]. MTs have been characterized in a number of fish species, including goldfish [105], flounder [106], carp [107], dogfish [108], and trout [109]. The liver is one of the primary sites of MT expression in fish, especially during reproduction and development [110]. In accordance with their vertebrate cousins, MT overexpression is associated with resistance to heavy metal toxicity in fish after exposure to metal ions.

3.3. Amphibian Metallothioneins

Amphibian MTs were first identified in the early 1980s, present as a single zinc and copper binding isoform across a number of frog species including *Rana catesbiana* [111], *Bombina orientalis, Bufo bufo japonicus,* and *Hyla arborea japonica* [112]. Since this time, levels of MT have been used as a measure of environmental toxicology from heavy metals. In these studies, MT is consistently upregulated in the liver and kidney, where it is involved in the handling of heavy metals in frogs [113,114].

Recent investigations have explored the molecular expression of MT in the model frog system, *Xenopus laevis*. There is a single *Xenopus* MT gene, XlMT-A, which displays a three exon, two intron structure similar to mammalian MTs [115]. The *Xenopus* MT is a 62 amino acid protein, more closely related to avian then fish and finally mammalian MTs [116]. During development, *Xenopus* MT is expressed in a range of sites from the tailbud stage onwards, including the liver and the CNS [117]. As with all vertebrate MTs, *Xenopus* MT expression is highly inducible by zinc, copper, and cadmium [116].

3.4. Reptile Metallothioneins

There is little information regarding MTs in reptiles. A protein with characteristics of MT was first identified in lizard species such as *Triturus pyrrhogaster* [118] or *Podarcis muralis* [119]. In a recent report, a cDNA encoding MT in the lizard *Podarcis sicula* was identified, producing a 63 amino acid protein which bears a strong resemblance to avian MTs, including preservation of the cysteine structure [120]. MT proteins have been found also in tortoises [121] and alligators [122]. Like all other vertebrate MTs, expression of vertebrate MTs seems to be inducible by cadmium.

4. METALLOTHIONEIN-1 AND -2 FUNCTIONAL ASPECTS

4.1. Metal Homeostasis

MT-1/-2 are generally considered to have a role in metal homeostasis, based primarily upon many years of study involving heavy metal toxicity experiments in animals. The general consensus of these studies is that heavy metal exposure in animals (either by diet or environment) results in a consistent elevation in MT-1/-2 expression, most prominently in the liver. Furthermore, similar exposure to metals in MT-1/-2 knockout mice leads to metal toxicity, while MT-1/-2 overexpressing mice are relatively protected from heavy metal toxicity (see review by Coyle et al. [123]). These studies indicate an important role for MT in maintaining heavy metal homeostasis within the body. This is most apparent for toxic metal species such as Cd^{2+} and Hg^{2+}. Indeed, in areas where people are exposed to high environmental levels of these metals, they exhibit elevated levels of heavy metal bound MT-1/-2 in their body. As a further example of the protective capacity of MT under physiological conditions of metal-induced stress, crossing a mouse model of Menkes disease (a copper efflux disease) with MT knockout mice results in embryonic lethality [124]. Whether metal detoxification represents a fortuitous property of MT or rather an evolutionary function is not clear [125].

It is also important to consider MT-1/-2 mediated metal homeostasis at the cellular level. In this regard, it is a consistent observation that cells isolated from MT-1/-2 knockout mice are more susceptible to metal exposure than those isolated from wild-type mice. Similarly, the exposed metal is generally found to be bound to MT-1/-2 within the cell. How does MT-1/-2 maintain metal homeostasis within the cell? When exposed to heavy metal, two possibilities exist for metal handling: firstly, the existing pool of MT-1/-2 can bind this metal, and secondly, the cell can produce new MT-1/-2.

Intracellularly, MT-1/-2 is commonly found as a mixture of Cu^{+}- and mostly Zn^{2+}-MT, although thionein (the metal-free form of MT) has also been detected in cells [126] and tissues [127]. Since it is well established that higher affinity binding metals can readily displace Cu^{+} and Zn^{2+} from MT, this suggests that at least initially the pre-existing pool of MT-1/-2 can buffer excess metal. However, this raises an important issue that by displacing zinc from MT, heavy metal exposure may increase the pool of free zinc in the cell, leading to cytoxicity or perturbation in cellular function (discussed in Krezel et al. [72]). Hence for effective heavy metal homeostasis, it seems likely that the production of new MT-1/-2 (in the form of thionein) is equally important.

4.2. Metal Transfer and Redox Potential

While MT-1/-2 expression is upregulated in response to heavy metal exposure, and heavy metal bound MT-1/-2 is observed in tissue, an important issue to consider is the redox potential of metal bound MT-1/-2, and in particular the role of MT in regulating the availability of metals (and particularly zinc) inside cells. In the case of Zn-MT-1/-2, the redox potential is low, meaning that mild biological oxidants can release zinc from the protein [70,128]. Furthermore, while binding metals such as cadmium and copper bind significantly stronger than zinc to MT-1/-2, they are also readily released by the actions of sulfur ligand oxidation [129,130]. This would suggest that particularly under oxidative stress conditions, MT-1/-2 may not be effective in maintaining heavy metal homeostasis. We should also consider the redox potential of thionein in this system, since there has recently been increased understanding that metal free MT-1/-2 may exist as an important form of the protein under a variety of physiological conditions within the cell (recently reviewed by Petering et al. [131]). In this regard, it appears that thionein reacts more readily to oxidants than Zn-MT [132,133]. As an example of this, addition of nitric oxide (in the form of diethylamine-nonoate) in zinc pre-treated TE671 cells (to induce intracellular levels of zinc-MT and thionein) resulted in an almost exclusive reaction with the thionein fraction of the MT pool [131]. This suggests the possibility that endogenous thionein present within the cell may act as a sacrificial respondent to nitric oxide, protecting the pool of zinc-MT and preventing a subsequent accumulation of free zinc (released from oxidized Zn-MT).

Further highlighting the potential redox role of MT is the cycling between zinc-MT and thionein coupled to glutathione (GSH)/glutathione disulfide (GSSG) [134]. Here, GSSG can oxidize MT leading to the release of zinc, while GSH reduces oxidized MT to become thionein, which can then bind zinc. Furthermore, this cycle appears to be able to control the removal and addition of zinc to inhibitory sites on enzymes involved in glycolysis and signal transduction [135]. While these processes are not particularly efficient, it has been found that selenium compounds catalyze this reaction [134], giving greater credence to the role of MT in redox-mediated function in the cellular environment.

Another contributor to the release of metals from MTs is reactive oxygen and nitrogen species (ROS), which is part of the MT-1/-2 antioxidant mechanism of action. In general, ROS attack and oxidize the metal-thiolate clusters, and subsequently MT-1/-2 release their metals, which may be complexed again in the case of reductive conditions and if the proper metals are provided [35]. However, it has also been reported by Feng et al. [136] that MT-1/-2 can form intramolecular disulfide bonds *in vivo*, which increase after the release of the metal ions.

4.3. Antioxidant Actions

Oxidative stress refers to the cytotoxic consequences of a mismatch between the formation of ROS and the ability of the cells to produce antioxidants. During any stressful or pathological condition, pro-inflammatory responses and activated leukocytes increase significantly ROS production, and accordingly, inflammation is often followed by oxidation and/or nitration of lipids, proteins, DNA, and carbohydrates [137]. Cells protect themselves from oxidative stress by producing antioxidant enzymes (catalase, superoxide dismutase, glutathione peroxidase among others), by using free radical scavengers such as glutathione, vitamin E, and β-carotene, as well as proteins that bind transition metals such as iron and copper (for reviews see [138–141]).

The first study suggesting that MT-1/-2 could be considered as antioxidant proteins was published in 1982 [142], a report which demonstrated that significant radiation resistance was provided by high MT-1/-2 levels in cultured cells. The study by Thornalley and Vašák [74] was a major breakthrough in the MT field, demonstrating that these proteins were efficient *in vitro* scavengers of free hydroxyl ($^{\bullet}OH$) and superoxide ($O_2^{-\bullet}$) radicals produced by the xanthine/xanthine oxidase reaction. Since this pioneering report, many studies have fully confirmed that MTs may react with a number of free radicals, and, from a physiological point of view, the most remarkable is that most, if not all, prooxidant conditions will induce MT-1/-2 synthesis, further underlining a role of these proteins in the host response to oxidative stress [90,143].

As could be expected, MT-1/-2 must be in the vicinity of the free radical generation site in order to exert a significant effect. In transfection experiments carried out in V79 Chinese hamster cells enriched in MT [144], MT conferred protection to DNA from hydroxyl radical attack by H_2O_2, while the mortality rate remained unaffected, which is consistent with its localization in the cell nucleus. Conversely, cytoplasmic presence of MT protects NIH 3T3 cells from the membrane permeant oxidant *tert*-butylhydroperoxide, while no protection of the DNA was evident [145]. More definitive evidence that MT-1/-2 proteins are antioxidants is provided by transgenic mice. Thus, MT-1-2 null embryonic cells show enhanced sensitivity to oxidative stress caused by *tert*-butylhydroperoxide and the redox cycling toxin paraquat [146], and MT-1/-2 null mice show increased liver and kidney damage caused by paraquat administration [147]. This is consistent with the reported functional substitution of yeast copper-zinc superoxide dismutase, a well-known antioxidant enzyme, by both yeast and mammalian MTs [148]. A flow of studies carried out in transgenic mice since then strongly suggest an antioxidant role for these proteins *in vivo* [149–155].

4.4. Antiapoptotic Actions

Numerous *in vitro*, *in vivo*, and human studies have shown antiapoptotic actions of MT-1/-2, although the actual mechanisms of action remain to be fully clarified.

MTs are stress proteins that are strongly upregulated following tissue injury. Inflammation and oxidative stress are prototypical hallmarks in those situations, and both are well-known for inducing apoptosis [139,156,157]; since MT-1/-2 have antioxidant actions and display significant antiinflammatory actions, this would be a mechanism underlying the antiapoptotic effects of MT. However, it is not likely that these effects of MT-1/-2 will provide all the answers, and particularly as mounting data have shown direct antiapoptotic actions of MT-1/-2 *in vitro* [158–163].

MT-1/-2 interact with a range of molecular pathways that directly or indirectly regulate the apoptotic cascade. The cytochrome *c* outflow from the mitochondria is critical as an initiator of ROS-triggered formation of the apoptosome, which is followed by pro-caspase-3 cleavage generating active caspase-3, the executioner of the apoptotic death process [164]. During pro-apoptotic conditions, MT-1/-2 cause a significant reduction in the cytochrome *c* leakage into the cytoplasm and reduced levels of caspases (e.g., caspase-1 and -3), although the precise molecular mechanism of action has not been clarified [161,162,165–169]. Another key inducer of apoptosis is the tumor suppressor protein p53, a protein that has been shown to be inhibited by MT-1/-2 in some cell lines by means of a direct interaction [170,171]. Another possible antiapoptotic mechanism of MT-1/-2 could be the reported interaction with the transcription factor nuclear factor KappaB (NFκB), which is implicated in both cell death and survival, as MT-1/-2 affect the cellular concentration and activity level of NFκB [159,165,172,173], albeit the mechanisms are by no means clear [174].

In a dose-dependent manner, MT-1/-2 also increase *de novo* synthesis and expression levels of common protooncogenes (e.g., c-myc), apoptosis-inhibitory genes (e.g., bcl-2) [175] and a number of growth/trophic factors (like FGF, TGFβ, VEGF, BDNF, GDNF, NTs) that are all well-known as promoters of cell survival [176,177].

4.5. Immunoregulation

A consensus is building up that MT may act as an intermediary in the stress-induced immunomodulatory response. For instance, MT-1/-2 expression can be induced by a number of acute phase cytokines such as IL-1, IL-6, TNF-α, IFN-α, and IFN-γ [178–182]. Furthermore, MT has been observed to interact directly with the plasma membrane of immune cells [183,184]. In further studies, Youn and Lynes [185] found that extracellular MT dramatically decreases both cytotoxic T lymphocyte (CTL) induction and cytolytic activity against allogenic target cells. Since CTLs are known to have an important role in protective immunity, this indicates that MT may have a significant role in the immune function. Indeed, it has been demonstrated that MT can modulate the *in vivo* humoral immune responses to T-dependent antigens, as observed by a decrease in immune response following injection of exogenous MT [186], and increased immune response to ovalbumin with simultaneous injection of a monoclonal anti-MT antibody [187]. Supporting this, genetic inability to express MT-1/-2 in MT knockout mice leads to elevated humoral response to T-dependent antigen challenge [172].

4.6. Other Potential Functions: A Potential Extracellular Role

There is now increasing evidence supporting the hypothesis that in addition to important intracellular roles (such as metal homeostasis and protection from oxidative stress), MTs have important extracellular roles. There are three lines of evidence from a number of different models that support this; namely the release of MT by cultured cells and detection *in vivo*, the uptake of extracellular MT by cells, and the action of extracellular MT upon cellular function.

It has been known for a number of years now that MT-1/-2 can be detected in the extracellular environment *in vivo* [188–190], and is actively secreted by adipocytes [191]. Similarly, it has been reported that MT-3 is secreted by cultured astrocytes [192]. Whether secretion of MT is regulated by cellular stress or other mechanisms has yet to be determined.

In more recent studies, it has been reported that extracellular MT-1/-2 is internalized by kidney tubule cells, with the endocytic receptor megalin responsible for between 30–60% of MT-1/-2 uptake by these cells [193,194]. Furthermore, it was found that megalin and exogenously applied MT-1/-2 colocalize within kidney cells, and that in the presence of the megalin ligand RAP or anti-megalin antibodies MT-1/-2 uptake is reduced. Surface plasma resonance (SPR) studies found consistent binding of approximately 0.7–0.9 mol of MT/mol of megalin, consistent with one binding site, with an estimated dissociation constant of 9.8×10^{-5} M. Finally, these authors demonstrated that the specific region of the MT molecule responsible for megalin binding resides in the α-domain of the protein.

A potential role for extracellular MT in modulating cellular function was revealed by Uchida and colleagues [47], who demonstrated that exogenously applied MT-3 inhibits neuronal survival of cultured cortical neurons, while other MT isoforms exhibit no such inhibitory activity. This biological property of MT-3 is commonly referred to as neuronal growth inhibitory activity. To date, a number of reports have confirmed the inhibitory activity of MT-3 upon neuronal survival, although in all cases, a crude brain homogenate was required for the bioactivity of MT-3 (e.g., see [48,195]). In all of these assays, MT-1/-2 exhibited no inhibitory effect. Interestingly, MT-1/-2 appear to have the opposite effect upon neurite outgrowth, with studies reporting that these proteins actually promote neurite outgrowth of cortical [196], and dopaminergic neurons [158]. Furthermore, MT-1/-2 have been observed to promote axon regeneration of injured neurons also following cortical injury to the rat brain [196].

In summary, there is an increasing understanding that in addition to important intracellular roles, extracellular MTs have important roles in regulating cellular function, including neuronal differentiation and survival, and immune response (see Section 4.5).

5. METALLOTHIONEINS IN THE CENTRAL NERVOUS SYSTEM

5.1. Metallothionein Expression in the Normal and Aging Adult Human Brain

The MT-1/-2 and MT-3 isoforms are found in varying levels throughout the adult mammalian brain. Within the uninjured adult brain, the levels of MT-1/-2 are relatively low; Nakajima and Suzuki [197] reported levels of 3.6 and 34.5 μg/g wet weight of MT for rat and human brain, respectively. Using immunohistochemical and *in situ* hybridization techniques it has been determined that the majority of this MT-1/-2 expression within the brain under basal conditions is by subsets of astrocytes, in particular protoplasmic astrocytes in the gray matter, and less so by astrocytes in the white matter [197–200]. However, MT-1/-2 are not exclusively expressed by astrocytes and indeed, other cells such as endothelial cells of blood vessels, meningeal cells, ependymal cells, and epithelial cells of the choroid plexus are also found to express MT-1/-2 [198]. In the spinal cord, MT-1/-2 is found particularly in gray matter astrocytes [201,202]. There are also some reports of neuronal expression of MT-1/-2 in the brain and spinal cord [201,203], although these are inconsistent and at considerably lower expression levels than those observed in astrocytes.

In 1991 a brain-specific MT isoform, MT-3, was identified in the human brain [47], whose expression was particularly prominent in hippocampal neurons within the CA3 region [204], which interestingly are known to contain zinc in their axon terminals. In addition to this, high levels of MT-3 mRNA have been observed in neurons within the cerebral cortex, amygdala, and cerebellum [204], and immunohistochemistry has confirmed this expression pattern of MT-3 in canine [205] and rat brains [206]. MT-3 is estimated to contribute approximately one third of the total MT levels within the adult brain. Interestingly, other studies suggest that astrocytes are the primary source of MT-3 expression within the brain, and particularly those in the cerebellum, neocortex, hippocampus, striatum, brain stem, and spinal cord [47,207–209]. It is not clear why there is such discord in the expression pattern of MT-3 in the mammalian brain, but it may relate to the experimental model investigated, the use of *in situ* hybridization *versus* immunohistochemical techniques and to the different antisera raised against different MT-3 antigens. It is also possible that some of the immunohistochemical detection of MT-3 in cells may not represent expression by that cell, but rather uptake of MT-3. In this regard, it has been demonstrated that MT-3 is secreted by cultured astrocytes [192] and that extracellular MT-3 interacts directly with neurons to inhibit neurite outgrowth [47,210].

It is well established that MT levels increase with age. In the CNS, MT-1/-2 is expressed at relatively low levels during human fetal development and increases significantly after birth [199]. For instance, elevated levels of MT-1/-2 mRNA and protein have been observed in aged humans, rats, and dogs. Suzuki et al. [200] reported a notable difference in MT-1/-2 expression levels and distribution in the aged brain when compared to those of the young human brain. Kojima et al. [205] found MT-1/-2 in the aged dog brain to be localized to astrocytes of the cerebral cortex and around the blood vessels where severe age-related morphological changes were observed.

Similarly, changes in MT-3 expression are also apparent within the aged brain. Miyazaki et al. [211] showed that the expression of MT-3, at both the mRNA and protein level, was up-regulated in the aged rat brain, and particularly in the cerebral cortices, hippocampus, hypothalamus, and amygdaloid nucleus. The consistent elevation of MT-1/-2 and -3 in the ageing brain has led to the proposal that MT-1/-2 can be used as a marker for ageing within the CNS [212].

5.2. Changes in Metallothionein Expression in the Neurodegenerative Brain

Levels of MT expression are also altered in neurodegenerative diseases. Indeed, the MT-3 isoform was first identified as a protein found in the normal adult brain but that it is deficient in the Alzheimer's disease brain [47]. While it is now not clear whether this is the case in all Alzheimer's disease brains, or particular to some of the AD populations, it is apparent that MT levels are very different in this and other neurodegenerative diseases.

MT expression levels have been well characterized in AD, and there are consistent reports of an increase in MT-1/-2 expression in the AD brain. For example, Duguid and colleagues demonstrated an increase in MT-1/-2 mRNA in the hippocampus of AD brain samples [213]. Immunohistochemistry has revealed that expression of MT-1/-2 is increased in layers II-VI of the gray matter and white matter, mostly within GFAP-positive astrocytes [214]; interestingly, pre-clinical AD cases also exhibit elevated levels of MT-1/-2 expression. In a further study, Zambenedetti and colleagues also noted an upregulation of MT-1/-2, particularly in astrocytes and microcapillaries of AD cerebral cortex and cerebral white matter, but differences between controls and AD cases were less distinct in the cerebellar cortex and white matter [215].

There has been intense interest in the role of MT-3 in AD since its discovery as a growth inhibitory factor with reduced expression level in this disease [47]. Using a MT-3 specific antibody this group established that MT-3 expression was reduced in AD cortex homogenates compared to that of normal, age matched brains, which was supported by immunohistochemical studies. In a subsequent study, northern blot analysis of AD and control brain homogenates revealed that MT-3 mRNA was reduced in AD [216]. In a later study involving 25 AD cases and 10 controls, immunocytochemistry, western blotting, and mRNA analysis found that MT-3 levels were lower in the AD cases [217]. However, other studies have reported little or no change in MT-3 mRNA [48,218] or protein levels [48] between AD and control cases. In a later study, Carrasco and colleagues found that MT-3 mRNA and protein levels are both doubled in the AD brain compared to controls [209]. It is possible that the reason for these discrepancies may relate to the populations involved in the studies and the different genetic or environmental components of these populations [48].

Not only are MT levels altered in AD, however, elevated levels of MT-1/-2 have also been reported in astrocytes within the spinal cord of patients with amyotrophic lateral sclerosis [202,219], a neurodegenerative disease causing progressive loss of spinal motor neurons. In multiple sclerosis, an inflammatory neurodegenerative disease, MT-1/-2 levels are elevated in astrocytes and activated monocytes/macrophages [220]. Interestingly, MT-1/-2 levels were mildly greater in inactive lesions than active lesions, suggesting that MT-1/-2 are involved in the remission periods of this disease. Rather surprisingly then, it has been reported that there is no change in MT-1/-2 expression in the Parkinson's disease brain [221].

The expression of MT-3 is affected by a number of human neurological diseases other than AD. MT-3 expression has been shown to be altered (up- or downregulated) in Down syndrome [222], Creutzfeld-Jakob disease [223], pontosubicular necrosis [224], Parkinson disease, meningitis, and amyotrophic lateral sclerosis (ALS) [207].

5.3. Metallothionein Expression in Experimental Animal Models of Human Diseases

Given that MT-1/-2 levels are elevated in the ageing or neurodegenerating human brain, there has been significant effort placed in understanding this mechanism, and particularly through the use of experimental animal models of human neurodegenerative diseases. In this regard, studies using animal models of Alzheimer's disease, amyotropic lateral sclerosis, and multiple sclerosis (MS) have been particularly insightful.

A range of different AD animal models exist, which produce either β-amyloid (Aβ) plaques or neurofibrillary tangles within the brain, two of the pathological hallmarks of AD. It has recently been reported that in mice expressing mutation bearing human β-amyloid precursor protein (resulting in the deposition of Aβ plaques within the brain) there are significantly increased levels of MT-1 expression in the brain [225]. In this study, three different transgenic models of AD, namely the Tg2576, TgCRND8, and Tg-SwD1 were found to have elevated levels of MT-1 mRNA, with induction greater and wider in the latter two strains of mice, consistent with the severity of Aβ deposition in these models.

Importantly, MT-1 upregulation was prominent in reactive astrocytes and microglia closely related to amyloid plaques, and not in areas away from plaques. This suggests a specific induction of MT-1 by Aβ plaque formation, likely associated with the elevated inflammation present surrounding plaques, or with the accumulation of zinc and copper within plaques. Interestingly, in the same study it was found that MT-3 expression was essentially unaltered in all three AD mouse models, with expression levels surrounding plaques similar to levels away from plaques. This suggests that those factors inducing MT-1 (gliosis, inflammation, and metals) are probably not important for the regulation of MT-3, and also that perhaps MT-3 is not specifically involved in the progression of AD.

A small number (approximately 1%) of all ALS cases are genetically linked to abnormalities in the superoxide dismutase (SOD1) gene (of which a number of mutations have been identified), and mice expressing human SOD1 bearing such a mutation develop muscular wasting and neurodegeneration similar to the human disease. In these SOD1 mutant mice, MT-1/-2 expression is elevated in astrocytes within the white and gray matter of the spinal cord [226]. Elevated MT-1/-2 levels (both mRNA and protein) are apparent prior to onset of motor deficits, and continue as neurodegeneration progresses [226–228], suggesting that MT-1/-2 serve an important function in the progression of this disease. In these mice, MT-3 levels are also elevated, although this occurs once motor deficits manifest, and changes from being expressed primarily by neurons in the dorsal and ventral horns to also being expressed by astrocytes throughout the white and gray matter [226,228]. These studies suggest the possibility of different functions for MT-3 in ALS.

In experimental autoimmune encephalomyelitis (EAE), an animal model of MS, MT-1 expression is strongly induced by reactive astrocytes and macrophages, with the level of expression directly proportional to the severity of the disease, in both mice [229] and rats [230,231]. Similar experiments using IFN-γ knockout mice revealed that this cytokine is not necessary for induction of MT-1 in this model [232], which is interesting given that this cytokine is involved in the regulation of oxidative stress in EAE [233].

Besides these three models, there are a number of studies showing MT-1/-2 induction in animal models of inflammation [8,11,182,234], stress [235–238], excitotoxicity [165,239–244], traumatic brain injury [209,245–247], stroke/ischemia [203,244,248–250], and gliodegeneration [251–253]. As for MT-3, up- or downregulation depending on the model, time, etc., has been observed in animal models of brain injury. MT-3 expression has been shown to be increased by stab wounds [254–256] and kainic acid administration [254], but decreased by cortical ablation of the somatosensory cortex [257], facial nerve transection [258], and middle cerebral artery occlusion [259]. A biphasic response of MT-3 to CNS injury, with initial downregulation followed by upregulation, was observed in response to N-methyl-D-aspartate (NMDA) [260] or to a cryolesion [176,246].

5.4. Functional Roles of Metallothioneins in the Central Nervous System: Transgenic Mice

We will limit the discussion here to the MT-1/-2 isoforms (see Chapter 11 for more insight into MT-3). As stated above, MT-1/-2 levels are increased in a number of animal models of human diseases and following tissue injury and inflammation. Increased MT-1/-2 expression, however, does not necessarily indicate an important role. The generation of genetically modified mice [261–263] has been tremendously useful, demonstrating that these proteins are indeed important in the CNS. Thus, mice overexpressing MT-1 showed lower infarcts and better functional recovery than wild-type mice following mild focal cerebral ischemia and reperfusion [203], and the opposite was observed in *Mt-1 & 2*-null mice [250]. Other studies with transgenic mice have equally involved MT-1/-2 as important proteins following damage elicited by kainic acid-induced seizures [165], gliotoxins [252,264], 6-hydroxydopamine [265], traumatic brain injury [176,247,266], and transgenic IL-6-induced neuropathology [267–269]. Despite the fact that these are quite different models of brain injury, MT-1/-2 exerted similar effects in all cases, decreasing oxidative stress, inflammation, and apoptosis, in other words, offering neuroprotection.

Similar conclusions have been drawn from studies using animal models of neurodegenerative diseases. Shedding light upon the possible roles of MTs in amyotrophic lateral sclerosis, Puttaparthi and colleagues [270] crossed SOD1 mutant mice with MT-1/-2 knockout mice, and observed significantly faster onset of motor dysfunction and faster mortality in these mice. A neuroprotective role of MT-1/-2 in ALS was also suggested independently [271]. These studies further support the neuroprotective role of MTs in ALS, and suggest that these functions are cell and isoform specific. Further investigation is required to determine how MTs act neuroprotectively in

ALS, but may be related to the handling of zinc (zinc-deficient SOD1 is highly toxic to motor neurons) or oxidative stress (which is greatly elevated in ALS).

A clear role for MT is emerging in the protection against myelin and neuronal damage associated with multiple sclerosis. When EAE was induced in MT-1/-2 knockout mice, the animals suffered significantly greater severity in neurodegeneration including increased inflammation and neuronal apoptosis [272], suggesting that MT-1/-2 are important protective proteins in this model. This was further highlighted by the demonstration that intraperitoneal injection of zinc-MT-2 was able to delay onset and progression of disease symptoms in EAE mice, including a reduction in inflammatory response of IL-6 and TNF-α [230,231], prevention of demyelination and axonal degeneration, and stimulation of oligodendrocyte regeneration and elevated expression of a number of trophic growth factors [273].

Independent studies by Canpolat and Lynes indicate that exogenous MT-1/-2 can act as negative regulators of immune response by effecting cell proliferation and macrophage and cytotoxic T lymphocyte function [187], suggesting that MT-1/-2 may act by suppressing the severity of autoimmune attacks in this disease. It will be interesting to investigate if similar MT-2 treatment can protect in MT-1/-2 knockout mice undergoing EAE. Given these extensive studies and comparative studies of MT-1/-2 expression in humans, it seems clear that MT-1/-2 are involved in delaying the neurodegenerative progression of MS and suggests MT-1/-2 as a possible therapy for MS. As a note, MT-3 expression appears unaltered in EAE mice, with no significant induction adjacent to inflammatory infiltrates [229].

Despite the strong correlation between Aβ plaque formation and MT-1 expression in AD mouse models [225], the role of MT-1 in AD is still not clear. Given the antiapoptotic and antiinflammatory properties (see Section 4) of MT, it is plausible to suggest that the role of MT-1 (and most likely MT-2 given that their regulation and function seem similar) in AD is again that of neuroprotection. Future studies investigating the phenotype of MT-1/-2 knockout mice crossed with AD mice will be particularly insightful.

6. CONCLUDING REMARKS AND FUTURE DIRECTIONS

In summary, MTs are a highly conserved family of proteins. The number of isoforms has grown with evolution, and there are reasons to believe that the different proteins may have similar but also specific functions; this deserves huge efforts in future studies.

Regardless of these issues, it seems likely that the structure of MT and its reactivity is intimately related to its function. In this regard, the three-dimensional structure of MT is dependent upon the formation of metal-thiolate clusters (indeed, thionein has a random coil structure), and the formation of significantly different peptide conformations between unmetallated forms and metallated forms with different metal ions may contribute to the variety of different functions that MT may serve in the cell.

MTs appear to have important protective functions in a range of neurodegenerative conditions. Further investigation is required to determine whether this represents a general protective effect of MTs (mediated by metal regulation, modulation of oxidative stress or inflammation, all of which are important mechanisms in neurodegenerative diseases) or a direct mechanism specific to each disease. Since the exogenous administration of these proteins exert significant effects which are very much in line with those concluded from the endogenous proteins, a therapeutic use of MTs is more than a possibility.

ACKNOWLEDGMENTS

The authors appreciate the funding obtained from Ministerio de Ciencia y Tecnología and Feder SAF2005-00671 and European Comission FP6 Integrated Project Exgenesis (Ref. LSHM-CT-2004-005272) (JH); the Australian Research Council (DP0556630, LP0774820) and the National Health and Medical Research Council of Australia (352623)(RC); the IMK Almene Fond, Scleroseforeningen, KAS Herlev's Forskningsråd, Kathrine og Vigo Skovgaards Fond, Vera og Carl Michaelsens Legat, Hørslev-fonden, Lundbeckfonden, Karen A Tolstrups Fond, Eva & Henry Frænkels Mindefond, Fonden til Lægevidenskabens Fremme, Th. Maigaard's Eftf. Fru Lily Benthine Lunds Fond, Dir. Leo Nielsens Legat, Kong Christian den Tiendes Fond, Lægeforeningens Forskningsfond/The Danish Medical Association Research Fund, Toyota-Fonden, Dansk Parkinsonforening, Dir. Ib Henriksens Fond & Dagmar Marshalls Fond (MP); and the Swiss National Science Foundation Grant 3100A0-111884 (MV).

ABBREVIATIONS

AD	Alzheimer's disease
ALS	amyotrophic lateral sclerosis
ARE	antioxidant response element
Aβ	β amyloid
ckMT	chicken metallothionein
CNS	central nervous system
CTL	cytotoxic T lymphocyte

DTNB	5,5′-dithiobis-(2-nitrobenzoic acid)
EAE	experimental autoimmune encephalomyelitis
EDTA	ethylenediamine-N,N,N′,N′-tetraacetic acid
EPR	electron paramagnetic resonance
EXAFS	extended X-ray absorption fine structure
GFAP	glial fibrillary acidic protein
GRE	glucocorticoid responsive element
GSH	glutathione
GSSG	glutathione disulfide
IFN-γ	interferon-γ
IL	interleukin
MBF-1	metal responsive element binding factor-1
MCD	magnetic circular dichroism
MRE	metal responsive element
MREBP	metal responsive element binding protein
MS	multiple sclerosis
MT	metallothionein
MTF-1	metal responsive element binding transcription factor
n.d.	not defined
NFκB	nuclear factor κB
NMDA	N-methyl-D-aspartate
NMR	nuclear magnetic resonance
NOE	nuclear Overhauser effect
NTA	nitrilotriacetate
RAP	receptor-associated protein
rmsd	root-mean-square deviation
ROS	reactive oxygen species
SOD	superoxide dismutase
SPR	surface plasma resonance
TNF-α	tumor necrosis factor-α
ZAP	zinc activated protein
ZiRF1	zinc regulated factor

REFERENCES

1. M. Margoshes and B. L. Vallee, *J. Amer. Chem. Soc.*, 1957, **79**, 4813–4814.
2. J. H. R. Kägi and B. L. Vallee, *J. Biol. Chem.*, 1960, **235**, 3460–3465.
3. J. H. R. Kägi and B. L. Valle, *J. Biol. Chem.*, 1961, **236**, 2435–2442.
4. P.-A. Binz and J. H. R. Kägi, *Metallothionein Molecular Evolution and Classification*, in *Metallothionein IV*, Ed. C. D. Klaassen, Birkhäuser Verlag, Basel, 1999, pp. 7–13.

5. M. Valls, R. Bofill, R. Gonzalez-Duarte, P. Gonzalez-Duarte, M. Capdevila and S. Atrian, *J. Biol. Chem.*, 2001, **276**, 32835–32843.
6. M. Karin, R. L. Eddy, W. M. Henry, L. L. Haley, M. G. Byers and T. B. Shows, *Proc. Natl. Acad. Sci. USA*, 1984, **81**, 5494–5498.
7. A. K. West, R. Stallings, C. E. Hildebrand, R. Chiu, M. Karin and R. I. Richards, *Genomics*, 1990, **8**, 513–518.
8. R. D. Palmiter, S. D. Findley, T. E. Whitmore and D. M. Durnam, *Proc. Natl. Acad. Sci. USA*, 1992, **89**, 6333–6337.
9. C. J. Quaife, S. D. Findley, J. C. Erickson, G. J. Froelick, E. J. Kelly, B. P. Zambrowicz and R. D. Palmiter, *Biochemistry*, 1994, **33**, 7250–7259.
10. D. R. Cox and R. D. Palmiter, *Hum. Genet.*, 1983, **64**, 61–64.
11. P. F. Searle, B. L. Davison, G. W. Stuart, T. M. Wilkie, G. Norstedt and R. D. Palmiter, *Mol. Cell Biol.*, 1984, **4**, 1221–1230.
12. J. Hidalgo, M. Aschner, P. Zatta and M. Vašák, *Brain Res. Bull.*, 2001, **55**, 133–145.
13. J. H. Kägi, S. R. Himmelhoch, P. D. Whanger, J. L. Bethune and B. L. Vallee, *J. Biol. Chem.*, 1974, **249**, 3537–3542.
14. J. H. Kägi and Y. Kojima, Experientia (suppl.), 1987, **52**, 25–61.
15. G. F. Nordberg, *Biol. Trace Elem. Res.*, 1989, 131–135.
16. J. H. Kägi and A. Schäffer, *Biochemistry*, 1988, **27**, 8509–8515.
17. D. G. Nettesheim, H. R. Engeseth and J. D. Otvos, *Biochemistry*, 1985, **24**, 6744–6751.
18. J. D. Otvos, H. R. Engeseth, D. G. Nettesheim and C. R. Hilt, Experientia (suppl.), 1987, **52**, 171–178.
19. D. R. Winge and K. A. Miklossy, *J. Biol. Chem.*, 1982, **257**, 3471–3476.
20. J. D. Otvos and I. M. Armitage, *Proc. Natl. Acad. Sci. USA*, 1980, **77**, 7094–7098.
21. M. Vašák, *J. Am. Chem. Soc.*, 1980, **102**, 3953–3955.
22. A. Arseniev, P. Schultze, E. Wörgötter, W. Braun, G. Wagner, M. Vašák, J. H. Kägi and K. Wüthrich, *J. Mol. Biol.*, 1988, **201**, 637–657.
23. P. Schultze, E. Wörgötter, W. Braun, G. Wagner, M. Vašák, J. H. Kägi and K. Wüthrich, *J. Mol. Biol.*, 1988, **203**, 251–268.
24. B. A. Messerle, M. Bos, A. Schäffer, M. Vašák, J. H. Kägi and K. Wüthrich, *J. Mol. Biol.*, 1990, **214**, 781–786.
25. K. Zangger, G. Oz, J. D. Otvos and I. M. Armitage, *Protein Sci.*, 1999, **8**, 2630–2638.
26. A. H. Robbins, D. E. McRee, M. Williamson, S. A. Collett, N. H. Xuong, W. F. Furey, B. C. Wang and C. D. Stout, *J. Mol. Biol.*, 1991, **221**, 1269–1293.
27. W. Braun, M. Vašák, A. H. Robbins, C. D. Stout, G. Wagner, J. H. Kägi and K. Wüthrich, *Proc. Natl. Acad. Sci. USA*, 1992, **89**, 10124–10128.
28. C. D. Berweger, W. Thiel and W. F. van Gunsteren, *Proteins: Structure, Function, and Genetics*, 2000, **41**, 299–315.
29. M. Good, R. Hollenstein and M. Vašák, *Eur. J. Biochem.*, 1991, **197**, 655–659.
30. D. L. Pountney and M. Vašák, *Eur. J. Biochem.*, 1992, **209**, 335–341.
31. K. S. Hagen, D. W. Stephan and R. H. Holm, *Inorg. Chem.*, 1982, **21**, 3928–3936.

32. M. Good, R. Hollenstein, P. J. Sadler and M. Vašák, *Biochemistry*, 1988, **27**, 7163–7166.
33. M. Vašák and J. H. Kägi, *Proc. Natl. Acad. Sci. USA*, 1981, **78**, 6709–6713.
34. I. Bertini, C. Luchinat, L. Messori and M. Vašák, *J. Am. Chem. Soc.*, 1989, **111**, 7296–7300.
35. N. Romero-Isart and M. Vašák, *J. Inorg. Biochem.*, 2002, **88**, 388–396.
36. A. Krezel and W. Maret, *J. Am. Chem. Soc.*, 2007, **129**, 10911–10921.
37. V. Calderone, B. Dolderer, H. J. Hartmann, H. Echner, C. Luchinat, C. Del Bianco, S. Mangani and U. Weser, *Proc. Natl. Acad. Sci. USA*, 2005, **102**, 51–56.
38. M. J. Stillman, *Coord. Chem. Rev.*, 1995, **144**, 461–511.
39. K. B. Nielson, C. L. Atkin and D. R. Winge, *J. Biol. Chem.*, 1985, **260**, 5342–5350.
40. D. L. Pountney, I. Schauwecker, J. Zarn and M. Vašák, *Biochemistry*, 1994, **33**, 9699–9705.
41. K. B. Nielson and D. R. Winge, *J. Biol. Chem.*, 1984, **259**, 4941–4946.
42. K. B. Nielson and D. R. Winge, *J. Biol. Chem.*, 1983, **258**, 13063–13069.
43. R. Bogumil, P. Faller, D. L. Pountney and M. Vašák, *Eur. J. Biochem.*, 1996, **238**, 698–705.
44. R. Bogumil, P. Faller, P. A. Binz, M. Vašák, J. M. Charnock and C. D. Garner, *Eur. J. Biochem.*, 1998, **255**, 172–177.
45. B. Roschitzki and M. Vašák, *J. Biol. Inorg. Chem.*, 2002, **7**, 611–616.
46. B. Roschitzki and M. Vašák, *Biochemistry*, 2003, **42**, 9822–9828.
47. Y. Uchida, K. Takio, K. Titani, Y. Ihara and M. Tomonaga, *Neuron*, 1991, **7**, 337–347.
48. J. C. Erickson, A. K. Sewell, L. T. Jensen, D. R. Winge and R. D. Palmiter, *Brain Res.*, 1994, **649**, 297–304.
49. P. Faller, D. W. Hasler, O. Zerbe, S. Klauser, D. R. Winge and M. Vašák, *Biochemistry*, 1999, **38**, 10158–10167.
50. D. W. Hasler, L. T. Jensen, O. Zerbe, D. R. Winge and M. Vašák, *Biochemistry*, 2000, **39**, 14567–14575.
51. G. Öz, K. Zangger and I. M. Armitage, *Biochemistry*, 2001, **40**, 11433–11441.
52. H. Wang, Q. Zhang, B. Cai, H. Li, K. H. Sze, Z. X. Huang, H. M. Wu and H. Sun, *FEBS Lett.* 2006, **580**, 795–800.
53. N. Romero-Isart, L. T. Jensen, O. Zerbe, D. R. Winge and M. Vašák, *J. Biol. Chem.*, 2002, **277**, 37023–37028.
54. P. Palumaa, E. Eriste, O. Njunkova, L. Pokras, H. Jornvall and R. Sillard, *Biochemistry*, 2002, **41**, 6158–6163.
55. P. Palumaa, I. Tammiste, K. Kruusel, L. Kangur, H. Jornvall and R. Sillard, *Biochim. Biophys. Acta*, 2005, **1747**, 205–211.
56. L. Liang, K. Fu, D. K. Lee, R. J. Sobieski, T. Dalton and G. K. Andrews, *Mol. Reprod. Dev.*, 1996, **43**, 25–37.
57. T. Schlake and T. Boehm, *Mech. Dev.*, 2001, **109**, 419–422.
58. L. Tio, L. Villarreal, S. Atrian and M. Capdevila, *J. Biol. Chem.*, 2004, **279**, 24403–24413.

59. G. Meloni, K. Zovo, J. Kazantseva, P. Palumaa and M. Vašák, *J. Biol. Chem.*, 2006, **281**, 14588–14595.
60. A. E. Martell, R. D. Hancock and R. J. Motekaitis, *Coord. Chem. Rev.*, 1994, **133**, 39–65.
61. J. D. Otvos, X. Liu, H. Li, G. Shen and M. Basti, *Dynamic Aspects of Metallothionein Structure*, in *Metallothionein III*, Ed. K.T. Suzuki, N. Imura and M. Kimura, Birkhäuser Verlag, Basel, 1993, pp. 57–74.
62. A. O. Udom and F. O. Brady, *Biochem. J.*, 1980, **187**, 329–335.
63. C. Jacob, W. Maret and B. L. Vallee, *Proc. Natl. Acad. Sci. USA*, 1998, **95**, 3489–3494.
64. J. Zeng, R. Heuchel, W. Schaffner and J. H. Kägi, *FEBS Lett*, 1991, **279**, 310–312.
65. G. Roesijadi, R. Bogumil, M. Vašák and J. H. Kägi, *J. Biol. Chem.*, 1998, **273**, 17425–17432.
66. C. F. Shaw, M. M. Savas and D. H. Petering, *Methods Enzymol.*, 1991, **205**, 401–414.
67. X. Yu, Z. Wu and C. Fenselau, *Biochemistry*, 1995, **34**, 3377–3385.
68. J. Zaia, L. Jiang, M. S. Han, J. R. Tabb, Z. Wu, D. Fabris and C. Fenselau, *Biochemistry*, 1996, **35**, 2830–2835.
69. A. Muñoz, F. Laib, D. H. Petering and C. F. Shaw III, *J. Biol. Inorg. Chem.*, 1999, **4**, 495–507.
70. W. Maret and B. L. Vallee, *Proc. Natl. Acad. Sci. USA*, 1998, **95**, 3478–3482.
71. L. J. Jiang, W. Maret and B. L. Vallee, *Proc. Natl. Acad. Sci. USA*, 1998, **95**, 3483–3488.
72. A. Krezel, Q. Hao and W. Maret, *Arch. Biochem. Biophys.*, 2007, **402**, 551–558.
73. G. Meloni, P. Faller and M. Vašák, *J. Biol. Chem.*, 2007, **282**, 16068–16078.
74. P. J. Thornalley and M. Vašák, *Biochim. Biophys. Acta*, 1985, **827**, 36–44.
75. K. D. Kroncke, K. Fehsel, T. Schmidt, F. T. Zenke, I. Dasting, J. R. Wesener, H. Bettermann, K. D. Breunig and V. Kolb-Bachofen, *Biochem. Biophys. Res. Commun.*, 1994, **200**, 1105–1110.
76. D. H. Petering, S. Krezoski, P. Chen, A. Pattaniak and C. F. Shaw, III, *Kinetic Reactivity of Metallothionein*, in *Metallothionein-Synthesis, Structure and Properties of Metallothioneins, Phytochelatins and Metal-Thiolate Complexes*, Ed. M. J. Stillman, C. F. Shaw, K. T. Suzuki III, VCH, Weinheim, New York, 1992, pp. 164–183.
77. T. Gan, A. Munoz, C. F. Shaw, III and D. H. Petering, *J. Biol. Chem.*, 1995, **270**, 5339–5345.
78. H. Li and J. D. Otvos, *J. Inorg. Biochem.*, 1998, **70**, 187–194.
79. I. Bremner, *Experientia* (suppl.), 1987, **52**, 81–107.
80. M. K. Yagle and R. D. Palmiter, *Mol. Cell Biol.*, 1985, **5**, 291–294.
81. N. W. Shworak, O. C. T N. C. Wong and L. Gedamu, *J. Biol. Chem.*, 1993, **268**, 24460–24466.
82. M. Karin, A. Haslinger, H. Holtgreve, R. I. Richards, P. Krauter, H. M. Westphal and M. Beato, *Nature*, 1984, **308**, 513–519.
83. F. Radtke, R. Heuchel, O. Georgiev, M. Hergersberg, M. Gariglio, Z. Dembic and W. Schaffner, *EMBO J.*, 1993, **12**, 1355–1362.

84. E. J. Kelly, E. P. Sandgren, R. L. Brinster and R. D. Palmiter, *Proc. Natl. Acad. Sci. USA*, 1997, **94**, 10045–10050.
85. S. L. Samson and L. Gedamu, *Prog. Nucleic Acid Res. Mol. Biol.*, 1998, **59**, 257–288.
86. R. D. Palmiter, *Proc. Natl. Acad. Sci. USA*, 1994, **91**, 1219–1223.
87. C. M. Tang, J. Westling and E. Seto, *Mol. Cell Biol.*, 1999, **19**, 680–689.
88. S. Koizumi, K. Suzuki and F. Otsuka, *J. Biol. Chem.*, 1992, **267**, 18659–18664.
89. G. Westin and W. Schaffner, *EMBO J.*, 1988, **7**, 3763–3770.
90. G. K. Andrews, *Biochem. Pharmacol.*, 2000, **59**, 95–104.
91. M. H. Vasconcelos, S. C. Tam, J. E. Hesketh, M. Reid and J. H. Beattie, *Toxicol. Appl. Pharmacol.*, 2002, **182**, 91–97.
92. M. H. Vasconcelos, S. C. Tam, J. H. Beattie and J. E. Hesketh, *Biochem. J.*, 1996, **315**, 665–671.
93. J. P. Bourdineaud, M. Baudrimont, P. Gonzalez and J. L. Moreau, *Biochimie*, 2006, **88**, 1787–1792.
94. G. K. Andrews, L. P. Fernando, K. L. Moore, T. P. Dalton and R. J. Sobieski, *J. Nutr.*, 1996, **126**, 1317S–1323S.
95. K. L. Shartzer, K. Kage, R. J. Sobieski and G. K. Andrews, *J. Mol. Evol.*, 1993, **36**, 255–262.
96. Y. J. Lee, Y. P. Chen, S. H. Wang, W. Y. Chow and L. Y. Lin, *Gene*, 1996, **176**, 85–92.
97. L. F. Maxfield, C. D. Fraize and J. M. Coffin, *Proc. Natl. Acad. Sci. USA*, 2005, **102**, 1436–1441.
98. C. C. McCormick, C. S. Fullmer and J. S. Garvey, *Proc. Natl. Acad. Sci. USA*, 1988, **85**, 309–313.
99. D. Y. Wei and G. K. Andrews, *Nucleic Acids Res.*, 1988, **16**, 537–553.
100. L. Y. Lin, W. C. Lin and P. C. Huang, *Biochim. Biophys. Acta*, 1990, **1037**, 248–255.
101. L. Villarreal, L. Tio, M. Capdevila and S. Atrian, *FEBS J.* 2006, **273**, 523–535.
102. U. Weser, H. Rupp, F. Donay, F. Linnemann, W. Voelter, W. Voetsch and G. Jung, *Eur. J. Biochem.*, 1973, **39**, 127–140.
103. L. P. Fernando, D. Y. Wei and G. K. Andrews, *J. Nutr.*, 1989, **119**, 309–318.
104. C. Capasso, V. Carginale, O. Crescenzi, D. Di Maro, E. Parisi, R. Spadaccini, P. A. Temussi, *Structure*, 2003, **11**, 435–434.
105. E. Carpene, P. Cortesi, S. Tacconi, O. Cattani, G. Isani and G. P. Serrazanetti, *Comp. Biochem. Physiol. C*, 1987, **86**, 267–272.
106. J. Overnell, R. McIntosh and T. C. Fletcher, *Experientia*, 1987, **43**, 178–181.
107. R. P. Cosson, *Biol. Trace Elem. Res.*, 1994, **46**, 229–245.
108. J. Hidalgo and R. Flos, *Comp. Biochem. Physiol. C*, 1986, **83**, 99–103.
109. H. L. d. Ley, M. L. Failla and D. S. Cherry, *Comp. Biochem. Physiol. B*, 1983, **74**, 507–513.
110. J. Price-Haughey, K. Bonham and L. Gedamu, *Environ. Health Perspect.*, 1986, 141–147.
111. K. T. Suzuki and H. Akitomi, *Comp. Biochem. Physiol. C*, 1983, **75**, 211–215.
112. K. T. Suzuki and R. Kawamura, *Comp. Biochem. Physiol. C*, 1984, **79**, 255–260.

113. A. K. Vogiatzis and N. S. Loumbourdis, *Arch. Environ. Contam. Toxicol.*, 1998, **34**, 64–68.
114. E. A. Papadimitriou and N. S. Loumbourdis, *Biometals*, 2003, **16**, 271–277.
115. E. Saint-Jacques, M. J. April and C. Seguin, *Gene*, 1995, **160**, 201–206.
116. J. P. Muller, D. Wouters-Tyrou, N. E. Erraiss, M. Vedel, N. Touzet, J. Mesnard, P. Sautiere and M. Wegnez, *DNA Cell Biol.*, 1993, **12**, 341–349.
117. M. Durliat, J. P. Muller, M. André and M. Wegnez, *Int. J. Dev. Biol.*, 1999, **43**, 575–578.
118. K. T. Suzuki, H. Akitomi and R. Kawamura, *Toxicol. Lett.*, 1984, **21**, 179–184.
119. R. Flos, J. Bas and J. Hidalgo, *Comp. Biochem. Physiol. C*, 1986, **83**, 93–98.
120. M. Riggio, F. Trinchella, S. Filosa, E. Parisi and R. Scudiero, *Mol. Reprod. Dev.*, 2003, **66**, 374–382.
121. M. Yamamura and K. T. Suzuki, *Comp. Biochem. Physiol. C*, 1984, **79**, 63–69.
122. J. U. Bell and J. M. Lopez, *Comp. Biochem. Physiol. C*, 1985, **82**, 123–128.
123. P. Coyle, J. C. Philcox, L. C. Carey and A. M. Rofe, *Cell Mol. Life Sci.*, 2002, **59**, 627–647.
124. E. J. Kelly and R. D. Palmiter, *Nat. Genet.*, 1996, **13**, 219–222.
125. R. D. Palmiter, *Proc. Natl. Acad. Sci. USA*, 1998, **95**, 8428–8430.
126. A. Pattanaik, C. F. Shaw III, D. H. Petering, J. Garvey and A. J. Kraker, *J. Inorg. Biochem.*, 1994, **54**, 91–105.
127. Y. Yang, W. Maret and B. L. Vallee, *Proc. Natl. Acad. Sci. USA*, 2001, **98**, 5556–5559.
128. W. Maret, *Biochemistry*, 2004, **43**, 3301–3309.
129. R. R. Misra, J. F. Hochadel, G. T. Smith, J. C. Cook, M. P. Waalkes and D. A. Wink, *Chem. Res. Toxicol.*, 1996, **9**, 326–332.
130. S. X. Liu, J. P. Fabisiak, V. A. Tyurin, G. G. Borisenko, B. R. Pitt, J. S. Lazo and V. E. Kagan, *Chem. Res. Toxicol.*, 2000, **13**, 922–931.
131. D. H. Petering, J. Zhu, S. Krezoski, J. Meeusen, C. Kiekenbush, S. Krull, T. Specher and M. Dughish, *Exp. Biol. Med.*, 2006, **231**, 1528–1534.
132. J. Ejnik, J. Robinson, J. Zhu, H. Forsterling, C. F. Shaw and D. H. Petering, *J. Inorg. Biochem.*, 2002, **88**, 144–152.
133. D. Krepkiy, W. E. Antholine and D. H. Petering, *Chem. Res. Toxicol.*, 2003, **16**, 750–756.
134. Y. Chen and W. Maret, *Eur. J. Biochem.*, 2001, **268**, 3346–3353.
135. W. Maret, C. Jacob, B. L. Vallee and E. H. Fischer, *Proc. Natl. Acad. Sci. USA*, 1999, **96**, 1936–1940.
136. W. Feng, F. W. Benz, J. Cai, W. M. Pierce and Y. J. Kang, *J. Biol. Chem.*, 2006, **281**, 681–687.
137. S. M. Allan and N. J. Rothwell, *Philos. Trans. R. Soc. Lond. B Biol. Sci.*, 2003, **358**, 1669–1677.
138. B. Halliwell, *J. Neurochem.*, 1992, **59**, 1609–1623.
139. J. Coyle and P. Puttfarcken, *Science*, 1993, **262**, 689–695.
140. C. Olanow, *Trends Neurosci.*, 1993, **16**, 439–444.
141. V. W. Pentreath and N. D. Slamon, *Human Exp. Toxicol.*, 2000, **19**, 641–649.
142. A. Bakka, A. S. Johnsen, L. Endresen and H. E. Rugstad, *Experientia*, 1982, **38**, 381–383.

143. M. Sato and I. Bremner, *Free Radical Biology Medicine*, 1993, **14**, 325–337.
144. L. S. Chubatsu and R. Meneghini, *Biochem. J.*, 1993, **291**, 193–198.
145. M. A. Schwarz, J. S. Lazo, J. C. Yalowich, I. Reynolds, V. E. Kagan, V. Tyurin, Y. M. Kim, S. C. Watkins and B. R. Pitt, *J. Biol. Chem.*, 1994, **269**, 15238–15243.
146. J. S. Lazo, Y. Kondo, D. Dellapiazza, A. E. Michalska, K. H. Choo and B. R. Pitt, *J. Biol. Chem.*, 1995, **270**, 5506–5510.
147. M. Sato, M. Apostolova, M. Hamaya, J. Yamaki, K. Choo, A. Michalska, N. Kodama and C. Tohyama, *Environ. Toxicol. Pharmacol.*, 1996, **1**, 221–225.
148. K. T. Tamai, E. B. Gralla, L. M. Ellerby, J. S. Valentine and D. J. Thiele, *Proc. Natl. Acad. Sci. USA*, 1993, **90**, 8013–8017.
149. Y. J. Kang, Y. Chen, A. Yu, M. Voss McCowan and P. N. Epstein, *J. Clin. Invest.*, 1997, **100**, 1501–1506.
150. Y. Kondo, J. Rusnak, D. Hoyt, C. Settineri, B. Pitt and J. Lazo, *Mol. Pharmacol.*, 1997, **52**, 195–201.
151. K. Fu, T. Tomita, M. P. Sarras Jr, R. C. De Lisle and G. K. Andrews, *Pancreas*, 1998, **17**, 238–246.
152. Y. J. Kang, G. Li and J. T. Saari, *Am. J. Physiol.*, 1999, **276**, H993–997.
153. Y. J. Kang, *Proc. Soc. Exp. Biol. Med.*, 1999, **222**, 263–273.
154. J. Liu, Y. Liu, D. Hartley, C. D. Klaassen, S. E. Shehin Johnson, A. Lucas and S. D. Cohen, *J. Pharmacol. Exp. Ther.*, 1999, **289**, 580–586.
155. S. Suemori, M. Shimazawa, K. Kawase, M. Satoh, H. Nagase, T. Yamamoto and H. Hara, *Invest. Ophthalmol. Vis. Sci.*, 2006, **47**, 3975–3982.
156. T. Buttke and P. Sandstrom, *Immunol. Today*, 1994, **15**, 7–10.
157. K. J. Barnham, C. L. Masters and A. I. Bush, *Nat. Rev. Drug Discov.*, 2004, **3**, 205–214.
158. L. B. Kohler, V. Berezin, E. Bock and M. Penkowa, *Brain Res.*, 2003, **992**, 128–136.
159. H. L. Butcher, W. A. Kennette, O. Collins, R. K. Zalups and J. Koropatnick, *J. Pharmacol. Exp. Ther.*, 2004, **310**, 589–598.
160. X. Yang, T. A. Doser, C. X. Fang, J. M. Nunn, R. Janardhanan, M. Zhu, N. Sreejayan, M. T. Quinn and J. Ren, *FASEB J.*, 2006, **20**, 1024–1026.
161. F. Reinecke, O. Levanets, Y. Olivier, R. Louw, B. Semete, A. Grobler, J. Hidalgo, J. Smeitink, A. Olckers and F. H. Van der Westhuizen, *Biochem. J.*, 2006, **395**, 405–415.
162. D. A. Wiseman, S. M. Wells, M. Hubbard, J. E. Welker and S. M. Black, *Am. J. Physiol. Lung Cell. Mol. Physiol.*, 2007, **292**, L165–L177.
163. Z. Peng, L. Peng, Y. Fan, E. Zandi, H. G. Shertzer and Y. Xia, *J. Biol. Chem.*, 2007, in press.
164. J. A. Potashkin and G. E. Meredith, *Antioxid. Redox Signal*, 2006, **8**, 144–151.
165. J. Carrasco, M. Penkowa, H. Hadberg, A. Molinero and J. Hidalgo, *Eur. J. Neurosci.*, 2000, **12**, 2311–2322.
166. G. W. Wang, J. B. Klein and Y. J. Kang, *J. Pharmacol. Exp. Ther.*, 2001, **298**, 461–468.
167. G. W. Wang, Z. Zhou, J. B. Klein and Y. J. Kang, *Am. J. Physiol. Heart Circ. Physiol.*, 2001, **280**, H2292–H2299.

168. Y. J. Kang, Y. Li, X. Sun and X. Sun, *Am. J. Pathol.*, 2003, **163**, 1579–1586.
169. L. Cai, Y. Wang, G. Zhou, T. Chen, Y. Song, X. Li and Y. J. Kang, *J. Am. Coll. Cardiol.*, 2006, **48**, 1688–1697.
170. C. Meplan, M. J. Richard and P. Hainaut, *Oncogene*, 2000, **19**, 5227–5236.
171. E. A. Ostrakhovitch, P. E. Olsson, S. Jiang and M. G. Cherian, *FEBS Lett.*, 2006, **580**, 1235–1238.
172. K. C. Crowthers, V. Kline, C. Giardina and M. A. Lynes, *Toxicol. Appl. Pharmacol.*, 2000, **166**, 161–172.
173. C. H. Kim, J. H. Kim, J. Lee and Y. S. Ahn, *Toxicol. Appl. Pharmacol.*, 2003, **190**, 189–196.
174. E. Papouli, M. Defais and F. Larminat, *J. Biol. Chem.*, 2002, **277**, 4764–4769.
175. S. Tekur and S. M. Ho, *Mol. Carcinog.*, 2002, **33**, 44–55.
176. M. Penkowa, J. Carrasco, M. Giralt, A. Molinero, J. Hernández, I. L. Campbell and J. Hidalgo, *J. Cerebral Blood Flow Metab.*, 2000, **20**, 1174–1189.
177. M. Penkowa, L. Tió, M. Giralt, A. Quintana, A. Molinero, S. Atrian, M. Vašák and J. Hidalgo, *J. Neurosci. Res.*, 2006, **83**, 974–984.
178. M. Karin, R. J. Imbra, A. Heguy and G. Wong, *Mol. Cell Biol.*, 1985, **5**, 2866–2869.
179. J. J. Schroeder and R. J. Cousins, *Proc. Natl. Acad. Sci. USA*, 1990, **87**, 3137–3141.
180. M. Sato, M. Sasaki and H. Hojo, *Res. Commun. Chem. Pathol. Pharmacol.*, 1992, **75**, 159–172.
181. R. L. Friedman and G. R. Stark, *Nature*, 1985, **314**, 637–639.
182. S. K. De, M. T. McMaster and G. K. Andrews, *J. Biol. Chem.*, 1990, **265**, 15267–15274.
183. J. Youn, L. A. Borghesi, E. A. Olson and M. A. Lynes, *J. Toxicol. Environ. Health*, 1995, **45**, 397–413.
184. L. A. Borghesi, J. Youn, E. A. Olson and M. A. Lynes, *Toxicology*, 1996, **108**, 129–140.
185. J. Youn and M. A. Lynes, *Toxicol. Sci.*, 1999, **52**, 199–208.
186. M. A. Lynes, L. A. Borghesi, J. Youn and E. A. Olson, *Toxicology*, 1993, **85**, 161–177.
187. E. Canpolat and M. A. Lynes, *Toxicol. Sci.*, 2001, **62**, 61–70.
188. J. S. Garvey, *Environ. Health Perspect*, 1984, 117–127.
189. I. Bremner, J. N. Morrison, A. M. Wood and J. R. Arthur, *J. Nutr.*, 1987, **117**, 1595–1602.
190. J. Hidalgo, M. Giralt, J. S. Garvey and A. Armario, *Am. J. Physiol.*, 1988, **254**, E71–78.
191. P. Trayhurn, J. S. Duncan, A. M. Wood and J. H. Beattie, *Am. J. Physiol. Regul. Integr. Comp. Physiol.*, 2000, **279**, R2329–R2335.
192. Y. Uchida, F. Gomi, T. Masumizu and Y. Miura, *J. Biol. Chem.*, 2002, **277**, 32353–32359.
193. R. B. Klassen, K. Crenshaw, R. Kozyraki, P. J. Verroust, L. Tio, S. Atrian, P. L. Allen and T. G. Hammond, *Am. J. Physiol. Renal Physiol.*, 2004, **287**, F393–F403.

194. N. A. Wolff, M. Abouhamed, P. J. Verroust and F. Thevenod, *J. Pharmacol. Exp. Ther.*, 2006, **318**, 782–791.
195. A. K. Sewell, L. T. Jensen, J. C. Erickson, R. D. Palmiter and D. R. Winge, *Biochemistry*, 1995, **34**, 4740–4747.
196. R. S. Chung, J. C. Vickers, M. I. Chuah and A. K. West, *J. Neurosci.*, 2003, **23**, 3336–3342.
197. K. Nakajima and K. Suzuki, *Neurochem. Int.*, 1995, **27**, 73–87.
198. H. G. Blaauwgeers, P. A. Sillevis Smitt, J. M. De Jong and D. Troost, *Glia*, 1993, **8**, 62–70.
199. K. Suzuki, K. Nakajima, N. Otaki and M. Kimura, *Biol. Signals*, 1994, **3**, 188–192.
200. K. Suzuki, K. Nakajima, N. Otaki, M. Kimura, U. Kawaharada, K. Uehara, F. Hara, Y. Nakazato and M. Takatama, *Pathol. Int.*, 1994, **44**, 20–26.
201. K. Kiningham, X. Bi and E. J. Kasarskis, *Neurochem. Int.*, 1995, **27**, 105–109.
202. P. A. Sillevis Smitt, H. G. Blaauwgeers, D. Troost and J. M. de Jong, *Neurosci. Lett.*, 1992, **144**, 107–110.
203. M. van Lookeren Campagne, H. Thibodeaux, N. van Bruggen, B. Cairns, R. Gerlai, J. T. Palmer, S. P. Williams and D. G. Lowe, *Proc. Natl. Acad. Sci. USA*, 1999, **96**, 12870–12875.
204. B. A. Masters, C. J. Quaife, J. C. Erickson, E. J. Kelly, G. J. Froelick, B. P. Zambrowicz, R. L. Brinster and R. D. Palmiter, *J. Neurosci.*, 1994, **14**, 5844–5857.
205. S. Kojima, A. Shimada, T. Morita, Y. Yamano and T. Umemura, *J. Vet. Med. Sci.*, 1999, **61**, 343–349.
206. S. Yanagitani, H. Miyazaki, Y. Nakahashi, K. Kuno, Y. Ueno, M. Matsushita, Y. Naitoh, S. Taketani and K. Inoue, *Life Sci.*, 1999, **64**, 707–715.
207. Y. Uchida, *Biol. Signals*, 1994, **3**, 211–215.
208. M. Yamada, S. Hayashi, I. Hozumi, T. Inuzuka, S. Tsuji and H. Takahashi, *Brain Res.*, 1996, **735**, 257–264.
209. J. Carrasco, M. Giralt, A. Molinero, M. Penkowa, T. Moos and J. Hidalgo, *J. Neurotrauma*, 1999, **16**, 1115–1129.
210. R. S. Chung, J. C. Vickers, M. I. Chuah, B. L. Eckhardt and A. K. West, *Exp. Neurol.*, 2002, **178**, 1–12.
211. I. Miyazaki, M. Asanuma, Y. Higashi, C. A. Sogawa, K. Tanaka and N. Ogawa, *Neurosci. Res.*, 2002, **43**, 323–333.
212. E. Mocchegiani, R. Giacconi, C. Cipriano, M. Muzzioli, P. Fattoretti, C. Bertoni-Freddari, G. Isani, P. Zambenedetti and P. Zatta, *Brain Res. Bull.*, 2001, **55**, 147–153.
213. J. R. Duguid, C. W. Bohmont, N. G. Liu and W. W. Tourtellotte, *Proc. Natl. Acad. Sci. USA*, 1989, **86**, 7260–7264.
214. P. A. Adlard, A. K. West and J. C. Vickers, *Neurobiol. Dis.*, 1998, **5**, 349–356.
215. P. Zambenedetti, R. Giordano and P. Zatta, *J. Chem. Neuroanat.*, 1998, **15**, 21–26.
216. S. Tsuji, H. Kobayashi, Y. Uchida, Y. Ihara and T. Miyatake, *EMBO J.*, 1992, **11**, 4843–4850.

217. W. H. Yu, W. J. Lukiw, C. Bergeron, H. B. Niznik and P. E. Fraser, *Brain Res.*, 2001, **894**, 37–45.
218. M. C. Amoureux, D. Van Gool, M. T. Herrero, R. Dom, F. C. Colpaert and P. J. Pauwels, *Mol. Chem. Neuropathol.*, 1997, **32**, 101–121.
219. H. G. Blaauwgeers, M. Anwar Chand, F. M. van den Berg, J. M. Vianney de Jong and D. Troost, *J. Neurol. Sci.*, 1996, **142**, 39–44.
220. M. Penkowa, C. Espejo, A. Ortega-Aznar, J. Hidalgo, X. Montalban and E. M. Martínez-Cáceres, *Cell Mol. Life Sci.*, 2003, **60**, 1258–1266.
221. B. Mirza, H. Hadberg, P. Thomsen and T. Moos, *Neuroscience*, 2000, **95**, 425–432.
222. Y. Arai, Y. Uchida and S. Takashima, *Pediatr. Neurol.*, 1997, **17**, 134–138.
223. T. Kawashima, K. Doh-ura, M. Torisu, Y. Uchida, A. Furuta and T. Iwaki, *Dement. Geriatr. Cogn. Disord.*, 2000, **11**, 251–262.
224. H. Isumi, Y. Uchida, T. Hayashi, S. Furukawa and S. Takashima, *Clin. Neuropathol.*, 2000, **19**, 77–84.
225. J. Carrasco, P. Adlard, C. Cotman, A. Quintana, M. Penkowa, F. Xu, W. E. Van Nostrand and J. Hidalgo, *Neuroscience*, 2006, **143**, 911–922.
226. Y. H. Gong and J. L. Elliott, *Exp. Neurol.*, 2000, **162**, 27–36.
227. S. Ono, Y. Endo, E. Tokuda, K. Ishige, K. Tabata, S. Asami, Y. Ito and T. Suzuki, *Biol. Trace Elem. Res.*, 2006, **113**, 93–104.
228. E. Tokuda, S. Ono, K. Ishige, A. Naganuma, Y. Ito and T. Suzuki, *Toxicology*, 2007, **229**, 33–41.
229. C. Espejo, J. Carrasco, J. Hidalgo, M. Penkowa, A. García, I. Sáez-Torres and E. M. Martínez-Cáceres, *Neuroscience*, 2001, **105**, 1055–1065.
230. M. Penkowa and J. Hidalgo, *Glia*, 2000, **32**, 247–263.
231. M. Penkowa and J. Hidalgo, *Exp. Neurol.*, 2001, **170**, 1–14.
232. C. Espejo, M. Penkowa, I. Sáez-Torres, J. Xaus, A. Celada, X. Montalban and E. M. Martínez-Cáceres, *Exp. Neurol.*, 2001, **172**, 460–468.
233. C. Espejo, M. Penkowa, I. Sáez-Torres, J. Hidalgo, A. García, X. Montalban and E. M. Martínez-Cáceres, *Exp. Neurol.*, 2002, **177**, 21–31.
234. Y. Itano, S. Noji, E. Koyama, S. Taniguchi, N. Taga, T. Takahashi, K. Ono and F. Kosaka, *Neurosci. Lett.*, 1991, **124**, 13–16.
235. J. Hidalgo, M. Borras, J. S. Garvey and A. Armario, *J. Neurochem.*, 1990, **55**, 651–654.
236. T. Gasull, M. Giralt, A. García and J. Hidalgo, *Glia*, 1994, **12**, 135–143.
237. J. Carrasco, J. Hernández, H. Bluethmann and J. Hidalgo, *Mol. Brain Res.*, 1998, **57**, 221–234.
238. S. T. Jacob, K. Ghoshal and J. F. Sheridan, *Gene Expr*, 1999, **7**, 301–310.
239. H. Zheng, N. E. Berman and C. D. Klaassen, *Neurochem. Int.*, 1995, **27**, 43–58.
240. T. Dalton, T. L. Pazdernik, J. Wagner, F. Samson and G. K. Andrews, *Neurochem. Int.*, 1995, **27**, 59–71.
241. J. Hidalgo, B. Castellano and I. L. Campbell, *Current Topics Neurochem.*, 1997, **1**, 1–26.
242. P. Montpied, F. de Bock, M. Baldy Moulinier and G. Rondouin, *Neuroreport*, 1998, **9**, 79–83.

243. L. Acarin, B. González, J. Hidalgo, A. J. Castro and B. Castellano, *Neuroscience*, 1999, **92**, 827–839.
244. Y. Tang, A. Lu, B. J. Aronow, K. R. Wagner and F. R. Sharp, *Eur. J. Neurosci.*, 2002, **15**, 1937–1952.
245. M. Penkowa and T. Moos, *Glia*, 1995, **13**, 217–227.
246. M. Penkowa, T. Moos, J. Carrasco, H. Hadberg, A. Molinero, H. Bluethmann and J. Hidalgo, *Glia*, 1999, **25**, 343–357.
247. M. Penkowa, J. Carrasco, M. Giralt, T. Moos and J. Hidalgo, *J. Neurosci.*, 1999, **19**, 2535–2545.
248. J. W. Neal, S. K. Singhrao, B. Jasani and G. R. Newman, *Neuropathol. Appl. Neurobiol.*, 1996, **22**, 243–247.
249. M. van Lookeren Campagne, H. Thiobodeaux, N. van Bruggen, B. Cairns and D. G. Lowe, *J. Neurosci.*, 2000, **20**, 5200–5207.
250. G. Trendelenburg, K. Prass, J. Priller, K. Kapinya, A. Polley, C. Muselmann, K. Ruscher, U. Kannbley, A. O. Schmitt, S. Castell, F. Wiegand, A. Meisel, A. Rosenthal and U. Dirnagl, *J. Neurosci.*, 2002, **22**, 5879–5888.
251. M. Penkowa, J. Hidalgo and T. Moos, *Brain Res.*, 1997, **774**, 256–259.
252. M. Penkowa, M. Giralt, T. Moos, P. S. Thomsen, J. Hernández and J. Hidalgo, *Exp. Neurol.*, 1999, **156**, 149–164.
253. M. Penkowa and J. Hidalgo, *Exp. Neurol.*, 2000, **163**, 72–84.
254. T. Anezaki, H. Ishiguro, I. Hozumi, T. Inuzuka, M. Hiraiwa, H. Kobayashi, T. Yuguchi, A. Wanaka, Y. Uda, T. Miyatake, K. Yamada, M. Tohyama and S. Tsuji, *Neurochem. Int.*, 1995, **27**, 89–94.
255. I. Hozumi, T. Inuzuka, M. Hiraiwa, Y. Uchida, T. Anezaki, H. Ishiguro, H. Kobayashi, Y. Uda, T. Miyatake and S. Tsuji, *Brain Res.*, 1995, **688**, 143–148.
256. I. Hozumi, T. Inuzuka, H. Ishiguro, M. Hiraiwa, Y. Uchida and S. Tsuji, *Brain Res.*, 1996, **741**, 197–204.
257. T. Yuguchi, E. Kohmura, K. Yamada, T. Sakaki, T. Yamashita, H. Otsuki, K. Kataoka, S. Tsuji and T. Hayakawa, *J. Neurotrauma*, 1995, **12**, 299–306.
258. T. Yuguchi, E. Kohmura, K. Yamada, T. Sakaki, T. Yamashita, H. Otsuki, A. Wanaka, M. Tohyama, S. Tsuji and T. Hayakawa, *Mol. Brain Res.*, 1995, **28**, 181–185.
259. T. Inuzuka, I. Hozumi, A. Tamura, M. Hiraiwa and S. Tsuji, *Brain Res.*, 1996, **709**, 151–153.
260. L. Acarin, J. Carrasco, B. González, J. Hidalgo and B. Castellano, *J. Neuropathol. Exp.. Neurol.*, 1999, **58**, 389–397.
261. R. D. Palmiter, E. P. Sandgren, D. M. Koeller and R. L. Brinster, *Mol. Cell Biol.*, 1993, **13**, 5266–5275.
262. A. E. Michalska and K. H. Choo, *Proc. Natl. Acad. Sci. USA*, 1993, **90**, 8088–8092.
263. B. A. Masters, E. J. Kelly, C. J. Quaife, R. L. Brinster and R. D. Palmiter, *Proc. Natl. Acad. Sci. USA*, 1994, **91**, 584–588.
264. M. Penkowa, M. Giralt, J. Camats and J. Hidalgo, *J. Comp. Neurol.*, 2002, **444**, 174–189.
265. M. Asanuma, I. Miyazaki, Y. Higashi, K. Tanaka, M. E. Haque, N. Fujita and N. Ogawa, *Neurosci. Lett.*, 2002, **327**, 61–65.

266. M. Giralt, M. Penkowa, N. Lago, A. Molinero and J. Hidalgo, *Exp. Neurol.*, 2002, **173**, 114–128.
267. M. Giralt, M. Penkowa, J. Hernández, A. Molinero, J. Carrasco, N. Lago, J. Camats, I. L. Campbell and J. Hidalgo, *Neurobiol. Dis.*, 2002, **9**, 319–338.
268. A. Molinero, M. Penkowa, J. Hernández, J. Camats, M. Giralt, N. Lago, J. Carrasco, I. L. Campbell and J. Hidalgo, *J. Neuropathol. Exp. Neurol.*, 2003, **62**, 315–328.
269. M. Penkowa, J. Camats, M. Giralt, A. Molinero, J. Hernández, J. Carrasco, I. L. Campbell and J. Hidalgo, *Glia*, 2003, **42**, 287–306.
270. K. Puttaparthi, W. L. Gitomer, U. Krishnan, M. Son, B. Rajendran and J. L. Elliott, *J. Neurosci.*, 2002, **22**, 8790–8796.
271. S. Nagano, M. Satoh, H. Sumi, H. Fujimura, C. Tohyama, T. Yanagihara and S. Sakoda, *Eur. J. Neurosci.*, 2001, **13**, 1363–1370.
272. M. Penkowa, C. Espejo, E. M. Martínez-Cáceres, X. Montalban and J. Hidalgo, *Cell. Mol. Life Sci.*, 2003, **60**, 185–197.
273. M. Penkowa and J. Hidalgo, *J. Neurosci. Res.*, 2003, **72**, 574–586.
274. M. Vašák and N. Romero-Isart, *Metallothioneins*, in Vol. V of *Encyclopedia of Inorganic Chemistry*, 2nd edn., Ed. R. B. King, John Wiley & Sons, Ltd., Chichester, UK, 2005, pp. 3208–3221.

Met. Ions Life Sci. **2009**, *5*, 319–351

11

Metallothionein-3, Zinc, and Copper in the Central Nervous System

Milan Vašák and Gabriele Meloni

Institute of Biochemistry, University of Zürich, Winterthurerstrasse, 190, CH-8057 Zürich, Switzerland

<mvasak@bioc.uzh.ch>

Metal Ions in Life Sciences, Volume 5 Edited by Astrid Sigel, Helmut Sigel, and Roland K. O. Sigel

Published by the Royal Society of Chemistry, www.rsc.org DOI: 10.1039/9781847558992-00319

ABSTRACT: Metallothionein-3 (MT-3), also known as the neuronal growth inhibitory factor, has been discovered by Uchida and coworkers in 1991 in their search for a cellular component responsible for antagonizing aberrant neuritic sprouting and increased survival of cultured neurons stimulated by Alzheimer's disease (AD) brain extract. Since this initial discovery further studies showed that MT-3 possesses peculiar structural and functional properties not shared by other members of the mammalian MT family. Several lines of evidence suggest that the metal-binding protein MT-3 plays a vital role in zinc and copper homeostasis in the brain. Although far from being understood, the unusual structural properties of MT-3 are responsible for its neuronal growth inhibitory activity, involvement in trafficking of zinc vesicles in the central nervous system, protection against copper-mediated toxicity in AD and in controlling abnormal metal-protein interactions in other neurodegenerative disorders.

KEYWORDS: Alzheimer's disease · brain metallothioneins · copper · growth inhibitory factor · metal binding · metallothionein-3 · neurodegeneration · oxidative stress · structure and reactivity · zinc

1. INTRODUCTION

The pathological characteristics of Alzheimer's disease (AD) are the progressive loss of neurons accompanied by the formation of extracellular amyloid plaques and intracellular neurofibrillary tangles [1]. An early hypothesis postulated that these alterations are caused by an imbalance of neurotrophic factors [2]. Performing *in vitro* cell culture studies, Uchida and coworkers discovered that protein extract from AD brain increased rat cortical neuron survival more effectively than extract from normal human brain, suggesting that AD brain possesses elevated neurotrophic activity [3]. The apparent increase in neurotrophic activity was found to be due to the loss of a growth inhibitory factor. This neuronal growth inhibitory factor (GIF) was subsequently purified from normal human brain [4]. Its characterization revealed a metalloprotein of 68 amino acids with the primary structure similar to mammalian metallothionein-1 and -2 (MT-1/-2), including the preserved array of 20 cysteine residues. These features together with molecular biological studies led to its classification as metallothionein-3 (MT-3) [5].

In contrast to the widely expressed MT-1/-2 isoforms, MT-3 is expressed mainly in the central nervous system (CNS). Since these initial reports, a number of studies revealed that MT-3 possesses peculiar structural properties not found in MT-1/-2 isoforms, which are responsible for its distinct biological activity. Mounting evidence suggests that MT-3 is critically involved in copper and zinc homeostasis in the CNS. Recently, its role in cell protection from pathological processes that occur after brain injury and in a number of neurodegenerative disorders has received particular attention. In the previous chapter structural and functional aspects of mammalian MTs, with a particular focus on MT-1/-2, have already been discussed (see Chapter 10). Since the present chapter is devoted to MT-3, the MT-1/-2 isoforms are only briefly discussed when required.

2. MAMMALIAN METALLOTHIONEINS IN THE BRAIN

2.1. Mammalian Metallothionein Family

In mammals, the MT gene family consists of four different subfamilies designated m1 (MT-1) to m4 (MT-4). Based on the classification that takes into account taxonomic parameters and the patterns of distribution of Cys residues along the MT sequence [6], mammalian MTs belong to the vertebrate family (family 1). This family is characterized by the consensus sequence: K-Xaa(1,2)-C-C-Xaa-C-C-P-Xaa(2)-C, where Xaa stands for an amino acid other than Cys. In *Homo sapiens*, a number of *MT* genes and pseudogenes exist in the genome out of which at least 10 are functional, including single *MT-3* and *MT-4* genes. All human isoforms are clustered on chromosome 16 and possess a similar gene organization [7] (see also Chapter 10). Besides MT-4, whose biosynthesis is restricted to cornified and stratified squamous epithelium [8], all other isoforms are present in the CNS.

MTs are composed of a single polypeptide chain of 61–68 amino acids. They are characterized by a conserved array of 20 cysteines and the absence of aromatic amino acids and histidine. Whereas MT-1/-2 contain 61–62 amino acids, MT-3 is a 68 amino acid residues metalloprotein which exhibits 70% sequence identity with the MT-1/-2 isoforms. The proteins bind 7 divalent (Zn^{2+}, Cd^{2+}) and up to 12 monovalent (Cu^{+}) metal ions through metal-thiolate coordinative bonds [9]. These metal ions are organized in two metal-thiolate clusters located in independent protein domains (for details see Chapter 10). While the isolated MT-1/-2 isoforms, which are expressed in almost all organs, usually contain only Zn^{2+} ions, MT-3 isolated from bovine or human brains contains both Cu^{+} and Zn^{2+} ions [4,10].

The present knowledge suggests that MT-1/-2 can have multiple biological functions which include a heavy metal detoxification, regulation of zinc-dependent cellular processes, protection from oxidative stress and apoptosis [10–13] (Chapter 10). The biosynthesis of these ubiquitous isoforms is inducible by various agents such as heavy metals, reactive oxygen species (ROS), hormones, and xenobiotics [14]. In contrast, MT-3 is a noninducible protein primarily confined to the CNS [5]. Low expression levels of this protein have also been reported in kidney, organs of the reproductive system, stomach, heart, pancreas, and salivary glands [15–18]. Several studies demonstrated that the MT-3 isoform possesses biological functions in the CNS that are not shared by the MT-1/-2 isoforms (see Section 2.3).

2.2. Expression Pattern of Metallothioneins in the Brain

In the human brain the MT-1/-2 and MT-3 isoforms occur [7]. Numerous studies investigated the localization of these three isoforms in the CNS. MT-1/-2 are present throughout the brain and spinal cord. Although the main cell type expressing these MT isoforms are the astrocytes, their expression was also found in ependymal cells, epithelial cells of choroid plexus, meningeal cells of the pia mater, and endothelial cells of blood vessels. Neurons rarely express MT-1/-2, although in a few reports their positive immunochemical staining has been shown (Chapter 10).

Conflicting results as to the cellular localization of MT-3 in the CNS have been reported based on the analysis of MT-3 mRNA and the protein. In the pioneering work of Uchida and coworkers, a clear immunoreactivity throughout the brain was prominent in astrocytes closely associated with neuronal perikarya and dendrites, while in neurons and other type of glial cells the MT-3 immunoreactivity was essentially absent [19]. Similar results were also obtained by *in situ* hybridization experiments in human brains and by MT-3 mRNA analysis of cultured rat astrocytes, neurons, and microglia [20]. However, a number of later reports clearly showed that the MT-3 expression is not restricted to astrocytes. Thus, *in situ* hybridization experiments in normal mice and the analysis of the expression pattern of the protein in transgenic mice convincingly demonstrated that MT-3 is expressed predominantly in neurons, while glial cells were essentially devoid of a MT-3 signal [21]. A significant signal was also obtained in ependymal cells and in endothelial cells of the choroid plexus. The studies also demonstrated that MT-3 is present in several brain regions at different concentrations [22]. Further *in situ* hybridization studies in rats and mice confirmed the predominant expression of MT-3 in neurons [23–35]. Besides the latter MT-3 mRNA studies, the immunochemical localization of the protein only in neurons, but not in glial cells has also been shown. It may be noted, however, that in a number of experimental animal models of the CNS injury, MT-3 mRNA levels were found increased in glial cells and presumably also in astrocytes. Moreover, the MT-3 expression in microglial cells could not be ruled out. An analysis for the protein by electron microscopic and confocal laser immunohistochemistry also showed higher amounts in astrocytes than in neurons [4,23,27,29,36–39]. In contrary to MT-1/-2, MT-3 has been reported to be secreted from astrocytes resulting in its comparable intra- and extracellular concentrations [40]. Overall, it would appear that MT-3 is expressed in both neurons and astrocytes [7].

The control of MT-3 expression levels in development and aging of the brain also differs from the MT-1/-2 isoforms. Thus, the MT-3 expression is not induced by classical MT inducers like lipopolysaccharides (LPS), Zn^{2+}, Cd^{2+}, ethanol or kainic acid (KA) [41]. However, increased levels of MT-3 mRNA have been observed upon the administration of drugs capable of generating ROS [41]. The binding of the metal-regulatory transcription factor (MTF-1) to the metal responsive elements (MREs) upstream of the MT-1/-2 genes regulates their expression. Although the upstream region of MT-3 gene also contains multiple MREs, they are thought to be non-functional as no *in vitro* binding of MTF-1 to these elements was observed [42]. The existence of a 25-times CTG repeat in the upstream region of mouse MT-3 gene has been suggested to be involved in the repression of the downstream gene [43].

In other studies, a DNA sequence similar to the John Cunningham (JC) virus silencer occurring downstream of the CTG repeat region has been suggested to contribute to the suppression of MT-3 gene expression, at least in primary glial cells [44]. However, other authors argued against the regulatory role of these sequences and suggested instead that the regulatory sequences for the MT-3 gene are located more than 2450-bp upstream from the transcription site [42]. Contrary to MT-1/-2, DNA methylation in the promoter region of MT-3 does not differ between permissive and not permissive cells [42,45]. However, in several gastric carcinomas DNA methylation of the CpG islands in the first intron of the *MT-3* gene has been correlated with its inducibility [45,46]. Thus it would appear that gene silencing by the higher order of chromatine may be involved in the tissue specific expression of the MT-3 gene. Overall, compared with the MT-1/-2 genes, the transcriptional regulation of MT-3 gene is poorly understood.

2.3. Function of Metallothionein-3 in the Brain

MT-3 displays biological properties not observed for MT-1/-2. This is clearly documented by *in vivo* studies, in which mice overexpressing MT-3 in most organs died as a result of pancreatic atrophy, whereas expression of similar amounts of MT-1 had no effect [47]. Although the reason for MT-3 toxicity is unknown, these results provide biological evidence that the MT isoforms have different functional properties. Other *in vivo* studies in support of this conclusion showed that in a mouse model of brain injury exogenously administered MT-3, in contrast to human MT-2, does not affect inflammation, oxidative stress, and apoptosis. Moreover, in this study opposite effects of MT-3 on several growth factors, neurotrophins, and markers of synaptic growth and plasticity and neuroinflammation have also been shown. These results highlight specific and divergent roles of exogenous MT-3 when compared with the MT-1/-2 isoforms. Overall, the *in vivo* studies clearly demonstrated a different biological function for this endogenously occurring protein [48].

In the CNS, MT-3 is expressed in both neurons and astrocytes, where it appears to play an important role in the homeostasis of copper and zinc [7]. Although MT-1/-2 are also expressed in astrocytes, only MT-3 has been reported to be secreted. This suggests that MT-3 may play a different biological role in the intra- and extracellular space [40]. MT-3 is present in several brain regions at different concentrations, being particularly concentrated in presynaptic vesicles of zinc-enriched neurons (ZEN) of cerebral cortex and hippocampus. ZEN belong to a subset of glutamergic neurons and as such contain presynaptic zinc vesicles in which 10–15% of the total Zn^{2+} in the brain is present [22]. Zn^{2+} is released from presynaptic neurons in a Ca^{2+}- and impulse-dependent fashion.

The co-localization of MT-3 and vesicular zinc in ZEN led to the suggestion that the protein may contribute to the utilization of Zn^{2+} as a neuromodulator. Since MT-3 is not required for a direct protection against exogenous Zn^{2+} or Cd^{2+}, it may play a role in protecting neurons from zinc toxicity by facilitating zinc transport or handling. Studies on MT-3 knockout mice revealed important MT-3 functions in zinc dependent processes [22]. While in MT-3 knockout mice the pool of histochemically reactive zinc was not affected, a decreased concentration of zinc in several brain regions including hippocampus occurred. After release, zinc is probably recovered by a specific uptake mechanism and recycled back into synaptic vesicles.

The studies of MT-3 and zinc transporter 3 (ZnT3) knockout mice revealed that ZnT3, a transporter that concentrates Zn^{2+} in presynaptic vesicles, and MT-3 function in the same pathway [49]. From these studies an important role for MT-3 in the recycling of Zn^{2+} has been suggested [50]. The demonstration of the direct interaction of MT-3 with Rab3A, a small GTPase involved in the regulation of the synaptic vesicle cycle, supports this role. The interaction indicates that MT-3 is not merely a cellular Zn^{2+} buffer, but actively participates in synaptic vesicle trafficking upstream of vesicle fusion [51]. The immunochemical identification of MT-3 as a component of a brain multiprotein complex with heat shock protein 84 (HSP84) and creatine kinase (CK) [52] suggests that MT-3 may play additional intracellular functions that remain to be elucidated.

The most distinctive function of MT-3 is its extracellular growth inhibitory activity in neuronal primary cultures. Thus MT-3, but not MT-1/-2 antagonizes the ability of Alzheimer's disease brain extract to stimulate survival and neuritic sprouting of cultured neurons [4,53]. The observation that constitutive expression of MT-3, but not MT-1 inhibited the growth of cultured kidney cells under zinc deficient conditions is in support of this activity [54]. The discovered bioactivity led to the hypothesis that MT-3 may be involved in pathogenic processes leading to AD. Independent studies in support of its role in AD showed that MT-3, but not MT-1/-2 protects the neuronal cells from the toxic effect of amyloid-β ($A\beta_{1-40}$) [55]. However, these two effects are functionally unrelated.

The higher expression of proinflammatory cytokines in MT-1/-2 knockout mice suggested their main regulatory functions in the brain's inflammatory response to injury. In this regard, the observation of MT-3 induction by cytokines remains unclear [7]. MT-3 appears to be synthesized in resting astrocytes and is thus a quiescent state-specific protein. However, its expression is increased by the epidermal growth factor (EGF), which stimulates biological responses such as astrocytic proliferation and neuronal differentiation [56]. These results led to the proposal that MT-3 may play a role in the repair of neuronal damage by inhibiting aberrant neuritic sprouting [56].

It is well established that MT-1/-2 protect cells against oxidative stress through efficient scavenging of free radicals [57–66]. In agreement with *in vitro* radical scavenging studies, cell culture experiments revealed that MT-3, similarly to MT-1/-2, protects cells from the ROS toxicity [40]. Exogenous MT-3 prevents neurite extension of cortical neurons at the early stage of differentiation and the death of differentiated neurons caused by an exposure to high oxygen concentrations. In agreement with this observation, a down-regulation of MT-3 in cortical neurons by antisense oligonucleotides promoted neuronal death under high oxygen conditions. In addition, EPR spin-trapping studies demonstrated that MT-3 scavenges generated free hydroxyl radicals more effectively than MT-1/-2 [40]. MT-3 can also scavenge free nitric oxide (NO) and that released from *S*-nitrosothiols more efficiently than MT-1/-2 [67]. Furthermore, in a cellular system, in which all three MT isoforms were similarly effective in protecting primary neuronal culture against hydrogen peroxide, MT-3 showed a much more pronounced protecting effect against *S*-nitrosothiols [67]. The *S*-nitrosylation of MT-3 occurs via an *S*-transnitrosation reaction. In this processes, the thiolate ligands are targets of both free NO and *S*-nitrosothiols leading to the protein modification and metal release [67]. The increased reactivity of MT-3 with free NO and *S*-nitrosothiols has led to the proposal that MT-3 may specifically convert NO signals to zinc signals [67]. Since the expression of MT-3 could be induced by oxidative stress in the acute phase following brain injury, a special function for MT-3 in the protection against reactive oxygen and nitrogen species has been suggested.

3. ZINC AND COPPER IN THE BRAIN

3.1. Zinc Physiology

Zinc is after iron the second most prevalent trace element in mammals. The body of a 70-kg human contains about 3 g of zinc. Large concentrations are present in the brain (~10 μg/g wet weight) [68]. Zinc is an essential component of numerous proteins where it plays a structural, catalytic or regulatory role [69]. Apart from its importance in protein complexes, the zinc ion is closely involved in intracellular signaling and neurotransmission. The ZIP (Zrt- and Irt-like proteins) family of membraneous transporter proteins is mainly involved in cellular zinc uptake, whereas the ZnT transporter family mediates the zinc efflux (Figure 1).

Zinc is primarily absorbed in the intestine and transported in the blood where it is bound mainly to albumin and the amino acids histidine and cysteine. Liver is essential for zinc removal from the plasma and distribution to other organs. The transport of zinc into the brain parenchyma via the blood-brain barrier system occurs by means of various carrier proteins [70]. The total Zn^{2+} concentration in the serum and extracellular fluids has been estimated to be 0.15 μM [71]. However, extracellular zinc-binding proteins and amino acids lower free Zn^{2+} concentrations to nanomolar or lower levels [71]. Neurons, like most cells, accumulate Zn^{2+} [72] with a total Zn^{2+} concentration in the cytosol reaching up to 150 μM [71]. However, also in this case zinc-binding proteins, mainly MTs, lower intracellular free Zn^{2+} concentrations below 1 nM [73–75].

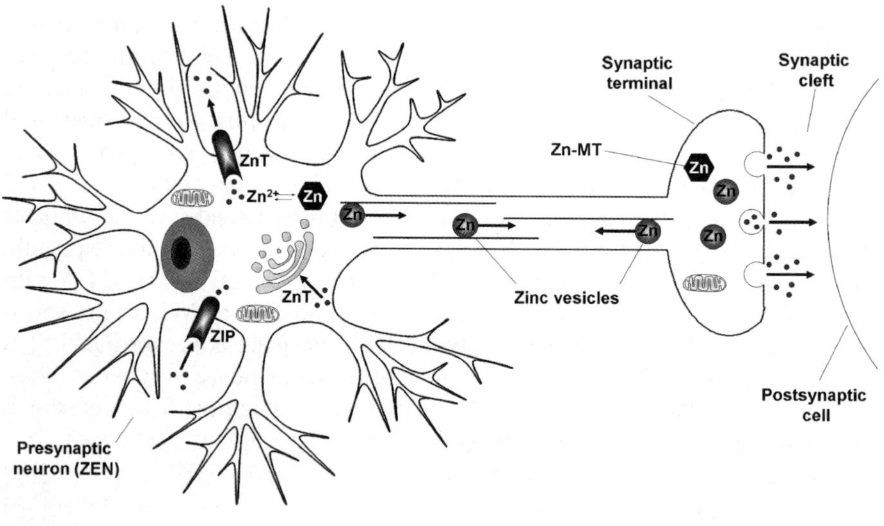

Figure 1. Zinc homeostasis in zinc-enriched neurons. Carrier-mediated influx and efflux of Zn^{2+} take place in the cell body. The intracellular free Zn^{2+} concentration is buffered by MT-3 or sequestered into cytoplasmic organelles. Zinc transporters are responsible for the organellar sequestration of free Zn^{2+}. These organelles could be mitochondria or maturing synaptic vesicles. Upon excitation Zn^{2+} is released from axonal terminal into the synaptic cleft along with glutamate.

In the CNS, a common intracellular homeostatic mechanism for zinc exists. All cell types, including neurons, possess two pathways for Zn^{2+} influx, i.e., a carrier-mediated transport and the Zn^{2+} flux through receptor-operated and voltage-gated channels. An excess of intracellular zinc can be exported into the extracellular space by the action of zinc transporters localized at the plasma membrane. Specific zinc transporters are also present at intracellular membranes, where they are involved in the sequestration of cytosolic Zn^{2+} and in its intracellular supply to Zn^{2+}-binding proteins during their biosynthesis. Under conditions when intracellular free Zn^{2+} concentrations rise to toxic levels, the cytosolic free Zn^{2+} is chelated by zinc-binding proteins or sequestered into cytoplasmic organelles. In neurons, the mitochondria are the major site for Zn^{2+} sequestration and the low intracellular free Zn^{2+} concentration in the cytosol is mainly maintained by the metallothionein/thionein pair [76–78].

MTs bind seven Zn^{2+} ions with a high affinity through the array of cysteine ligands. However, the binding is redox-sensitive as under oxidative conditions Zn^{2+} is released concomitant with cysteine oxidation. MTs possess not only a high Zn^{2+} sequestration capacity, but also acts as a Zn^{2+} depot, providing a source of Zn^{2+} for intra- and extracellular Zn^{2+} signals. Little is known how Zn^{2+} upon its entry becomes bound to MTs and how Zn^{2+} transporters contribute to Zn^{2+} compartmentalization and trafficking. Moreover, no evidence for intracellular zinc chaperones is currently available.

Besides cytoplasmic organelles involved in Zn^{2+} sequestration, 10–15% of the total brain zinc is localized in presynaptic vesicles of zinc-enriched neurons [79]. In ZEN three pools of Zn^{2+} occur: free Zn^{2+}, protein-bound Zn^{2+}, and compartmentalized Zn^{2+} (Figure 1) [80]. Like in other cells, the free Zn^{2+} concentration in ZEN is extremely low. A unique feature of these neurons is the existence of an exchangeable pool of compartmentalized Zn^{2+} inside synaptic vesicles with zinc concentrations in the millimolar range. This Zn^{2+} is termed "labile" Zn^{2+} being sensitive to metal chelators. Because of the neuron architecture, the presynaptic zincergic neuron is challenged with the difficult task of moving Zn^{2+} from sites of carrier-mediated uptake in the cell body to distant sites of Zn^{2+} release from synaptic vesicles at axon terminals. The mechanisms and structures involved in the cellular trafficking of vesicular Zn^{2+} are still enigmatic. However, that Zn^{2+} is sequestered during this journey into vesicles to protect the cell from the deleterious effects of free Zn^{2+} is well established. In ZEN, vesicular Zn^{2+} is released from presynaptic terminals in a Ca^{2+}- and stimulation-dependent manner through exocytosis. Although the physiological significance of synaptic Zn^{2+} release is not fully understood, it probably plays a modulator role in synaptic transmission via modulation of neurotransmitter receptors and transporters localized at the postsynaptic membrane. Additionally, released

Zn^{2+} is translocated into the postsynaptic neuron, entering the cell via Ca^{2+}-permeable channels.

The excitotoxic stimulation occurring under certain acute conditions, including epilepsy and transient global ischemia, leads to the release of large amounts of glutamate and Zn^{2+}. In this case, a massive influx of Zn^{2+} into the postsynaptic cell gives rise to a large sustained increase in intracellular free Zn^{2+} levels, thereby contributing to the selective neuronal loss [68,81–83]. In neuronal cell culture studies intracellular free Zn^{2+} concentrations of 400–600 nM have been reported to trigger a widespread neuronal death [84,85]. Under these pathological conditions Zn^{2+} is taken up into mitochondria with an impairment of mitochondrial structure and function [85–91]. This process contributes to the cell death via apoptotic pathways [92]. However, the mechanism of Zn^{2+}-induced cell death likely comprises several pathways. Released Zn^{2+} may be recycled by Zn^{2+} transporters restoring Zn^{2+} levels in synaptic vesicles. Alternatively, released Zn^{2+} may be taken up by glia. A dysregulation of Zn^{2+} homeostasis appears to be involved in the progression of a number of metal-linked degenerative disorders such as AD [93].

3.2. Copper Physiology

Copper is an essential trace element which plays a crucial catalytic role in a number of important copper enzymes. The body of a 70-kg human contains <110 mg of Cu with less than 10% of it present in the brain (≈ 9 mg) [94]. Since copper is redox active, it can catalyze the production of ROS through Fenton-type reactions. Therefore, its intracellular concentration must be tightly controlled. An upper limit of 10^{-18} M for the concentration of free Cu^{2+} in unstressed cells has been estimated [95].

In contrast to zinc homeostasis that of copper is better understood. The maintenance of cellular copper concentrations requires membrane copper transporters and copper chaperones, which deliver copper to specific sites in the cell (Figure 2). Copper transporters and chaperones identified in lower eukaryotes are also present in mammals, indicating evolutionary conservation of the systems involved in copper trafficking within cells. The membrane proteins encoded by *ATP7A* and *ATP7B* genes play an essential role in the physiological regulation of copper, controlling systemic and cellular copper concentrations. Both genes encode copper ATPases that are members of the P-type ATPase family of cation transporters. These proteins are copper efflux pumps that regulate the amount of copper leaving the cell and supply copper to copper enzymes in the trans-Golgi network.

Copper is primarily absorbed in the small intestine and most of the newly absorbed copper initially enters the liver. In the liver, copper is supplied to endogenous enzymes, incorporated into ceruloplasmin, and secreted into the blood [96]; if in excess, it is secreted in the bile through the involvement of

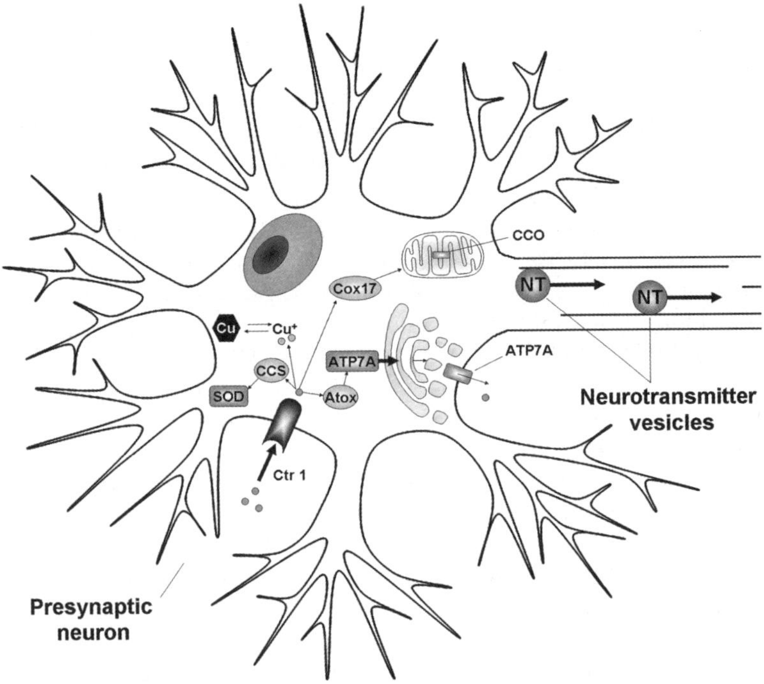

Figure 2. Model for the homeostatic regulation of copper in the central nervous system. For details see text.

the Cu-ATPase ATP7B. The transport of copper to the brain requires copper transport across the blood-brain barrier. Since this step is blocked in Menkes' disease, which is characterized by a copper deficiency caused by the mutation in the Cu-ATPase ATP7A, this suggests that the same copper pump is involved. Copper is then transported as Cu^{+} into neurons and astrocytes by the protein Ctr1 (Figure 2) [97–99].

However, the mechanism of Cu^{2+} reduction prior to its uptake remains unknown. The delivery of copper to target enzymes relies on at least three metallochaperone systems: (a) the Cox17 system, which delivers copper to mitochondria for its insertion into cytochrome *c* oxidase (CCO) under participation of the co-chaperones Cox11 and Sco-1/-2 [100], (b) the copper chaperone for the superoxide dismutase (CCS) system, which delivers copper to superoxide dismutase (SOD-1) in the cytoplasm [101], and (c) the Atox1 or HAH1 chaperone system, which delivers copper to ATP7A in the trans-Golgi network for its incorporation into other copper proteins such as ceruloplasmin [102]. In the case of increased copper levels, ATP7A moves from the trans-Golgi network to the plasma membrane for the copper efflux

[103]. In concert with Atox1, ATP7A also provides copper to secretory copper enzymes such as peptidylglycine α-amidating monooxygenase (PAM) [104]. The presence of nonfunctional ATP7A results in the altered activity of PAM, which leads to an impaired maturation of specific neuropeptides [105]. The activity of dopamine-β-monooxygenase (DBM) also depends on the ATP7A expression [106]. ATP7A is likely required for the transfer of copper to other copper enzymes.

The mechanisms for the delivery of copper to membrane-associated enzymes such as the extracellular superoxide dismutase and the amine oxidase vascular adhesion protein 1 are still to be established. Although copper may also bind to MTs, the role of MTs in cerebral copper homeostasis is far from being understood. Extensive research on neurodegenerative disorders is providing increasing evidence that other neuronal membrane proteins such as the amyloid precursor protein (APP) and the prion protein (PrP) are also involved in copper trafficking in the brain. Alterations in the tight copper regulation are considered to contribute to several neurodegenerative disorders. Free or abnormally bound copper may act as a catalyst for the generation of the ROS. Therefore, an improper copper compartmentalization, elevation of copper levels in the cell and abnormal binding of Cu^{2+} to cellular proteins are currently being explored as sources of pathological oxidative stress in neurodegenerative disorders like AD, amyotrophic lateral sclerosis (ALS), transmissible spongiform encephalopathies (TSEs), Parkinson (PD), and Huntington diseases.

3.3. Zinc and Copper in Neurodegenerative Disorders

A predisposing risk factor associated with neurodegenerative diseases is age. In the normal brain a high concentration of transition metal ions such as Zn, Cu, and Fe is present. As a consequence of normal aging, a rise in the levels of transition metals in the brain tissue occurs. Recent evidence suggests that a breakdown of metal homeostasis and the associated oxidative stress represent a key factor in a variety of neurodegenerative disorders [107].

Oxidative stress is the result of an imbalance between the free radical production and opposing antioxidant defense. Physiological amounts of ROS modulate intracellular biological processes such as signal transduction, transcription, and apoptosis. However, due to their high reactivity, the excessive production of ROS is extremely dangerous for the structural and functional integrity of the cell. ROS can either directly modify cellular DNA, proteins and lipids or initiate radical chain reactions, yielding an extensive oxidative damage. Although cells possess a variety of defense mechanisms and repair systems, the production of ROS can overwhelm the antioxidant defenses leading to oxidative stress. A chronic oxidative stress is being implicated in a number of neurodegenerative diseases [94]. The brain is particularly vulnerable to oxidative stress, since despite accounting for 2% of the body weight it consumes approximately one fifth of the total oxygen inhaled, which is required for the high rate turnover of large quantities of ATP [108].

These neurodegenerative processes are best documented for AD. Oxidative stress associated with the metal-catalyzed transformation of amyloid-β plays a significant role in normal brain and in AD pathology [93]. The Aβ peptide, which aggregates into extracellular amyloid plaques in the AD brain, is a proteolytic fragment of 39–43 amino acid residues generated from the substantially larger amyloid precursor protein by β- and γ-secretases. In AD brains, the metabolic control of Zn and Cu levels is impaired. Aβ shows a high affinity to Zn^{2+} and Cu^{2+} and binding of these metals is critical for the formation of amyloid plaques [109]. Both metals are also found in senile plaques ($\sim$1 mM Zn^{2+} and 0.3 mM Cu^{2+}). Despite a high concentration in plaques, iron is not likely to interact directly with Aβ *in vivo*. $A\beta_{1-40}$ and $A\beta_{1-42}$ can bind copper in two binding sites of different affinities, but only the high affinity binding site appears to be physiologically relevant. In the high affinity binding site ($K_d \approx$ 10–100 nM), Cu^{2+} is coordinated in a square planar geometry by three histidine imidazole rings (His_6, His_{13}, and His_{14}) [93,110–112]. The amine nitrogen or carboxylate oxygen of the N-terminal aspartate, backbone amide NHs of Glu_3-Val_{18} or phenolic oxygen of Tyr_{10} have been suggested as the fourth ligand [93,110–112].

The binding of transition metals to Aβ modulate its toxicity in cell cultures, with Cu^{2+} increasing toxicity and Zn^{2+} attenuating it. Since the addition of Zn^{2+} protected cultured neurons from Aβ-Cu^{2+} toxicity, it has been suggested that this effect represents a potential antioxidant response enabling the redox-silencing and entombment of Aβ by zinc [113]. The Aβ peptide efficiently reduces Cu^{2+} to Cu^{+} in a process which generates ROS through a Fenton-type reaction [114]. The reactions involved in the generation of ROS by Aβ in the presence of redox-active transition metals are summarized as follows:

$$A\beta + M^{(n+1)+} \rightarrow A\beta + M^{n+} \quad (1)$$

$$M^{n+} + O_2 \rightarrow M^{n+1} + O_2^{\bullet -} \quad (2)$$

$$2\,O_2^{\bullet -} + 2H^+ \rightarrow H_2O_2 + O_2 \quad (3)$$

$$M^{n+} + H_2O_2 \rightarrow M^{(n+1)} + OH^{\bullet} + OH^{-} \quad \text{(Fenton reaction)} \quad (4)$$

$$O_2^{\bullet -} + H_2O_2 \rightarrow OH^{\bullet} + OH^{-} + O_2 \quad \text{(Haber-Weiss reaction)} \quad (5)$$

The initial reduction of metal ions (Cu^{2+}) requires the oxidation of an electron donor. The sulfur of Met_{35} in Aβ through its oxidation to sulfoxide has been shown to represent the source of reducing equivalents. In addition, the redox cycling of this metal in the presence of oxygen can be accomplished by biological reducing agents such as ascorbate, dopamine, or cholesterol. The CNS has the highest concentration of ascorbate of all organs. Ascorbate is actively transported to the brain extracellular fluid to a concentration of 200–400 μM and accumulates further in neurons where it reaches a concentration of around 10 mM [115]. The formation of a Tyr_{10} radical during the redox cycling process has been shown to be responsible for the formation of dityrosine cross-linked soluble Aβ aggregates.

The formation of dityrosine adducts *in vivo* is a general sign of oxidative stress in AD. A number of other oxidative adducts of Aβ, originating from the redox cycling of the bound redox-active copper, have also been observed *in vitro* and *in vivo* [108]. Emerging evidence suggests that the early formation of soluble Aβ aggregates plays a critical role in the neurotoxicity of this peptide [116]. This is reflected by a significant amount of ROS generated during early stages of peptide aggregation. The ROS production is also responsible for the formation of insoluble SDS resistant aggregates [117]. Thus, the soluble and oligomeric Aβ-metal complexes appear to be the dominant toxic species [118,119]. This conclusion is supported by the observation that stable Aβ oligomers, but not monomers or fibrils inhibit hippocampal long-term potentiation.

Similar phenomena have been observed in other neurodegenerative pathologies where metal binding and associated oxidative stress appear to be potentiating agents of the neurodegenerative processes. These include copper binding to synuclein in PD, prion protein in prion diseases, SOD-1 in ALS, huntigtin in Huntington disease and probably also in other less characterized pathologies [120–122]. Indeed, neurodegenerative diseases such as AD [93], prion disease [94], and PD [123] share common pathophysiological hallmarks, i.e., misfolding of Aβ, prion and α-synuclein, the formation of protein aggregates, abnormal metal-protein interactions and oxidative stress. Molecules able to modulate metal homeostasis in the CNS, including metal-binding proteins like MT-3, may thus play important protective roles in the control of these processes.

4. METALLOTHIONEIN-3 STRUCTURE AND REACTIVITY

4.1. Structure of Metallothionein-3 with Divalent Metal Ions

Compared with the amino acid sequences of mammalian MT-1/-2 and MT-4 (see Chapter 10, Figure 1), the MT-3 sequence shows two inserts, an acidic

```
             1        10        20        30        40        50        60
MT-3 Human   MDPETCPCPSGGSCTCADSCKCEGCKCTSCKKSCCSCCPAECEKCAKDCVCKGGEAAEAEAEKCSCCQ
MT-3 Bovine  MDPETCPCPTGGSCTCSDPCKCEGCTCASSKKSCCSCCPAECEKCAKDCVCKGGEGAEAEEKKCSCCQ
MT-3 Horse   MDPETCPCPTGGSCTCSGECKCEGCKCTSCKKSCCSCCPAECEKCAKDCVCKGGEGAEAEAEKCSCCQ
MT-3 Mouse   MDPETCPCPTGGSCTCSDKCKCKGCKCTNCKKSCCSCCPAGCEKCAKDCVCKGEEGAKAEAEKCSCCQ
MT-2 Human   MDPN.CSCAAGDSCTCAGSCKCKECKCTSCKKSCCSCCPVGCAKCAQGCICKGA......SDKCSCCA
```

Figure 3. KALIGN amino acid sequence alignment of MT-3 isoforms from four different mammalian species compared with the human metallothionein-2 sequence with the conserved residues highlighted. The figure was generated with the program ESPript version 2.2.

hexapeptide in the C-terminal region and a Thr in position 5 in the N-terminal region followed by a conserved C_6PCP_9 motif. These features are conserved among known mammalian MT-3 sequences (Figure 3). The C_6PCP_9 motif is essential for the growth inhibitory activity of Zn_7MT-3, as the mutation of two prolines to Ser and Ala, amino acids found in MT-2, abolished the biological activity [53]. Accordingly, it has been suggested that the specific structural features introduced by the two unique Pro residues are responsible for the extracellular bioactivity of Zn_7MT-3 [124]. However, how this protein exerts its neuroinhibitory activity is currently unknown.

While the metal-free protein (apoMT-3) possesses a predominantly disordered structure, a well-defined structure develops upon metal binding. The structural studies on recombinant Zn_7MT-3 and Cd_7MT-3 established the presence of two mutually interacting protein domains, resembling those reported for MT-1 and MT-2, with each domain encompassing a metal-thiolate cluster (see Chapter 10) [125–128]. A 3-metal cluster is located in the N-terminal β-domain (residues 1–31) and a 4-metal cluster in the C-terminal α-domain (residues 32–68) of M^{II}_7MT-3 [126,127]. The distinct biological activity of MT-3 has been linked to the structural differences between MT-1/-2 and MT-3. The structure of MTs is determined by interplay between the polypeptide chain and metal ions. In this regard, the structure of metal-thiolate clusters formed plays an important role in the final polypeptide fold.

Structural information on the metal-thiolate clusters in M^{II}_7MT-3 was obtained from spectroscopic investigations of the recombinant protein and chemically synthesized single protein domains [126,127]. Based on the first shell X-ray absorption fine structure (EXAFS) data of Zn_7MT-3 the seven Zn^{2+} are tetrahedrally coordinated by terminal and μ_2-bridging thiolate ligands. However, in both Zn_7MT-3 and its isolated Zn_3-β-domain outer shell zinc-backscattering and sulfur-backscattering interactions at about 3.28 and 2.9 Å, respectively, were detected (Figure 4) [125]. Such a short Zn···Zn distance of 3.28 Å is inconsistent with the cyclohexane-like Zn_3Cys_9 cluster present in the β-domain of mammalian MT-2, where a Zn···Zn distance of about 3.8 Å exists [129]. In EXAFS studies of the transcriptional activator

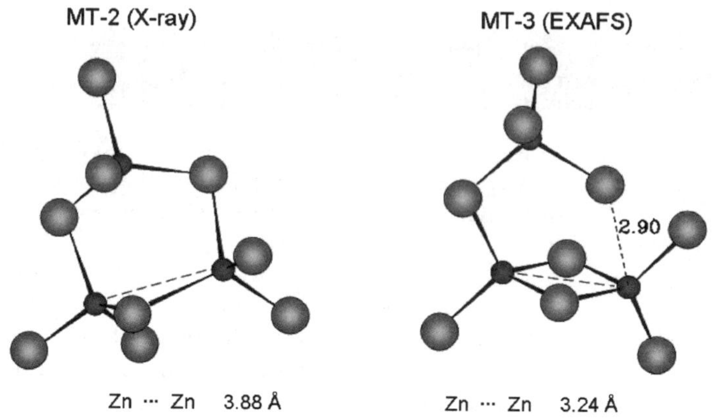

Figure 4. Proposed structural model of the $Zn_3(Cys)_9$ cluster in Zn_7MT-3 (right) in comparison with the cyclohexane-like Zn_3Cys_9 cluster present in M^{II}_7MT-2 (left). Metal ions are shown as balls and cysteine thiolates as shaded spheres. Adapted from [125] with permission from *Eur. J. Biochem.*, copyright (1998).

PPR1, which contains a binuclear $Zn_2(Cys)_6$ cluster, a Zn···Zn distance of 3.16 Å has been determined [130]. The presence of two Cys bridges between the metal ions in this cluster is responsible for the short Zn···Zn distance. Using this information a tentative structure of the Zn_3-thiolate cluster in Zn_7MT-3 was proposed. In this model a binuclear cluster, as found in PPR1, is linked to a third tetrahedral Zn site (Figure 4) [125]. Thus, an unusual Zn_3-thiolate cluster is apparently formed in the biologically active β-domain of Zn_7MT-3.

The three-dimensional (3D) NMR structure of mouse and human Cd_7MT-3 has been reported [131,132]. The two protein domains in human Cd_7MT-3 are connected by a flexible hinge region of a conserved Lys-Lys sequence in the middle of the polypeptide chain (Figure 5). However, because of dynamic processes in Cd_7MT-3 only the 3D structure of the C-terminal α-domain (residues 32–68), containing an adamantane-like Cd_4Cys_{11} cluster, could be determined by NMR [132]. The structure of this domain reveals a peptide fold and cluster organization very similar to that found in Cd_7MT-1/-2, with the exception of an extended flexible loop encompassing the acidic hexapeptide insert [133].

Evidence for dynamic processes in the MT-3 structure has been obtained from the ^{113}Cd NMR studies of human $^{113}Cd_7$MT-3. This investigation revealed the presence of unprecedented dynamic processes within the metal-thiolate clusters. The observed significant broadening of all ^{113}Cd resonances, the absence of homonuclear ^{113}Cd-^{113}Cd couplings and the very low

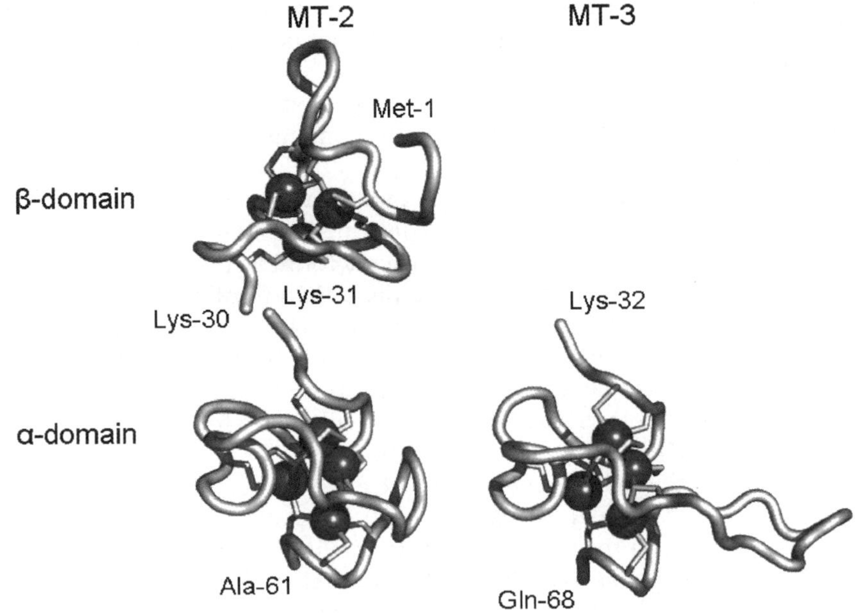

Figure 5. Three-dimensional structure of human Cd_7MT-2 [133] and the α-domain of human $^{113}Cd_7$-MT-3 [132] determined by nuclear magnetic resonance spectroscopy. Cd^{2+} ions are shown as shaded spheres connected to the protein backbone by cysteine thiolate ligands. The models were generated with the program PyMOL v0.99 (http://www.delanoscientific.com/) using the PDB coordinates 1mhu, 2mhu, and 2f5 h.

and temperature-independent intensity of the Cd_3Cys_9-cluster resonances were taken to indicate the presence of dynamic processes acting on two different NMR time scales: (i) fast exchange processes among conformational cluster substates giving rise to broad, weight-averaged resonances and (ii) additional very slow exchange processes between configurational cluster substates in the β-domain [128]. The conformational cluster substates may be visualized as minor dynamic fluctuations of the metal coordination environment and the configurational cluster substates as major structural alterations brought about by temporarily breaking and reforming of the metal-thiolate bonds. Whereas conformational cluster substates are present in both clusters, the configurational cluster substates were only observed in the Cd_3Cys_9 cluster of the β-domain, precluding its structure determination by NMR. To account for slow dynamic events centered at the 3-metal cluster of MT-3, a partial unfolding of the β-domain, whose kinetics could be determined by the *cis/trans* interconversion of Cys-Pro amide bonds in the

C_6PCP_9 motif, was suggested [134]. The unfolding of the β-domain ought to be accompanied by the rearrangement of metal-thiolate coordinative bonds. This effect may account for the EXAFS data obtained on freeze-dried samples of Zn_7MT-3 and its isolated Zn_3-β-domain at low temperature (77 K). Under these conditions mainly one configurational cluster substate may be populated.

The existence of interchanging configurational cluster substates of comparable stability have already been demonstrated for inorganic adamantane-like metal-thiolate clusters with the general formula $[M_4(SPh)_{10}]^{2-}$ (M = Cd^{II}, Zn^{II}, Co^{II}, and Fe^{II}) [135]. Since the mutation of conserved proline residues in the C_6PCP_9 motif to the amino acids present in MT-2, Ala and Ser, abolished the neuroinhibitory activity and dynamics of the 3-metal cluster, this suggests that alterations of the surface topology and the greater flexibility of the β-domain could be an important factor for the bioactivity of MT-3 in protein and/or receptor recognition.

Additional insights into the dynamic of the β-domain of Cd_7MT-3 were provided by a molecular dynamics (MD) simulation [136]. The studies revealed that, due to the structural constraints introduced by the T_5CPCP_9 motif, an unusual conformation of the N-terminal fragment (aa 1–13) is formed when compared with Cd_7MT-2 and that the formation of a *trans*/*trans* isomer is energetically more favorable. Further simulation of the partial unfolding supported the proposed role for *cis*/*trans* interconversion of Cys-Pro amide bonds in the folding/unfolding process of the β-domain. Other studies, using the chimeric MT forms generated by swopping the MT-3 and MT-1 domains, showed that the α-domain via domain-domain interactions modulates the bioactivity of the β-domain of MT-3 and that, besides Thr_5, Pro_7 and Pro_9 mutations [124], also the mutation of Glu_{23} to Lys abolishes the growth inhibitory activity of MT-3 [137,138]. Although the mechanisms underlining these effects remain to be elucidated, the so far obtained results suggest that the structure of the β-domain of MT-3 is subjected to a fine tuning.

As a result of these structural properties, the metal-binding affinity of MT-3 towards divalent metal ions is weaker than the one of MT-1/-2. The average apparent binding constants for Cd^{2+} and Zn^{2+} ions in MT-3 are $2.0 \times 10^{14}\,M^{-1}$ and $6.2 \times 10^{10}\,M^{-1}$, respectively, while the corresponding constants in MT-2 are $7.0 \times 10^{14}\,M^{-1}$ and $3.1 \times 10^{11}\,M^{-1}$ [124]. In addition, in contrast to the cooperative metal binding to MT-1/-2, the metal binding to MT-3 is noncooperative and should therefore proceed through a different folding pathway [134,139].

In summary, the structural studies together with the presence of a unique T_5CPCP_9 motif led to the conclusion that both the specific structural features and the structure dynamics are necessary prerequisites for the extracellular biological activity of Zn_7MT-3 [124,140].

4.2. Structural Features of Cu(I)-Bound Metallothionein-3

Structural studies have also been carried out on isolated MT-3 containing Cu^+ and Zn^{2+} ions. The EXAFS studies on isolated $Cu(I)_4Zn_{3-4}$MT-3 revealed the presence of two homometallic clusters, a Zn_{3-4}-thiolate cluster and a $Cu(I)_4$-thiolate cluster [125]. In contrast to tetrahedrally coordinated Zn^{2+} ions, Cu^+ ions are digonally and/or trigonally coordinated by two or three cysteine ligands [125]. This observation, also seen for the $Cu(I)_4$-thiolate clusters in MT-1/-2, signifies that different coordination geometries for the binding of monovalent and divalent metal ions to MT-3 exist. To accommodate metal-thiolate clusters with different coordinating geometries, the MT-3 structure should possesses a high degree of flexibility [141].

Detailed information on the interaction of Cu^+ with MT-3 was forthcoming from the studies of Cu^+ binding into the metal-free synthetic domains and the full length protein. The stepwise Cu^+ filling of both synthetic MT-3 domains showed that two well-defined Cu(I)-thiolate cluster forms are generated during this process. In the case of the β-domain (residues 1–32), the titration with Cu^+ ions resulted in the successive formation of two cluster forms involving all 9 cysteine ligands, i.e., Cu_4S_9 and Cu_6S_9 clusters [126]. Similar studies on the α-domain (residues 32–68), containing 11 cysteine ligands, resulted in the formation of the Cu_4S_{8-9} cluster followed by the Cu_6S_{11} cluster [127]. The major differences in the respective spectroscopic features of both cluster forms were observed in the low-temperature Cu^+ luminescence emission spectra. Thus, while the $Cu(I)_6$ clusters exhibited only a single emissive band at 600 nm, in the case of the $Cu(I)_4$ clusters two emissive bands at 420 and 610 nm were discerned. The presence of two emissive bands in $Cu(I)_4$ clusters has been correlated with short intranuclear Cu···Cu distances (<2.8 Å), allowing metal-metal interactions due to a d^{10}-d^{10} orbital overlap. Accordingly, the high energy emissive band has been assigned to ^{3}CC (cluster centered) origin and the one at low energy to triplet charge transfer [126,127]. The presence of metal-metal interactions appears to be important for the stability of the $Cu(I)_4$-thiolate cluster towards oxidation by air oxygen. This is supported by the observation that the expansion of the metal core, generating a $Cu(I)_6$-thiolate cluster, results in increased Cu···Cu distances and susceptibility to oxidation [126]. By immunochemical methods the $Cu(I)_4$-thiolate cluster was found located in the N-terminal β-domain [142]. Similarities between the spectral features of isolated Cu_4,Zn_{3-4}MT-3 and those of the $Cu(I)_4$-thiolate cluster in the β-domain suggested its presence also in the isolated protein.

The generation of a native-like MT-3 structure through a direct Cu^+ and Zn^{2+} incorporation into the recombinant apoprotein revealed monomeric $Cu(I)_4$,Zn_4MT-3 in which, besides the $Cu(I)_4$ cluster in the β-domain, a Zn_4 cluster in the α-domain was present. While the $Cu(I)_4$ cluster in

$Cu(I)_4Zn_4MT$-3 possesses the already mentioned unprecedented stability in air, the Zn_4 cluster was found to be air-sensitive. Its oxidation resulted in the disulfide formation and the release of one Zn^{2+} yielding $Cu(I)_4Zn_3MT$-3. This process could be prevented or reversed under reducing conditions [143]. Growth inhibitory activity in neuronal assays has been established for $Cu(I)_4Zn_{3-4}MT$-3 isolated from human [4] and bovine [144] brains and for recombinant human Zn_7MT-3 [145]. However, comparative biological studies on well-defined metalloforms are currently lacking.

4.3. Reactivity of Metallothionein-3

MT-3, like other mammalian MTs, shows a highly unusual reactivity. Despite a high thermodynamic stability of the bound metal ions, they are kinetically very labile, permitting (i) a rapid metal transfer to and from other metalloproteins, (ii) metal substitution reactions, which follow the order of increasing affinity of inorganic thiolates, i.e., Cu(I) ≫ Cd(II) > Zn(II), (iii) reactions of thiolate ligands with electrophiles, reflected by a high reactivity observed with alkylating and oxidizing agents [9,11,146], and (iv) reactions with radical species such as hydroxyl ($OH^{\bullet}$), superoxide ($O_2^{\bullet -}$), and nitric oxide (NO). The free radical attack occurs at the metal-bound thiolates and results in the thiolate oxidation and/or modification with a subsequent metal release [40,67,147].

Although the affinity of metal ions to MT-3 follows the order $Cu^+ > Cd^{2+} > Zn^{2+}$ seen also with MT-1/-2 (Chapter 10), the affinity of Cu^+ to MT-3 appears to be significantly higher than that to MT-1/-2. This information has been forthcoming from the differences in initial rates of the reaction of these MT isoforms with 5,5′-dithio-bis(2-nitrobenzoic acid) (DTNB) [148]. The larger affinity of Cu^+ to MT-3 might be related to the structural features of its β-domain in which copper is preferentially bound. All MT-3 sequences contain a hexapeptide insert in the α-domain of MT-3 (Figure 3). To gain an insight as to its effect on the MT-3 structure, a series of MT-3 variants of the EAAEAE hexapeptide were prepared by site-directed mutagenesis and their properties and reactivity investigated as function of pH, EDTA and DTNB concentrations. The studies revealed that the EAAEAE insert, by increasing the solvent exposed surface, is in part responsible for rendering the α-domain looser. As a result, the overall stability of the metal-thiolate cluster is reduced [149].

The generation of ROS through the redox cycling of free or abnormally bound Cu^{2+} in a Fenton-type reaction is harmful in a number of neurodegenerative disorders. In view of the extracellular occurrence of MT-3, the reaction of Zn_7MT-3 with free Cu^{2+} has been investigated. Zn_7MT-3 was able to scavenge free Cu^{2+} ions through their reduction to Cu^+ and binding to the protein. In this reaction thiolate ligands are oxidized to disulfides concomitant with Zn^{2+} release. The binding of the first four Cu^{2+} is cooperative, forming a $Cu(I)_4$-thiolate cluster in the β-domain of $Cu(I)_4,Zn_4$ MT-3 together with two disulfide bonds, presumably present in the same domain. The $Cu(I)_4$-thiolate cluster in partially oxidized $Cu(I)_4,Zn_4$MT-3 exhibits an unusual stability toward air oxygen. Evidence for metal-metal interactions in this cluster may be in part responsible for its stability. Since this process completely quenched the copper-catalyzed free hydroxyl radical production, this suggests that Zn_7MT-3, by efficiently silencing the free redox-active Cu^{2+} ions, would play an important protective role from Cu^{2+}-mediated toxicity in the brain [150].

MT-3, like other MTs, efficiently scavenges ROS. EPR spin-trapping studies demonstrated that MT-3 scavenges hydroxyl radicals generated by a Fenton-type reaction or the photolysis of hydrogen peroxide much more effectively than the same concentration of MT-1/-2 [40]. In another study, using the isolated α- and the β-domain, both domains efficiently reacted with hydrogen peroxide. While the β-domain of Zn_7MT-3 was more reactive than the β-domain of MT-2, an opposite order of reactivity was observed for the α-domains. In all these instances, the reaction resulted in thiolate oxidation and zinc release [67]. Besides the reaction with the ROS, MT-3 also efficiently scavenges reactive nitrogen species (RNS) like NO and *S*-nitrosothiols. Studies of the full-length Zn_7MT-3 and both individual zinc-reconstituted domains demonstrated that in the reaction with *S*-nitrosothiols zinc is released from both domains of MT-3. In these processes, the zinc-thiolate bonds are targets for both free NO and *S*-nitrosothiols leading to their modification and metal release. *S*-Nitrosylation of thiols in MT-3 occurs via *S*-transnitrosation in which the direct transfer of NO^+ equivalents between *S*-nitrosothiol and cysteine ligands in MT-3 takes place [67].

5. ROLES OF METALLOTHIONEIN-3 IN ZINC AND COPPER PHYSIOLOGY AND PATHOLOGY

A protective role for MT-3 in pathological situations such as seizures has been reported. During seizures, zinc and glutamate are released from synaptic vesicles and both activate kainic acid and α-amino-3-hydroxy-5-methylisoxazole-4-propionic acid (AMPA) receptors on postsynaptic neurons in a synergistic manner. In this regard, MT-3 knockout mice were more sensitive to seizures induced by KA and exhibited a greater neuronal injury in the CA3 field of the hippocampus [22]. In this brain region, large amounts of zinc are usually released during seizures from mossy fibers in the dentate gyrus. These results are in agreement with the observation that MT-3 transgenic mice with elevated MT-3 levels were more resistant to seizure-induced neuronal damage [22].

Several studies investigated the role of MT-3 and the MT-1/-2 isoforms in animal models of brain injury. The brain damage by stab wound injury produced a significant increase in MT-3 expression in reactive astrocytes around the wound, while neurons appeared mostly unresponsive as judged by no significant change in MT-3 levels [24,151,152]. An up-regulation of MT-3 in reactive astrocytes around the degenerated neurons in the CA3 field was also seen upon the administration of KA into the ventricles [24]. In contrast, cortical ablation of the somatosensory cortex decreased MT-3 expression in the cortex ipsilateral to the injury one day after the ablation, but its level increased transiently at the fourth day after which it returned to normal level, the exception being the area surrounding the injury, where increased MT-3 expression was observed between 2–3 weeks [32]. These studies demonstrated that, besides astrocytes, also neurons responded to the ablation with increased MT-3 expression. The topic administration of basic fibroblast growth factor (bFGF) can potentiate the decrease of MT-3 mRNA levels. Since bFGF facilitates neurite extension and given the inhibitory effect of MT-3 on neuronal survival *in vitro*, the initial down-regulation of MT-3 after cortical ablation would be a physiological attempt to facilitate neurite extension in response to injury [32]. Such effect of bFGF on MT-3 expression might be indirect, as in other studies no significant effect of bFGF on MT-3 expression was found in confluent astrocytes *in vitro*. The initial MT-3 downregulation followed by an increase of MT-3 expression would suppress oversprouting promoted by neurotrophic factors produced in response to tissue injury.

The above suggestion is supported by results obtained following facial nerve transaction. In this case, MT-3 mRNA levels decreased significantly in the ipsilateral facial nucleus, presumably to facilitate the regeneration of axons [33]. Other model studies of rat brain injury have shown that in ischemia caused by a middle cerebral artery occlusion, the MT-3 expression decreases progressively until day seven after the brain damage and returns to normal control levels between 21–28 days [153]. Later studies have shown a biphasic response of MT-3 to CNS injury with initial decrease and later increase of its expression in response to the injection of the glutamate analogue NMDA [23] or to a cryolesion [29,154]. Taken together, the results indicate that brain damage is associated with significant alterations in MT-3 expression, and that the type and temporal pattern of the MT-3 response depends on the nature of the insult used to inflict brain damage.

Several studies suggest that MT-3 may play an important role in the progression of a number of neurodegenerative diseases including AD, Creutzfeldt-Jakob disease (CJD), PD, ALS, and spinocerebellar degeneration (SCD) [155]. In these neurodegenerative diseases, the expression of

MT-3 has been found down-regulated or altered and metals such as zinc and copper have been implicated as possible etiological factors [107,120,121].

The protective role of MT-3 in these diseases is best documented for AD. The growth inhibitory activity of MT-3, which abolishes aberrant neuritic sprouting stimulated by the AD brain extract, led to the hypothesis that MT-3 may be involved in pathogenic processes leading to AD [4]. Independent studies in support of its role in AD showed that MT-3, but not MT-1 and MT-2, protects neuronal cells from toxic effects of $A\beta_{1-40}$ [55,156]. However, these two effects appear to be functionally unrelated [156]. Despite several contrasting reports, in AD patients the down-regulation of MT-3 appears to be correlated with neuronal loss or disease duration. *In vitro* studies indicated that the MT-3 levels are influenced by the state of the cell cycle, but not by neuroglial interactions. However, the molecular mechanism leading to MT-3 down-regulation is so far unknown [155].

The protective effect exerted by MT-3 in AD brains could be related to its metal binding capacity and reactivity. In this regard, it has been reported that Zn_7MT-3 can efficiently remove copper not only from soluble Aβ-Cu(II) aggregates, but also from insoluble aggregates through the reduction of Cu^{2+} to Cu^{+} and binding to the protein forming $Cu(I)_4Zn_4$MT-3 which contains two disulfide bonds. At the same time, the released Zn^{2+} from the protein binds to $A\beta_{1-40}$ forming the non-toxic $A\beta_{1-40}$-Zn(II) species. This metal swap completely quenches the ROS production mediated by Cu^{2+} bound to $A\beta_{1-40}$. The metal swap occurred not only *in vitro*, but also in human neuroblastoma cell cultures whereby the toxic effect of $A\beta_{1-40}$-Cu(II) was abolished. Overall, these studies indicate a distinct protective role of this protein in AD [157].

A similar protective effect of MT-3 can be envisaged in other metal-linked neurodegenerative pathologies such as PD, transmissible spongiform encephalopathies (also known as the prion diseases), and ALS in which altered MT-3 levels and the dysregulated Cu^{2+} metabolism have been reported [121]. In both PD and prion disorders MT-3 was found to be downregulated [41,158]. Studies conducted on hemi-Parkinsonian rats suggest that the free radical scavenging potency, including that of MT-3, is reduced in the Parkinson brain and that levodopa treatment fails to induce MT-3 expression, effects accelerating the progression of PD [155].

In the pathology of PD, the fibrillation and aggregation of α-synuclein is a key process in the formation of intracellular inclusions, Lewy bodies, in neurons of substantia nigra [159]. At present, several studies suggest that copper binding and oxidative stress might contribute to an abnormal aggregation of this molecule. The Cu^{2+} binding site in α-synuclein is located at the N-terminus in which Cu^{2+} is anchored through the residue H_{50} and other ill defined nitrogen/oxygen donor atoms in a square planar or distorted tetragonal geometry [123]. Evidence for the direct involvement of

Cu^{2+} in α-synuclein aggregation and in the generation of ROS comes from studies showing that Cu^{2+} induces self-oligomerization of α-synuclein, the formation of hydrogen peroxide by α-synuclein in a Fenton-type reaction and that lipids, in the presence of Cu^{2+} and hydrogen peroxide, enhance the metal-catalyzed oxidative self-oligomerization of α-synuclein. The latter is an important event as it leads to the dissociation of α-synuclein from the cell membrane and its aggregation in the cytosol [121,160–162].

The prion diseases are a group of progressive neurodegenerative disorders that affect the brain and nervous system of humans and animals and are transmitted by prions, proteinaceous infective agents [163]. Prion diseases result from the accumulation of a misfolded form of the endogenous prion protein (PrP^C) and represent an ongoing threat to human health. The transition from natively folded PrP^C to misfolded PrP^{Sc} is a crucial pathogenic event [164]. This fatal neurodegenerative disease shares hallmarks with AD and PD, such as are regional spongiform degeneration, neuronal loss, astrocytosis, and the amyloid plaque formation. The prion protein is the major constituent of the extracellular TSEs plaques. Currently, the process leading to their formation is the subject of intense research. For a long time, the prion structure conversion and aggregation have been considered a key parameter [163]. Increasing experimental evidence is mounting to show that oxidative stress, associated with the copper-catalyzed transformations of prion protein, plays an important role in this disease. The mainly disordered part of PrP structure can bind up to six Cu^{2+} ions. In this structure four octarepeats containing the residues HGGGW are present. Each octarepeat is involved in the coordination of one Cu^{2+} by the imidazole nitrogen of the histidine, two glycine amide nitrogens and a carbonyl oxygen in a square planar geometry. Two additional copper binding sites are present outside of the octarepaet region.

In ALS, contrasting reports as to the alteration of MT-3 levels in the spinal cord of ALS model mice exist. However, it has been suggested that copper-mediated oxidative stress contributes to the pathogenesis of ALS caused by the mutation in *SOD-1* gene. This is in agreement with the reported alterations of zinc and copper concentrations and increased levels of lipid peroxidation in a mice model of ALS. As a response to the increased oxidative stress load, increased levels of MT-3 were found. A protective role of MTs in ALS is supported by the fact that crossing ALS mice model with either MT-1/-2 or MT-3 knockout mice was found to reduce survival time and accelerate the onset and progression of ALS [165]. Taken together, the finding that MT-3 exerts multiple functions, including zinc and copper homeostasis and protection against ROS, supports possible protective roles of this protein in the progression of different metal-linked neurodegenerative disorders.

As discussed above, the regulation of the basal MT-3 expression is poorly understood. Nevertheless, in recent years overexpression of MT-3 in certain cancer types such as bladder [166], prostate [167] and breast [168] cancer have been described. The biological significance of this phenomenon is so far unknown. The overexpression in breast cancer is associated with tumors having a poor prognosis and that in bladder, where no basal expression in healthy tissue is present, was suggested to be used as a cancer marker. That MT-3, similarly to MT-1/-2, may confer cellular resistance in platinum-based chemotherapy in these cell types has also been shown [169].

6. CONCLUDING REMARKS

The unusual biological and structural properties of tissue-specific MT-3 clearly distinguish it from the widely expressed MT-1/-2 isoforms. The biological functions of MT-3 include the neuronal growth inhibitory activity, involvement in trafficking of zinc vesicles in the CNS, protection against copper-mediated toxicity in Alzheimer's disease, and the control of abnormal metal-protein interactions in other neurodegenerative disorders. The majority of these functions are closely related to the structure of its β-domain. However, the 3D structure of this domain, encompassing the Zn_3- and $Cu(I)_4$-thiolate cluster in Zn_7MT-3 and $Cu(I)_4,Zn_4$MT-3, respectively, is so far unknown. The structure of the $Cu(I)_4$-thiolate clusters in the fully reduced and partially oxidized $Cu(I)_4,Zn_4$MT-3 species would shed a light into the underlying structural features responsible for their unusual stability to oxidation in air. Moreover, understanding of the ability of Zn_7MT-3 to redox silence the toxic Cu^{2+} ions through their reduction and binding into the protein structure may be of importance for its emerging protective role in metal-linked neurodegenerative disorders.

ACKNOWLEDGMENTS

The authors appreciate the funding obtained from Swiss National Science Foundation Grant 3100A0-100246/1, Swiss Academy of Engineering Sciences (SATW) "PAI Germaine de Staël" Grant 08345VK, Hartmann-Müller-Stiftung, and the Forschungskredit der Universität Zürich Grant 54043901.

ABBREVIATIONS

3D	three-dimensional
AD	Alzheimer's disease
ALS	amyotrophic lateral sclerosis

AMPA	α-amino-3-hydroxy-5-methylisoxazole-4-propionic acid
APP	amyloid precursor protein
ARE	antioxidant response element
Aβ	amyloid-β
bFGF	basic fibroblast growth factor
CC	cluster centered
CCO	cytochrome *c* oxidase
CCS	copper chaperone for superoxide dismutase
CJD	Creutzfeldt-Jakob disease
CK	creatine kinase
CNS	central nervous system
DBM	dopamine-β-monooxygenase
DTNB	5,5′-dithiobis-(2-nitrobenzoic acid)
EDTA	ethylenediamine-N,N,N′,N′-tetraacetic acid
EGF	epidermal growth factor
EPR	electron paramagnetic resonance
EXAFS	extended X-ray absorption fine structure
GIF	growth inhibitory factor
HSP84	heat shock protein 84
JC virus	John Cunningham virus
KA	kainic acid
LPS	lipopolysaccharides
MD	molecular dynamics
MREs	metal responsive elements
MT	metallothionein
MTF-1	metal-regulatory transcription factor -1
NMDA	N-methyl-D-aspartic acid
NMR	nuclear magnetic resonance
NO	nitric oxide
($O_2^{\bullet -}$)	superoxide radical
$OH^{\bullet}$	hydroxyl radical
PAM	peptidylglycine α-amidating monooxygenase
PD	Parkinson disease
PDB	Protein data bank
PPR1	pyrimidine pathway regulatory protein 1
PrP	prion protein
RNS	reactive nitrogen species
ROS	reactive oxygen species
SCD	spinocerebellar degeneration
SDS	sodium dodecyl sulfate
SOD	superoxide dismutase
TSEs	transmissible spongiform encephalopathies
ZEN	zinc enriched neurons

ZIP	Zrt- and Irt-like proteins
ZnT	zinc transporter

REFERENCES

1. M. P. Mattson, *Nature*, 2004, **430**, 631–639.
2. S. H. Appel, *Ann. Neurol.*, 1981, **10**, 499–505.
3. Y. Uchida, Y. Ihara and M. Tomonaga, *Biochem. Biophys. Res. Commun.*, 1988, **150**, 1263–1267.
4. Y. Uchida, K. Takio, K. Titani, Y. Ihara and M. Tomonaga, *Neuron*, 1991, **7**, 337–347.
5. R. D. Palmiter, S. D. Findley, T. E. Whitmore and D. M. Durnam, *Proc. Natl. Acad. Sci. USA*, 1992, **89**, 6333–6337.
6. P. A. Binz and J. H. R. Kägi, in *Metallothioenein IV;* Ed. C. Klaassen, Birkhäuser Verlag, Basel, 1999, p. 7–13.
7. J. Hidalgo, M. Aschner, P. Zatta and M. Vašák, *Brain Res. Bull.*, 2001, **55**, 133–145.
8. C. J. Quaife, S. D. Findley, J. C. Erickson, G. J. Froelick, E. J. Kelly, B. P. Zambrowicz and R. D. Palmiter, *Biochemistry*, 1994, **33**, 7250–7259.
9. M. Vašák and N. Romero-Isart, in *Encyclopedia of Inorganic Chemistry*, 2nd edn, Ed. R. B. King, J. Wiley & Sons Ltd, New York, 2005, p. 3208–3221.
10. R. Bogumil, P. Faller, D. L. Pountney and M. Vašák, *Eur. J. Biochem.*, 1996, **238**, 698–705.
11. M. Vašák and D. W. Hasler, *Curr. Opin. Chem. Biol.*, 2000, **4**, 177–183.
12. J. H. R. Kägi, *Methods Enzymol.*, 1991, **205**, 613–626.
13. J. H. R. Kägi, in *Metallothionein III*, Ed. K. T. Suzuki, N. Imura and M. Kimura, Birkhäuser Verlag, Basel, 1993, p. 29–56.
14. J. H. Kägi, *Methods Enzymol.*, 1991, **205**, 613–626.
15. P. Moffatt and C. Seguin, *DNA Cell. Biol.*, 1998, **17**, 501–510.
16. Y. Irie, F. Mori, W. M. Keung, Y. Mizushima and K. Wakabayashi, *Physiol. Res.*, 2004, **53**, 719–723.
17. S. H. Garrett, M. A. Sens, J. H. Todd, S. Somji and D. A. Sens, *Toxicol. Lett.*, 1999, **105**, 207–214.
18. K. S. Clifford and M. J. MacDonald, *Diabetes Res. Clin. Pract.*, 2000, **49**, 77–85.
19. Y. Uchida, *Biol. Signals*, 1994, **3**, 211–215.
20. H. Kobayashi, Y. Uchida, Y. Ihara, K. Nakajima, S. Kohsaka, T. Miyatake and S. Tsuji, *Brain Res. Mol. Brain Res.*, 1993, **19**, 188–194.
21. B. A. Masters, C. J. Quaife, J. C. Erickson, E. J. Kelly, G. J. Froelick, B. P. Zambrowicz, R. L. Brinster and R. D. Palmiter, *J. Neurosci.*, 1994, **14**, 5844–5857.
22. J. C. Erickson, G. Hollopeter, S. A. Thomas, G. J. Froelick and R. D. Palmiter, *J. Neurosci.*, 1997, **17**, 1271–1281.
23. L. Acarin, J. Carrasco, B. Gonzalez, J. Hidalgo and B. Castellano, *J. Neuropathol. Exp. Neurol.*, 1999, **58**, 389–397.

24. T. Anezaki, H. Ishiguro, I. Hozumi, T. Inuzuka, M. Hiraiwa, H. Kobayashi, T. Yuguchi, A. Wanaka, Y. Uda, T. Miyatake, K. Yamada, M. Tohyama and S. Tsuji, *Neurochem. Int.*, 1995, **27**, 89–94.
25. J. Carrasco, J. Hernandez, H. Bluethmann and J. Hidalgo, *Brain Res. Mol. Brain Res.*, 1998, **57**, 221–234.
26. J. Carrasco, J. Hernandez, B. Gonzalez, I. L. Campbell and J. Hidalgo, *Exp. Neurol.*, 1998, **153**, 184–194.
27. J. Carrasco, M. Giralt, M. Penkowa, A. K. Stalder, I. L. Campbell and J. Hidalgo, *Exp. Neurol.*, 2000, **163**, 46–54.
28. S. Kojima, A. Shimada, A. Kodan, K. Kobayashi, T. Morita, Y. Yamano and T. Umemura, *Can. J. Vet. Res.*, 1998, **62**, 148–151.
29. M. Penkowa, T. Moos, J. Carrasco, H. Hadberg, A. Molinero, H. Bluethmann and J. Hidalgo, *Glia*, 1999, **25**, 343–357.
30. R. A. Velazquez, Y. Cai, Q. Shi and A. A. Larson, *J. Neurosci.*, 1999, **19**, 2288–300.
31. T. Yuguchi, E. Kohmura, T. Sakaki, M. Nonaka, K. Yamada, T. Yamashita, T. Kishiguchi, T. Sakaguchi and T. Hayakawa, *J. Cereb. Blood Flow Metab.*, 1997, **17**, 745–752.
32. T. Yuguchi, E. Kohmura, K. Yamada, T. Sakaki, T. Yamashita, H. Otsuki, K. Kataoka, S. Tsuji and T. Hayakawa, *J. Neurotrauma*, 1995, **12**, 299–306.
33. T. Yuguchi, E. Kohmura, K. Yamada, T. Sakaki, T. Yamashita, H. Otsuki, A. Wanaka, M. Tohyama, S. Tsuji and T. Hayakawa, *Brain Res. Mol. Brain Res.*, 1995, **28**, 181–185.
34. H. Zheng, N. E. Berman and C. D. Klaassen, *Neurochem. Int.*, 1995, **27**, 43–58.
35. S. Choudhuri, K. K. Kramer, N. E. Berman, T. P. Dalton, G. K. Andrews and C. D. Klaassen, *Toxicol. Appl. Pharmacol.*, 1995, **131**, 144–154.
36. J. Carrasco, M. Giralt, A. Molinero, M. Penkowa, T. Moos and J. Hidalgo, *J. Neurotrauma*, 1999, **16**, 1115–1129.
37. I. Hozumi, T. Inuzuka, H. Ishiguro, M. Hiraiwa, Y. Uchida and S. Tsuji, *Brain Res.*, 1996, **741**, 197–204.
38. M. Penkowa, M. Giralt, P. S. Thomsen, J. Carrasco and J. Hidalgo, *J. Neurotrauma*, 2001, **18**, 447–463.
39. M. Yamada, S. Hayashi, I. Hozumi, T. Inuzuka, S. Tsuji and H. Takahashi, *Brain Res.*, 1996, **735**, 257–264.
40. Y. Uchida, F. Gomi, T. Masumizu and Y. Miura, *J. Biol. Chem.*, 2002, **277**, 32353–32359.
41. C. A. Sogawa, M. Asanuma, N. Sogawa, I. Miyazaki, T. Nakanishi, H. Furuta and N. Ogawa, *Acta. Med. Okayama*, 2001, **55**, 1–9.
42. R. Faraonio, P. Moffatt, O. Larochelle, H. M. Schipper, R. S. Arnaud and C. Seguin, *Eur. J. Biochem.*, 2000, **267**, 1743–1753.
43. M. Imagawa, Y. Ishikawa, H. Shimano, S. Osada and T. Nishihara, *J. Biol. Chem.*, 1995, **270**, 20898–20900.
44. M. Watabe, S. Gross, C. Lawyer, G. J. Brewer, T. Mashimo and K. Watabe, *Cell. Mol. Neurobiol.*, 1997, **17**, 235–243.
45. D. Deng, W. El-Rifai, J. Ji, B. Zhu, P. Trampont, J. Li, M. F. Smith and S. M. Powel, *Carcinogenesis*, 2003, **24**, 25–29.

46. W. El-Rifai, H. F. Frierson Jr., J. C. Harper, S. M. Powell and S. Knuutila, *Int. J. Cancer*, 2001, **92**, 832–838.
47. C. J. Quaife, E. J. Kelly, B. A. Masters, R. L. Brinster and R. D. Palmiter, *Toxicol. Appl. Pharmacol.*, 1998, **148**, 148–157.
48. M. Penkowa, L. Tio, M. Giralt, A. Quintana, A. Molinero, S. Atrian, M. Vašák and J. Hidalgo, *J. Neurosci. Res.*, 2006, **83**, 974–984.
49. T. B. Cole, C. A. Robbins, H. J. Wenzel, P. A. Schwartzkroin and R. D. Palmiter, *Epilepsy Res.*, 2000, **39**, 153–169.
50. M. Aschner, M. G. Cherian, C. D. Klaassen, R. D. Palmiter, J. C. Erickson and A. I. Bush, *Toxicol. Appl. Pharmacol.*, 1997, **142**, 229–242.
51. M. Knipp, G. Meloni, B. Roschitzki and M. Vašák, *Biochemistry*, 2005, **44**, 3159–3165.
52. I. El Ghazi, B. L. Martin and I. M. Armitage, *Exp. Biol. Med. (Maywood)*, 2006, **231**, 1500–1506.
53. A. K. Sewell, L. T. Jensen, J. C. Erickson, R. D. Palmiter and D. R. Winge, *Biochemistry*, 1995, **34**, 4740–4747.
54. R. D. Palmiter, *Toxicol. Appl. Pharmacol.*, 1995, **135**, 139–146.
55. Y. Irie and W. M. Keung, *Biochem. Biophys. Res. Commun.*, 2001, **282**, 416–420.
56. Y. Uchida, *J. Neurochem.*, 1999, **73**, 1945–1953.
57. K. Hanada, D. Sawamura, K. Tamai, T. Baba, I. Hashimoto, T. Muramatsu, N. Miura and A. Naganuma, *J. Invest. Dermatol.*, 1998, **111**, 582–5.
58. Y. J. Kang, *Proc. Soc. Exp. Biol. Med.*, 1999, **222**, 263–273.
59. Y. J. Kang, G. Li and J. T. Saari, *Am. J. Physiol.*, 1999, **276**, H993–997.
60. Y. J. Kang, Y. Chen, A. Yu, M. Voss-McCowan and P. N. Epstein, *J. Clin. Invest.*, 1997, **100**, 1501–1506.
61. J. S. Lazo, Y. Kondo, D. Dellapiazza, A. E. Michalska, K. H. A. Choo and B. R. Pitt, *J. Biol. Chem.*, 1995, **270**, 5506–5510.
62. J. Liu, Y. Liu, D. Hartley, C. D. Klaassen, S. E. Shehin-Johnson, A. Lucas and S. D. Cohen, *J. Pharmacol. Exp. Ther.*, 1999, **289**, 580–586.
63. Y. Kondo, J. M. Rusnak, D. G. Hoyt, C. E. Settineri, B. R. Pitt and J. S. Lazo, *Mol. Pharmacol.*, 1997, **52**, 195–201.
64. G. W. Wang, D. A. Schuschke and Y. J. Kang, *Am. J. Physiol.*, 1999, **276**, H167–75.
65. P. J. Thornalley and M. Vašák, *Biochim. Biophys. Acta.*, 1985, **827**, 36–44.
66. M. Sato and I. Bremner, *Free Radic. Biol. Med.*, 1993, **14**, 325–337.
67. Y. Chen, Y. Irie, W. M. Keung and W. Maret, *Biochemistry*, 2002, **41**, 8360–8367.
68. C. J. Frederickson, *Int. Rev. Neurobiol.*, 1989, **31**, 145–238.
69. B. L. Vallee and K. H. Falchuk, *Physiol. Rev.*, 1993, **73**, 79–118.
70. A. Takeda, *Biometals*, 2001, **14**, 343–351.
71. A. Takeda, *Brain. Res. Brain. Res. Rev.*, 2000, **34**, 137–148.
72. R. A. Colvin, *Am. J. Physiol. Cell. Physiol.*, 2002, **282**, 317–329.
73. L. M. Canzoniero, S. L. Sensi and D. W. Choi, *Neurobiol. Dis.*, 1997, **4**, 275–279.
74. R. B. Thompson, D. Peterson, W. Mahoney, M. Cramer, B. P. Maliwal, S. W. Suh, C. Frederickson, C. Fierke and P. Herman, *J. Neurosci. Methods*, 2002, **118**, 63–75.

75. L. A. Finney and T. V. O'Halloran, *Science*, 2003, **300**, 931–936.
76. R. D. Palmiter, *Proc. Natl. Acad. Sci. USA*, 1998, **95**, 8428–8430.
77. S. C. Burdette and S. J. Lippard, *Proc. Natl. Acad. Sci. USA*, 2003, **100**, 3605–3610.
78. W. Maret, *Antioxid. Redox Signal.*, 2006, **8**, 1419–1441.
79. C. J. Frederickson, J. Y. Koh and A. I. Bush, *Nat. Rev. Neurosci.*, 2005, **6**, 449–462.
80. R. A. Colvin, C. P. Fontaine, M. Laskowski and D. Thomas, *Eur. J. Pharmacol.*, 2003, **479**, 171–185.
81. S. W. Suh, R. B. Thompson and C. J. Frederickson, *Neuroreport*, 2001, **12**, 1523–1525.
82. Y. Li, C. J. Hough, S. W. Suh, J. M. Sarvey and C. J. Frederickson, *J. Neurophysiol.*, 2001, **86**, 2597–2604.
83. C. J. Frederickson, M. D. Hernandez, S. A. Goik, J. D. Morton and J. F. McGinty, *Brain Res.*, 1988, **446**, 383–386.
84. L. M. Canzoniero, D. M. Turetsky and D. W. Choi, *J. Neurosci*, 1999, **19**, RC31.
85. S. L. Sensi, H. Z. Yin, S. G. Carriedo, S. S. Rao and J. H. Weiss, *Proc. Natl. Acad. Sci. USA*, 1999, **96**, 2414–2419.
86. S. L. Sensi, H. Z. Yin and J. H. Weiss, *Eur. J. Neurosci*, 2000, **12**, 3813–3818.
87. S. L. Sensi, D. Ton-That and J. H. Weiss, *Neurobiol. Dis.*, 2002, **10**, 100–108.
88. S. L. Sensi, D. Ton-That, P. G. Sullivan, E. A. Jonas, K. R. Gee, L. K. Kaczmarek and J. H. Weiss, *Proc. Natl. Acad. Sci. USA*, 2003, **100**, 6157–6162.
89. J. Wudarczyk, G. Debska and E. Lenartowicz, *Arch. Biochem. Biophys.*, 1999, **363**, 1–8.
90. D. Jiang, P. G. Sullivan, S. L. Sensi, O. Steward and J. H. Weiss, *J. Biol. Chem.*, 2001, **276**, 47524–47529.
91. K. E. Dineley, T. V. Votyakova and I. J. Reynolds, *J. Neurochem.*, 2003, **85**, 563–570.
92. S. L. Budd, L. Tenneti, T. Lishnak and S. A. Lipton, *Proc. Natl. Acad. Sci. USA*, 2000, **97**, 6161–6166.
93. P. A. Adlard and A. I. Bush, *J. Alzheimers Dis.*, 2006, **10**, 145–163.
94. E. Gaggelli, H. Kozlowski, D. Valensin and G. Valensin, *Chem. Rev.*, 2006, **106**, 1995–2044.
95. T. D. Rae, P. J. Schmidt, R. A. Pufahl, V. C. Culotta and T. V. O'Halloran, *Science*, 1999, **284**, 805–808.
96. J. F. Mercer and R. M. Llanos, *J. Nutr.*, 2003, **133**, 1481S–1484S.
97. B. Zhou and J. Gitschier, *Proc. Natl. Acad. Sci. USA*, 1997, **94**, 7481–7486.
98. J. Lee, J. R. Prohaska, S. L. Dagenais, T. W. Glover and D. J. Thiele, *Gene*, 2000, **254**, 87–96.
99. J. Lee, J. R. Prohaska and D. J. Thiele, *Proc. Natl. Acad. Sci. USA*, 2001, **98**, 6842–6847.
100. P. A. Cobine, F. Pierrel and D. R. Winge, *Biochim. Biophys. Acta.*, 2006, **1763**, 759–772.
101. V. C. Culotta, L. W. Klomp, J. Strain, R. L. Casareno, B. Krems and J. D. Gitlin, *J. Biol. Chem.*, 1997, **272**, 23469–23472.

102. L. W. Klomp, S. J. Lin, D. S. Yuan, R. D. Klausner, V. C. Culotta and J. D. Gitlin, *J. Biol. Chem.*, 1997, **272**, 9221–9226.
103. S. Lutsenko and M. J. Petris, *J. Membr. Biol.*, 2003, **191**, 1–12.
104. R. El Meskini, V. C. Culotta, R. E. Mains and B. A. Eipper, *J. Biol. Chem.*, 2003, **278**, 12278–12284.
105. T. C. Steveson, G. D. Ciccotosto, X. M. Ma, G. P. Mueller, R. E. Mains and B. A. Eipper, *Endocrinology*, 2003, **144**, 188–200.
106. J. R. Prohaska and T. L. Smith, *J. Nutr.*, 1982, **112**, 1706–1717.
107. K. J. Barnham, C. L. Masters and A. I. Bush, *Nat. Rev. Drug. Discov.*, 2004, **3**, 205–14.
108. D. G. Smith, R. Cappai and K. J. Barnham, *Biochim. Biophys. Acta.*, 2007, **1768**, 1976–1990.
109. X. Huang, C. S. Atwood, R. D. Moir, M. A. Hartshorn, R. E. Tanzi and A. I. Bush, *J. Biol. Inorg. Chem.*, 2004, **9**, 954–960.
110. L. Hou and M. G. Zagorski, *J. Am. Chem. Soc.*, 2006, **128**, 9260–9261.
111. C. D. Syme, R. C. Nadal, S. E. Rigby and J. H. Viles, *J. Biol. Chem.*, 2004, **279**, 18169–18177.
112. L. Guilloreau, L. Damian, Y. Coppel, H. Mazarguil, M. Winterhalter and P. Faller, *J. Biol. Inorg. Chem.*, 2006, **11**, 1024–1038.
113. M. P. Cuajungco, L. E. Goldstein, A. Nunomura, M. A. Smith, J. T. Lim, C. S. Atwood, X. D. Huang, Y. W. Farrag, G. Perry and A. I. Bush, *J. Biol. Chem.*, 2000, **275**, 19439–19442.
114. X. Huang, C. S. Atwood, M. A. Hartshorn, G. Multhaup, L. E. Goldstein, R. C. Scarpa, M. P. Cuajungco, D. N. Gray, J. Lim, R. D. Moir, R. E. Tanzi and A. I. Bush, *Biochemistry*, 1999, **38**, 7609–7616.
115. M. E. Rice, *Trends Neurosc.*, 2000, **23**, 209–216.
116. D. P. Smith, D. G. Smith, C. C. Curtain, J. F. Boas, J. R. Pilbrow, G. D. Ciccotosto, T. L. Lau, D. J. Tew, K. Perez, J. D. Wade, A. I. Bush, S. C. Drew, F. Separovic, C. L. Masters, R. Cappai and K. J. Barnham, *J. Biol. Chem.*, 2006, **281**, 15145–15154.
117. B. J. Tabner, O. M. El-Agnaf, M. J. German, N. J. Fullwood and D. Allsop, *Biochem. Soc. Trans.*, 2005, **33**, 1082–1086.
118. J. Hardy and D. J. Selkoe, *Science*, 2002, **297**, 353–356.
119. R. Kayed, E. Head, J. L. Thompson, T. M. McIntire, S. C. Milton, C. W. Cotman and C. G. Glabe, *Science*, 2003, **300**, 486–489.
120. K. J. Barnham, R. Cappai, K. Beyreuther, C. L. Masters and A. F. Hill, *Trends Biochem. Sci.*, 2006, **31**, 465–472.
121. A. Gaeta and R. C. Hider, *Br. J. Pharmacol.*, 2005, **146**, 1041–1059.
122. J. H. Fox, J. A. Kama, G. Lieberman, R. Chopra, K. Dorsey, V. Chopra, I. Volitakis, R. A. Cherny, A. I. Bush and S. Hersch, *PLoS ONE*, 2007, **2**, e334.
123. R. M. Rasia, C. W. Bertoncini, D. Marsh, W. Hoyer, D. Cherny, M. Zweckstetter, C. Griesinger, T. M. Jovin and C. O. Fernandez, *Proc. Natl. Acad. Sci. USA*, 2005, **102**, 4294–4299.
124. D. W. Hasler, L. T. Jensen, O. Zerbe, D. R. Winge and M. Vašák, *Biochemistry*, 2000, **39**, 14567–14575.
125. R. Bogumil, P. Faller, P. A. Binz, M. Vašák, J. M. Charnock and C. D. Garner, *Eur. J. Biochem.*, 1998, **255**, 172–177.

126. P. Faller and M. Vašák, *Biochemistry*, 1997, **36**, 13341–13348.
127. D. W. Hasler, P. Faller and M. Vašák, *Biochemistry*, 1998, **37**, 14966–14973.
128. P. Faller, D. W. Hasler, O. Zerbe, S. Klauser, D. R. Winge and M. Vašák, *Biochemistry*, 1999, **38**, 10158–10167.
129. W. Braun, M. Vašák, A. H. Robbins, C. D. Stout, G. Wagner, J. H. Kägi and K. Wüthrich, *Proc. Natl. Acad. Sci. USA*, 1992, **89**, 10124–10128.
130. L. J. Ball, G. P. Diakun, P. L. Gadhavi, N. A. Young, E. M. Armstrong, C. D. Garner and E. D. Laue, *FEBS Lett.*, 1995, **358**, 278–282.
131. G. Öz, K. Zangger and I. M. Armitage, *Biochemistry*, 2001, **40**, 11433–11441.
132. H. Wang, Q. Zhang, B. Cai, H. Li, K. H. Sze, Z. X. Huang, H. M. Wu and H. Sun, *FEBS Lett.*, 2006, **580**, 795–800.
133. B. A. Messerle, A. Schaffer, M. Vašák, J. H. Kägi and K. Wüthrich, *J. Mol. Biol.*, 1990, **214**, 765–779.
134. D. W. Hasler, L. T. Jensen, O. Zerbe, D. R. Winge and M. Vašák, *Biochemistry*, 2000, **39**, 14567–14575.
135. K. S. Hagen, D. W. Stephan and R. H. Holm, *Inorg. Chem.*, 1982, **21**, 3928–3936.
136. F. Y. Ni, B. Cai, Z. C. Ding, F. Zheng, M. J. Zhang, H. M. Wu, H. Z. Sun and Z. X. Huang, *Proteins*, 2007, **68**, 255–266.
137. Z. C. Ding, X. C. Teng, B. Cai, H. Wang, Q. Zheng, Y. Wang, G. M. Zhou, M. J. Zhang, H. M. Wu, H. Z. Sun and Z. X. Huang, *Biochem. Biophys. Res. Commun.*, 2006, **349**, 674–682.
138. Z. C. Ding, Q. Zheng, B. Cai, W. H. Yu, X. C. Teng, Y. Wang, G. M. Zhou, H. M. Wu, H. Z. Sun, M. J. Zhang and Z. X. Huang, *J. Biol. Inorg. Chem.*, 2007, **12**, 1173–1179.
139. P. Palumaa, E. Eriste, O. Njunkova, L. Pokras, H. Jørnvall and R. Sillard, *Biochemistry*, 2002, **41**, 6158–6163.
140. N. Romero-Isart, L. T. Jensen, O. Zerbe, D. R. Winge and M. Vašák, *J. Biol. Chem.*, 2002, **277**, 37023–37028.
141. B. Dolderer, H. Echner, A. Beck, H. J. Hartmann, U. Weser, C. Luchinat and C. Del Bianco, *FEBS J.*, 2007, **274**, 2349–2362.
142. B. Roschitzki and M. Vašák, *J. Biol. Inorg. Chem.*, 2002, **7**, 611–616.
143. B. Roschitzki and M. Vašák, *Biochemistry*, 2003, **42**, 9822–9828.
144. A. Bruinink, P. Faller, C. Sidler, R. Bogumil and M. Vašák, *Chem. Biol. Interact.*, 1998, **115**, 167–174.
145. J. C. Erickson, A. K. Sewell, L. T. Jensen, D. R. Winge and R. D. Palmiter, *Brain Res.*, 1994, **649**, 297–304.
146. N. Romero-Isart and M. Vašák, *J. Inorg. Biochem.*, 2002, **88**, 388–396.
147. Y. Shi, W. Wang, J. Mo, L. Du, S. Yao and W. Tang, *Biometals*, 2003, **16**, 383–389.
148. S. Toriumi, T. Saito, T. Hosokawa, Y. Takahashi, T. Numata and M. Kurasaki, *Basic. Clin. Pharmacol. Toxicol.*, 2005, **96**, 295–301.
149. Q. Zheng, W. M. Yang, W. H. Yu, B. Cai, X. C. Teng, Y. Xie, H. Z. Sun, M. J. Zhang and Z. X. Huang, *Protein. Eng.*, 2003, **16**, 865–870.
150. G. Meloni, P. Faller and M. Vašák, *J. Biol. Chem.*, 2007, **282**, 16068–16078.
151. I. Hozumi, T. Inuzuka, M. Hiraiwa, Y. Uchida, T. Anezaki, H. Ishiguro, H. Kobayashi, Y. Uda, T. Miyatake and S. Tsuji, *Brain Res.*, 1995, **688**, 143–148.

152. I. Hozumi, T. Inuzuka, H. Ishiguro, M. Hiraiwa, Y. Uchida and S. Tsuji, *Brain. Res.*, 1996, **741**, 197–204.
153. T. Inuzuka, I. Hozumi, A. Tamura, M. Hiraiwa and S. Tsuji, *Brain Res.*, 1996, **709**, 151–131.
154. M. Penkowa, J. Carrasco, M. Giralt, A. Molinero, J. Hernandez, I. L. Campbell and J. Hidalgo, *J. Cereb. Blood. Flow. Metab.*, 2000, **20**, 1174–1189.
155. I. Hozumi, M. Asanuma, M. Yamada and Y. Uchida, *J. Health Sci.*, 2004, **50**, 323–331.
156. Y. Irie and W. M. Keung, *Brain Res.*, 2003, **960**, 228–234.
157. G. Meloni, V. Sonois, T. Delaine, L. Guilloreau, A. Gillet, J. Teissie, P. Faller and M. Vašák, *Nature Chem. Biol.*, 2008, **4**, 366–372.
158. T. Kawashima, K. Doh-ura, M. Torisu, Y. Uchida, A. Furuta and T. Iwaki, *Dement. Geriatr. Cogn. Disord.*, 2000, **11**, 251–62.
159. T. M. Dawson and V. L. Dawson, *Science*, 2003, **302**, 819–822.
160. S. R. Paik, H. J. Shin, J. H. Lee, C. S. Chang and J. Kim, *Biochem. J.*, 1999, **340**, 821–828.
161. S. R. Paik, H. J. Shin and J. H. Lee, *Arch. Biochem. Biophys.*, 2000, **378**, 269–277.
162. S. Turnbull, B. J. Tabner, O. M. El-Agnaf, S. Moore, Y. Davies and D. Allsop, *Free Radic. Biol. Med.*, 2001, **30**, 1163–1170.
163. A. Aguzzi and M. Heikenwalder, *Nat. Rev. Microbiol.*, 2006, **4**, 765–775.
164. S. B. Prusiner, *Science*, 1982, **216**, 136–144.
165. K. Puttaparthi, W. L. Gitomer, U. Krishnan, M. Son, B. Rajendran and J. L. Elliott, *J. Neurosci.*, 2002, **22**, 8790–8796.
166. M. A. Sens, S. Somji, D. L. Lamm, S. H. Garrett, F. Slovinsky, J. H. Todd and D. A. Sens, *Environ. Health Perspect.*, 2000, **108**, 413–418.
167. S. H. Garrett, M. A. Sens, D. Shukla, S. Nestor, S. Somji, J. H. Todd and D. A. Sens, *Prostate*, 1999, **41**, 196–202.
168. M. A. Sens, S. Somji, S. H. Garrett, C. L. Beall and D. A. Sens, *Am. J. Pathol.*, 2001, **159**, 21–26.
169. R. Dutta, D. A. Sens, S. Somji, M. A. Sens and S. H. Garrett, *Prostate*, 2002, **52**, 89–97.

Met. Ions Life Sci. **2009**, *5*, 353–397

12

Metallothionein Toxicology: Metal Ion Trafficking and Cellular Protection

David H. Petering, [1] ***Susan Krezoski,*** [1] ***and Niloofar M. Tabatabai*** [2]

[1] Department of Chemistry and Biochemistry, University of Wisconsin-Milwaukee, Milwaukee, WI 53201, USA
<petering@uwm.edu>
<suek@uwm.edu>

[2] Division of Endocrinology, Metabolism and Clinical Nutrition and Kidney Disease Center, Medical College of Wisconsin, Milwaukee, WI 53226, USA
<ntabatab@mcw.edu>

Metal Ions in Life Sciences, Volume 5 Edited by Astrid Sigel, Helmut Sigel, and Roland K. O. Sigel

Published by the Royal Society of Chemistry, www.rsc.org DOI: 10.1039/9781847558992-00353

ABSTRACT: The literature is replete with reports about the involvement of metallothionein in host defense against injurious chemical, biological, and physical agents. Yet, metallothionein's functional roles are still being debated. This review addresses the issues that have left the physiological significance of metallothionein in doubt and moves on to assess the MT's importance in cell toxicology. It is evident that the protein is broadly involved in protecting cells from injury due to toxic metal ions, oxidants, and electrophiles. Attention is focused on MT's structural and chemical properties that confer this widespread role in cell protection. Particular emphasis is placed on the implications of finding that metal ion unsaturated metallothionein is commonly present in many cells and tissues and the question, how does selectivity of reaction with metallothionein take place in the cellular environment that includes large numbers of competing metal binding sites and high concentrations of protein and glutathione sulfhydryl groups?

KEYWORDS: electrophile · metal ions · metallothionein · metallothionein-null · oxidant · reactivity · toxicity

1. INTRODUCTION

Metallothionein (MT) was discovered during an exploratory project to analyze metal ion content and distribution in mammalian tissues. Having obtained samples of horse kidney, Margoshes and Vallee discovered that in addition to essential transition metal ions, it contained a readily measurable concentration of Cd^{2+} that was bound to a single protein, MT [1]. Shortly thereafter, Vallee and Kägi carried out the basic physico-chemical characterization of the protein, showing that it was sulfhydryl-rich, that it could bind Cd^{2+} or Zn^{2+} with an SH/M^{2+} ratio of 2.9, and that the pH dependence of metal ion binding favored Cd^{2+} over Zn^{2+} [2]. The clear hypothesis emanating from these studies was that MT apo-protein successfully sequesters Cd^{2+} in competition with the rest of the cell's constituents and in the process either protects the cell or serves as a site of toxicity.

These two papers set in motion the intense study of metallothionein that continues to the present. Originally, the striking observation that a metal ion, Cd^{2+}, known to be toxic to humans was largely associated with only one cellular protein propelled the field of metallothionein research toward the study of MT's role in Cd^{2+} toxicology and its potential as a participant in the cellular reactions of other heavy metal ions [3]. Subsequently, it was recognized that MT contributes to Zn^{2+} metabolism and that MT's cellular activities had to be extended to include essential as well as toxic metals [4].

The focus of metallothionein studies expanded again as investigators inquired about the reactivity of MT's 20 SH groups with xenobiotic oxidants and electrophiles [5,6]. In this context, the view of the protein was turned on its head: it became a sulfhydryl-rich protein that acts as a potent reductant or nucleophile. *In vitro*, metallothionein was all of these things – a robust binder of multiple essential and toxic/therapeutic metal ions, a strong reductant, and a stellar nucleophile [2,7–11]. In model *in vivo* systems, a similar range of reactivity could be inferred [12–17]. But in the midst of such an array of reactions, the question has persisted: what are the authentic physiological and pathological functions of metallothionein?

In this review, attention will be given to possible reasons for the lack of clarity about metallothionein's functional roles after nearly a half century of intensive study. Nevertheless, this chapter's aim is to answer this question as clearly as possible as it relates to MT's role in the cell's response to toxic conditions. In the midst of the huge literature on MT in relation to cell and tissue injury, the approach will be to draw selectively on *in vivo* research and to use the body of chemical knowledge about MT structure and reactivity to illuminate these studies at a molecular level.

It is recognized that metallothioneins or MT-like proteins or peptides exist in virtually all species. Mammalian metallothioneins are the subjects of this review. Chapters 3–8 of this volume consider metallothioneins in other life forms.

2. ANIMAL METALLOTHIONEINS

The primary structure of metallothionein is conserved throughout the animal kingdom [18]. Based on peptide sequences of approximately 60–70 amino acids that include 18–20 cysteine residues, 6 or 7 Zn^{2+} or Cd^{2+}(M^{2+}) ions bind to the protein [19,20]. Four major metallothionein isoforms have been discovered and characterized. MT-1 and MT-2 are found throughout the organism; MT-3 and MT-4 display more confined distributions [21,22]. Originally discovered and thought to be specific for brain neurons, MT-3 has also been observed in primary cultures of human proximal tubule cells and various cancers [23,24]. MT-4 has received the least attention and has been observed in various epithelial tissues [22].

2.1. MT-1 and MT-2

Metallothioneins-1 and -2 have very similar amino acid sequences and structures. Cd_7-MT-2 and Cd_5,Zn_2-MT-2, two species for which 3-dimensional structures have been determined, exist as 2-domain structures, each enclosing a metal-thiolate cluster with stoichiometry M_3S_9 or M_4S_{11} (Figure 1) [19,25,26]. In crustacean MTs that contain two 3-metal clusters, analogous metal-thiolate connectivities exist, minus the fourth metal ion and 2 cysteine sulfhydryl groups in the C-terminal cluster [20,27,28]. Each cluster contains

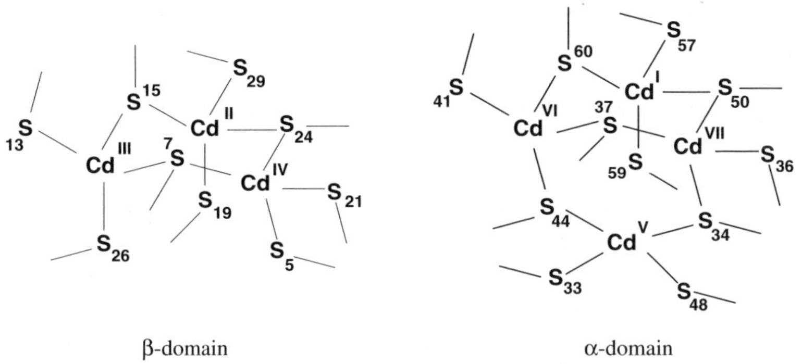

Figure 1. Cd^{2+}-thiolate connectivities in the metallothionein Cd_3S_9 and Cd_4S_{11} clusters, located in the N-terminal β-domain and C-terminal α-domain, respectively. Roman numerals refer to numbers of $^{113}Cd^{2+}$ NMR peaks [19]; numbers indicate cysteine positions in amino acid sequence.

multiple metal ions knit together by bridging thiolate ligands and linked to the peptide backbone through bridging and terminal sulfhydryl groups. It is intriguing that with 18–20 sulfhydryl groups and multiple metal ions available for interaction, only 3 discrete clusters with fixed metal ion-thiolate connectivities are observed.

2.2. MT-3

Neuronal growth inhibitory factor, a protein isolated on the basis of its capacity to inhibit the growth of neurons in culture has been identified as MT-3 [29,30]. The sequence of MT-3 contains a 6 amino acid insert in the α-domain that draws one's attention [31]. Although it exerts a temperature-dependent effect on the $^{113}Cd_4S_{11}$ NMR spectrum, it is the $^{113}Cd_3S_9$ resonances that are substantially perturbed and broadened. The latter property has been traced to the presence of 2 proline residues at positions 7 and 9 in the β-domain [31]. Such sequence perturbations translate into a reduction in metal ion binding affinity in MT-3 [32]. But according to one cursory study, they do not affect cluster reactivity [33]. However, relatively few structure-reactivity correlation studies have been undertaken to investigate how substitutions among non-cysteine residues affect cluster structure and reactivity [34].

The remainder of this review is devoted to the roles of MT-1 and MT-2 in cell toxicology. Chapter 11 of this volume addresses the activities of MT-3 in the brain.

3. METALLOTHIONEIN AND TOXICOLOGY. AN OVERVIEW

3.1. Experimental Approaches Considered

The number of papers that support the involvement of metallothionein in organismic, tissue, or cellular response to toxic agents is immense and repetitive but not definitive. Depending on the study, investigators may have resorted to non-physiological modes of chemical exposure such as subcutaneous injection, to grossly large applications of toxic reagent to test whether metallothionein is important for cell protection, or to studies in the presence of artificially inflated concentrations of metallothionein. In turn, it has been common to measure endpoints such as animal mortality, overt tissue injury, or cell death that are not typical of chronic human diseases associated with Cd^{2+} exposure. Thus, their relevance to the human context has been difficult to assess.

Additionally, there is an element of experiment-specific variability to metallothionein studies that has left the field unable to reach definitive conclusions about the physiological roles of MT. For example, after decades of experiments that generally reveal MT as a cellular protective molecule against at least the acute toxicity of Cd^{2+}, papers still appear which confound this conclusion [35]. When compared with the relative uniformity of results emanating from experiments inquiring about the role of glutathione in cellular protection, the seemingly slippery nature of MT may well signal a more complicated cellular chemistry, embracing metal-ion binding, thiol reactivity, and even protein-protein interactions [36–38].

Experiments implicating metallothionein in cellular protection reveal that MT tissue concentrations rise after exposure to toxic or stressful reagents [4,39–41]. MT protein increases after physiological or pathological manipulations such as exercise, exposure to cold, heat, bacteria, and other stresses, including oxygen deprivation and exposure to chemicals, provide clear indication that MT participates in the normal organismic response to a bewildering variety of conditions.

Other experiments with a number of chemicals show that toxicity diminishes when MT concentration is increased through pre-induction with Zn^{2+} or a tolerable concentration of Cd^{2+} [42–44]. Recent variations of this experiment yielding similar results involve cells containing plasmids that overexpress MT or MT-transgenic animals that constitutively express elevated concentrations of MT throughout the organism or in selected tissues [45–47]. However, the hypothesis that MT protects against chemical injury through chemical reaction with the offending reagent commonly has not been tested by direct measurement of toxicant sequestration by MT.

Some of these experiments are subject to the criticism that when protection by artificially enhanced levels of MT is observed, such concentrations may not be relevant under normal physiological/pathological conditions. Nevertheless, some normal tissues contain substantial concentrations of constitutively expressed MT [48–50]. In addition, as mentioned above, MT protein synthesis is induced in some tissues under a variety of stress conditions [4,40,41]. Furthermore, in pathological conditions such as cancer, tumors commonly contain large concentrations of MT protein [51,52]. Thus, toxic reagents may encounter substantial concentrations of MT, depending on the tissue and the physiological condition.

A second issue complicates the interpretation of studies of MT's role in cell protection. It has been recognized recently that MT expressed in normal tissues, tissues from stressed animals, normal and transformed cells in culture, cells containing transgenic MT, cells induced to synthesized extra MT, etc., contain metal-ion unsaturated MT [51,53,54]. Thus, in considering the reactions of the MT pool with injurious chemicals, two types of reactions need to be considered – those involving thiolate sites saturated with metal ion, usually Zn^{2+}, and others in which metal ion-free thiolates undergo reaction. The chemical reactivity of these two classes of sites is expected to be substantially different. In the first case, thiolate nucleophilicity or oxidation potential is relatively deactivated through binding to a positively charged metal ion and the reduction in sulfur atom solvent accessibility in the folded, native protein. In the second instance, the multiple free sulfhydryl groups associated with the unstructured polypeptide are powerful nucleophilic and redox reagents that may act in concert as part of the same MT molecule [54].

In this complicated setting, the most secure strategy to test the involvement of MT in protection against specific chemicals or physical agents is to compare the susceptibility of wild-type and MT-1- and -2-null animals or cells [55,56]. Effects in wild-type animals represents the native, unaugmented response that the organism mounts in reaction to toxic exposure. Differences observed in MT-null mice are traced directly to the absence of the protein. Even with this model, however, questions of interpretation may exist. Although the MT-null phenotype (survival without obvious compromised development) suggests that the protein is not required for normal cell function, mice lacking MT differ from wild-type counterparts. For example, pregnancies of MT-null mice are not routine. Even mild zinc deficiency results in substantial fetal resorption and malformation that is not seen in the wild-type organism [57]. Moreover, the physiology of MT-null animals is abnormal; for instance, they display deficiencies in spatial learning and memory [58,59]. Similarly, MT-null cells may have surprisingly different phenotypes from control mice. Their macrophage inducible nitric oxide synthase produces distinctly lower amounts of NO than control cells [60]. They also display less NF-κB activation in response to lipopolysaccharide [61].

Nevertheless, with these caveats, experiments with MT-null mice offer the simplest model to address whether MT participates in processes that modulate the toxicity of trial toxicants. Section 4 examines the results of studies with MT-null mice.

3.2. Mechanisms of Cell Protection by Metallothionein Considered

The metal ion binding capacity of MT stands out as an obvious mechanism by which the protein can protect cells against the toxicity of various metal ions such as Cd^{2+}, e.g.:

$$Zn_7\text{-MT} + 7\,Cd^{2+} \rightleftharpoons Cd_7\text{-MT} + 7\,Zn^{2+} \tag{1}$$

In contrast, estimating the potential for MT's sulfhydryl groups to contribute to cellular protection against non-metal ion oxidants and electrophiles is more complicated (reactions 2 and 3).

$$2\ \text{RS(H)} + \text{oxidant} \rightleftharpoons \text{RSSR} + \text{reductant} \tag{2}$$

$$\text{RS(H)} + \text{R}'\text{-X} \rightarrow \text{RS-R}' + \text{X}^- \tag{3}$$

Although an induced pool of MT may contain a substantial concentration of SH groups, glutathione concentrations are typically in the millimolar range as is the sum of the proteomic sulfhydryl groups [36,37,54]. Thus, the question immediately arises: among the plethora of thiols associated with the tripeptide, glutathione, and the proteome, are there properties of MT's complement of sulfhydryl groups that render them selectively reactive with oxidants and electrophiles?

With some exceptions, chemical studies of the capacity of MT to undergo examples of reactions (2) and (3) have not compared the reactivity of its sulfhydryl groups (metal-bound and metal-free) with those of GSH or general proteomic thiols. Moreover, most cellular studies have not directly assessed sulfhydryl group status among these three classes of thiols in studies of putative cell protection by MT.

Other less direct mechanisms might exist whereby basal or induced concentrations of MT affect cellular response to injurious chemicals or physical agents. A persistent suggestion has been that in reactions (1)–(3), one of the products is Zn^{2+} and that released Zn^{2+} might act as an intracellular signaling agent that secondarily modifies other molecules and reactions [62,63]. For example, it is hypothesized that Cd^{2+} induces the synthesis of MT by displacing Zn^{2+} from preexistent Zn-MT [64]. In turn, released Zn^{2+} binds to the inactive transcription factor, apo-MTF-1, converting it into Zn-MTF-1 that binds to metal response elements of the MT promoter and stimulates MT mRNA synthesis.

Recently, it has become apparent that MT has intracellular protein binding partners [38,65]. Conceivably, formation of specific MT-protein adducts could contribute to metabolic regulation as well.

4. METALLOTHIONEIN'S ROLE IN TOXICOLOGY: RESULTS WITH THE MT-1- AND MT-2-NULL MICE AND DERIVED CELLS

Table 1 summarizes studies [66–87] showing that MT-1- and MT-2-null mice are significantly more sensitive than wild-type, control mice to a broad variety of agents applied in different ways that are classified as metal ions or metalloids, inflammatory or stress inducing agents that may act through oxidative stress mechanisms, and electrophiles. Only a few of the studies in Table 1 characterize the role of metallothionein. For example, the study of mercury vapor exposure of pregnant mice revealed that placental metallothionein in wild-type animals contained most of that tissue's Hg [68]. Thus, MT probably acts to prevent fetal mercury accumulation by binding and trapping Hg^{2+} during its transport from the mother into the fetus.

Most reports presume that MT acts in its protective role as a metal ion binder, antioxidant, or nucleophile without supporting evidence. Generally, these assumptions make sense. Still, there are cases in which straight forward chemical hypotheses fall short. For instance, the role of MT in Pb^{2+} carcinogenesis is murky [69]. Lead does not appear to associate directly with MT in cells or animals, though Pb^{2+} is certainly capable of binding with sulfhydryl groups. Nor does the chronic, low level exposure regime of the experiment immediately suggest that oxidative stress is involved. The authors argue that MT is necessary for Pb inclusion body formation, a protective response of kidney cells to the presence of Pb^{2+}, and that MT binds to these microscopically visible concentrations of Pb and protein. Perhaps, MT-protein interactions play a role in their synthesis. If so, the novel area of MT protein binding would take on toxicological significance [38,65]. Nevertheless, with all these caveats, the overwhelming impression from these studies is that MT is an important participant in cell protection against a variety of agents that exhibit a wide range of cellular chemistry.

The general questions emerging from these results center on the chemical mechanisms by which metallothionein contributes to cell protection.

Table 1. Survey of toxicity of chemical, biological, and physical agents in metallothionein wild-type and null mice.

Toxic Agent	Effects in Null versus Wild-type Mice	Exposure Route	Reference
Metals and Metalloids			
Cd^{2+}	Elevated nephrotoxicity	Intraperitoneal injection	66
	Increased bone injury	Subcutaneous injection	66,67
Hg°	Elevated Hg in fetus	Inhalation by pregnant dams	68
Pb^{2+}	Increased proliferative lesions and renal carcinogenesis	Chronic oral exposure	69
Arsenite, arsenate, dimethylarsinic acid	Increased lung, liver, and kidney pathology and serum and urinary 8-hydroxy-2′-deoxyguanosine	Oral exposure and injection	70,71
Inflammatory and Oxidative Stress			
Bacterial endotoxin	Heightened susceptibility to lung inflammation	Intratracheal installation	72
Ovalbumin antigen	Increased lung inflammation	Intratracheal installation	73
Helicobacter-induced infection	More gastritis	Innoculation	74
Brain cryo-injury	Increased inflammatory response		75,76
Acetaminophen	Increased hepatotoxicity and lethality	Intraperitoneal injection	77
t-Butyl-peroxide	Elevated apoptosis	Embryonic cell incubation	78,79
Paraquat	Reduced cell survival	Embryonic cell incubation	80
UV-B radiation	More cell injury	Exposure to skin and skin explants	81,82
γ-Irradiation	Enhanced apoptosis in thymus	Whole body radiation	83

Table 1. (*Continued*).

Toxic Agent	Effects in Null versus Wild-type Mice	Exposure Route	Reference
Electrophiles			
CCl_4	Elevated liver damage	Oral exposure	84
7,12-Dimethylbenz[a]-anthracene	Increased carcinogenesis	Skin application	85
cis-Dichlorodiammine Platinum(II)	Increase in hepatocellular carcinoma	Intraperitoneal injection	86
Melphalan	Enhanced apoptosis	Embryonic cell incubation	78,79
N-Butyl-N-(4-hydroxybutyl)-nitrosamine	Increased carcinogenicity, decreased malignant potential	Oral exposure	87

The sections below address this issue in terms of what is known about the structure and reactivity of metallothionein.

5. METAL ION TOXICOLOGY IN RELATION TO METALLOTHIONEIN CHEMISTRY

5.1. Cadmium Toxicity

Cadmium ion has been recognized as a major global industrial and commercial pollutant with established human health effects for many decades. Its organ specific toxicity includes nephrotoxicity characterized as a Fanconi syndrome in which the kidney tubule fails to resorb multiple nutrients from the glomerular filtrate [88]. Bone targeted toxicity was first observed in Itai-Itai disease in Japan [89]. Cd^{2+}-related lung disease has also been recognized [90]. Despite epidemiological studies linking Cd to particular organ-system toxicity, the majority of mechanistic studies have utilized model organ or cell systems other than kidney, bone, or lung. Moreover, many investigators' primary interest has been in the role of metallothionein in relation to Cd^{2+} toxicity not in the underlying mechanisms of Cd^{2+}-induced cell injury. Considering the generality of mammalian cell biology, it makes sense to employ a variety of convenient models to understand how Cd^{2+} damages cell function. At some point, however, such studies need to address the targeted nature of Cd^{2+} toxicity and its relation to metallothionein.

5.1.1. *Cadmium Trafficking*

An early, unusual study of the time-dependent distribution of Cd^{2+} intraperitoneally injected into rodents revealed in liver that the metal ion first associated with other proteins. Then, as MT was induced, the metal ion shifted into the MT pool [91]. Thereafter, the Cd-MT protein pool acquired Zn^{2+} to become Cd,Zn-MT.

These observations suggest the following series of reactions:

$$Cd^{2+}_{OUT} \rightleftharpoons Cd^{2+}_{IN} \tag{4}$$

$$Cd^{2+}_{IN} + (Zn)\text{-proteome} \rightleftharpoons Cd\text{-proteome} + (Zn^{2+}) \tag{5}$$

$$Cd\text{-proteome} + (\text{apo-or Zn-})MT \rightleftharpoons Cd\text{-}MT + \text{proteome} + (Zn^{2+}) \tag{6}$$

$$Cd\text{-}MT + Zn\text{-}MT \rightleftharpoons Cd,Zn\text{-}MT \tag{7}$$

Cd^{2+} is transported into the cell and binds adventitiously to members of the proteome including any extant metallothionein. Among the outcomes of these reactions, MT mRNA transcription is induced and apo-MT is synthesized. With the appearance of additional MT, reaction (6), involving either induced apo-MT or apo-MT that has acquired Zn^{2+}, shifts the Cd^{2+} distribution toward MT. The formation of mixed metal, Cd,Zn-MT follows, requiring both the synthesis of extra MT beyond what is necessary for Cd^{2+} binding and the transport of Zn^{2+} into the cell.

The sections that follow develop the chemical foundation supporting the operation of reactions (6) and (7) in the cell's response to Cd^{2+}.

5.1.1.1. Metallothionein Binding Stoichiometry with Cd^{2+} or Zn^{2+}. Both Cd^{2+} and Zn^{2+} react with apo-MT in titration experiments to form M_7-MT species in which the 7 metal ions are distributed among 2 metal-thiolate clusters, M_3S_9 and M_4S_{11}, that stabilize the folding of each domain about the cluster through metal ion-thiolate bonding (see Figure 1 in Section 2.1). Except for cysteines 50, 57, 59, and 60, which coordinate to a single Cd^{2+} in the α-domain cluster, there are no other examples of 4 consecutive sulfhydryl ligands binding to the same metal ion in either cluster

[25,26]. The fact that thiolates bridge almost all pairs of Cd^{2+} ions in each cluster suggest that the clusters may act as units in some of their chemical reactions.

5.1.1.2. Cooperative Metal Ion Binding to Form Clusters. Experiments from two decades ago examined species that formed during the titration of apo-MT with Zn^{2+} and Cd^{2+} [92]. After each addition, a non-specific proteinase was added to digest unreacted protein with the result that after addition of 1–4 equivalents of M^{2+}, only fully occupied α-domain was isolated that was resistant to hydrolysis by pronase. It was concluded that Zn^{2+} and Cd^{2+} (M^{2+}) prefer binding to the α-domain and that "cooperative" all or nothing binding is observed (reaction 8) with $K_4 \gg K_{1-3}$ (reactions 9–12):

$$nM^{2+} + \alpha\text{-domain} \rightleftharpoons n/4M_4\text{-}\alpha\text{-domain} \quad (8)$$

However, the presence of the protease in the system complicates the interpretation. Perhaps, M^{2+} initially binds independently to the protein and only after the addition of the fourth M^{2+} is the native cluster structure achieved (reactions 9–12, $K_4 \sim K_{1-3}$).

$$M^{2+} + \text{apo-MT} \rightleftharpoons M_1\text{-MT} \qquad K_1 \quad (9)$$

$$M^{2+} + M_1\text{-MT} \rightleftharpoons M_2\text{-MT} \qquad K_2 \quad (10)$$

$$M^{2+} + M_2\text{-MT} \rightleftharpoons M_3\text{-MT} \qquad K_3 \quad (11)$$

$$M^{2+} + M_3\text{-MT} \rightleftharpoons M_4\text{-}\alpha\text{-domain} \qquad K_4 \quad (12)$$

But as the enzyme degrades MT, protease-resistant, intact domains are favored, resulting in the sole observation of M_4-α-domain at each stage of the titration.

In fact, the latter description represents the pathway of reaction of Co^{2+} with apo-MT [93]. Vašák and co-workers showed that in the step-wise titration, 1–3 equivalents of paramagnetic Co^{2+} associate with apo-MT with linearly increasing intensity of paramagnetic resonance, indicative of non-interacting Co-thiolate binding. Upon addition of the fourth Co^{2+}, the intensity collapses as spin coupling occurs in the Co_4S_{11} cluster. A recent extension of this experiment fixed the location of Co^{2+} within the protein by titrating the free SH groups with iodoacetamide [94]. In the Co_3-MT species,

SH groups from both domains were utilized to form 3 independent CoS_4 sites. By substituting Co(II) with $^{111}Cd^{2+}$ (I = 1/2) in the iodoacetamide-modified Co-MT species and determining the resultant 1-dimensional ^{111}Cd NMR spectrum, it was further concluded that the isolated Co(II) sites are non-specific.

A similar experiment involving the titration of apo-MT with $^{111}Cd^{2+}$ also showed that the signature ^{111}Cd NMR spectrum of the α-domain was absent until addition of the third and fourth $^{111}Cd^{2+}$ ion. The α-domain gained full occupancy only after 5 Cd^{2+} equivalents were added and some β-domain resonant intensity was observed, suggesting that α-domain residence is only marginally preferred relative to the β-domain. As with Co^{2+}, native cluster structures are favored in the presence of sufficient Cd^{2+} to constitute native domains, but other structures can exist at substoichiometric concentrations of Cd^{2+}, in the sequence Cd_1-MT, Cd_2-MT, Cd_3-MT, Cd_4-α-domain + Cd_1-β-domain [94,95]. Little effort has been devoted to testing whether cellular MT may contain such species.

The finding that metal-thiolate species other than fully occupied clusters can exist during titration experiments needs to be reconciled with the repeated observations described below that various reactions of metal ion-saturated MT, such as thiol modification or ligand substitution, display biphasic kinetics and represent the unitary reaction of individual clusters not parallel or sequential reactions of individual sites within the clusters (e.g., [96,97]). The simplest explanation is that the clusters have kinetic integrity such that the initial reaction on the cluster is rate limiting for reaction of the entire complement of sulfhydryl groups or metal ions in the cluster.

5.1.1.3. Stability Constants of M_7-Metallothionein. Stability constants for Zn_7-MT have directly been determined through equilibrium competition with ligands of known stability constant with Zn^{2+}[31,98–100]. These range from $10^{11.2}$ per Zn^{2+} at pH 7.4 to $10^{11.5}$ at pH 8, assuming cooperative, all or nothing reaction of clusters with competing ligands. Interestingly, such measurements have not, heretofore, detected a significant difference in equilibrium binding affinity of the two domain clusters for Zn^{2+}. This is consistent with the above titration experiments that show an overlap in the formation of the α- and β-domain clusters during the titration (see Section 5.1.1.2). Hydrogen ion titrations of Zn_7-MT and Cd_7-MT done by Kägi and Vallee, also failed to detect substantial differences in stability of α- and β-domains [2]. These titrations demonstrated that Cd^{2+} binds to MT three orders of magnitude more strongly than Zn^{2+}. Thus, at pH 7.4 the stability constant of Cd_7-MT per Cd^{2+} is on the order of 10^{14}.

A recent paper concluded that Zn_7-MT undergoes ligand competition in three steps, involving Zn_3-β-, Zn_2-β-, and Zn_4-α-domains [101]. These reactions were linked to four stability constants at pH 7.4, $10^{11.8}/Zn^{2+}$ for cooperative formation of Zn_4-α-, $10^{7.7}/Zn^{2+}$ for β-converting into Zn_1-β-, and $10^{10.0}$ and $10^{10.4}/Zn^{2+}$ for the further step-wise formation of Zn_3β-domains. These results strongly contrast with previous findings. Reasons for the differences remain to be elucidated.

The apparent stability constants at pH 7.4 for Zn-MT are large but less than, for example, the constant for Zn-carbonic anhydrase of 10^{12} [102]. In contrast, the binding constant for Cd_7-MT domains of $10^{8.6}$ is such that apo-MT stoichiometrically competes for Cd^{2+} bound to Cd-CA [103]. This difference reflects the fact that Cd^{2+} displays a greater preference for sulfhydryl ligands in comparison with imidazole nitrogen ligands in carbonic anhydrase than does Zn^{2+}. In turn, the results provide a rationale for the repeated observation in cells that the great majority of the intracellular Cd^{2+} is associated with MT.

5.1.1.4. Kinetics of M_7-Metallothionein Formation. Stopped flow experiments demonstrate that apo-MT reacts with Zn^{2+}, Cd^{2+}, or simple amino acid complexes within the time of mixing (ca. 4 ms) and with an excess of metal ions forms clusters within milliseconds [94]. Cluster formation may be considered as analogous to hydrophobic side chain coalescence in typical protein folding [104]. In that context, one can view the intermediate non-specific Cd-thiolate complexes described above in Section 5.1.1.2 as mechanistically similar to the intermediate molten globule state between the typical random coil peptide to its 3-dimensional structure. In both, the key property is greatly reduced conformational flexibility that limits further conformational options for the peptide backbone as it folds [104].

5.1.1.5. Formation of Mixed Metal Ion Metallothioneins. It is necessary to move beyond homogeneous metal ion metallothioneins because virtually all Cd^{2+}-containing MTs isolated from biological sources are mixed metal ion species, primarily $Cd_n,Zn_{(7-n)}$-MT [19,91]. Using $^{113}Cd^{2+}$ NMR spectroscopy, Nettesheim, Engeseth, and Otvos made the remarkable discovery that native $^{113}Cd_n,Zn_{(7-n)}$-MT is constituted through a unique interprotein metal ion exchange reaction that produces a defined population of mixed metal ion species for each ratio of starting materials [105]:

$$n\ ^{113}Cd_7\text{-MT} + (7-n)Zn_7\text{-MT} \rightleftharpoons 7\ ^{113}Cd_{n},Zn_{(7-n)}\text{-MT} \qquad (13)$$

Notably, Cd^{2+} tends to move from the β- to the α-domain and Zn^{2+} shifts in the other direction. As a result, Cd^{2+} tends to be located in the α-domain and Zn^{2+} in the β-domain. The segregation of metal ions supports the hypothesis that the domains have different functions, one for toxic metal ion storage and the other for functional Zn^{2+} metabolism [98]. Were there simply one cluster that favored Cd^{2+} over Zn^{2+} binding, this opportunity for bifunctionality would not exist.

Reaction (13) must be bimolecular and would seem to require interprotein thiolate interactions between α- and β-domains that facilitate the metal ion exchange process. The plausibility of such specific interactions was suggested by the X-ray crystallographic structural analysis of Cd_5,Zn_2-MT [26]. MT molecules exist as quasi-dimers in the crystal unit cell, with α- and β-domains of adjacent molecules proximate to one another. That such interactions do occur in solution may be inferred (a) from the observation of the dimer state by column chromatography and (b) through the detection of differences in the Cd_4-α-domain conformation of the holoprotein dimer when compared with the isolated domain [106,107]. The 2-dimensional $^{111}Cd^{2+}$ NMR study in (b) indicated that in the dimer, the Cd_4-α-domain peptide and cluster are perturbed at sites at the dimer interface revealed in the X-ray crystallographic analysis [107].

The cellular occurrence of reaction (13) necessitates that Cd^{2+} ions first induce MT synthesis resulting in Cd_7- and Zn_7-MT and that these homogeneous metal proteins subsequently react with one another to generate mixed metal ion metallothioneins. Further, as the mixed metal ion protein is biodegraded, releasing metal ions, and is synthesized to recapture them in the steady state, these homogeneous metal ion species must continue to be made as transient intermediates. How all of these steps are choreographed in the cell is unknown.

The peculiarity of this process is underscored by considering a conceptually much simpler pathway to mixed metal metallothioneins. An alternative reaction results in the generation of

$$Zn_7\text{-MT} + n\ Cd^{2+} \rightleftharpoons Cd_{n,}Zn_{(7-n)}\text{-MT} + n\ Zn^{2+} \tag{14}$$

mixed metal ion protein according to $^{113}Cd^{2+}$ NMR spectroscopy but with completely different Cd^{2+} and Zn^{2+} distributions than observed in reaction (13) [105]. Later, Stillman, Cai, and Zelazowski showed that mild heating of the product mixture caused it to rearrange to that formed through the interprotein metal ion exchange reaction [108]. Thus, reaction (14) achieved a kinetically accessible product that with some additional energy of activation converted to the equilibrium product reached in reaction (13).

Reaction (14) might be the pathway to formation of Cd_7-MT. Stopped flow studies of the conversion of Zn_7-MT to Cd_7-MT reveal that the β-domain (Zn_3S_9) reacts with Cd^{2+} faster and in a single step, whereas the α-domain (Zn_4S_{11}) displays triphasic kinetics [109]. The three steps from fastest to slowest involve the substitution of 2, 1, and 1 equivalent of Zn^{2+} by Cd^{2+}. Each step begins with a Cd^{2+}-binding step, followed by exchange of Cd^{2+} for Zn^{2+} as in the initial reaction (reaction 15):

$$Zn_4\text{-}\alpha + Cd^{2+} \rightleftharpoons Cd\text{-}Zn_4\text{-}\alpha \rightarrow Cd,Zn_3\text{-MT} + Zn^{2+} \tag{15}$$

It is hypothesized that Cd^{2+} establishes bridging Cd-S-Zn linkages in the adduct species.

5.1.1.6. Models of Cd^{2+} Trafficking in Relation to Metallothionein. Metallothionein participates in Cd^{2+} ion trafficking by serving as its major intracellular binding site. Initially, as Cd^{2+} enters cells, it binds to constitutively expressed MT and then distributes adventitiously among other binding sites until new MT synthesis has been induced [91]. Thus, as described above, basal or induced apo-MT could react with Cd^{2+} or Cd-L (L, cellular ligand) to form Cd-MT (reactions 9–12). Alternatively, Zn-MT might undergo metal ion exchange with either of these species:

$$\text{apo-MT} + Cd^{2+} \rightleftharpoons \text{Cd-MT} \tag{16}$$

$$\text{apo-MT} + \text{Cd-L} \rightleftharpoons \text{Cd-MT} + \text{L} \tag{17}$$

$$\text{Zn-MT} + Cd^{2+} \rightleftharpoons \text{Cd-MT} + Zn^{2+} \tag{18}$$

$$\text{Zn-MT} + \text{Cd-L} \rightleftharpoons \text{Cd-MT} + \text{Zn-L} \tag{19}$$

Each of these reactions sequesters the injurious metal ion as Cd-MT, consistent with the general cellular picture in which MT induction results in the great majority of cellular Cd^{2+} becoming bound to the protein. Considering the plethora of adventitious metal binding sites that exist in proteins and other molecules, it is likely that the concentration of free Cd^{2+} is small. Thus, reactions (17) and (19) are the important ones to examine.

Section 6.2 reviews studies showing that metal ion unsaturated MT or apo-MT is widespread in cells under a variety of conditions. That being so, ligand substitution reactions involving apo-MT (reaction 16) probably play a significant role in Cd^{2+} trafficking. As written, this reaction is presumed to be central to MT's observed sequestration of Cd^{2+} as it distributes among various intracellular binding sites. Cd-carbonic anhydrase (Cd-CA) has been used to examine this reaction [110]. Like Zn-CA, the Cd^{2+} protein displays a large stability constant (10^9) but one that is several orders of magnitude less than that of Cd-MT [104]. Apo-MT readily competes to remove Cd^{2+} from Cd-CA with a rate constant of 6 $M^{-1} s^{-1}$ at pH 7 and 25°. A hypothesis to explain apo-MT's reactivity with Cd-CA is that the 4 cysteine residues at the C-terminus, which chelate to a single Cd^{2+} in the native structure, preferentially react with Cd-CA because they present less steric hindrance to reaction than interior sulfhydryl ligand sets. Indeed, the peptide 49–61, containing these sulfhydryl residues, reacts in the time of mixing with Cd-CA [111].

Besides ligand substitution reactions, Zn-MT might react directly with Cd-L to carry out metal ion exchange (reaction 20). If L normally exists as Zn-L, then this reaction would restore L to its native state. Could such reactions contribute to the intracellular distribution of Cd^{2+}?

To address this question, several Cd-substituted proteins have been examined. For example, Cd-CA exchanges metals with Zn-MT [110]. The Cd^{2+} complex of tramtrack, normally a C_2H_2 zinc finger transcription factor, has lost the specific DNA binding activity of Zn-tramtrack [112]. When reacted with Zn-MT, metal ion exchange occurs and, in the process,

$$\text{Zn-MT} + \text{Cd-tramtrack} \rightleftharpoons \text{Cd-MT} + \text{Zn-tramtrack} \tag{20}$$

its selective DNA binding property is recovered. Zn-MT not only serves as a Cd^{2+} sequestration site but, under the presumption that favored Cd^{2+} binding loci are Zn proteins, also provides Zn^{2+} to reconstitute the native protein. The thermodynamic favorability of this reaction follows from the greater stability constant of Cd-MT than Zn-MT and probable similar constants for the two tramtrack species [113].

Efforts to demonstrate the same reaction of Zn-MT with Cd-transcription factor IIIA and Cd-Sp1 (unpublished information) have not been successful although both have C_2H_2 ligand sets like tramtrak [114]. A possible reason is that in these structures Cd^{2+} binding causes a major ligand and conformational rearrangement resulting in the formation of a Cd-S_4 site from the sulfhydryl groups in adjacent fingers. Bound more firmly to four thiolate ligands, Cd^{2+} reaction with apo- or Zn-MT may be much less favorable.

Conceivably, Cd-MT, itself, might react to transfer Cd^{2+} to other sites in the reverse of reaction (18). However, the extent of such reactions may be limited as MT's cellular concentration rises into the 400 μM range, the K_D of the Cd-MT dimer [106]. Once in the dimer form, for example, Cd-MT's reactivity with EDTA declines precipitously due to the reduced solvent exposure of its Cd-thiolate clusters [97,107].

More generally, release of Cd^{2+} from cellular Cd,Zn-MT can be envisioned under conditions in which the protein is modified by oxidants or electrophiles. An unpublished study showed, for example, that cells pre-exposed to Cd^{2+} resulting in Cd-MT formation were sensitized to the toxicity of *cis*-dichlorodiammine Pt(II). According to measurements of Cd^{2+} and Pt distribution, the drug preferentially competed for MT thiolate groups, releasing Cd^{2+} into the cell. Although the formation of Pt-MT is protective (see below), the liberation of Cd^{2+} appeared to result in enhanced toxicity.

5.1.1.7. Metallothionein Induction by Cd^{2+}. Cadmium ion entering cells is expected to associate with constitutively expressed MT. As its intracellular concentration exceeds MT's binding capacity, Cd^{2+} efficiently induces additional MT synthesis. The metallothionein gene and its promoter were among the first eukaryotic genes to be sequenced and intensively studied [115,116]. Using cloning and DNA sequence deletion methods, Palmiter and coworkers established that the MT promoter contained a number of metal response elements (MREs) that were responsive to Cd^{2+}, Zn^{2+}, Cu, Co, Hg^{2+}, and Bi^{3+} and dependent on the activity of the metal ion responsive transcription factor, MTF-1 [117,118]. Later, MTF-1 was characterized by Schaffner and coworkers [119,120]. MTF-1 was demonstrated to be a requirement for MT induction by Zn^{2+} and Cd^{2+} by showing that MTF-1 knockout cells are unresponsive to these metal ions [121].

The primary structure of MTF-1 includes six tandem Zn-finger motifs located toward the N-terminus that serve as its MT promoter binding site [122]. Each is a canonical ββα Zn-finger domain, in which Zn^{2+} coordinates to 2 cysteine thiolate groups and 2 histidine imidazole nitrogens [107,123]. Experiments show that the DNA binding function is responsive to Zn^{2+} in the micromolar range, implying stability constants for the Zn^{2+} sensitive fingers on the order of 10^6 [121,124]. This result leads to a simple equilibrium hypothesis for Zn^{2+} switching between inactive and active conformations: perturbations of intracellular Zn^{2+} shift the ratio of active and inactive MTF-1 as in reaction (21) where MTF-1 is likely to be partially saturated with Zn^{2+}.

$$Zn_m\text{-MTF-1}_{INACTIVE} + n\,Zn^{2+} \rightleftharpoons Zn_{(m+n)}\text{-MTF-1}_{ACTIVE} \quad K \sim \mu M \quad (21)$$

Considering that the native 3-dimensional conformation of Zn-fingers requires the presence of Zn^{2+}, MTF-1 activation results from shifting reaction (21) toward the right. MTF-$1_{INACTIVE}$ is largely confined to the cytoplasm and upon activation, migrates into the nucleus to associate with cognate DNA [125]. The structural details of this reaction such as the identity of the fingers that are responsive to Zn^{2+} and its relationship to nuclear localization and DNA binding remain the subject of investigation [126–128].

Initially, it seemed possible to extend this hypothesis to include Cd^{2+}-dependent activation of MTF-1:

$$Zn_m\text{-MTF-1}_{INACTIVE} + n\,Cd^{2+} \rightleftharpoons Zn_m, Cd_n\text{-MTF-1}_{ACTIVE} \quad K \sim \mu M \qquad (22)$$

However, Andrews et al. showed that *in vitro* addition of Cd^{2+} to cell lysate containing MTF-$1_{INACTIVE}$ did not enhance binding to a test MRE [124]. In more chemically oriented studies, Petering and colleagues demonstrated that Cd^{2+} binding to Zn-finger transcription factor IIIA (TFIIIA) or TFIIIA finger 3 inhibited adduct formation with cognate DNA [113]. Cd^{2+} replacement of Zn^{2+} in a modified TFIIIA finger 3 retained the general ββα conformation but resulted in subtle reorganization of the DNA binding α-helical region consistent with hypothetical loss of DNA binding affinity (Figure 2). Similarly, *in vivo* or *in vitro* exposure to Cd^{2+} depresses binding of Zn-finger transcription factor Sp1 to its native DNA binding sites [129,130]. Thus, direct association of Cd^{2+} with C_2H_2 zinc-finger transcription factors as in reaction (22) does not result in Zn-finger transcription factors that are competent to bind to DNA.

An alternative hypothesis to explain the connection between Cd^{2+} and MTF-1 has been proposed [64]. Upon uptake by cells, Cd^{2+} initially binds to members of the Zn-proteome, including Zn-MT (reaction 23), displacing Zn^{2+}:

$$Cd^{2+} + \text{Zn-proteome} \rightleftharpoons \text{Cd-proteome} + Zn^{2+}_{FREE} \qquad (23)$$

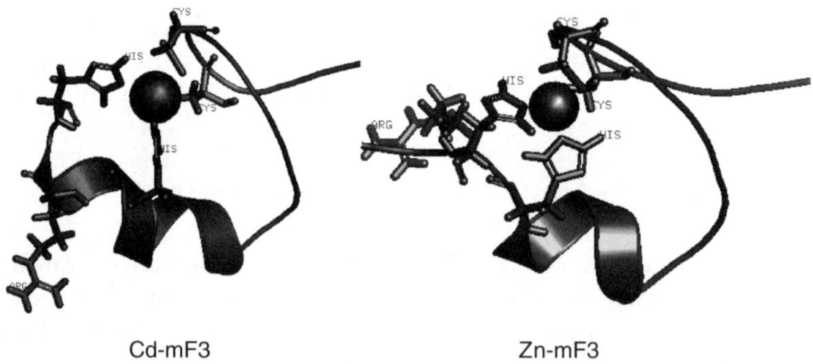

Figure 2. Comparative NMR structures of Cd- and Zn-mF3, a mutant zinc finger 3 of transcription factor IIIA [113]. Cd-mF3 displays changes in orientation of histidine ligands that perturb the recognition helix, shifting the position of an arginine side chain that is important for specific DNA binding.

Zn^{2+}_{FREE} becomes the surrogate or second messenger for Cd^{2+} in relation to MTF-$1_{INACTIVE}$.

$$Zn_m\text{-MTF-}1_{INACTIVE} + n\, Zn^{2+}_{FREE} \rightleftharpoons Zn_{(m+n)}\text{-MTF-}1_{ACTIVE} \quad K \sim \mu M \qquad (24)$$

Support for the hypothesis came from an experiment in which Cd^{2+} and Zn_7-MT were mixed with Zn_m-MTF-$1_{INACTIVE}$. As expected, Cd^{2+} displaced Zn^{2+} from MT making it available to switch MTF-1 to its active conformation (reaction 24). Whether this mechanism operates in cells, where many adventitious ligands compete with MTF-$1_{INACTIVE}$ for Zn^{2+}, is unknown.

5.1.2. *Cd^{2+}, Nephrotoxicity, and Metallothionein*

The fortuitous discovery of Cd-MT in horse kidney offered a molecular foothold in the study of the mechanism of Cd toxicity [1]. Since then, nearly stoichiometric localization of Cd^{2+} in MT has been observed in human kidney tissue as well as countless other animal and cellular model systems involving kidney and other organ systems. Generally, MT protein is induced by the presence of Cd^{2+} and proceeds to efficiently sequester the metal ion in a process that is understood as a mechanism that protects cells and tissues from its deleterious effects. Klaassen, Liu, and Choudhuri estimated that MT affords 10-fold protection such that in its absence background exposure to humans would cause nephrotoxicity [131].

5.1.2.1. Cd-Metallothionein as a Toxic Agent. An enormous number of studies have focused on the MT-based mechanism by which MT protects cells against Cd^{2+}. Fewer have asked how Cd^{2+} causes toxicity in the kidney. A persistent hypothesis during the 1980s and 1990s was that Cd-MT was the toxic form of Cd^{2+} for the kidney [132–134]. This view arose because of the particular model of Cd^{2+} toxicity that was commonly in use at the time. Basically, animals were injected with an overtly toxic Cd^{2+} concentration. This route of exposure results in rapid accumulation of most of the dose in the liver not the kidney. In turn, hepatic Cd^{2+} induced the synthesis of apo-MT and formation of Cd-MT even as it also quickly caused acute liver injury and hepatocyte cell death. Cd-MT released into the plasma was filtered by the kidney glomerulus and reabsorbed in the proximal tubule, where it underwent lysosomal degradation. In effect, a bolus of Cd^{2+} injected into the animal was delivered to the kidney by an indirect mechanism. Once there, it caused severe nephrotic damage.

The problem with the model was twofold: first, it focused on acute toxicity not chronic toxicity as seen in humans. Second, the route of Cd^{2+} administration delivered the metal ion to the liver not the kidney, where primary toxicity occurred. In contrast, ingestion, for example, in humans results in toxicity confined to the kidney [135]. In order to test the role of hepatic Cd-MT as a key source of toxic Cd^{2+} in the kidney, the injection experiment above was repeated in MT-null mice [136,137]. Despite the inability to make hepatic Cd-MT, the experiment still produced acute kidney toxicity. The concentration of Cd^{2+} in the kidney was much lower than in liver but the absence of a means to make endogenous tubular MT greatly increased the proximal tubule's vulnerability to lower levels of Cd^{2+}.

5.1.2.2. Cd^{2+}, Kidney Tubular Cell Toxicity, and Metallothionein. Inquiry into the mechanisms of sub-acute Cd^{2+} nephrotoxicity has been pursued by Petering and co-workers, using as the model system primary kidney cortical cells that have proximal tubular characteristics. At concentrations of Cd^{2+} that cause no overt toxicity, two transport defects have been observed, inhibition of Na^+-dependent glucose and phosphate resorption [138,139]. In each case, reduction in Na^+-nutrient cotransporter mRNA occurred in a Cd^{2+} concentration-dependent manner. Focusing on glucose transport, facilitated by Na^+-glucose cotransporters 1 and 2 (SGLT 1 and 2), subsequent studies showed that Cd^{2+} down-regulated the promoter binding activity of an essential transcription factor, Sp1 [129,130]. Sp1 is a Zn-finger protein, containing 3 tandem Zn-finger motifs [140]. *In vitro*, Cd^{2+} substitutes for Zn^{2+} under nearly stoichiometric concentrations with the loss of DNA-binding activity [130]:

$$Cd^{2+} + \text{Zn-Sp1} \rightleftharpoons \text{Cd-Sp1} + Zn^{2+} \qquad (25)$$

Whether this reaction accounts for Zn-Sp1 downregulation in cells treated with Cd^{2+} remains to be established. Interestingly, comparative exposure to Cd^{2+} and Cd-MT at concentrations that delivered the same amount of Cd^{2+} into the cells resulted in a much greater inhibitory effect of inorganic Cd^{2+} on glucose uptake [141].

The role of MT in this model of cell Cd^{2+} exposure has been investigated. Induction of MT protein synthesis by Cd^{2+}, first noticeable at 4 h post addition of Cd^{2+}, does not prevent depression of SGLT 1 and 2 activity even as it acquires nearly all of the Cd^{2+} that enters the cell over a 24 h period of observation [142]. Moreover, an additional 48 h incubation in Cd^{2+}-free medium does not reverse the damage. In contrast, when MT is preinduced, subsequent exposure of cells to Cd^{2+} largely prevents inhibition of glucose transport [142]. Apparently, the initial, relatively rapid reaction of Cd^{2+} with the proteome described by reaction (5) (see also Section 5.1.1)

$$Cd^{2+}_{IN} + (Zn)\text{-proteome} \rightleftharpoons Cd\text{-proteome} + (Zn^{2+}) \quad (5)$$

$$Cd\text{-proteome} + (\text{apo- or Zn-})MT \rightleftharpoons Cd\text{-}MT + \text{proteome} + (Zn^{2+}) \quad (6)$$

is sufficient to affect sodium-glucose cotransporter synthesis even after cellular MT is elevated. The results of this study indicate that successful competition for Cd^{2+} by MT in reaction (6) may not be sufficient to ablate cell injury. Alternatively, key sites of proteomic Cd^{2+} binding that are involved in the injury process either may not be accessible to or may not undergo reaction with MT.

5.2. Copper Toxicology

Copper is an essential metal ion that is subject to tightly controlled intracellular trafficking mechanisms throughout eukaryotic organisms, involving chaperone proteins that move Cu through specific pathways terminating in binding to functional metalloproteins [143,144]. Cu ion can exist in 2 oxidation states. In order to protect cells from excess Cu that has escaped normal metabolic trafficking routes, MT serves as a relatively inert binding site for Cu(I) [145].

Cu(I) binding by MT has general physiological significance because during fetal development, Cu(I) is stored safely in MT for later dispersal to the rest of the organism during neonatal development [49,146,147]. Genetic diseases related to defects in copper trafficking exist, including Wilson's and Menkes diseases [148]. In both, deranged Cu metabolism leads to tissue increases in Cu that are reflected in the formation of elevated concentrations of Cu-MT [149–152].

The properties of Cu(I)-MT are complicated and controversial. Early titration results showed that each domain of MT could accept 6 Cu(I) ions and it was hypothesized that Cu^{1+} binds with different thiolate coordination and cluster geometry than Zn^{2+} and Cd^{2+} [7,153]. In contrast, assessment of the metal ion-binding stoichiometry of naturally occurring Cu_m,Zn_n-MTs from bovine calf liver showed both that (m + n) could equal 7 over a wide range of m/n ratios [147], indicative of similar coordination by Cu(I) and Zn^{2+}. Nevertheless, species containing more than 7 Cu(I) ions were also evident in some liver samples. Moreover, it was demonstrated that Zn,Cu(I) clusters, like their Cd,Zn cluster counterparts, could be formed by inter-protein metal ion exchange between homogeneous Cu- and Zn-MT proteins (as in reaction 13) [154].

Recent studies on the stoichiometry question resulted in yet another stoichiometry of Cu_8-MT for the homogeneous metal ion protein, with both domains hosting 4 Cu(I) ions [155]. One conclusion from these reports is that MT domains bind Cu(I) in a range of stoichiometries that result in a variety of Cu(I)-thiolate cluster configurations. The ligand substitution reactivity of Cu_m-MT, where m varies from 7 to 12, has been studied using the strong binding Cu(I) ligand, 4,7-ρ-sulfonylphenyl-2,9-dimethyl-1,10-phenantholine (bathophenanthroline disulfonate, BCS) [147,156].

$$\mathrm{Cu(I)_m\text{-}MT + 2m(BCS) \rightleftharpoons m\,Cu(I)(BCS)_2 + apo\text{-}MT} \tag{26}$$

The kinetics of the reaction are biphasic and consistent with relatively slow, rate limiting reactions of BCS with one Cu(I) in each cluster, followed by rapid acquisition of the rest of the Cu(I) by BCS. Despite the huge stability constant of $Cu(I)(BCS)_2$ of approximately 10^{19} at pH 7.4, 200-fold excess ligand relative to Cu(I) needs to be employed to drive the reaction of Cu_7-MT with BCS to completion [156].

The same reaction carried out with Cu_{12}-MT displays quite different properties (reaction 26) [156]. Four to five Cu(I) ions undergo rapid, stoichiometric reaction with BCS, leaving a Cu-MT species that behaves like Cu_7-MT, in that a large excess of BCS is needed to induce further reaction. This study confirms in one experiment the existence of 2 sets of Cu-thiolate clusters derived from Cu_{12}-MT (Cu_6S_9, Cu_6S_{11}) that are more labile and less stable and Cu_{7-8}-MT ($Cu_{3-4}S_9$, Cu_4S_{11}) that are much less reactive with BCS.

An important observation from the study of naturally occurring Cu,Zn-MT was that the copper-containing protein is stable to oxidation [148]. This is a critical property, considering MT's role as a innocuous storage site for redox active Cu. It is also an intriguing property deserving of more study because many thiol compounds are readily oxidized in the presence of catalytic concentrations of Cu and molecular oxygen. Yet, Cu-MT incorporating multiple Cu(I) ions and thiolate groups resists such oxidation.

Lastly, recent studies in cell culture suggest that MT is not necessarily the terminal binding site for MT-bound Cu [54]. Multiple cell types exposed to Cu^{2+} rapidly synthesize MT to form Cu-MT. Once extracellular Cu is removed, MT-bound Cu begins to efflux from the cells, leaving behind apo-MT. The features of the reaction suggest that direct Cu transfer from MT may be a key step in the process.

$$\mathrm{Cu\text{-}MT_{IN\,CELL} \rightarrow apo\text{-}MT_{IN} + Cu_{OUT}} \tag{27}$$

However, the large binding affinity of much of the Cu associated with MT weights against this hypothesis. In light of these results, the possibility that metal ions bound to MT can be exported from mammalian cells *in vivo* needs to be explored.

6. OXIDANT TOXICOLOGY IN RELATION TO METALLOTHIONEIN CHEMISTRY

A classic paper demonstrated that hepatic metallothionein was induced by a diversity of stresses such as heat, cold, bacterial infection, exposure to the irritant, turpentine, etc. [4]. These results suggested that MT was part of the generalized host stress response that includes the response to inflammatory agents. Its induction is coordinated with the rapid depletion of loosely bound plasma Zn^{2+} and its simultaneous accumulation in the liver as Zn-MT [4,157].

The ability of metallothionein to bind and sequester metal ions is complemented by its wide-spread reactivity with oxidants and electrophiles (see Table 1 in Section 4). With its multitude of sulfhydryl groups, MT has both strong reductant capability and more general nucleophilic reactivity. These activities are also suggestive of MT's role as a stress protein that participates in the organism's response to inflammatory agents and the oxidant/electrophilic agents that they induce.

6.1. Oxidant Metabolism and Metallothionein

The interaction of oxidants with metallothionein shifts the focus from MT's metal binding capability to the reactivity of its thiol groups. Table 1 lists a number of studies that implicate MT in organismal response to and protection from oxidant species. Relatively few experiments have been directed toward understanding the chemical reactivity of the sulfhydryl groups in metallothionein. Nevertheless, this is an important subject. One needs to be able to rationalize how this reductant pool that supplies only a fraction of the cell's thiol complement manages to contribute significantly to protection against oxidants as well as electrophiles in the presence of mM concentrations of glutathione and proteomic sulfhydryl groups [54].

6.2. Apo-Metallothionein

Most chemical studies of MT's range of reactivity with oxidant and electrophilic species have been conducted with metal-saturated MT, commonly Zn-MT, before the presence and potential biological significance of apo-MT were well known. Its presence provides a pool of high density sulfhydryl groups that are much more reactive than those participating in metal-thiolate clusters.

Apo-MT is hypothetically the initial product of metallothionein protein synthesis on the ribosome. With a stability constant of $10^{11.2}$ per Zn^{2+} in each domain at pH 7, Zn_7-MT or Cd_7-MT formation rapidly occurs [94,98].

$$7\,Zn^{2+} + \text{apo-MT} \rightleftharpoons Zn_7\text{-MT} \tag{28}$$

Indeed, it has been assumed that in the absence of such a reaction, apo-MT rapidly degrades because the apo-peptide should be unstructured in the absence of Zn^{2+} and susceptible to intracellular proteolysis [158]. Nevertheless, it was shown in 1994 that many cancer cell lines and solid tumors contain large, basal, steady-state concentrations of Zn^{2+}-unsaturated metallothionein [52]. Later, experiments established that normal mammalian tissues that may not contain much MT nevertheless include significant fractions as the metal ion-unsaturated species [53]. Recently, it was reported that a variety of conditions leading to the overexpression of MT such as exposure to dexamethasone, plasmid expression of MT, and even induction of MT by Zn^{2+}, result in metallothionein protein that contains a large fraction of free binding sites [54]. Thus, under basal or elevated conditions of expression of MT, apo-MT is likely to be a significant form of the protein.

This has enormous implications for understanding the physiological functions of metallothionein. Unsaturated binding sites can bind exogenous metal ions and potentially compete for endogenous Zn^{2+} and Cu^{1+}. Metal-free sulfhydryl groups are expected to be much more reactive with oxidants and electrophiles than thiolate residues bound to metal ions in clusters [94].

The apparent steady-state stability of apo-MT is surprising, considering its hypothetical unfolded conformation. Nevertheless, measurement of the rate of apo-MT biodegradation in Zn-deprived Ehrlich cells showed that its half time for turnover was not particularly fast but, instead, was similar to that of the general protein pool [159]. A recent inquiry into the conformation of apo-MT using a computational approach suggested that the protein may adopt a folded structure [160]. The repeated observation that apo-MT migrates like metal ion-saturated MT over Sephadex G-75 as a 10 kDa protein not as a larger, unstructured peptide, gives credence to this hypothesis [161].

6.2.1. Zn-Proteomic Requirements for Apo-Metallothionein's Steady-State Existence in Cells

The persistence of cellular apo-MT, a potent thermodynamic sink for Zn^{2+}, under diverse conditions including induction by Zn^{2+}, itself, is surprising in light of the presence of a reservoir of external nutrient Zn^{2+} and hundreds of μM intracellular Zn^{2+} distributed among thousands of Zn proteins [54,162]. The implications are manifold: (a) either the Zn-proteome is kinetically inert to reaction with apo-MT or the Zn^{2+} stability constants of members of the Zn-proteome exceed that of apo-MT. (b) Extracellular Zn^{2+} is not passively/thermodynamically linked to cellular apo-MT. (c) By extension, the plasma membrane Zn^{2+} transporters, importers, and exporters that control net uptake of Zn^{2+}, are not tightly coupled to the MT thermodynamic sink, such that MT remains under-saturated with Zn^{2+}[54].

The chemical basis for the maintenance of apo-MT in the face of the Zn-proteome has been investigated by observing the extent of reaction of Zn-proteome with a series of competing ligands, L [163]:

$$\text{Zn-proteome} + \text{L} \rightleftharpoons \text{apo-proteome} + \text{Zn-L} \tag{29}$$

Isolated Zn-proteome is unreactive with apo-MT (log $K = 11.2$, pH 7.4) within the uncertainty of the measurement (stability constant). In marked contrast, similar concentrations of several small multidentate ligands, TPEN (15.6), EDTA (13.4), and EGTA (8.8) extract 20–30% of proteomic Zn^{2+}. The difference in reaction of EGTA and apo-MT demonstrates that apo-MT is kinetically inert to reaction with Zn-proteome within the error of the measurement. Another experiment showing that Zn-proteome is relatively unreactive with glutathione (GSH) at a physiological concentration of 2 mM GSH, underscores the peculiarity that a large fraction of the Zn-proteome is recalcitrant to reaction with the two major intracellular competing ligands for Zn^{2+}, apo-MT and GSH. The lack of reactivity of apo-MT with sources of intracellular Zn^{2+} is striking when compared with MT's ability to compete for intracellular Cd^{2+} (Section 5.1.1).

6.2.2. Redox State of Cellular Apo-Metallothionein

Repeated isolation of apo-MT in the presence of minimal O_2 and the absence of thiol reducing agents has demonstrated that the metal-free species exists in reduced form in a variety of cells [52,54]. Another report indicates that little of the metal ion-unsaturated form of MT exists in rat liver as oxidized protein [164]. Considering either apo-MT or Zn-MT as participants in reactions with toxic oxidants, it must be hypothesized that apo-$MT_{OXIDIZED}$ can be re-reduced by a cellular reductant so that it can continue to contribute to redox cycling, much like glutathione [165].

6.3. Metallothionein Reaction with 5,5′-Dithio-bis(2-nitrobenzoate) and Glutathione Disulfide

The thiol-disulfide interchange reaction of metal-bound and apo-MT with the disulfide, 5,5′-dithio-bis(2-nitrobenzoate), DTNB, illustrates much of what is known about MT's sulfhydryl reactivity. Apo-MT reacts with DTNB within the time of mixing [94]. In strong contrast, the rates of reactions of Zn- and Cd-MT with DTNB are much slower and biphasic [96,166,167]. With Zn_7- and Cd_7-MT, each phase is indicative of the independent, cooperative reaction of α- and β-domain clusters with DTNB [96]. In turn, for each phase the rate is dependent on dissociative (k_1) and associative (k_2) processes.

$$k = k_1 + k_2[\mathrm{DTNB}] \tag{30}$$

Thus, in either pathway, an initial event, thought to be metal ion-thiolate bond dissociation for k_1, and initial nucleophile attack of a cluster thiolate on a single DTNB molecule (k_2), is rate-determining. As a result, each cluster reacts as a cooperative unit with DTNB. In the thiol-disulfide interchange written below without Zn^{2+}, the initial step results in mixed disulfide formation and the release of the yellow TNB (5-thiol-2-nitrobenzoate) species:

$$\mathrm{TNB\text{-}S\text{-}S\text{-}TNB} + \mathrm{MT\text{-}S^-} \rightleftharpoons \mathrm{MT\text{-}S\text{-}TNB} + \mathrm{TNB\text{-}S^-} \tag{31}$$

However, the reaction proceeds further because of the intramolecular proximity of 20 thiol groups dispersed along the protein backbone [168]:

$$\mathrm{MT(S^-)\text{-}S\text{-}TNB} \rightleftharpoons \mathrm{MT(S\text{-}S)} + \mathrm{TNB\text{-}S^-} \tag{32}$$

The overall reaction,

$$\mathrm{M_7\text{-}MT(S^-)_{20}} + 10\ \mathrm{TNB\text{-}S\text{-}S\text{-}TNB} \rightleftharpoons \mathrm{MT(S\text{-}S)_{10}} + 20\ \mathrm{TNB\text{-}S^-} + 7\ \mathrm{Zn^{2+}} \tag{33}$$

immediately shows this to be a redox reaction, which is driven by the stronger reduction potential of M_7-MT in comparison with DTNB. Considering that the reaction also involves the unfavorable dissociation of Zn^{2+}, Cd^{2+} ($K \sim 10^{11}$ and 10^{14}, respectively) or even Cu^{1+} from MT, the reduction potential of apo-MT must be strongly negative [147]. The intramolecular oxidation of MT sulfhydryl groups as in reaction (32) favors its low reduction potential.

The kinetics of this reaction are relatively slow because the target thiol groups are wrapped within a peptide structure that together with the clusters shields the sulfur atoms to varying degrees from contact with molecules in the solvent [107,169]. Notably, however, the unimolecular process (k_1)

insures that even at low DTNB concentration, there remains significant thiolate reactivity. That this dissociative step is seen in ligand substitution as well as redox reactions suggests that the clusters undergo rate-limiting thiol-metal ion dissociation processes that are important in multiple reactions [97].

The DTNB reaction described above has been used as a model for the hypothesis that intracellular oxidants, particularly disulfides, might control the metal ion content of metallothionein [170,171]. Thus, it has been suggested that glutathione disulfide (GSSG) might serve this role and release Zn^{2+} for essential cellular processes under oxidative conditions that favor the formation of GSSG. If the reaction works with Zn-MT, it might also take place with Cd,Zn-MT as occurs with DTNB, releasing Cd^{2+} and causing toxicity. However, an earlier study showed that even at mM concentrations of GSSG, the rate of reaction of Zn-MT is exceedingly slow (25% complete in 72 h with 200 μM GSSG) and, therefore, unlikely to be physiologically relevant [168].

6.4. Oxidant Reactivity with Metallothionein: Cluster Thiolate Solvent Accessibility

The details of domain structure provide additional insights into the connection between structure, reactivity, and function. The metal ion-thiolate cluster essentially occupies the interior of each domain. Thus, as the peptide backbone folds about the cluster, the solvent exposure of both metal ions and sulfur ligands is reduced [26,107,169]. Adding the side chains further decreases the solvent accessibility of the clusters (Figure 3, Table 2). The result is that in each domain, only some of the sulfhydryl groups are in contact with water and readily able to undergo reaction. Interestingly, with either mammalian (3 and 4 metal clusters) or crustacean (two 3 metal clusters) MT, the two domains differ substantially in aggregate thiolate solvent exposure (Figure 3). Although the β-domain cluster (M_3S_9) tends to be more reactive with some reagents, this is not uniformly true [96,97]. A hypothesis linking structure and reactivity is that in so far as reactions are bimolecular, differences in cluster solvent accessibility contribute to the rates of such reactions [169].

6.5. Reactions of Oxygen Species with Metallothionein

A handful of studies has shown that Zn- or Cd-MT is highly reactive with superoxide anion, hydrogen peroxide, and hydroxyl radical, three partially reduced forms of O_2 that are implicated in various forms of cell injury, as well as peroxynitrite, another highly toxic species [10,172,173]. With each

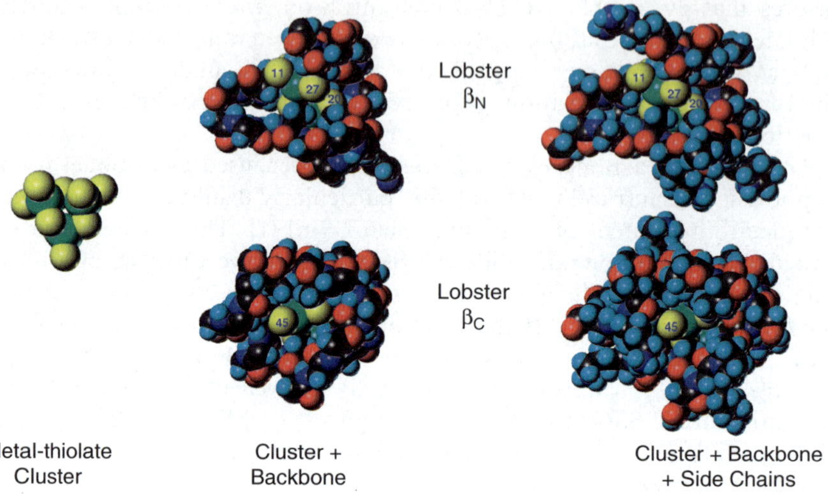

Figure 3. Metallothionein cluster thiolate surface accessible to the solvent in the naked cluster, the cluster surrounded by the peptide backbone, and in the holo-domain. Color code: sulfur atoms (yellow), cadmium ions (green), nitrogens (dark blue), oxygens (red), hydrogens (aqua-blue). Numbering on sulfurs refer to sequence positions of cysteines.

Table 2. Sulfur accessible surface area (Å^2) of metallothionein domains.

	Cluster (A[a])	A + methylene and immediate backbone (B)[b]	A + B + rest of backbone (C)	A + B + C + side chains (D)
Cd_3S_9 Domains				
Lobster, β_N	512	104	62	45
Lobster, β_C	539	108	42	17
Cd_4S_{11} Domains				
Rabbit, α_C holoprotein dimer	600	68	40	22
Rabbit, α_C	638	120	83	50

[a] Naked cluster core.

[b] Core plus -$CH_2CH(NH-)CO$-.

reactant, efficient reaction has been demonstrated with metal ion-bound thiolate groups in the MT clusters.

Only one paper has compared the relative cellular and *in vitro* reactivity of MT with GSH and other protein thiols with a model oxidant [172].

In that report, the effects of H_2O_2 on intracellular thiol pools were determined in the absence and presence of DMSO, added to minimize the contribution of hydroxyl radical to the observed results. Notably, in control cells exposed to 200 μM H_2O_2 for 30 min, the proteomic and GSH thiol pools were reduced 60 and 80%, respectively. In Zn-MT induced cells, the impact of H_2O_2 on these pools declined to 29 and 35%, respectively, with the reactivity of the thiol groups in the MT fraction responsible for this difference. It was clear that selective reaction of hydrogen peroxide with MT protected both proteomic and glutathione sulfhydryl groups. Since MT is a major site of reaction of reactive oxygen species, consideration must be given to the nature of the products and how they might be reduced in the cell to restore MT's sulfhydryl content. The existence of such a thiol redox cycling mechanism would provide a pathway for the ongoing activity of the protein in cellular redox chemistry (Section 6.2.2).

Virtually all of the chemical work in this area has been done with metal ion-saturated MT. The properties of reaction of apo- or metal ion-unsaturated MT with reactive oxygen species have not been described. Nor have cellular studies recognized that the MT is likely to be significantly undersaturated with metal ions, leading to the strong possibility that free MT thiol groups are preferentially reactive within the MT pool.

6.6. Nitric Oxide Species

In contrast to the paucity of studies on the reactions of MT with reactive oxygen species, the reactivity of MT with nitric oxide-related molecules has been the subject of a number of investigations. According to various studies, chemical reactions of Zn- or Cd-MT with NO, NO oxidation products, or S-nitrosyl thiolates has resulted in thiol modification and Zn^{2+} release [174–177]. These results have been used to rationalize rapid NO-dependent mobilization of intracellular Zn^{2+} by NO in terms of specific interaction of NO species with Zn-MT [178,179]. Direct support for this hypothesis was derived from an experiment investigating the effect of NO on cells containing a MT chimera assembled from MT with cyan fluorescence protein attached to its N-terminus and enhanced yellow fluorescence protein to its C-terminal end [180]. The native, Zn-containing protein displays fluorescence resonance energy transfer (FRET) between the two fluorophores. However, upon exposure to NO, the FRET property was lost, indicative of the destruction of the cluster structure and the unfolding of the protein.

In other experiments aimed at clarifying the reactivity of Zn-, Cd-, and apo-MT with the variety of nitric oxide species, Zhu et al. found that Zn-MT was unreactive with NO under anaerobic conditions and RS-NO (S-nitrosyl-penicillamine and S-nitrosyl-glutathione) in the absence of photochemical dissociation of the S-nitrosyl bond [54,181]. NO in the presence of O_2 forms NO_2 and N_2O_3; NO plus PTIO (2-phenyl-4,4,5,5-tetramethylimidazoline-1-oxyl-3-oxide), an oxygen atom donor, converts NO to NO_2. Either reaction mixture slowly oxidizes the sulfhydryl groups of Zn-MT to the level of disulfides.

$$\text{Zn-MT}(\text{S}^-)_2 + 2\,\text{NO}_2 \rightarrow \text{Zn}^{2+} + \text{MT(S-S)} + 2\,\text{NO}_2^- \qquad (34)$$

Notably, glutathione competes effectively with Zn-MT for reactive NO species [181,182]. In contrast to the sluggish reactivity of Zn-MT, apo-MT reacts within the time of mixing with S-nitrosyl-penicillamine and anaerobic NO [54,181].

A cellular study examined the comparative reactivity of aerobic NO with thiol groups of the general proteome, Zn- and apo-MT, and glutathione [54]. Within the MT pool, only apo-MT underwent reaction. Strikingly, proteomic and glutathione sulfhydryl groups also were oxidized and the extent of reaction of the three thiol sources was directly proportional to their intracellular relative concentrations. Thus, with this oxidizing agent, no preference for MT was observed. The results underscore the reactivity of the proteome pool of SH groups, collectively a larger concentration of sulfhydryl groups than either MT or GSH. In light of these results, a report that plasmid driven overexpression of MT protects cells from NO exposure suggests that MT concentration was so large that it became the dominant thiol pool that reacts with NO oxidation products [183].

6.7. Arsenic and Chromate Compounds

Environmental exposure to arsenic is a worldwide problem because of geological contamination of groundwater [184]. Arsenite is classified as a human carcinogen [185]. Chromate is a common contaminant resulting from industrial activity and concern exists about occupational as well as general population exposure because of its carcinogenic potential [186]. The recent identification of arsenic trioxide as an effective antineoplastic drug against some blood cancers broadens the toxicological significance of arsenic compounds [187]. Chemical forms of each element exist in different oxidation states and display reactivity with sulfhydryl containing compounds. Thus, consideration of their reactions with metallothionein need to be part of the program to understand the cellular chemistry of these toxic agents.

6.7.1. *Arsenic Species*

Arsenic displays a rich chemistry characterized by multiple oxidation states and the involvement of organic species [188,189]. Many studies show that AsO_3^{3-} and AsO_4^{3-} induce oxidative stress in exposed cells [190,191]. Thus, metallothionein may act as a protective agent by intercepting reactive oxygen species. Alternatively, the strong reactivity of arsenite with sulfhydryl groups suggests that arsenic-thiol adducts may contribute to toxicity as well as protection [192,193]. Therefore, MT might protect cells by forming thiol adducts with arsenic species.

Two studies have investigated the reactions of MT with arsenite and its methylated metabolites [194,195]. Apo-MT SH groups react with arsenite in a succession of substitution reactions leading to binding stoichiometries of As_3S_{11} and As_3S_9 for the α- and β-domains, respectively [194]. Most likely, the resultant species involve As bound to three thiolate groups:

$$AsO_3^{3-} + 3\ MT(S^-)_3 + 3\ H^+ \rightleftharpoons AsS_3\text{-}MT + 3\ OH^- \qquad (35)$$

This would be consistent with the conclusion that As binds to independent sites and not within As-thiolate clusters [194].

A second study included the examination of the reaction of methylated species, $CH_3As(OH)_2$ and $(CH_3)_2AsOH$ [195]. Apo-MT can bind as many as 10 monomethyl and 20 dimethyl arsenical molecules as expected based on the substitution chemistry of reaction (35). Neither report [194,195] investigated the reactivity of Zn-MT with these reagents. Indeed, each study utilized low pH in order to conduct electrospray mass spectrometry on the products. As a result, MT's thiol groups were protonated and deactivated in comparison with physiological pH, where these substitution reactions would be more favorable.

6.7.2. *Chromate*

Chromate is activated to cause cell injury through reduction [186]. Thiol species such as glutathione have been considered potential intracellular reductants [196,197]. MT might serve this role as well. Chromate is very slowly reduced by Zn-MT's sulfhydryl groups [197]. By comparison, apo-MT reacts rapidly with CrO_4^{2-}, beginning with a binding step considered to be the formation of an apo-MT-Cr(VI) adduct [197]. In analogy to the reaction of other thiol compounds with CrO_4^{2-}, it is presumed that thiol groups have replaced oxygens in the Cr coordination sphere:

$$MT(SH)_3 + CrO_4^{2-} \rightleftharpoons MT\text{-}S_3Cr(VI)(O)^{1+} + 3\ OH^- \qquad (36)$$

Accompanying adduct formation is the simultaneous appearance of a Cr(V) species, indicative of thiolate-Cr(VI) oxidation-reduction. As Cr(VI) is converted to Cr(III), the ultimate product of reaction, superoxide and hydroxyl radicals are spin-trapped. A Cr(III)-MT adduct is detected initially by ESR spectroscopy but gradually dissociates as MT sulfhydryl groups undergo oxidation.

In all respects, rapidity of formation of the initial adduct species, the appearance of Cr(V), and the overall rate of Cr(III) production, apo-MT is much more potent than glutathione in its reaction with chromate. Thus, the possibility needs to be considered that apo-MT may contribute to the reductive activation not inhibition of this toxic agent.

7. ELECTROPHILE TOXICOLOGY AND METALLOTHIONEIN CHEMISTRY

Besides oxidants, metallothionein undergoes reaction with a variety of electrophiles including N-ethylmaleimide, acrolein, acetaldehyde, and nitrogen mustards [198–201]. Such reactions provide one rationale for the protection that MT affords against polycyclic aromatic hydrocarbon carcinogenesis (see Table 1 in Section 4), namely that it intercepts activated electrophilic epoxides. However, it might also antagonize the promotional process that is widely thought to involve inflammation and the production of oxy radical species [202].

Among electrophilic reagents, the reaction of various MT species with *cis*-dichlorodiammine Pt(II) (*cis*-DDP) has received the greatest attention [203–206]. Following the demonstration that a large fraction of intracellular Pt becomes bound to MT upon reaction with Zn-MT, metallothioncin has been considered part of the cancer cell's defense against electrophilic chemotherapeutic agents such as *cis*-DDP and nitrogen mustards [200,207].

7.1. Metallothionein and Cancer Pathogenicity and Chemotherapy

The exquisitely simple square planar complex *cis*-DDP has been used intensively for decades to treat a variety of solid cancers [208]. Accompanying its strong anticancer activity are serious side effects that include nephro- and ototoxicity [208]. It is thought that its principal mechanism of action involves *cis*-complexation with N^7 nitrogens on proximate guanine residues of DNA [209]. The relatively weak nucleophilicity of these ring nitrogens makes it necessary for *cis*-DDP to dissociate and undergo aquation in the rate-limiting reaction for DNA binding:

$$Cl_2(NH_3)_2Pt(II) + H_2O \rightleftharpoons (H_2O)Cl(NH_3)_2Pt(II) + Cl^- \quad k_{-1} \tag{37}$$

$$(H_2O)Cl(NH_3)_2Pt(II) + H_2O \rightleftharpoons (H_2O)_2(NH_3)_2Pt(II) + Cl^- \quad k_{-2} \tag{38}$$

$$(H_2O)_2(NH_3)_2Pt(II) + N^7\text{-}N^7(DNA) \rightleftharpoons (N^7)_2(NH_3)_2Pt(II) + 2\,H_2O \quad \text{fast} \tag{39}$$

Because of the necessity to form the aquo-Pt(II) species in the overall reaction, it is possible for stronger cellular nucleophiles to compete with DNA for Pt(II) binding by undergoing direct substitution [209]. The obvious choices are glutathione, a ubiquitous thiol-containing tripeptide that is found in mM concentration in most cells and metallothionein with its high density of sulfhydryl groups. In both cases, their reactions with *cis*-DDP are bimolecular even with the dichloro form of the complex and occur much faster than the reaction of DNA with the drug [203,204]. Thus, cluster thiolate groups directly react with *cis*-DDP. Direct comparison of their quantitative rate expressions reveals somewhat surprisingly that either Cd_7- or Zn_7-MT reacts faster than GSH with *cis*-DDP. The products of reaction with the metallothioneins appear to be Pt_{10}- and Pt_7-MT, respectively. In either case, bridging thiolate ligands are necessary to accommodate the S_4 ligand set required of each Pt(II). Considering that Pt(II) forms rigorously square planar complexes, the interesting question arises how the peptide folds about the presumed Pt-thiolate clusters.

7.2. Bi^{3+} and Metallothionein

Bi^{3+} intersects with cellular metallothionein chemistry in a peculiar way. As a hard, +3 metal ion that prefers oxygen ligands, little thought was given to its reaction with metallothionein until reports appeared showing that oral administration of insoluble bismuth subnitrate resulted in organ specific protection against the principal, non-tumor toxicity of *cis*-dichlorodiammine Pt(II) (kidney), doxorubicin (heart), and X-radiation (bone marrow) [210–212]. These studies demonstrated that metallothionein had been induced in each target tissue and hypothesized that elevated concentrations of MT provided protection through direct reaction with the platinum drug or secondary reaction with products such as reactive oxygen species generated by doxorubicin and γ-radiation. Remarkably, the protection afforded by MT did not compromise the antitumor activity of these agents.

These unexpected findings stimulated an inquiry into the possibility that Bi^{3+} can undergo complexation with MT. *In vitro* titration of Zn_7-MT with

Bi-citrate, a soluble form of Bi^{3+}, resulted in the formation of Bi_7-MT [213]:

$$7\ \text{Bi-citrate} + \text{Zn}_7\text{-MT} \rightleftharpoons \text{Bi}_7\text{-MT} + 7\ \text{Zn-citrate} \qquad (40)$$

The resultant Bi-substituted MT is highly stable as judged by its stability at pH 1. That being the case, whether Bi-MT can readily react with *cis*-DDP or other electrophilic or oxidant species as implied from *in vivo* studies remains to be determined.

Incubation of cells with Bi-citrate results in MT induction in several cell types [54,214]. In human U373 glioblastoma cells, Bi-MT is the principal product. In contrast, pig kidney LLC-PK_1 cells accumulate Bi^{3+} but not in its MT pool. As a result, a large concentration of apo-MT is observed. Since none of the above mouse studies examined the metallation state of MT induced by Bi-subnitrate exposure, it is possible that substantial concentrations of apo-MT were produced that were highly reactive with *cis*-DDP and other agents.

8. GENERAL CONCLUSIONS

Metallothionein is a key member of the cell's defense mechanisms against multiple toxic agents, ranging from metal ions to oxidants and electrophiles. MT's unique array of thiol groups strung along its short peptide backbone supports highly efficient metal ion binding and robust nucleophilic and redox reaction with toxic agents that may not be duplicated by other monothiol compounds such as glutathione.

Structure-reactivity studies have begun to link the chemistry of the protein molecule to empirical information about its protective role in cells. However, considerably more effort needs to be devoted to understanding the variety and structural basis of its chemistry. The unexpected, widespread existence of cellular apo- or unsaturated MT complicates this task since the apoprotein contains a complement of much more reactive thiol groups that may play a dominant role in reaction with toxicants. More generally, the presence of apo-MT in cells offers researchers a bridge between studies of MT in relation to cell toxicology and inquiries into its participation in the trafficking mechanisms of biologically essential Zn^{2+} and Cu^{1+}.

ACKNOWLEDGMENTS

Current research described in this chapter was supported by the National Institute of Environmental Health Sciences through grants ES-04026 and ES-04184.

ABBREVIATIONS

Apo-MT	metal ion unsaturated or metal ion free metallothionein
BCS	bathocuproine disulfonate = 4,7-ρ-sulfonylphenyl-2,9-dimethyl-1,10-phenantholine
CA	carbonic anhydrase
Cd-MT	metallothionein with Cd^{2+} bound without reference to stoichiometry of metal ion binding. The same convention applies to all metal ion-metallothionein species.
cis-DDP	*cis*-dichlorodiammine platinum (II)
DMSO	dimethylsulfoxide
DTNB	5,5′-dithio-bis(2-nitrobenzoate)
EDTA	ethylenediammine-N,N,N′,N′-tetraacetate
EGTA	[2,2′-oxypropylene-dinitrilo]tetraacetate
FRET	fluorescence resonance energy transfer
GSH	glutathione
GSSH	glutathione disulfide
MRE	metal response element
MT	metallothionein
MTF-1	metal ion responsive transcription factor
PTIO	2-phenyl-4,4,5,5-tetramethylimidazoline-1-oxyl-3-oxide
SH	sulfhydryl group
Sp1	common transcription factor
TFIIIA	transcription factor IIIA
TNB	5-thio-2-nitrobenzoate
TPEN	N,N,N′,N′-(2-pyridylethyl)ethylenediammine

REFERENCES

1. M. Margoshes and B. L. Vallee, *J. Am. Chem. Soc.*, 1957, **79**, 4813–4814.
2. J. H. Kägi and B. L. Vallee, *J. Biol. Chem.*, 1961, **236**, 2435–2442.
3. L. Friberg, M. Piscator, G. F. Nordberg and T. Kjellstrom, *Cadmium in the Environment*, C.R.C. Press, Cleveland, 1974.
4. S. H. Oh, J. T. Deagen, P. D. Whanger and P. H. Weswig, *Am. J. Physiol.*, 1978, **234**, E282–285.
5. C.D. Klaassen and S. Z. Kagen, *Adv. Exp. Med. Biol.*, 1982, **136A**, 663–646.
6. L. Endresen, A. Bakke and H. E. Rugstad, *Cancer Res.*, 1983, **43**, 2918–2926.
7. K. B. Nielson and D. R. Winge, *J. Biol. Chem.*, 1984, **259**, 4941–4946.
8. D. Holt, L. Magos and M. Webb, *Chem. Biol. Interact.*, 1980, **32**, 125–135.
9. J. E. Laib, C. F. Shaw 3rd, D. H. Petering, M. K. Eidsness, R. C. Elder and J. S. Garvey, *Biochemistry*, 1985, **24**, 1977–1986.
10. P. J. Thornalley and M. Vašák, *Biochim. Biophys. Acta*, 1985, **827**, 36–44.

11. W. R. Bernhard, M. Vašák and J. H. Kägi, *Biochemistry*, 1986, **25**, 1975–1980.
12. M. G. Cherian, *Toxicology*, 1980, **17**, 225–231.
13. K. S. Squibb, R. J. Cousins and S. L. Feldman, *Biochem. J.*, 1977, **164**, 223–228.
14. D. R. Winge, B. L. Geller and J. Garvey, *Arch. Biochem. Biophys.*, 1981, **208**, 160–166.
15. J. W. Baumann, J. Liu, Y. P. Liu and C. D. Klaassen, *Toxicol. Appl. Pharmacol.*, 1991, **110**, 47–54.
16. A. C. Mello-Filho, L. S. Chubatsu and R. Meneghini, *Biochem. J.*, 1988, **256**, 475–479.
17. S. Z. Cagen and C. D. Klaassen, *Toxicol. Appl. Pharmacol.*, 1979, **51**, 107–116.
18. http://beta.uniprot.org/docs/metallo and http://ca.expasy.org/cgi-bin/sprot-search-de?metallothionein.
19. J. D. Otvos and I. M. Armitage, *Proc. Natl. Acad. Sci. USA*, 1980, **77**, 7094–7098.
20. J. D. Otvos, R. W. Olafson and I. M. Armitage, *J. Biol. Chem.*, 1982, **257**, 2427–2431.
21. B. A. Masters, C. J. Quaife, J. C. Erickson, E. J. Kelly, G. J. Froelick, B. P. Zambrowicz, R. L. Brinster and R. D. Palmiter, *J. Neurosci.*, 1994, **14**, 5844–5857.
22. C. J. Quaife, S. D. Findley, J. C. Erickson, G. J. Froelick, E. J. Kelly, B. P. Zambrowicz and R. D. Palmiter, *Biochemistry*, 1994, **33**, 7250–7259.
23. S. Somji, S. H. Garrett, M. A. Sens, V. Gurel and D. A. Sens, *Toxicol. Sci.*, 2004, **80**, 358–366.
24. M. A. Sens, S. Somji, S. H. Garrett, C. L. Beall and D. A. Sens, *Am. J. Pathol.*, 2001, **159**, 21–26.
25. A. Arseniev, P. Schultze, E. Wörgötter, W. Braun, G. Wagner, M. Vašák, J. H. Kägi and K. Wüthrich, *J. Mol. Biol.*, 1988, **201**, 637–657.
26. A. H. Robbins, D. E. McRee, M. Williamson, S. A. Collett, N. H. Xuong, W. F. Furey, B. C. Wang and C. D. Stout, *J. Mol. Biol.*, 1991, **221**, 1269–1293.
27. Z. Zhu, E. F. DeRose, G. P. Mullen, D. H. Petering and C. F. Shaw 3rd, *Biochemistry*, 1994, **33**, 8858–8865.
28. S. S. Narula, M. Brouwer, Y. Hua and I. M. Armitage, *Biochemistry*, 1995, **34**, 620–631.
29. Y. Uchida and M. Tomonaga, *Brain Res.*, 1989, **481**, 190–193.
30. Y. Uchida, K. Takio, K. Titani, Y. Ihara and M. Tomonaga, *Neuron*, 1991, **7**, 337–347.
31. D. W. Hasler, L. T. Jensen, O. Zerbe, D. R. Winge and M. Vašák, *Biochemistry*, 2000, **39**, 14567–14575.
32. P. Palumaa, I. Tammiste, K. Kruusel, l. Kangur, H. Jörnvall and R. Sillard, *Biochim. Biophys. Acta*, 2005, **1747**, 205–211.
33. A. K. Sewell, L. T. Jensen, J. C. Erickson, R. D. Palmiter and D. R. Winge, *Biochemistry*, 1995, **34**, 4740–4747.
34. A. Muñoz, D. H. Petering and C. F. Shaw 3rd, *Inorg. Chem.*, 2000, **39**, 6114–6123.
35. W. Kennette, O. M. Collins, R. K. Zalups and J. Koropatnick, *Toxicol. Sci.*, 2005, **88**, 602–613.

36. A. Pastore, G. Federici, E. Bertini and F. Piemonte, *Clin. Chim. Acta*, 2003, **333**, 19–39.
37. R. Dringen, *Prog. Neurobiol.*, 2000, **62**, 649–671.
38. M. Knipp, G. Meloni, B. Roschitzki and M. Vašák, *Biochemistry*, 2005, **44**, 3159–3165.
39. D. Grebić, H. Jakovac, I. Mrakovcić-Sutić, J. Tomac, A. Bulog, V. Micović and B. Radosević-Stasić, *Histol. Histopathol.*, 2007, **22**, 593–601.
40. M. Penkowa, P. Keller, C. Keller, J. Hidalgo, M. Giralt and B. K. Pedersen, *Exp. Physiol.*, 2005, **90**, 477–486.
41. G. Trendelenburg, K. Prass, J. Priller, K. Kapinya, A. Polley, C. Muselmann, K. Ruscher, U. Kannbley, A. O Schmitt, S. Castell, F. Wiegand, A. Meisel, R. Rosenthal and U. Dirnagl, *J. Neurosci.*, 2002, **15**, 5879–5888.
42. A. Ajjimaporn, J. Swinscoe, S. Shavali, P. Govitrapong and M. Ebadi, *Brain Res. Bull.*, 2005, **67**, 466–475.
43. M. J. Pérez and A. I. Cederbaum, *Free Rad. Biol. Med.*, 2003, **34**, 443–455.
44. L. Cai, S. Iskander, M. G. Cherian and R. R. Hammond, *Toxicol. Lett.*, 2004, **146**, 217–226.
45. C. P. Yao, J. W. Allen, D. R. Conklin and M. Aschner, *Brain Res.*, 1999, **818**, 414–420.
46. Y. Liu, J. Liu, M. B. Iszard, G. K. Andrews, R. D. Palmiter and C. D. Klaassen, *Toxicol. Appl. Pharmacol.*, 1995, **135**, 222–228.
47. Y. J. Kang, *Cardiovasc. Toxicol.*, 2007, **7**, 95–100.
48. S. Onosaka and M. G. Cherian, *Toxicology*, 1982, **23**, 11–20.
49. R. Mason, A. Bakka, G. P. Samarawickrama and M. Webb, *Br. J. Nutr.*, 1981, **45**, 375–389.
50. D. H. Petering, J. Loftsgaarden, J. Schneider and B. Fowler, *Environ. Health Perspect.*, 1984, **54**, 73–81.
51. A. Pattanaik, C. F. Shaw 3rd, D. H. Petering, J. Garvey and A. J. Kraker, *J. Inorg. Biochem.*, 1994, **54**, 91–105.
52. M. G. Cherian, A. Jayasurya and B. H. Bay, *Mutat. Res.*, 2003, **533**, 201–209.
53. Y. Yang, W. Maret and B. L. Vallee, *Proc. Natl. Acad. Sci. USA*, 2001, **98**, 5556–5565.
54. D. H. Petering, J. Zhu, S. Krezoski, J. Meeusen, C. Kiekenbush, S. Krull, T. Specher and M. Dughish, *Exp. Biol. Med. (Maywood)*, 2006, **231**, 1528–1534.
55. A. E. Michalska and K. H. Choo, *Proc. Natl. Acad. Sci. USA*, 1993, **90**, 8088–8092.
56. E. J. Kelly, C. J. Quaife, G. J. Froelick and R. D. Palmiter, *J. Nutr.*, 1996, **126**, 1782–1790.
57. G. K. Andrews and J. Geiser, *J. Nutr.*, 1999, **129**, 1643–1648.
58. J. J. McAuliffe, B. Joseph, E. Hughes, L. Miles and C. V. Vorhees, *Brain Res.*, 2008, **1190**, 175–185.
59. E. D. Levin, C. Perraut, N. Pollard and J. H. Freedman, *Physiol. Behav.*, 2006, **87**, 513–518.
60. N. Itoh, H. Shibayama, M. Kanekiyo, D. Namphung, T. Nakanishi, A. Matsuyama, T. Odani and K. Tanaka, *Toxicology*, 2005, **216**, 188–196.

61. M. Kanekiyo, N. Itoh, A. Kawasaki, A. Matsuyama, K. Matsuda, T. Nakanishi and K. Tanaka, *Biochem. J.*, 2002, **361**, 363–369.
62. Z. L. Tang, K. J. Wasserloos, X. Liu, M. S. Stitt, I. J. Reynolds, B. R. Pitt and C. M. St. Croix, *Mol. Cell. Biochem.*, 2002, **234–235**, 211–217.
63. D. A. Wiseman, S. M. Wells, M. Hubbard, J. E. Welker and S. M. Black, *Am. J. Physiol. Lung Cell. Mol. Physiol.*, 2007, **292**, L165–L177.
64. B. Zhang, O. Georgiev, M. Hagmann, C. Günes, M. Cramer, P. Faller, M. Vašák and W. Schaffner, *Mol. Cell. Biol.*, 2003, **23**, 8471–8485.
65. I. El Ghazi, B. L. Martin and I. M. Armitage, *Exp. Biol. Med. (Maywood)*, 2006, **231**, 1500–1506.
66. Y. Liu, J. Liu, S. M. Habeebu, M. P. Waalkes and C. D. Klaassen, *Toxicol. Sci.*, 2000, **57**, 167–176.
67. S. S. Habeebu, J. Liu, Y. Liu and C. D. Klaassen, *Toxicol. Sci.*, 2000, **56**, 211–219.
68. M. Yoshida, M. Satoh, A. Shimada, E. Yamamoto, A. Yasutake and C. Tohyama, *Toxicology*, 2002, **175**, 215–222.
69. M. P. Waalkes, J. Liu, R. A. Goyer and B. A. Diwan, *Cancer Res.*, 2004, **64**, 7766–7772.
70. J. Liu, Y. Liu, R. A. Goyer, W. Achanzar and M. P. Waalkes, *Toxicol. Sci.*, 2000, **55**, 460–467.
71. G. Jia, H. Sone, N. Nishimura, M. Satoh and C. Tohyama, *Int. J. Oncol.*, 2004, **25**, 325–333.
72. T. Takano, K. Inoue, R. Yanagisawa, M. Sato, A. Shimada, T. Morita, M. Sawada, K. Nakamura, C. Sanbongi and T. Yoshikawa, *Thorax*, 2004, **59**, 1057–1062.
73. K. Inoue, H. Takano, R. Yanagisawa, M. Sakurai, T. Ichinose, K. Sadakane, K. Hiyoshi, M. Sato, A. Shimada, M. Inoue and T. Yoshikawa, *Exp. Biol. Med. (Maywood)*, 2005, **230**, 75–81.
74. C. D. Tran, H. Huynh, M. van den Berg, M. van der Pas, M. A. Campbell, J. C. Philcox, P. Coyle, A. M. Rofe and R. N. Butler, *Helicobacter*, 2003, **8**, 533–541.
75. J. Carrasco, M. Penkowa, M. Giralt, J. Camats, A. Molinero, I. L. Campbell, R. D. Palmiter and J. Hidalgo, *Neurobiol. Dis.*, 2003, **13**, 22–36.
76. M. Penkowa, M. Caceres, R. Borup, F. C. Nielsen, C. B. Poulsen, A. Quintana, A. Molinero, J. Carrasco, S. Florit, M. Giralt and J. Hidalgo, *J. Neurosci. Res.*, 2006, **84**, 1452–1474.
77. J. Liu, Y. Liu, D. Hartley, C. D. Klaassen, S. E. Shehin-Johnson, A. Lucas and S. D. Cohen, *J. Pharmacol. Exp. Ther.*, 1999, **289**, 580–586.
78. Y. Kondo, E. S. Woo, A. E. Michalska, K. H. Choo and J. S. Lazo, *Cancer Res.*, 1995, **55**, 2021–2023.
79. Y. Kondo, J. M. Rusnak, D. G. Hoyt, C.E. Settineri, B. R. Pitt and J. S. Lazo, *Mol. Pharmacol.*, 1997, **52**, 195–201.
80. J. S. Lazo, Y. Kondo, D. Dellapiazza, A. E. Michalska, K. H. Choo and B. R. Pitt, *J. Biol. Chem.*, 1995, **270**, 5506–5510.
81. K. Hanada, *J. Dermatol. Sci.*, 2000, **23**(Suppl. 1), S51–S56.
82. W. H. Wang, L. F. Li, B. X. Zhang and X. Y. Lu, *Clin. Exp. Dermatol.*, 2004, **29**, 57–61.

83. D. X. Deng, L. Cai, S. Chakrabarti and M. G. Cherian, *Toxicology*, 1999, **134**, 39–49.
84. S. R. Davis, D. A. Samuelson and R. J. Cousins, *J. Nutr.*, 2001, **131**, 215–222.
85. B. Zhang, M. Satoh, N. Nishimura, J. S. Suzuki, H. Sone, Y. Aoki and C. Tohyama, *Cancer Res.*, 1998, **58**, 4044–4046.
86. M. P. Waalkes, J. Liu, K. S. Kasprzak and B. A. Diwan, *Int. J. Cancer*, 2006, **119**, 28–32.
87. Y. Kondo, S. Himeno, W. Endo, M. Mita, Y. Suzuki, K. Nemoto, M. Akimoto, J. S. Lazo and N. Imura, *Carcinogenesis*, 1999, **20**, 1625–1627.
88. O. Barbier, G. Jacquillet, M. Tauc, M. Cougnon and P. Poujeol, *Nephron. Physiol.*, 2005, **99**, 105–110.
89. G. Kazantzis, *Biometals*, 2004, **17**, 493–498.
90. *Cadmium, Environmental Health Criteria 134*, World Health Organization, Geneva, 1992, pp. 152–154.
91. D. R. Winge, R. Premakumar and K. V. Rajagopalan, *Arch. Biochem. Biophys.*, 1975, **170**, 242–252.
92. K. B. Nielson and D. R. Winge, *J. Biol. Chem.*, 1983, **258**, 13063–13069.
93. M. Vašák and J. H. Kägi, *Proc. Natl. Acad. Sci. USA*, 1981, **78**, 6709–6713.
94. J. Ejnik, J. Robinson, J. Zhu, H. Försterling, C. F. Shaw and D. H. Petering, *J. Inorg. Biochem.*, 2002, **88**, 144–152.
95. K. E. Rigby Duncan and M. J. Stillman, *FEBS J.*, 2007, **274**, 2253–2261.
96. A. Muñoz, D. H. Petering and C. F. Shaw 3rd, *Inorg. Chem.*, 1999, **38**, 5655–5659.
97. T. Gan, A. Muñoz, C. F. Shaw 3rd and D. H. Petering, *J. Biol. Chem.*, 1995, **270**, 5339–5345.
98. J. D. Otvos, D. H. Petering and C. F. Shaw, *Comments Inorg. Chem.*, 1989, **9**, 1–35.
99. D. H. Petering and C. F. Shaw 3rd, *Methods Enzymol.*, 1991, **205**, 475–484.
100. B. Roschitzki and M. Vašák, *Biochemistry*, 2003, **42**, 9822–9828.
101. A. Krezel and W. Maret, *J. Am. Chem. Soc.*, 2007, **129**, 10911–10921.
102. S. Lindskog and P. O. Nyman, *Biochem. Biophys. Acta*, 1964, **85**, 464–474.
103. K. A. McCall and C. A. Fierke, *Biochemistry*, 2004, **43**, 3979–3986.
104. K. A. Dill, *Biochemistry*, 1990, **29**, 7133–7155.
105. D. G. Nettesheim, H. R. Engeseth and J. D. Otvos, *Biochemistry*, 1985, **24**, 6744–6751.
106. J. D. Otvos, X. Liu, H. Li, G. Shen and M. Basti, in *Metallothionein-III*, Ed. K. I. Suzuki, N. Imura and M. Kimura, Birkhäuser Verlag, Basel, 1993, pp. 57–74.
107. J. W. Ejnik, A. Muñoz, E. DeRose, C. F. Shaw 3rd and D. H. Petering, *Biochemistry*, 2003, **42**, 8403–8410.
108. M. J. Stillman, W. Cai and A. J. Zelazowski, *J. Biol. Chem.*, 1987, **262**, 4538–4548.
109. J. Ejnik, C. F. Shaw III and D. H. Petering, submitted for publication.
110. J. Ejnik, A. Muñoz, T. Gan, C. F. Shaw, 3rd, and D. H. Petering, *J. Biol. Inorg. Chem.*, 1999, **4**, 784–90.
111. A. Muñoz, F. Laib, D. H. Petering and C. F. Shaw 3rd, *J. Biol. Inorg. Chem.*, 1999, **4**, 495–507.
112. G. Roesijadi, R. Bogumil, M. Vašák and J. H. Kägi, *J. Biol. Chem.*, 1998, **273**, 17425–17432.

113. D. Krepkiy, F. H. Försterling and D. H. Petering, *Chem. Res. Toxicol.*, 2004, **17**, 863–870.
114. M. Huang, D. Krepkiy, W. Hu and D. H. Petering, *J. Inorg. Biochem.*, 2004, **98**, 775–785.
115. N. Glanville, D. M. Durnam and R. D. Palmiter, *Nature*, 1981, **292**, 267–369.
116. G. W. Stuart, P. F. Searle, H. Y. Chen, R. L. Brinster and R. D. Palmiter, *Proc. Natl. Acad. Sci. USA*, 1984, **81**, 7318–7322.
117. G. W. Stuart, P. F. Searle and R. D. Palmiter, *Nature*, 1985, **317**, 828–831.
118. R. D. Palmiter, *Proc. Natl. Acad. Sci. USA*, 1994, **91**, 1219–1223.
119. G. Westin and W. Schaffner, *EMBO J.*, 1988, **7**, 3763–3770.
120. F. Radtke, R. Heuchel, O. Georgiev, M. Hergersberg, M. Gariglio, Z. Dembic and W. Schaffner, *EMBO J.*, 1993, **12**, 1355–1362.
121. R. Heuchel, F. Radtke, O. Georgiev, G. Stark, M. Aguet and W. Schaffner, *EMBO J.*, 1994, **13**, 2870–2875.
122. F. Radtke, O. Georgiev, H. P. Müller, E. Brugnera and W. Schaffner, *Nucleic Acids Res.*, 1995, **23**, 2277–2286.
123. N. P. Pavletich and C. O. Pabo, *Science*, 1991, **252**, 809–817.
124. D. Bittel, T. Dalton, S. L. Samson, L. Gedamu and G. K. Andrews, *J. Biol. Chem.*, 1998, **273**, 7127–7133.
125. I. V. Smirnova, D. C. Bittel, R. Ravindra and H. Jiang, *Andrews J. Biol. Chem.*, 2000, **275**, 9377–9384.
126. X. Chen, A. Agarwal and D. P. Giedroc, *Biochemistry*, 1998, **37**, 11152–11161.
127. T. P. Dalton, D. Bittel and G. K. Andrews, *Mol. Cell. Biol.*, 1997, **17**, 2781–2789.
128. B. M. Potter, L. S. Feng, P. Parasuram, V. A. Matskevich, J. A. Wilson, G. K. Andrews and J. H. Laity, *J. Biol. Chem.*, 2005, **280**, 28529–28540.
129. N. M. Tabatabai, S. S. Blumenthal and D. H. Petering, *Toxicology*, 2005, **207**, 369–382.
130. N. M. Tabatabai, A. B. Blodgett, R. K. Kothinti and D. H. Petering, submitted for publication.
131. C. D. Klaassen, J. Liu and S. Choudhuri, *Ann. Rev. Pharmacol. Toxicol.*, 1999, **39**, 267–294.
132. G. F. Nordberg, R. Goyer and M. Nordberg, *Arch. Pathol.*, 1975, **99**, 192–197.
133. R. E. Dudley, L. M. Gammal and C. D. Klaassen, *Toxicol. Appl. Pharmacol.*, 1985, **77**, 414–426.
134. J. Sudo, T. Hayashi, S. Kimura, K. Kakuno, J. Terui, K. Takashima and M. Soyama, *J. Toxicol. Environ. Health*, 1996, **48**, 333–348.
135. S. Satarug, J. R. Baker, S. Urbenjapol, M. Haswell-Elkins, P. E. Reilly, D. J. Williams and M. R. Moore, *Toxicol. Lett.*, 2003, **137**, 65–83.
136. J. Liu, Y. Liu, S. S. Habeebu and C. D. Klaassen, *Toxicol. Sci.*, 1998, **46**, 197–203.
137. J. Liu, S. S. Habeebu, Y. Liu and C. D. Klaassen, *Toxicol. Appl. Pharmacol.*, 1998, **153**, 48–58.
138. S. S. Blumenthal, D. L. Lewand, M. A. Buday, J. G. Kleinman, S. K. Krezoski and D. H. Petering, *Am. J. Physiol.*, 1990, **258**, F1625–1633.
139. D. H. Petering, M. Dughish, M. Huang, S. Krezoski, S. Krull, D. Lewand, A. Muñoz, L. Ren, S. Venkatexh, S. Blumenthal and C. F. ShawIII, in *Metallothionein-IV*, Ed. C. D. Klaassen, Birkhäuser Verlag, Basel, 1999, pp. 459–466.

140. V. A. Narayan, R. W. Kriwacki and J. P. Caradonna, *J. Biol. Chem.*, 1997, **272**, 7801–7809.
141. S. Blumenthal, D. Lewand, S. K. Krezoski and D. H. Petering, *Toxicol. Appl. Pharmacol.*, 1996, **136**, 220–228.
142. S. Blumenthal, D. Lewand, A. Sochanik, S. Krezoski and D. H. Petering, *Toxicol. Appl. Pharmacol.*, 1994, **129**, 177–187.
143. V. C. Culotta, S. J. Lin, P. Schmidt, L. W. Klomp, R. L. Casareno and J. Gitlin, *Adv. Exp. Med. Biol.*, 1999, **448**, 247–254.
144. S. La Fontaine and J. F. Mercer, *Arch. Biochem. Biophys.*, 2007, **463**, 149–167.
145. L. Tapia, M. González-Agüero, M. F. Cisternas, M. Suazo, V. Cambiazo, R. Uauy and M. González, *Biochem. J.*, 2004, **378**, 617–624.
146. A. Bakka and M. Webb, *Biochem. Pharmacol.*, 1981, **30**, 721–725.
147. P. Chen, P. Onana, C. F. Shaw 3rd and D. H. Petering, *Biochem. J.*, 1996, **317**, 389–394.
148. P. de Bie, P. Muller, C. Wijmenga and L. W. Klomp, *J. Med. Genet.*, 2007, **44**, 673–688.
149. N. O. Nartey, J. V. Frei and M. G. Cherian, *Lab. Invest.*, 1987, **57**, 397–401.
150. K. T. Suzuki, *Res. Commun. Mol. Pathol. Pharmacol.*, 1995, **89**, 221–240.
151. H. Kodama and Y. Murata, *Pediatr. Int.*, 1999, **41**, 430–435.
152. E. J. Kelly and R. D. Palmiter, *Nat. Genet.*, 1996, **13**, 219–222.
153. K. B. Nielson, C. L. Atkin, D. R. Winge, *J. Biol. Chem.*, 1985, 260, 5342–50.
154. P. Chen, A. Muñoz, D. Nettesheim, C. F. Shaw 3rd and D. H. Petering, *Biochem. J.*, 1996, **317**, 395–402.
155. D. L. Pountney, I. Schauwecker, J. Zarn and M. Vašák, *Biochemistry*, 1994, **33**, 9699–9705.
156. M. Dughish and D. H. Petering, unpublished information.
157. B. Ujjani, G. Krakower, G. Bachowski, S. Krezoski, C. F. Shaw 3rd and D. H. Petering, *Biochem. J.*, 1986, **233**, 99–105.
158. S. R. Davis and R. J. Cousins, *J. Nutr.*, 2000, **130**, 1085–1088.
159. S. K. Krezoski, J. Villalobos, C. F. Shaw 3rd and D. H. Petering, *Biochem. J.*, 1988, **255**, 483–491.
160. K. E. Rigby Duncan and M. J. Stillman, *J. Inorg. Biochem.*, 2006, **100**, 2101–2107.
161. D. H. Petering, A. Quesada, M. Dughish, S. Eisch, T. Gan, D. Lemkuil, A. Pattanaik, R. W. Byrnes, M. Savas, H. Whelan and C. F. Shaw III, in *Metallothionein-III*, Ed. K. Suzuki, Birkhäuser Verlag, Basel, 1993, pp. 329–346.
162. C. Andreini, L. Banci, I. Bertini and A. Rosato, *J. Proteome Res.*, 2006, **5**, 196–201.
163. U. Rana, R. Kothinti, J. Meeusen, N. M. Tabatabai, S. Krezoski and D. H. Petering, *J. Inorg. Biochem.*, 2008, **102**, 489–499.
164. A. Krezel and W. Maret, *Biochem. J.*, 2007, **402**, 551–558.
165. D. A. Dickinson and H. J Forman, *Ann. NY Acad. Sci.*, 2002, **973**, 488–504.
166. T. Y. Li, D. T. Minkel, C. F. Shaw 3rd and D. H. Petering, *Biochem. J.*, 1981, **193**, 441–446.
167. C. F. Shaw 3rd, M. M. Savas and D. H. Petering, *Methods Enzymol.*, 1991, **205**, 401–414.

168. M. M. Savas, C. F. Shaw 3rd and D. H. Petering, *J. Inorg. Biochem.*, 1993, **52**, 235–249.
169. A. Muñoz, F. H. Försterling, C. F. Shaw 3rd and D. H. Petering, *J. Biol. Inorg. Chem.*, 2002, **7**, 713–724.
170. W. Maret, *Proc. Natl. Acad. Sci. USA*, 1994, **91**, 237–241.
171. L. J. Jiang, W. Maret and B. L. Vallee, *Proc. Natl. Acad. Sci. USA*, 1998, **95**, 3483–3488.
172. A. R. Quesada, R. W. Byrnes, S. O. Krezoski and D. H. Petering, *Arch. Biochem. Biophys.*, 1996, **334**, 241–250.
173. L. Cai, J. B. Klein and Y. J. Kang, *J. Biol. Chem.*, 2000, **275**, 38957–38960.
174. K. D. Kroncke, K. Fehsel, T. Schmidt, F. T. Zenke, I. Dasting, J. R. Wesener, H. Bettermann, K. D. Breunig and V. Kolb-Bachofen, *Biochem. Biophys. Res. Commun.*, 1994, **200**, 1105–1110.
175. C. T. Aravindakumar, J. Ceulemans and M. De Ley, *Biochem. J.*, 1999, **344**, 253–258.
176. C. T. Aravindakumar, J. Ceulemans and M. De Ley, *Biophys. Chem.*, 2000, **85**, 1–6.
177. R. R. Misra, J. F. Hochadel, G. T. Smith, J. C. Cook, M. P. Waalkes and D. A. Wink, *Chem. Res. Toxicol.*, 1996, **9**, 326–232.
178. L. L. Pearce, R. E. Gandley, W. Han, K. Wasserloos, M. Stitt, A. J. Kanai, M. K. McLaughlin, B. R. Pitt and E. S. Levitan, *Proc. Natl. Acad. Sci. USA*, 2000, **97**, 477–482.
179. C. M. St Croix, K. J. Wasserloos, K. E. Dineley, I. J. Reynolds, E. S. Levitan and B. R. Pitt, *Am. J. Physiol. Lung Cell. Mol. Physiol.*, 2002, **282**, L185–192.
180. C. M. St Croix, M. S. Stitt, S. C. Watkins and B. R. Pitt, *Methods Enzymol.*, 2005, **396**, 317–326.
181. J. Zhu and D. H. Petering, submitted for publication.
182. L. Khatai, W. Goessler, H. Lorencova and K. Zangger, *Eur. J. Biochem.*, 2004, **271**, 2408–2416.
183. M. A. Schwarz, J. S. Lazo, J. C. Yalowich, W. P. Allen, M. Whitmore, H. A. Bergonia, E. Tzeng, T. R. Billiar, P. D. Robbins and J. R. Lancaster Jr., *Proc. Natl. Acad. Sci. USA*, 1995, **92**, 4452–4456.
184. D. Das, A. Chatterjee, G. Samantra, B. Mandal, T. R. Chowdhury, P. P. Chowdhury, C. Chanda, G. Basu, D. Lodh, S. Nandi, T. Chakraborty, S. Mandal, S. M. Bhattacharya and D. Chakraborti, *Analyst*, 1994, **119**, N168–N170.
185. M. F. Hughes, *Toxicol. Lett.*, 2002, **133**, 1–16.
186. T. J. O'Brien, S. Ceryak and S. R. Patierno, *Mutat. Res.*, 2003, **533**, 3–36.
187. M. A. Sekeres, *Curr. Hematol. Rep.*, 2005, **4**, 59–63.
188. D. J. Thomas, M. Styblo and S. Lin, *Toxicol. Appl. Pharmacol.*, 2001, **176**, 127–144.
189. H. V. Aposhian, R. A. Zakharyan, M. D. Avram, A. Sampayo-Reyes and M. L. Wollenberg, *Toxicol. Appl. Pharmacol.*, 2004, **198**, 327–335.
190. R. C. Lantz and A. M. Hays, *Drug Metab. Rev.*, 2006, **38**, 791–804.
191. Y. Kumagai and D. Sumi, *Ann. Rev. Pharmacol. Toxicol.*, 2007, **47**, 243–262.
192. M. J. Han, X. Meng and L. Lippincott, *Toxicol. Lett.*, 2007, **175**, 57–63.

193. P. Kapahi, T. Takahashi, G. Natoli, S. R. Adams, Y. Chen, R. Y. Tsien and M. Karin, *J. Biol. Chem.*, 2000, **275**, 36062–36066.
194. T. T. Ngu and M. J. Stillman, *J. Am. Chem. Soc.*, 2006, **128**, 12473–12483.
195. G. Jiang, Z. Gong, X.-F. Li, W. R. Cullen and X. C. Le, *Chem. Res. Toxicol.*, 2003, **16**, 873–880.
196. H. J. Wiegand, H. Ottenwalder and H. M. Bolt, *Toxicology*, 1984, **33**, 341–348.
197. D. Krepkiy, W. E. Antholine and D. H. Petering, *Chem. Res. Toxicol.*, 2003, **16**, 750–756.
198. C. F. Shaw III, L. He, A. Muñoz, M. M. Savas, S. Chi, C. Fink, T. Gan and D. H. Petering, *J. Biol. Inorg. Chem.*, 1997, **2**, 65–73.
199. Q. Hao and W. Maret, *FEBS J.*, 2006, **273**, 4300–4310.
200. S. L. Kelley, A. Basu, B. A. Teicher, M. P. Hacker, D. H. Hamer and J. S. Lazo, *Science*, 1988, **241**, 1813–1815.
201. T. He, D. Wei, D. Fabris and C. Fenselau, *Cell. Mol. Biol. (Noisy-le-Grand)*, 2000, **46**, 383–392.
202. M. Karin, *Cell. Res.*, 2008, **18**, 334–342.
203. A. Pattanaik, G. Bachowski, J. Laib, D. Lemkuil, C. F. Shaw 3rd, D. H. Petering, A. Hitchcock and L. Saryan, *J. Biol. Chem.*, 1992, **267**, 16121–16128.
204. D. C. Lemkuil, D. Nettesheim, C. F. Shaw 3rd and D. H. Petering, *J. Biol. Chem.*, 1994, **269**, 24792–24797.
205. D. Hagrman, J. Goodisman, J. C. Dabrowiak and A. K. Souid, *Drug Metab. Dispos.*, 2003, **31**, 916–923.
206. M. Knipp, A. V. Karotki, S. Chesnov, G. Natile, P. J. Sadler, V. Brabec and M. Vašák, *J. Med. Chem.*, 2007, **50**, 4075–4086.
207. A. Kraker, J. Schmidt, S. Krezoski and D. H. Petering, *Biochem. Biophys. Res. Commun.*, 1985, **130**, 786–792.
208. T. Boulikas and M. Vougiouka, *Oncol. Rep.*, 2003, **10**, 1663–1682.
209. D. H. Petering, W. E. Antholine and L. A. Saryan, in *Anticancer and Interferon Agents, Synthesis and Properties*, Ed. R. M. Ottenbrite and G. B. Butler, Marcel Dekker, New York, 1984, Ch. 7.
210. Y. Kondo, M. Satoh, N. Imura and M. Akimoto, *Anticancer Res.*, 1992, **12**, 2303–2307.
211. M. Satoh, A. Naganuma and N. Imura, *Life Sci.*, 2000, **67**, 627–634.
212. N. Miura, M. Satoh, N. Imura and A. Naganuma, *J. Pharmacol. Exp. Ther.*, 1998, **286**, 1427–1430.
213. H. Sun, H. Li, I. Harvey and P. J. Sadler, *J. Biol. Chem.*, 1999, **274**, 29094–29101.
214. T. Kaji, M. Suzuki, C. Yamamoto, Y. Imaki, A. Mishima, Y. Fujiwara, M. Sakamoto and H. Kozuka, *Res. Commun. Mol. Pathol. Pharmacol.*, 1994, **86**, 25–35.

Met. Ions Life Sci. **2009**, *5*, 399–412

13

Metallothionein in Inorganic Carcinogenesis

Michael P. Waalkes and Jie Liu

Inorganic Carcinogenesis Section, Laboratory of Comparative Carcinogenesis, National Cancer Institute at NIEHS, 111 Alexander Drive, Mail Drop F0-09, Research Triangle Park, NC 27709, USA

<waalkes@niehs.nih.gov>

<liu6@niehs.nih.gov>

Metal Ions in Life Sciences, Volume 5 Edited by Astrid Sigel, Helmut Sigel, and Roland K. O. Sigel

Published by the Royal Society of Chemistry, www.rsc.org DOI: 10.1039/9781847558992-00399

ABSTRACT: Metallothionein (MT) is a cysteine-rich, metal-binding protein that plays an important role in the detoxication of heavy metals and in the homeostasis of essential metal ions. Deficiency in MT makes animals highly sensitive to toxicity of some metals, and may predispose to inorganic carcinogenesis. For instance, poor expression of MT in regions of rat prostate is a susceptibility factor in cadmium-induced prostate cancer. Similarly, *MT-1/2* null mice, where the major forms of MT are knocked out, are more sensitive than wild-type mice to the carcinogenic effects of cadmium, arsenic, lead, and cisplatin. On the other hand, the carcinogenic potential of nickel is unchanged in *MT-1/2* null mice or in *MT-1* overexpressing transgenic mice, suggesting a minimal role for this protein in nickel carcinogenesis. Several mechanisms have been proposed for the inhibitory role of MT in inorganic carcinogenesis, including metal sequestration, reduced oxidative stress, adaptation response, acquired apoptosis resistance, and compromised DNA repair. In mice a clear inability to form inclusion bodies is implicated in enhanced lead-induced renal carcinogenesis in *MT-1/2* null mice, while downregulation of MT occurs during hepatocarcinogenesis induced by transplacental arsenic. There is a great variation in human MT expression and polymorphisms of the *MT* gene exist that may affect individual response to toxic metal insult, and poor ability to produce MT in response to metal exposure clearly may predispose individuals to carcinogenesis, by some, but not all, inorganic carcinogens.

KEYWORDS: adaptation · apoptosis resistance · arsenic · cadmium · cisplatin · inclusion body formation · lead · metal carcinogenesis · metallothionein

1. INTRODUCTION

Metallothionein (MT) is a low-molecular-weight, metal-binding protein of incompletely defined function [1]. MT probably plays important roles in the detoxication of metals, in the homeostasis of essential metals, and in the scavenging of free radicals [2]. Moreover, MT expression can be greatly increased by exposure to a variety of stimuli, particularly metals (see Chapter 2). There are at least four major mammalian MT isoforms. The MT-1 and MT-2 isoforms are widely expressed [3], while MT-3 is largely brain-specific, and MT-4 is mainly located in stratified squamous epithelium [4].

The possible roles of MT in carcinogenesis were explored early on in a National Cancer Institute Workshop in 1992 [5] and in a US Society of Toxicology Symposium in 1994 [6]. MT can play a role in both tumor pathobiology and chemotherapy. For instance, the expression of MT in human tumors varies greatly depending on the precise tumor type and even the stage of an individual tumor [3]. MT has also been implicated in anticancer drug resistance, such as resistance to cisplatin [6]. In humans, for reasons that are not fully understood, there is great individual variation in MT expression [7]. For instance, in one study in human livers without any pathology MT protein varied from 0 to 104 μg/g tissue [8]. Various other studies show wide-ranging discrepancies in MT expression in human populations [9]. It also appears that polymorphism for human *MT-2A* gene can significantly affect MT expression [10].

The production of *MT-1* transgenic mice [11], which highly overexpress MT, and *MT-1/2* null mice [12], which poorly express the major MT forms (see Chapters 10 and 12), have greatly facilitated research on the possible role of MT in carcinogenesis, including metal carcinogenesis. *MT-1/2*-null mice are more sensitive than wild-type mice to the toxicity of various inorganics, including cadmium [2], mercury [13], arsenic [14], cisplatin [15], zinc, and copper [16]. *MT*-null mice are also more susceptible to carcinogenic effects of lead [17], cisplatin [18], and cadmium (see below). Thus, the expression of MT appears to be a key factor in determining sensitivity to toxicity and carcinogenicity for various inorganics. In this chapter, the role of MT in cadmium, arsenic, lead, cisplatin, and nickel carcinogenesis, as well as the possible mechanisms are discussed.

2. METALLOTHIONEIN IN METAL CARCINOGENESIS

2.1. Cadmium

Cadmium(II) is a toxic heavy metal ion and MT can clearly protect against cadmium toxicity in the kidney and elsewhere (See Chapter 12). *MT-1/2* null mice are more susceptible than wild-type mice to cadmium-induced acute lethality [16], and chronic toxicity to the kidney [19], liver [20], bone [21], hematopoietic and immune system [22]. Thus, accumulating evidence clearly indicates that MT is a major cellular protein for protection against cadmium toxicity [2].

An association between human or rodent cadmium exposure and prostate cancer has long been suspected [23]. There are indications that MT, the primary cellular protective mechanism against cadmium toxicity, is poorly expressed in the specific lobe of the rat prostate in which cadmium induces tumors [24], potentially indicating a basis for regional sensitivity. Repository injections of cadmium will also induce local sarcomas and repeated cadmium injections enhance the malignant progression of ensuring sarcomas in rats [25]. Immunohistochemically, the primary injection site sarcoma showed high levels of MT, while metastases were essentially devoid of MT, indicating that suppressed MT production is important for metastasis of cadmium-induced sarcomas [25]. The pathogenesis of cadmium-induced prostatic carcinogenesis might include aberrant gene expression resulting in stimulation of cell proliferation or blockage of apoptosis. Activation of the MT gene and activation of protooncogenes may enhance cell proliferation with damaged DNA. Suppression of DNA repair would add to the population of cells with damaged DNA, as apoptotic resistance could facilitate aberrant cell accumulation in prostate and testes [26].

Exposure of human urinary bladder UROtsa cells to cadmium results in malignant transformation, producing cells capable of tumor formation when inoculated into nude mice [27]. MT protein expression in these heterotransplant tumors is focal in pattern, varied in intensity, and highest in the less differentiated cells, suggesting a role in bladder carcinogenesis [27].

We thus hypothesize that the poor expression of MT should enhance the carcinogenic potential of cadmium. In order to directly test this hypothesis *in vivo* we have studied the carcinogenic response to cadmium using a MT-deficient mouse model. Adult male *MT-1/2* null or wild-type mice were injected subcutaneously with a single dose of cadmium at 0 (control), 1.0 or 5.0 μmoles Cd as $CdCl_2$/kg. The mice were then observed for up to 2 years. Survival was similar and these doses were not acutely toxic to mice of either phenotype. No difference occurred in spontaneous tumors between *MT-1/2* null and wild-type mice. However, cadmium-treated *MT-1/2* null mice showed a clear, dose-related increase in liver tumors (Figure 1) that did not occur in wild-type mice. Thus, mice deficient in MT are predisposed to cadmium carcinogenesis in the liver, a potential human target site of cadmium [23].

2.2. Arsenic

Arsenic is a toxic metalloid. In animal studies, arsenic effectively activates *MT* gene expression [28]. The induction of MT by arsenicals can be envisioned, at least in part, as an adaptive response to overcome toxic insult from the metalloid. For instance, *MT-1/2* null mice are clearly more sensitive than wild-type mice to arsenic-induced acute lethality [16], and are more susceptible to inorganic arsenic-induced hepatotoxicity and nephrotoxicity after long-term arsenic exposure [14]. These data support a protective role of MT against arsenical toxicity, at least in part, by binding arsenic to MT, and thus sequestering arsenic from critical cellular organelles [29].

The involvement of MT in arsenic carcinogenesis comes from several lines of evidence. Deficiency of MT makes *MT-1/2* null mice highly sensitive to the genotoxic effects of arsenic as assessed in peripheral blood cells produced by exposure to dimethylarsinic acid, a metabolite of inorganic arsenic [30]. In transplacental arsenic carcinogenesis studies, the expression of MT in transplacental arsenic-induced tumors and tumor surrounding tissues is downregulated when mice reach adulthood [31]. On the other hand, MT is induced by *in utero* arsenic exposure in fetal liver cells [32], or in rat liver cells chronically exposed to inorganic arsenic [33]. In the areas of Guizhou, China, where arsenicosis is endemic, poor expression of *MT-1A* and *MT-2A* in peripheral blood and buccal cells is associated with enhanced sensitivity to chronic arsenic intoxication [34]. Although the exact role of MT in arsenic carcinogenesis remains unknown, the wide variations in human MT

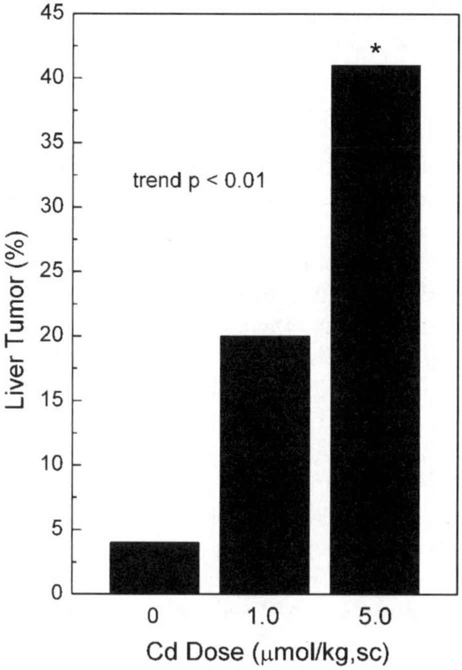

Figure 1. Cadmium-induced liver tumors in *MT-1/2* null mice. Adult *MT-1/2* null mice and wild-type mice were injected subcutaneously with a single dose of 1.0 or 5.0 μmoles Cd as $CdCl_2$/kg. The mice were observed for up to 2 years. Cadmium induced a clear dose-related increase in liver tumors that did not occur in wild-type mice. Trend $p < 0.01$ for liver tumor incidence. The asterisk (*) on top of the third column indicates that these results are significantly different from the control experiments at the dose of 5 μmol/kg.

expression [7] could well be a major factor in susceptibility to the carcinogenic effects of this metalloid.

2.3. Lead

Lead is a naturally occurring, ubiquitous environmental toxicant and potentially carcinogenic metal [35]. MT can be induced by lead after parental injections [36], but not after direct exposure of cultured hepatocytes [37]. Lead avidly binds to MT [38], and sequestration of lead in the cytosol by Zn-induced MT protects against lead toxicity in primary cultured hepatocytes [39]. However, *MT-1/2* null mice do not show increased sensitivity to acute lead-induced lethality [16]. A remarkable characteristic of lead poisoning is the production of a protein-lead complex, which appears in the renal cells of poisoned animals and humans, especially during chronic intoxication [40]. After subchronic exposure (10 weeks) of *MT-1/2* null mice to lead acetate, even though less lead is accumulated in kidney as compared to wild-type mice, MT deficient mice are more susceptible to lead-induced renal toxicity and proliferative lesions [41]. The inability to form lead-inclusion bodies was evident in *MT-1/2* null kidneys or cells exposed to the metal, in sharp contrast to their wild-type counterparts [41].

The susceptibility of *MT*-null mice to subchronic lead nephrotoxicity prompted the hypothesis that inability to form inclusion bodies in the *MT-1/2* null mice could impact the carcinogenic potential of lead. To test this hypothesis, adult male *MT-1/2* null and wild-type mice received drinking water with 0, 1000, 2000, and 4000 ppm lead acetate for up to 2 years. Renal adenoma and cystic tubular atypical hyperplasia (preneoplasia) were much more common and severe after lead exposure in *MT-1/2* null mice than in wild-type mice. A metastatic renal cell carcinoma also occurred in a lead-treated *MT-1/2* null mouse [17]. Lead-induced renal neoplastic lesions showed cyclin D1 overexpression, a common feature of human renal tumors [17]. Renal lead-containing nuclear inclusion bodies were frequently observed in wild-type mice, and MT was often found in the association with the outer portion of these bodies [17]. In contrast, none of the *MT*-null mice formed lead inclusion bodies after chronic lead exposure. Thus, the *MT-1/2* null phenotype was unable to form renal inclusion bodies, even with protracted lead exposure, and *MT-1/2* null mice show increased sensitivity to lead. It is possible that poor production of MT may also predispose humans to lead carcinogenicity, although this will require further investigation.

2.4. Cisplatin

Cisplatin is a well-known metallic chemotherapeutic agent and an inducer of MT that, in turn, is a major cellular mediator of anticancer drug resistance [6] and a probable carcinogen. Cellular MT can bind significant amounts of cisplatin [42], and there are multiple interactions of cisplatin with MT. For instance, preinduction of MT can protect against cisplatin-induced nephrotoxicity [15], clastogenicity and DNA damage [43], as well as cisplatin-induced lung carcinogenicity [44]. On the other hand, induction of MT blocks cisplatin efficacy in chemotherapy, and modulation or inhibition of MT induction by cisplatin alters drug resistance [45]. To kill cancer cells, chemotherapeutic agents are generally used at toxic levels. One long-term manifestation of these toxic effects could be secondary tumor formation. Cisplatin is an effective initiator of renal and dermal cancers in rodents. The mechanism of cisplatin as a chemotherapeutic agent is likely the same as a carcinogen [6,18]. Thus, cellular MT acts as a "double-edged sword" in regard to therapeutic/carcinogenic mode of action for cisplatin.

MT-1/2 null cells are more susceptible to cisplatin-induced cytotoxicity and DNA damage [46]. *MT-1/2* null mice are also more sensitive than wild-type animals to cisplatin-induced hepatotoxicity following acute exposure [47], and to cisplatin nephrotoxicity following subchronic exposures [48]. The hypothesis that MT deficiency might impact the carcinogenic effects of cisplatin was tested using *MT-1/2* null mice. Male *MT-1/2* null mice or wild-type mice were exposed to a single treatment of cisplatin (5 or 10 mg/kg, ip), or left untreated (control) and observed up to 2 years [18]. The doses of cisplatin used were equal to only a fraction of the total dose typically used in clinical settings. In cisplatin-treated *MT-1/2* null mice, a dose-related increase in hepatocellular carcinoma occurred (control, 0%; 5 mg/kg, 17%; 10 mg/kg, 36%) that was not seen in wild-type mice. Liver carcinoma multiplicity (HCC/liver) was markedly increased by cisplatin, but only in *MT-1/2* null mice and not wild-type mice, indicating the formation of multiple primaries in mice deficient in MT synthesis. Harderian gland carcinoma incidence was also increased by cisplatin treatment in *MT-1/2* null mice but not in wild-type mice [18]. Thus, deficiency of MT predisposes to cisplatin carcinogenesis in animals. The potential for an enhanced susceptibility to secondary tumor formation with cisplatin in persons poorly expressing MT should be explored.

2.5. Nickel

Nickel compounds have carcinogenic potential in humans and animals, possibly by production of oxidative stress among other effects. Nickel is also a metallic inducer of MT in animals [36], in rat primary hepatocyte cultures [37], in human peripheral lung epithelial cells [49], and in human peripheral lymphocytes [50]. However, high concentrations or doses of nickel are often required for MT induction. MT can bind nickel, but with a relatively low affinity [38]. Pre-induction of MT by Zn has been shown to protect against nickel-induced acute cytotoxicity [39] and hepatotoxicity [51]. *MT-1/2* null mice are more susceptible than wild-type mice to nickel-induced inflammation and lethality [52]. All these studies suggest a role of MT in acute nickel toxicity, but little was known about the role of MT in chronic nickel toxicity and carcinogenicity.

The impact of MT deficiency on the carcinogenic effects of nickel was studied using *MT-1/2* null and *MT-1* transgenic mouse models. Male adult *MT-1/2* null or corresponding wild-type mice were exposed to a single treatment of nickel (0.5 or 1.0 mg Ni as nickel sulfate/site, intramuscularly into both hind legs), or left untreated (control) and observed over the next 2 years. Nickel induced injection site fibrosarcomas in a dose-related fashion, but there was no clear difference between *MT-1/2* null and wild-type mice in incidence, suggesting *MT-1/2* null mice do not have any increased sensitivity to the carcinogenic effects of nickel [53]. This result was further fortified by observations in *MT-1* transgenic mice using the similar treatment protocol with injection site tumors. Fibrosarcomas started occurring 45 weeks after nickel injection in a dose-dependent fashion, but again no difference occurred in incidence between *MT-1* transgenic and corresponding wild-type mice [54]. Thus, MT does not appear to play a major role in altering the carcinogenic effects of nickel.

Taken together, MT plays a major protective role for cadmium toxicity, and important roles in the toxicity of cisplatin, arsenic, and lead. The carcinogenic potentials of these metals follow the protective role of MT against their chronic toxicity, while MT does not appear to play a role in nickel carcinogenesis.

3. MECHANISMS BY WHICH METALLOTHIONEIN MAY REDUCE METAL CARCINOGENESIS

3.1. Oxidative Stress

In Chapters 10 and 12, the ability of MT to act against metal-induced oxidative stress, in the regulation of cellular redox potentials, and as a cellular antioxidant component, have been discussed. Oxidative stress is the major mechanism for toxicity of various metals, but its role in metal toxicity varies during chronic exposure and the stage of metal carcinogenesis [55]. For example, oxidative stress is implicated in acute cadmium toxicity [56], but only minimally involved in chronic cadmium-induced malignant transformation [33]. Oxidative stress is implicated in arsenic toxicity [57], but in transplacental arsenic carcinogenesis, the role of oxidative stress appears to be minimal [58]. MT plays a protective role against nickel-induced oxidative damage following acute exposure [52], but MT does not protect against nickel carcinogenesis using the same *MT-1/2 null* mouse model [53]. Thus, the role of MT in metal carcinogenesis cannot be solely explained based on the antioxidant role of this metal-binding protein.

3.2. Adaptation and Apoptotic Resistance

Induction of MT is an important cellular adaptive mechanism affecting the magnitude and progression of toxicity from repeated toxic insults from metals [59]. Induction of MT in response to metal exposure may, at first, appear to be beneficial. However, recent evidence suggests that the metal-adapted phenotype could have deleterious consequences and may represent a double-edged sword in metal carcinogenesis [60]. For example, it has been shown that cadmium-adapted alveolar epithelial cells have a reduced ability to repair DNA damage due, in part, to the inhibition of two base excision repair enzymes (8-oxoguanine-DNA glycosylase and endonuclease III) [61]. Cells with genetic aberrations resulting from unrepaired DNA lesions would normally be removed from the tissues by apoptosis. However, induction of MT confers apoptotic resistance, and may allow damaged cells to survive and to proliferate [60]. Induction of MT is associated with diminished cellular oxidative stress in cadmium transformants [33]. Similar observations on MT induction and acquired apoptotic resistance were also evident in rat liver cells malignantly transformed by chronic arsenic exposure [62] and in similarly transformed human prostate epithelial cells [63]. Thus, the acquired apoptotic resistance in association with induction of MT could be an important mechanism in metal carcinogenesis in cells and in animals.

3.3. Inclusion Body Formation

A remarkable characteristic of lead poisoning is the production of a protein-lead complex, which appears in various cells of poisoned humans or animals as inclusion bodies [35]. Lead-induced inclusion bodies are often nuclear, and approximately spherical with an electron-dense core and a peripheral fibrillary network. Inclusion bodies first form in the cytoplasm and then often migrate to the nucleus [40]. Whether inclusion bodies are a permanent structure or decay with cessation of lead exposure is unknown. Similarly, the origin of the protein component of the inclusion bodies remains uncertain. Lead is highly concentrated in the inclusion bodies, and as much as half of lead in the kidney may be found there. It is suspected that these complexes are protective in that they may render lead toxicologically inert, thereby blocking interactions with more sensitive cellular targets. Thus, lead inclusion bodies may play a role in the intracellular inert storage of lead and thereby reduce its toxicity.

MT-1/2 null mice do not produce lead inclusion bodies in the kidney following subchronic lead exposure (10 weeks), and were sensitive to lead-induced renal toxicity, manifested as hypertrophy, diminished function and aberrant gene expression, as compared with wild-type mice [41]. In the two year bioassays, these *MT-1/2* null mice were more sensitive to lead-induced renal proliferative lesions (adenoma and cystic tubular atypical hyperplasia), and showed an inability to form lead inclusion bodies [17]. Poor production of MT clearly predisposes *MT-1/2* null mice to renal carcinogenic effects of lead, and would likely predispose human populations to lead carcinogenicity, although the later requires direct testing.

Table 1. Proposed mechanisms by which metallothionein reduces metal carcinogenesis.

Proposed Mechanisms	Proposed Roles of Metallothionein
Oxidative stress	Free radical scavenging
Apoptotic resistance	Damaged cell survival, proliferation, adaptation
Inability of forming inclusion bodies	Essential for forming lead IBs for lead detoxication
Metallothionein downregulation	Enhanced metastasis (?), aberrant gene expression

3.4. Downregulation of Metallothionein in Inorganic Carcinogenesis

Downregulation of MT is associated with spontaneous transformation in cultured rat liver cells [64] and murine and human hepatocellular carcinoma [65]. Low expression of *MT* in certain tumors could also be due to alterations in cell proliferation, or cell differentiation, such as the events seen during tumor progression [4], or due to indirect methylation of the *MT* genes [66]. *MT* gene hypermethylation is associated with reduced expression in various human or rodent tissues and cells [7]. Alterations in *MT* gene methylation significantly alter expression potential [66]. MT expression is decreased in transplacental arsenic-induced tumors and tumor surrounding tissues in mice [31] and in blood and buccal cells from arsenicosis patients [34]. The reduced *MT* gene expression would likely attenuate the body's defense mechanism against the toxic and carcinogenic metal, but may also alter gene expression perhaps impacting the carcinogenic process.

The above proposed roles for MT in inorganic carcinogenesis are summarized in Table 1, and these mechanisms could act together in an integrated manner during inorganic carcinogenesis.

4. CONCLUDING REMARKS AND FUTURE DIRECTIONS

MT plays various important complex roles in inorganic carcinogenesis, and impacts inorganics such as cadmium, arsenic, lead, and cisplatin. MT appears to play a minimal role in the carcinogenic potential of nickel. The hypothesis that individuals with a low ability for MT expression may be susceptible to inorganic carcinogenesis definitely warrants further investigation.

There is great individual variation in MT expression [9]. It also appears that polymorphisms for human *MT* gene can significantly limit MT expression [10]. Altered MT expression is clearly associated with carcinogenesis and can be used as a potential tumor progression biomarker [3]. Thus, defining the precise role of MT in inorganic carcinogensis is an important step in the protection of humans at risk, since metals are often difficult to remove from the environment.

ACKNOWLEDGMENTS

The authors thank Drs. Erik Tokar, Wei Qu, and Larry Keefer for their critical review of this book chapter. Research was supported in part by the Intramural Research Program of the NIH, National Cancer Institute, Center for Cancer Research, and National Institute of Environmental Sciences. The authors have no competing financial interest and the content of this publication does not necessarily reflect the views or policies of the Department of Health and Human Services.

ABBREVIATIONS

IBs	inclusion bodies
ip	intraperitoneally
MT	metallothionein
sc	subcutaneous

REFERENCES

1. J. H. Kägi, *Methods Enzymol.*, 1991, **205**, 613–626.
2. C. D. Klaassen, J. Liu and S. Choudhuri, *Annu. Rev. Pharmacol. Toxicol.*, 1999, **39**, 267–294.
3. M. G. Cherian, A. Jayasurya and B. H. Bay, *Mutat. Res.*, 2003, **533**, 201–209.
4. C. J. Quaife, S. D. Findley, J. C. Erickson, G. J. Froelick, E. J. Kelly, B. P. Zambrowicz and R. D. Palmiter, *Biochemistry*, 1994, **33**, 7250–7259.
5. M. G. Cherian, P. C. Huang, C. D. Klaassen, Y. P. Liu, D. G. Longfellow and M. P. Waalkes, *Cancer Res.*, 1993, **53**, 922–925.
6. M. G. Cherian, S. B. Howell, N. Imura, C. D. Klaassen, J. Koropatnick, J. S. Lazo and M. P. Waalkes, *Toxicol. Appl. Pharmacol.*, 1994, **126**, 1–5.
7. M. T. Wu, B. Demple, R. A. Bennett, D. C. Christiani, R. Fan and H. Hu, *J. Toxicol. Environ. Health A.*, 2000, **61**, 553–567.
8. P. A. Sillevis Smitt, H. van Beek, A. J. Baars, D. Troost, E. S. Louwerse, A. C. Krops-Hermus, F. A. de Wolff and J. M. de Jong, *Arch. Neurol.*, 1992, **49**, 721–724.

9. S. Onosaka, K. S. Min, C. Fukuhara, K. Tanaka, S. Tashiro, I. Shimizu, M. Furuta, T. Yasutomi, K. Kobashi and K. Yamamoto, *Toxicology*, 1986, **38**, 261–268.
10. K. Kita, N. Miura, M. Yoshida, K. Yamazaki, T. Ohkubo, Y. Imai and A. Naganuma, *Hum. Genet.*, 2006, **120**, 553–560.
11. R. D. Palmiter, E. P. Sandgren, D. M. Koeller and R. L. Brinster, *Mol. Cell Biol.*, 1993, **13**, 5266–5275.
12. B. A. Masters, E. J. Kelly, C. J. Quaife, R. L. Brinster and R. D. Palmiter, *Proc. Natl. Acad. Sci. USA*, 1994, **91**, 584–588.
13. M. Yoshida, C. Watanabe, M. Satoh, A. Yasutake, M. Sawada, Y. Ohtsuka, Y. Akama and C. Tohyama, *Toxicol. Sci.*, 2004, **80**, 69–73.
14. J. Liu, Y. Liu, R. A. Goyer, W. Achanzar and M. P. Waalkes, *Toxicol Sci.*, 2000, **55**, 460–467.
15. M. Satoh, Y. Aoki and C. Tohyama, *Cancer Chemother. Pharmacol.*, 1997, **40**, 358–362.
16. J. D. Park, Y. Liu and C. D. Klaassen, *Toxicology*, 2001, **163**, 93–100.
17. M. P. Waalkes, J. Liu, R. A. Goyer and B. A. Diwan, *Cancer Res.*, 2004, **64**, 7766–7772.
18. M. P. Waalkes, J. Liu, K. S. Kasprzak and B. A. Diwan, *Int. J. Cancer*, 2006, **119**, 28–32.
19. J. Liu, Y. Liu, S. S. Habeebu and C. D. Klaassen, *Toxicol. Sci.*, 1998, **46**, 197–203.
20. S. S. Habeebu, J. Liu, Y. Liu and C. D. Klaassen, *Toxicol. Sci.*, 2000, **56**, 211–219.
21. S. S. Habeebu, J. Liu, Y. Liu and C. D. Klaassen, *Toxicol. Sci.*, 2000, **55**, 223–232.
22. J. Liu, Y. Liu, S. S. Habeebu and C. D. Klaassen, *Toxicol. Appl. Pharmacol.*, 1999, **159**, 98–108.
23. M. P. Waalkes, *Mutat. Res.*, 2003, **533**, 107–120.
24. M. P. Waalkes and S. Rehm, *J. Toxicol. Environ. Health*, 1994, **43**, 251–269.
25. M. P. Waalkes, S. Rehm and M. G. Cherian, *Toxicol. Sci.*, 2000, **54**, 110–120.
26. R. A. Goyer, J. Liu and M. P. Waalkes, *Biometals*, 2004, **17**, 555–558.
27. X. D. Zhou, D. A. Sens, M. A. Sens, V. B. Namburi, R. K. Singh, S. H. Garrett and S. Somji, *Toxicol. Sci.*, 2006, **91**, 467–475.
28. H. Kreppel, J. W. Bauman, J. Liu, J. M. McKim Jr. and C. D. Klaassen, *Fundam. Appl. Toxicol.*, 1993, **20**, 184–189.
29. T. T. Ngu and M. J. Stillman, *J. Am. Chem. Soc.*, 2006, **128**, 12473–12483.
30. G. Jia, H. Sone, N. Nishimura, M. Satoh and C. Tohyama, *Int. J. Oncol.*, 2004, **25**, 325–333.
31. J. Liu, Y. Xie, D. M. Ducharme, J. Shen, B. A. Diwan, B. A. Merrick, S. F. Grissom, C. J. Tucker, R. S. Paules, R. Tennant and M. P. Waalkes, *Environ. Health Perspect.*, 2006, **114**, 404–411.
32. J. Liu, L. Yu, E. Tokar, C. Bortner, M. I. Sifre, Y. Sun and M. P. Waalkes, *Anal. New York Acad. Sci.* 2008, in press.
33. W. Qu, B. A. Diwan, J. M. Reece, C. D. Bortner, J. Pi, J. Liu and M. P. Waalkes, *Int. J. Cancer*, 2005, **114**, 346–355.

34. J. Liu, M. L. Cheng, Q. Yang, K. R. Shan, J. Shen, Y. Zhou, X. Zhang, A. L. Dill and M. P. Waalkes, *Environ. Health Perspect.*, 2007, **115**, 1101–1106.
35. E. K. Silbergeld, *Mutat. Res.*, **533**, 121–133.
36. M. P. Waalkes and C. D. Klaassen, *Fundam. Appl. Toxicol.*, 1985, **5**, 473–477.
37. W. M. Bracken and C. D. Klaassen, *J. Toxicol. Environ. Health.*, 1987, **22**, 163–174.
38. M. P. Waalkes, M. J. Harvey and C. D. Klaassen, *Toxicol. Lett.*, 1984, **20**, 33–39.
39. J. Liu, W. C. Kershaw and C. D. Klaassen, *Toxicol. Appl. Pharmacol.*, 1991, **107**, 27–34.
40. R. A. Goyer, *Environ. Health Perspect.*, 1996, **104**, 1050–1054.
41. W. Qu, B. A. Diwan, J. Liu, R. A. Goyer, T. Dawson, J. L. Horton, M. G. Cherian and M. P. Waalkes, *Am. J. Pathol.*, 2002, **160**, 1047–1056.
42. D. Hagrman, J. Goodisman, J. C. Dabrowiak and A. K. Souid, *Drug Metab. Dispos.*, 2003, **31**, 916–923.
43. (a) I. Nakagawa, E. Nishi, A. Naganuma and N. Imura, *Mutat. Res.*, 1995, **348**, 37–43; (b) O. Hosokawa, M. Okabe, S. Saito, T. Saito and M. Kurasaki, *Pharmacol. Toxicol.*, 2000, **86**, 276–282.
44. M. Satoh, Y. Kondo, M. Mita, I. Nakagawa, A. Naganuma and N. Imura, *Cancer Res.*, 1993, **53**, 4767–4768.
45. (a) M. Satoh, M. G. Cherian, N. Imura and H. Shimizu, *Cancer Res.*, 1994, **54**, 5255–5257; (b) Y. Saga, H. Hashimoto, S. Yachiku, T. Iwata and M. Tokumitsu, *Int. J. Urol.*, 2004, **11**, 407–415.
46. Y. Kondo, E. S. Woo, A. E. Michalska, K. H. Choo and J. S. Lazo, *Cancer Res.*, 1995, **55**, 2021–2013.
47. J. Liu, Y. Liu, S. S. Habeebu and C. D. Klaassen, *Toxicol. Appl. Pharmacol.*, 1998, **149**, 24–31.
48. M. Satoh, A. Shimada, B. Zhang and C. Tohyama, *Biochem. Pharmacol.*, 2000, **60**, 1729–1734.
49. R. Y. Cheng, A. Zhao, W. G. Alvord, D. A. Powell, R. M. Bare, A. Masuda, T. Takahashi, L. M. Anderson and K. S. Kasprzak, *Toxicol. Appl. Pharmacol.*, 2003, **191**, 22–39.
50. H. Yamada and S. Koizumi, *Chem. Biol. Interact.*, 1991, **78**, 347–354.
51. P. Sidhu, M. L. Garg and D. K. Dhawan, *Chem. Biol. Interact.*, 2004, **150**, 199–209.
52. S. C. Wesselkamper, S. A. McDowell, M. Medvedovic, T. P. Dalton, H. S. Deshmukh, M. A. Sartor, L. M. Case, L. N. Henning, M. T. Borchers, C. R. Tomlinson, D. R. Prows and G. D. Leikauf, *Am. J. Respir. Cell Mol. Biol.*, 2006, **34**, 73–82.
53. M. P. Waalkes, J. Liu, K. S. Kasprzak and B. A. Diwan, *Int. J. Toxicol.*, 2005, **24**, 215–20.
54. M. P. Waalkes, J. Liu, K. S. Kasprzak and B. A. Diwan, *Toxicol. Lett.*, 2004, **153**, 357–364.
55. J. Liu, R. A. Goyer and M. P. Waalkes, in *Casarett and Doull's Toxicology – The Basic Science of Poisons*, 7th edit., Ed. C. D. Klaassen, McGraw Hill, New York, 2007, pp. 931–975.

56. J. Liu, M. B. Kadiiska, J. C. Corton, W. Qu, M. P. Waalkes, R. P. Mason, Y. Liu and C. D. Klaassen, *Free Radic. Biol. Med.*, 2002, **32**, 525–535.
57. IARC, *Some Drinking Water Disinfectants and Contaminants, including Arsenic*, Vol. 84, IARC Press, Lyon, 2004, pp. 269–477.
58. M. P. Waalkes, J. Liu and B. A. Diwan, *Toxicol. Appl. Pharmacol.*, 2007, **222**, 271–280.
59. C. D. Klaassen and J. Liu, *Environ. Health Perspect.*, 1998, **106**(Suppl 1), 297–300.
60. M. P. Waalkes, D. A. Fox, J. C. States, S. R. Patierno and M. J. McCabe Jr., *Toxicol. Sci.*, 2000, **56**, 255–261.
61. B. A. Hart, R. J. Potts and R. D. Watkin, *Toxicology*, 2001, **160**, 65–70.
62. J. Liu, L. Benbrahim-Tallaa, X. Qian, L. Yu, Y. Xie, J. Boos, W. Qu and M. P. Waalkes, *Toxicol. Appl. Pharmacol.*, 2006, **216**, 407–415.
63. W. E. Achanzar, E. M. Brambila, B. A. Diwan, M. M. Webber and M. P. Waalkes, *J. Natl. Cancer Inst.*, 2002, **94**, 1888–1891.
64. T. P. Coogan, W. E. Achanzar and M. P. Waalkes, *J. Environ. Pathol. Toxicol. Oncol.*, 2000, **19**, 261–273.
65. M. P. Waalkes, B. A. Diwan, S. Rehm, J. M. Ward, M. Moussa, M. G. Cherian and R. A. Goyer, *J. Pharmacol. Exp. Ther.*, **277**, 1026–1033.
66. S. T. Jacob, S. Majumder and K. Ghoshal, *Environ. Health Perspect.*, 2002, **110**(Suppl 5), 827–830.

Met. Ions Life Sci. **2009**, *5*, 413–439

14

Thioredoxins and Glutaredoxins. Functions and Metal Ion Interactions

Christopher Horst Lillig[1, 2] ***and Carsten Berndt***[2]

[1]Department of Clinical Cytobiology and Cytopathology, Phillips University, D-35037 Marburg, Germany
<horst@lillig.de>

[2]The Medical Nobel Institute for Biochemistry, Department of Medical Biochemistry and Biophysics, Karolinska Institutet, S-17177 Stockholm, Sweden
<carsten.berndt@ki.se>

Metal Ions in Life Sciences, Volume 5 Edited by Astrid Sigel, Helmut Sigel, and Roland K. O. Sigel

Published by the Royal Society of Chemistry, www.rsc.org DOI: 10.1039/9781847558992-00413

ABSTRACT: Thioredoxins and glutaredoxins represent the major cellular systems for the reduction of protein disulfides and protein de-glutathionylation, respectively. These two systems are involved in many aspects of human health, for instance as electron donors of metabolic enzymes and by controlling and maintaining the cellular redox state. The members of this protein family are characterized by a common structural motif, the thioredoxin fold. This basic architecture consists of a central four-stranded β-sheet surrounded by three α-helices. During the past few years accumulating evidence suggests a close relationship between these redoxins, most of all the glutaredoxins, and the cellular iron pool. Today we know that the thioredoxin fold cannot only be utilized for specific protein-protein interactions but also for interactions with metals, for instance iron-sulfur centers. Within this chapter, we summarize these recent findings and discuss the potential physiological implications of these metal interactions.

KEYWORDS: glutaredoxin · iron-sulfur · metal binding · oxidoreductase · thioredoxin

1. INTRODUCTION

1.1. The Thioredoxin Family of Proteins

All members of the thioredoxin (Trx) family of proteins are characterized by a common structural motif with βαβαββα topology, in which the three helices surround a central four-stranded sheet [1–3] (Figure 1). This basic architecture is found in bacterial glutaredoxins (Grxs), while most Trxs contain an additional α-helix and β-sheet at the N-terminus [4]. In spite of considerable variation in overall structure, the Trx fold is present in a variety of proteins: thiol-disulfide oxidoreductases like Trxs [1], Grxs [5], DsbA [6], protein disulfide isomerases [7,8], and peroxiredoxins (Prxs) [9], but also functionally different proteins, such as glutathione transferases [10], glutathione peroxidases [11], proteins involved in cytochrome *c* oxidase assembly [12], and chloride intracellular channels [13]. If circular permutations are considered, a variety of other proteins including the C-terminal domain of tubulin, cytidine deaminase, and phospholipase D can be viewed as Trx fold proteins [14].

Hallmarks of the thiol-disulfide oxidoreductases from the Trx fold family are a Cys-X-X-Cys active site motif, located before and at the beginning of

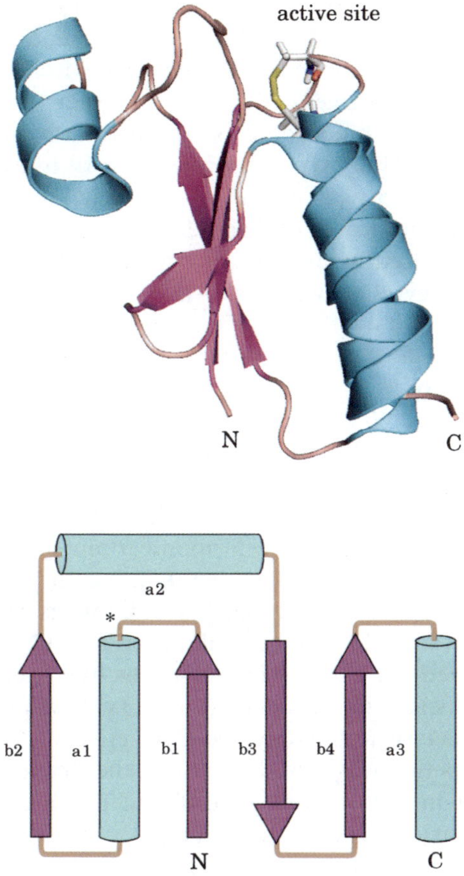

Figure 1. The thioredoxin fold. The basic motif of the thioredoxin fold, in which a central four-stranded sheet is surrounded by the three helices, is adopted only by bacterial Grxs. The presented structure shows oxidized *E. coli* Grx1 (PDB code 1EGO). The asterisk marks the position of the Cys-X-X-Cys active site.

α-helix 1, and a *cis*-proline adjacent to β-strand 3 [3,15]. Several variations of the original Cys-X-X-Cys motif have been described such as Ser-X-X-Cys, Thr-X-X-Cys, Cys-X-X-Ser, Cys-X-X-Thr, or Cys-X-X-X-Cys [12,16]. The nature and composition of the two central (X-X) residues of the active site motif, together with other flanking residues, are important determinants of the physico-chemical properties and specific activity of the individual oxidoreductase family members [17,18]. As an example, the redox potential of a mutant of the reductase Trx1 from *E. coli* harboring the active site of the disulfide oxidase DsbA ($\Delta E'_0 = -122$ mV [19]) increased from $\Delta E'_0 = -270$ mV [20] to $\Delta E'_0 = -204$ mV [21]. In reverse direction, the redox potential of a DsbA mutant with the Trx active site decreased by 92 mV [22]. In a similar manner, replacement of the Trx active site for the corresponding active site of protein disulfide isomerase resulted in an increase of its standard midpoint potential [20,21] accompanied by a 10-fold increase in protein disulfide isomerase activity [20,23]. The *cis*-proline, which has been described as essential residue for thiol-disulfide oxidoreductase activity [24–26] was found to be replaced by arginine in Prxs [27], histidine in Sco proteins [28], and cysteine in cytidine deaminases [29].

1.2. Thioredoxin

Trxs are small (12–15 kDa) ubiquitous proteins present in all forms of life; *E. coli*, yeast, and mammalian cells contain two Trxs each [30]. The mammalian Trx1 is mainly localized in the cytosol, but can also be translocated in the nucleus or exported from the cell upon certain stimuli [31,32]; Trx2 is targeted to mitochondria [33]. Plants contain a rich variety of Trxs with sometimes highly specialized functions [34–36], for instance the *A. thaliana* genome contains at least 19 genes encoding for Trxs [37].

Trx was discovered more than 40 years ago as hydrogen donor for *E. coli* ribonucleotide reductase (RNR) [38]. *E. coli* Trx1 was sequenced in 1968 revealing the characteristic Cys-Gly-Pro-Cys active site motif [39]. In 1975 the first crystal structure was solved disclosing the Trx fold for the first time [1]. Today, more than 200 structures of different Trxs are available including structures of both oxidized and reduced Trxs, for instance, from *E. coli* Trx1 [40] and human Trxs 1 [41] and 2 [42]. These structures confirmed an early observation from 1967 that reduction induces local conformational changes in and around the active site [43], which can have a dramatic effect on the binding affinity of Trx to other proteins (see Section 2.2.). Today, Trxs are primarily recognized as general protein disulfide reductases. Target disulfides are reduced in a thiol-disulfide exchange reaction leaving a disulfide in the active site of the Trx. This disulfide is subsequently reduced by Trx reductase (TrxR), a dimeric flavo-enzyme that utilizes NADPH as electron donor (Figure 2; for a detailed description see [44]).

Compared to their bacterial counterparts, mammalian cytosolic Trxs (Trxs1) contain additional cysteine residues, that have been suggested as important for regulating the proteins' activity in response to alterations in the general redox state. Human Trx1, for instance, contains three additional cysteines which can undergo reversible disulfide formation [45], *S*-nitrosylation [46,47] and/or *S*-glutathionylation [48]. Human Trx1 can also be processed *in vivo* yielding truncated variants (Trx80) that are secreted and

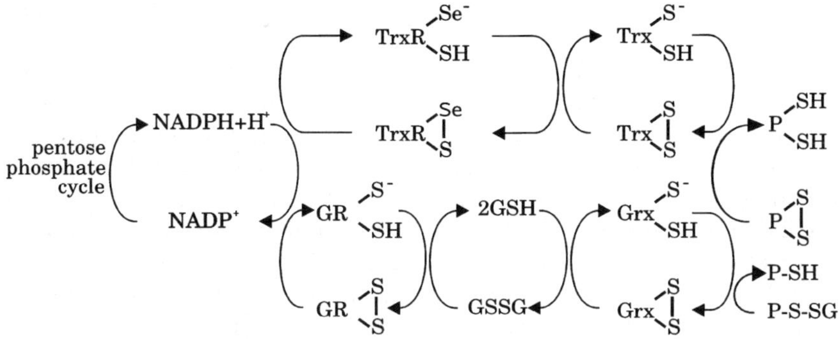

Figure 2. Electron flow from NADPH to target proteins of the thioredoxin and glutaredoxin systems. Thioredoxin (Trx) and glutaredoxin (Grx) reduce protein disulfides (P-S-S-P). In addition Grx reduces mixed disulfides between proteins and glutathione (P-S-SG). Oxidized Trx is reduced by thioredoxin reductase (TrxR), oxidized Grx by glutathione (GSH) yielding glutathione disulfide (GSSG), which is subsequently reduced by glutathione reductase (GR). Both, TrxR and GR, use electrons provided by NADPH. Mammalian TrxRs contain an active site selenol-thiol motif, while most other species utilize a dithiol motif for the reduction of Trx.

present in plasma where they were proposed to be an early signal in innate immune response [49,50].

1.3. Glutaredoxin

Grxs are small proteins of around 9–16 kDa existing in basically all glutathione (GSH)-containing forms of life. Similar to Trx, Grx was first discovered as a GSH-dependent hydrogen donor for RNR in *E. coli* cells lacking Trx [51]. The determination of the *E. coli* Grx1 amino acid sequence revealed a structural and functional relationship to Trx [4,52], as a matter of fact the basic representation of the Trx fold is present only in Grxs (Figure 1). Trxs and Grxs can compensate for each other's functions to some extent, but at the same time, both proteins display many unique features and functions [53].

Depending on their active site motif, Grxs can be categorized into (1) dithiol Grxs (consensus sequence: Cys-Pro-Tyr-Cys) and (2) monothiol Grxs (Cys-Gly-Phe-Ser). *E. coli* contains three dithiol Grxs [54,55] and one monothiol Grx [56], yeast and mammalian cells contain two dithiol Grxs [57–60] and two to five monothiol Grxs [61–64]. Mammalian dithiol Grx1 is a cytosolic protein, dithiol Grx2 and monothiol Grx5 are located inside the mitochondrion. Again, the situation in plants is far more complex.

The genome of *A. thaliana* encodes at least 14 dithiol Grxs and 17 monothiol Grxs [65].

Grxs use exclusively GSH as electron donor. The product of this reaction, glutathione disulfide (GSSG), is subsequently reduced by glutathione reductase (GR) at the expense of NADPH (Figure 2). As an exception from this general rule, Grx from the bacteriophage T4 and human Grx2 can be reduced by both GSH and TrxR, thus combining characteristics of both Trxs and Grxs [15,66]. Although some Grxs can reduce protein disulfides as efficiently as Trxs, Grxs are primarily seen as highly specific reductants of mixed disulfides formed between protein thiols and GSH (de-glutathionylation) [44,67]. Similar to Trxs, the structural comparison of reduced [5] and oxidized [68] *E. coli* Grx1 revealed small but significant changes in the area of the active site [69]. The reduced form displays a higher flexibility in this particular area leading to an increased binding affinity for substrates.

1.4. Related Proteins

Various other proteins from the Trx family of thiol-disulfide oxido-reductases have been described, see for instance refs. [70] and [71]. Excluding the endoplasmatic reticulum, nine such proteins have been described in mammals. These proteins are often tissue and/or organelle specific. SpTrxs 1-3 are three spermatocyte/spermatid-specific Trxs. SpTrx1 contains a C-terminal Trx domain, SpTrx2 a N-terminal Trx domain followed by three consecutive nucleoside diphosphate kinase domains. SpTrx3 is a single domain Trx-like protein [72].

The proteins Txl-1 and Txl-2 (Trx-like 1 and 2) both contain a N-terminal Trx domain [72,73], whereas nucleoredoxin contains a central Trx-domain [74]. Proteins with Grx domains include the thioredoxin-glutathione reductase [75], and a testis-specific splice variant of TrxR1 [76]. PICOT (protein kinase C interacting cousin of thioredoxin) contains a N-terminal Trx domain and two consecutive monothiol Grx domains [77]. From these proteins, only two exhibited oxidoreductase activity in standard Trx or Grx assays, i.e., Txl-1 and nucleoredoxin [73,74].

2. FUNCTIONS OF THIOREDOXINS AND GLUTAREDOXINS

Both Trx and Grx were first discovered for their ability to serve as electron donor for metabolic enzymes such as RNR. Over the past three decades the proteins turned out to be general protein disulfide reductases (Trx) and effective catalysts of reversible protein glutathionylation (Grx), respectively [71].

2.1. Thioredoxin and Glutaredoxin as Electron Donors

Trx was first described as electron donor for RNR and sulfate reduction in *E. coli* and yeast, respectively [38,78–81]. RNRs catalyze the conversion of nucleotides to deoxynucleotides, the building blocks for DNA synthesis [82,83]. The mammalian and aerobic *E. coli* enzymes belong to the class I RNRs. Enzymatic turn-over of ribonnucleotides requires the reduction of a disulfide in the RNR R1 subunit by Trxs or Grxs [38,51,54,84,85]. In *E. coli,* Grx1 may be the primary electron donor for RNR *in vivo* [86,87]. Yeast, on the other hand, seems to utilize both Trxs and Grxs as electron donor for RNR *in vivo* [88,89]. In mammalian cells the importance of both proteins as electron donor for RNR *in vivo* is less well understood, because the distribution of Trx/TrxR and Grx in tissues is not related to cell proliferation or distribution of RNR [90–92].

Bacteria, fungi, and plants are able to satisfy their need for reduced sulfur by assimilation of inorganic sulfate. Reduction of sulfate (SO_4^{2-}) to sulfide (S^{2-}) requires eight electrons and takes place in two steps. First, sulfate is activated to adenylylsulfate (APS) or phoshoadenylylsulfate (PAPS) and subsequently reduced to sulfite (SO_3^{2-}) by APS and PAPS reductase, respectively. Secondly, sulfite is reduced to sulfide using six electrons provided by NADPH in bacteria and fungi or ferredoxin in photosynthetic organisms [93,94]. The requirement for a low molecular weight dithiol reductand in the first step was originally described by Wilson et al. [79]. In parallel to the history of RNR, Grx and Trx were identified to be the alternative electron donors for PAPS reductase in *E. coli* [95–97].

Next to metabolic enzymes, Trx and Grx have also been described as electron donors for other antioxidant enzymes. For instance, Prxs are an ubiquitous family of thiol-dependent peroxidases that fall into three major classes. The typical (in human Prx1 to Prx4) and atypical 2-Cys Prxs (human Prx5), which utilize Trx as electron donor, and the 1-Cys Prxs (human Prx6), the electron donor of which is not yet clear [98–100]. Methionine sulfoxides that may form during oxidative stress are reduced by methionine sulfoxide reductases using Trx as electron donor [101]. This modification is discussed to be important in the regulation of protein function and aging [102,103].

2.2. Thiol Redox Control

Trxs and Grxs keep a reduced environment inside the cell by reducing protein disulfides and protein-GSH mixed disulfides, respectively, during oxidative challenges. By modulating the redox state of critical protein thiols both Trxs and Grxs function as redox regulators of various signaling molecules and transcription factors. Many of these redox-regulated transcription factors contain critical cysteines in their DNA binding domain. Trxs and Grxs provide the regulatory switch to modulate the binding activity of these factors in response to the cellular redox state. As an example, NF-κB p50 contains a single cysteine (Cys 62) in its DNA binding domain that is susceptible to oxidation [104]. Reduction of Cys62 is necessary for the binding of NF-κB to its target site in the DNA. Trx1 reduces Cys62 disulfides formed between two monomers of p50 in the nucleus enabling binding of NF-κB to its target site [105,106]. Grx1 may be part of this redox regulon as well because Cys62 of p50 can also undergo reversible glutathionylation [107,108].

Apoptosis signal-regulating kinase 1 (ASK1) is a mitogen-activated protein kinase which activates downstream kinases like c-Jun N-terminal kinase and p38 MAP kinase, for instance, in response to apoptotic stimuli like tumor necrosis factor α [109]. In humans, both Trx1 and Grx1 regulate the activity of this kinase by binding reversibly to ASK1 dependent on their own redox state. Remarkably, this regulation is independent of the redox activity of the redoxins. Reduced Trx forms a complex with the N-terminal portion of ASK1 in which the kinase activity of ASK1 is suppressed. Oxidation of Trx leads to dissociation of the complex and activation of ASK1 [110]. Grx1, on the other hand, binds to the C-terminal domain of ASK1 [111] and may regulate its kinase activity in response to the GSH redox state [112].

Recently, Casadei et al. [113] reported the reversible glutathionylation of metallothioneins (MTs) under conditions of nitrosative and oxidative stress *in vitro*. The authors proposed conserved cysteines in the N-terminal domain of the MTs, that were shown to undergo reversible S-nitrosylation before, as target for S-glutathionylation [113] (and references therein). MTs tend to aggregate under conditions of oxidative stress [114,115] and glutathionylation appeared to promote this behavior both *in vitro* and *in vivo* [113]. These initial findings suggest a possible regulation of MT function by Grxs.

2.3. Diseases Related to Dysfunction of the Thioredoxin and Glutaredoxin Systems

Trx and Grx have been implied in various aspects of human health and disease and this topic has been reviewed exhaustively before, see for instance refs. [71] and [116]. A detailed description of all these aspects is out of the focus of this chapter. We therefore provide a brief overview and summarize the major review articles and some of the key references therein.

The role of the Trx system in the nervous system has been reviewed in detail in [117,118]. For instance, both Trx1 and Grx1 exhibit a potential beneficial role during ischemia-reperfusion injury in animal models of focal cerebral ischemia. Trx staining decreases in ischemic regions but increases in not directly affected regions [119,120]. In a similar manner, Grx1 is reduced in the area of the ischemic insult [121]. The beneficial roles of both endogenous as well as exogenous Trx and Grx in the cardiovascular system have been summarized, for instance, in refs. [44], [122], and [123]. The role of Trxs, Grxs and GSH in protecting the lens against exogenous and endogenous oxidative stress have been discussed extensively in [124,125]. Secreted Trx1 and its truncated sibling Trx80 are powerful cytokines and chemokines involved in the activation of T cells and lymphocytes [49,50]. Detailed aspects of their function in immunity and during viral infections have been discussed before, see for instance refs [71] and [126].

A well balanced thiol-disulfide redox homeostasis is crucial for the maintenance of airway function and most of all Trxs play a prominent role in these regulatory circuits [127–129]. The roles of the Trx system in the promotion of cancer and as therapeutic target have recently been summarized [130]. In testis and cancer cells, the normally mitochondrial Grx2 is present in two alternative cytosolic/nuclear isoforms, implying a role for these proteins in the regulation of cell differentiation/de-differentiation [131]. Aging correlates with a continuous increase in oxidative damage of multiple cellular components. The role of cellular antioxidants such as the Trx and Grx systems in the aging process have been discussed, for instance, in refs. [132], [133], and [134]. The strongest evidence for a protective role of the proteins are Trx1 transgenic mice produced by Mitsui et al. [135] who exhibit a significantly increased lifespan [133,135]. Last but not least, recent evidence for a crucial role of Grxs in iron homeostasis point to potential roles in iron-related diseases, such as microcytic anemia [136].

3. METAL BINDING MEMBERS OF THE THIOREDOXIN FAMILY OF PROTEINS

The Cys-X-X-Cys active site motif of Trx family members is also well known as ligand motif for the coordination of iron, zinc, copper or cadmium in a variety of metal-binding proteins like MTs, zinc-finger proteins, and Hsp33 chaperones [137–139]. This basic similarity raises the question what mechanisms prevent the binding of metals in most Trx-related thiol-disulfide oxidoreductases. Recent reports based on mutagenesis of various Trxs and Grxs highlighted the importance of two features: (1) The *cis*-proline efficiently precludes metal binding by the active site [140]. Replacement of the *cis*-proline for a histidine residue in some members of the Trx family, e.g., Sco1, allows metal coordination [28]. (2) The two central residues in between the two active cysteine residues are another important factor distinguishing between metal-binding and metal-free proteins of the Trx family [17,141,142]. Next to these features, a number of proteins utilize alternative metal ligands outside the classical active site to convert the Trx fold into a scaffold for metal binding [143,144].

3.1. Iron Binding

3.1.1. Thioredoxins

As of today, no iron-coordinating wild-type Trx has been documented. This is not surprising, since all Trxs contain the *cis*-proline residue that, together with the Cys-Gly-Pro-Cys active site, efficiently precludes metal binding. However, the wealth of structural information on Trxs as well as their high expression levels and stability predestines them as design platform for the genetic and *in silico* engineering of metallocenters.

Exchange of the *cis*-proline in human Trx1 for serine, threonine, arginine or alanine resulted in the ability of the proteins to coordinate an [FeS] cluster [140]. Next to the *cis*-proline, the occurrence of a proline in the active site seems to preclude metal binding (see also Section 3.1.2). The group of Bardwell was able to genetically select *E. coli* Trx1 mutants in which the Cys-Gly-Pro-Cys active site was changed to Cys-Gly-Cys-Ala or Cys-Gly-Cys-Cys, respectively, that were able to coordinate an [FeS] cluster [141,145]. These mutations converted the reductase Trx1 into an O_2-dependent sulfhydryl oxidase allowing the proteins to catalyze oxidative protein folding in the bacterial periplasm and rescue strains lacking DsbA and DsbB [141], which are normally essential for disulfide bond formation (reviewed in [146]). Iron and acid labile sulfide determination, gel filtration, UV/VIS and CD spectroscopy, as well as crystal structures revealed the presence of a [2Fe2S] cluster in these mutants. The single [FeS] cluster was coordinated between two monomers by the two more N-terminal active site cysteine residues [140–141,145]. *E. coli* Trx1 mutants coordinating different iron centers were constructed by rational protein design, i.e., mononuclear non-heme iron superoxide dismutase-like iron centers [147], mononuclear rubredoxin-like iron centers [148], [4Fe4S] clusters [149], and Fe-His_3-O_2 centers [150].

3.1.2. Glutaredoxins

It came to a big surprise when in 2005 human Grx2 was described as the first native [FeS] protein from the Trx family of thiol-disulfide oxidoreductases [151]. This primarily mitochondrial Grx differs in many aspects from other "classical" dithiol Grxs. The active site prolyl residue is exchanged for a seryl residue (Cys-Ser-Tyr-Cys) [59,60], the protein can use electrons not only from GSH, but also from TrxR [66] and it is not inactivated by oxidative conditions [152]. The [2Fe2S] cluster bound to Grx2 bridges two Grx2 monomers to form a dimeric holo-Grx2 complex. Astonishingly, it is the holo-complex that lacks enzymatic activity while dissociation of the cluster yields the active monomeric apo-protein. Oxygen-dependent degradation of the holo-complex was efficiently prevented by GSH. GSSG and other redox active compounds promoted degradation [151].

Subsequent biochemical analysis revealed that the iron-sulfur cluster is complexed by the two N-terminal active site thiols of two Grx2 monomers and the thiol groups of two molecules of GSH non-covalently bound to the two Grx2 monomers in the holo-complex [17]. The structure of the dimeric holo Grx2 complex (Figure 3) was solved by X-ray diffraction [153]. Remarkably, most molecular interactions contributing to the holo complex involve the GSH molecules and the [FeS] center. Hardly any direct

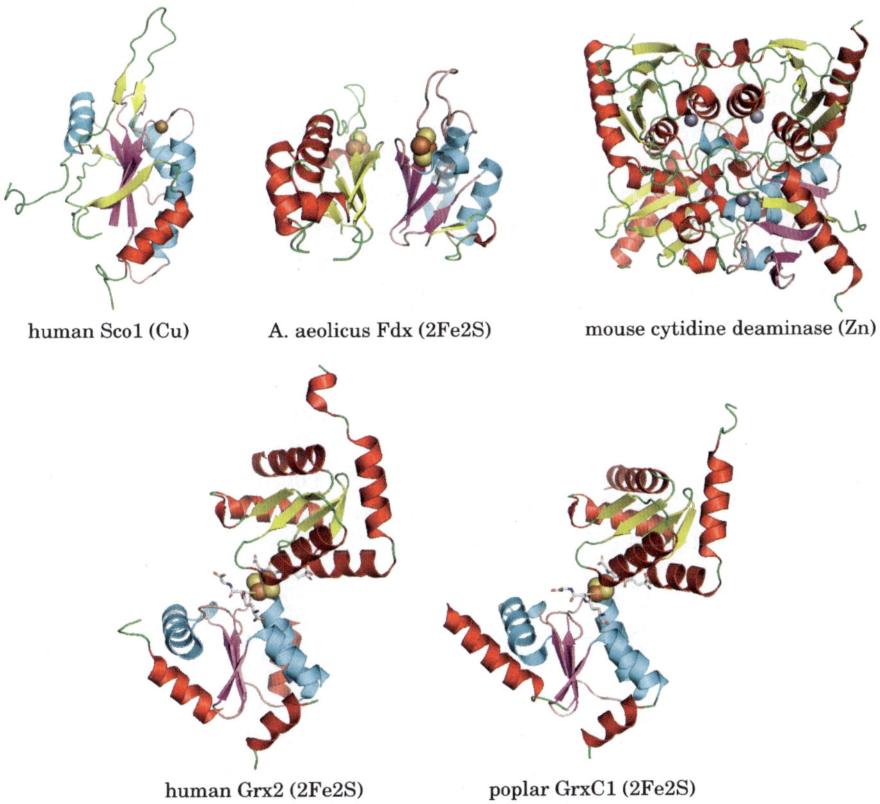

Figure 3. Different thioredoxin-fold proteins harboring metal binding sites. The basic Trx-fold is highlighted in blue and purple ribbon representation, in case of multimeric proteins, the motif is highlighted in only one subunit. Human Sco 1 (PDB code 2 gqm), dimeric *A. aeolicus* Fdx (1f37), tetrameric mouse cytidine deaminase (1fr6), dimeric human Grx2 (2ht9), and dimeric poplar GrxC1 (2e7p) are shown.

interactions between the two protein monomers are formed. Similar to human Grx2, the cytosolic GrxC1 from poplar with a glycyl residue instead of the consensus prolyl residue in the active site can form essentially the same dimeric iron-sulfur-containing holo-complex [142,154] (Figure 3).

Moreover, recent data from a number of groups indicate that most monothiol Grxs (active site Cys-Gly-Phe-Ser) are able to form the dimeric iron-sulfur cluster bridged holo-complex *in vitro* [63,155]. The main property that enables human Grx2, poplar GrxC1, and the various monothiol Grxs to form the [Fe-S] bridged dimeric holo complex is the exchange of the active site proline present in "classical" dithiol Grxs for a serine or glycine. This exchange allows a higher flexibility of the main chain in the active site area and thereby non-covalent binding of GSH [142,153]. As a matter of fact, when the active site of human Grx1 harbouring the classical Cys-Pro-Tyr-Cys active site, was changed to the corresponding Cys-Ser-Tyr-Cys sequence of Grx2, Grx1 became able to complex the [2Fe2S] cluster *in vitro* as well [17]. Mutagenesis of different poplar glutaredoxins suggests that the incorporation of an iron-sulfur cluster could be a general feature of various plant glutaredoxins possessing a Cys-Gly-Tyr-Cys active site [142].

Despite of the various Grxs that have been shown to bind an [2Fe2S] center *in vitro*, the physiological importance of these metal centers have yet to be demonstrated. At the time of writing, only human Grx2 was shown to bind iron *in vivo*, i.e., in cultured cells [151].

3.1.3. Other Thioredoxin Fold Proteins

Since about 40 years, ferredoxins (Fdxs) are well known as [FeS] proteins. In 2000 the structure of the [2Fe2S] Fdx from *Aquifex aeolicus* revealed that this Fdx belongs to the Trx fold family [143]. The [2Fe2S] cluster is coordinated by four cysteines, which are located outside but near the classical Trx active site region [143,156] (Figure 3).

3.2. Zinc and Other Metals

E. coli Trx2 is the only metal binding Trx described so far [144]. This protein has an additional N-terminal domain of 32 amino acids including two more Cys-X-X-Cys motifs. These four cysteine residues are able to coordinate zinc. Since the oxidoreductase activity of Trx2 is not affected by zinc binding, the function of the metal remains elusive. Several zinc binding sites [157] and a copper binding site found in blue copper proteins [158,159] were introduced by rational protein design into Trxs. Some of the copper binding sites were also able to bind cobalt(II) and mercury(II) [159]. A systematic mutational approach revealed that mutation of Glu30 and Gln62 to histidines allow *E. coli* Trx1 to coordinate nickel and copper, a feature that has been utilized to purify recombinant Trx fusion proteins via immobilized metal ion affinity chromatography [160]. Trx, covalently linked to the protein of interest, avoids inclusion body formation and thereby increases the levels of soluble proteins for multiple heterologously expressed proteins [161].

The Sco family of proteins is involved in the copper-dependent assembly of cytochrome *c* oxidase [162,163], the terminal enzyme complex of the respiratory chain [164]. Leary et al. proposed that Sco1 and Sco2 are also responsible for maintenance of the cellular copper homeostasis [165], a function that requires copper-binding [28,166,167]. These proteins were characterized as Trx fold proteins [12,168–172], in which the cysteine residues of a Cys-X-X-X-Cys motif and a single histidine in place of the Cys-X-X-Cys active site and the *cis*-proline, respectively, coordinate Cu(I), Cu(II), Ag(I), Co(II), Ni(II), or Zn(II) [28,170,173,174] (Figure 3). Noteworthy, the insertion of a histidine in place of the *cis*-proline, converted even human Trx1 to a copper and zinc binding protein [140].

Cytidine deaminases, zinc-binding proteins involved in the pyrimidine salvage pathway which enables certain organisms to utilize exogenous pyrimidine bases and nucleosides, exhibit a circular-permutated Trx fold [14]. Each monomer binds one zinc with the help of three ligands: the two cysteine residues of the Cys-X-X-Cys active site, and a histidine in the homodimeric forms found in *E. coli* [175] and *A. thaliana* [176], or a third cysteine in the homotetrameric forms present in *B. subtilis* [177], *M. musculus* [29], and *H. sapiens* [178]. Again, the third ligand is located in a position structurally similar to that of the *cis*-proline (Figure 3).

The uncharacterized COG3019 protein family is described as "predicted metal-binding protein" in the COG database [179] and several members of this family functionally cluster with multicopper oxidases, copper transporting ATPases, and cobalt-zinc-cadmium resistance proteins [180]. These proteins contain a conserved Cys-Gly-Cys-Cys active site motif and a histidine in place of the *cis*-proline [181]. In theory, this combination of features allows several metal ion interactions similar to Sco proteins, cytidine deaminases, and the Cys-X-Cys-X active site mutants of *E. coli* Trx1 (see above).

The C-terminal domains of tubulin α- and β-subunits show a more distantly related Trx-fold [14]. A putative zinc binding site has been proposed at the lateral contacts between α-tubulin C-terminal subunits, which consists of one histidine residue and one glutamate residue per subunit [182].

4. METAL ION INTERACTIONS AND PHYSIOLOGY

The high reactivity of iron with different oxygen species demands tightly controlled regulatory circuits during the synthesis of iron-containing cofactors, such as [FeS] centers, to avoid toxic side effects of free intracellular iron ions.

4.1. Iron-Sulfur Cluster Biogenesis

[FeS] centers are multipurpose structures found in all life forms. They can undergo reversible redox reactions, determine protein structure, act as catalytic centers, and as sensitive sensors of iron and various oxygen species [183,184]. In eukaryotic cells, iron-sulfur cluster synthesis is an essential function of mitochondria and thought to take place on the scaffold protein Isu (IscU/NifU/SufU in bacteria) from where the [FeS] units are transferred to apo [FeS] proteins with the help of DnaK and DnaJ type chaperones (overviews in [185] and [186]). Absence of the mitochondrial monothiol Grx5 in yeast led to constitutive oxidative damage, iron accumulation in the cell and inactivation of iron-sulfur containing enzymes. These defects could be complemented by over-expression of two proteins involved in [FeS] assembly, the molecular chaperon Ssq1 and the potential alternative scaffold Isa. Hence, a function of Grx5 in iron-sulfur cluster synthesis or repair was suggested [187].

Based on structural models, Alves et al. predicted specific complexes between Grx5 and early components of the yeast ISC (iron-sulfur cluster) machinery, i.e. the cysteine desulfurase and the scaffold proteins [188]. Two-hybrid analysis confirmed possible molecular interactions between Grx5 and Isa1 [189]. Mühlenhoff et al. demonstrated in their groundbreaking work from 2003 [190] that depletion of Grx5 from yeast increased the amount of iron, most likely in the form of [FeS] centers, bound to the scaffold protein Isu1. These results strongly suggest a function of Grx5 in a step following the initial [FeS] cluster synthesis on Isu1 when the [FeS] clusters have to be inserted into apo proteins with the help of the molecular chaperones Ssq1 and Atm1 [190] (Figure 4).

The importance of the mitochondrial monothiol Grxs for iron homeostasis was most impressively demonstrated when the hypochromic anemic zebrafish mutant Shiraz was shown to be caused by a deficiency in Grx5 leading to an impaired [FeS] cluster assembly and as a result to defects in iron homeostasis and heme synthesis [64]. In addition, a human counterpart of the Shiraz mutant has been identified by Camaschella et al. caused by a homozygous silent mutation in the human Grx gene that decreases splicing efficiency [136]. Corroboratively, the middle-aged patient displayed sideroblastic-like microcytic anemia and iron overload. Monothiol Grxs from various species including bacteria, protozoans, plants, and vertebrates are able to rescue the phenotype of the yeast Grx5^{-} mutant when targeted to mitochondria [64,191,192]. These findings indicate that the yet to be

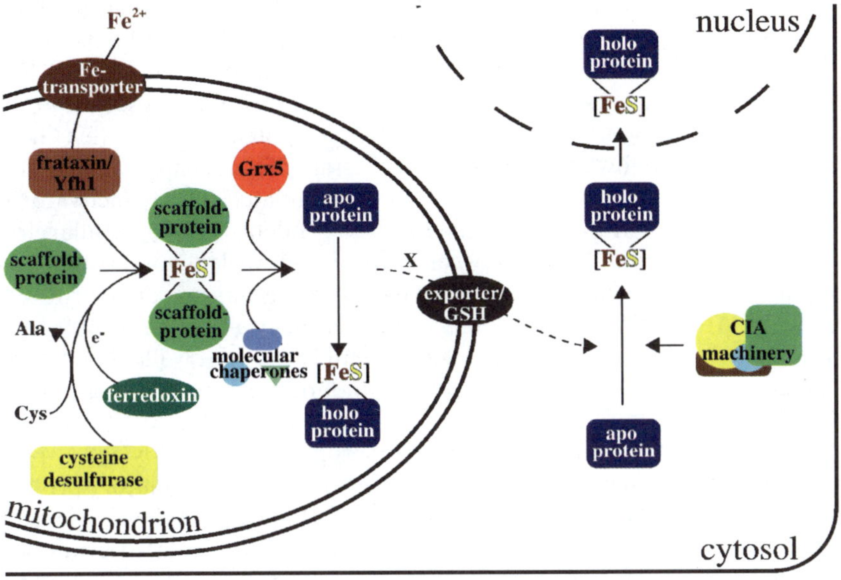

Figure 4. Model for the biosynthesis of iron-sulfur clusters in eukaryotic cells. The mitochondrial [FeS] cluster biosynthesis machinery is essential for assembly of mitochondrial, cytosolic, and nuclear [FeS] proteins. A hitherto unknown factor X, provided by the mitochondrial ISC machinery exported in a mechanism that requires glutathione (GSH), is required for the cytosolic iron-sulfur cluster assembly complex (CIA) for the biosynthesis of cytosolic and nuclear [FeS] proteins. In mitochondria, iron is delivered by frataxin and homologue proteins to the scaffold protein, which coordinates the newly synthesized [FeS] cluster before transferring it to apo proteins. Sulfide is provided by cysteine desulfurase and ferredoxin. For the transfer of [FeS] clusters to apo proteins several molecular chaperones and the monothiol Grx5 are required. For a detailed overview see reference [214].

established biochemical functions of monothiol Grxs in [FeS] assembly and iron homeostasis are conserved throughout evolution.

The physiological functions of IscA-like proteins (Isa, IscA, and SufA) are still being controversially discussed. IscA and SufA have been described as alternative scaffold proteins [193–195] and, more recently, as potential iron donors for the formation of an iron-sulfur cluster in the scaffold IscU [196–198]. IscA binds iron with high affinity [197,199] and releases iron in the presence of cysteine [200], the sulfur donor for iron-sulfur cluster biosynthesis. In *E. coli*, the Trx system was shown to mediate iron binding of IscA [201] and SufA [198].

The regulation of iron uptake, intracellular iron levels and iron-related proteins is, unlike [FeS] cluster synthesis, not conserved between different evolutionary distant groups. In *E. coli* and several other bacteria, the ferric uptake regulator (Fur) is the main sensor of iron. Ferrous iron-loaded Fur is a transcriptional repressor of multiple iron homeostasis-related genes [202]. Depletion of iron, especially in the case of *fur*$^{-}$ strain, caused dramatic elevation in levels of monothiol Grx4, implying a potential role of this Grx in iron-dependent pathways [56]. The expression of genes involved in iron homeostasis in *Saccharomyces cerevisiae* are regulated by Aft (activator of ferrous transport) transcription factors. Dependent on the cellular iron status, these proteins shuttle between cytosol and nucleus to activate transcription of iron regulon genes [203,204]. The mechanism of how iron is sensed by Aft is still unknown, however, this function depends on a functional mitochondrial iron sulfur cluster assembly machinery [205] and direct interactions of Aft with the two cytosolic monothiol Grxs, Grx3 and Grx4 [206,207].

Vertebrate cells regulate intracellular iron levels employing a post-transcriptional mechanism based on iron-dependent iron regulatory proteins (IRPs) [208–210]. Activated IRP1 and IRP2 bind to iron regulatory elements (IRE), secondary structure elements in the mRNA of regulated genes, and regulate the translation of these IRE-containing mRNAs. Loss of Grx5 in the zebrafish Shiraz mutant promoted the activation of IRP1. To some extend, knockdown of IRP1 restored hemoglobin synthesis in the Grx5 mutant, demonstrating a crosstalk between hemoglobin production and mitochondrial [FeS] cluster assembly [64].

Both IRP1 and IRP2 have long been known as redox-sensitive proteins. Their activation may be induced by oxidation, however, full IRE-binding activity is only achieved when IRP cysteinyl thiols are present in their reduced form. *In vitro*, 2-mercaptoethanol is required to obtain maximal IRP1 activity [211]. IRP1 can be activated by S-nitrosylation, but this activation *in vitro* is considerably lower compared to the one observed in cells [212]. Activation *in vitro* is significantly increased in the presence of reduced Trx and *in vivo* NO-mediated IRP1 activation is effectively prevented by anti-Trx antibodies [213]. These observations point to a potential physiological role of the Trx system in vertebrate iron homeostasis.

4.2. Oxidative Stress and Redox Regulation

The physiological and molecular function of the iron-sulfur centers in both monothiol and certain dithiol Grxs have, at the time of writing, yet to be established. However, it is tempting to speculate about a role of the [2Fe2S] cluster binding activity of Grx5 during the transfer of [FeS] cofactors from the scaffold proteins onto newly synthesized apo-proteins. Functions currently discussed for the other Grxs are (1) that of a sensor of iron levels, as in

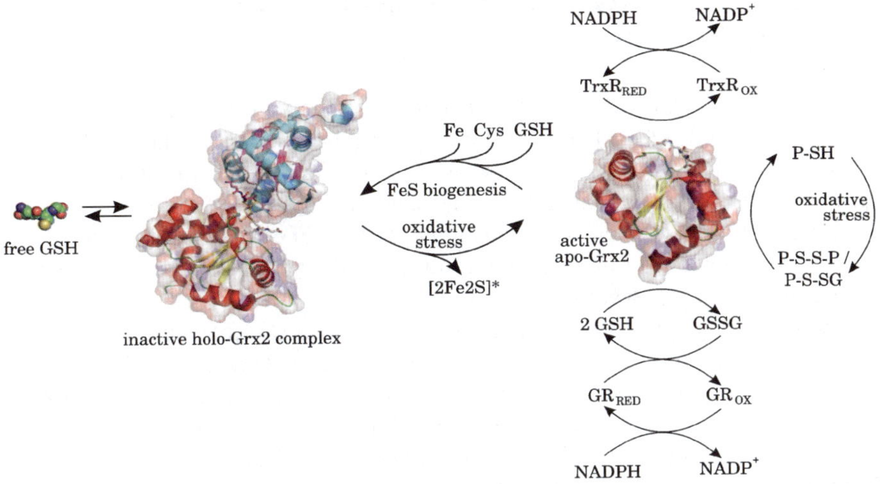

Figure 5. Model for the redox regulation of human glutaredoxin 2. In the holo-form of human glutaredoxin 2 (hGrx2) two monomers are bridged via an iron-sulfur cluster. This cluster is coordinated by the N-terminal active site cysteines of the protein monomers and the cysteine residues of non-covalently bound glutathione (GSH) molecules, which are in constant exchange with GSH in solution. Under conditions of oxidative stress, when GSH becomes the limiting factor for cluster coordination, the dimeric holo hGrx2 complex dissociates. Apo-hGrx2 is enzymatically active reducing protein disulfides and mixed disulfides between proteins and GSH using electrons provided by NADPH via thioredoxin reductase (TrxR), or glutathione reductase (GR) and GSH.

the case of yeast Grx3 and Grx4, and/or (2) as sensor of reactive oxygen species, a role first suggested for human Grx2. This mitochondrial dithiol Grx is a potent inhibitor of oxidative stress-induced apoptosis. Because Grx2′s [FeS] cluster is vulnerable to oxidative destruction yielding active apo-monomers, a role of the [FeS] cluster as sensor of oxidative stress has been proposed. In this model (Figure 5), Grx2 is primarily present in its inactive dimeric holo-form under normal conditions.

Intracellular oxidative stress is characterized by an increase in the ratio of GSSG to GSH and a decrease in total glutathione. Reduced GSH is required for the coordination of the [FeS] cofactor in the holo-Grx2 complex and this GSH is in dynamic equilibrium with free GSH. If reduced GSH becomes the limiting factor, the holo-Grx2 complex will dissociate and activate Grx2. In support of this speculative scenario is the fact that the activity of Grx2 is highly resistant to oxidative conditions, while other Grxs, such as human Grx1, that lack [FeS] cluster binding activity are inactivated by oxidative conditions [152].

5. CONCLUDING REMARKS AND FUTURE DIRECTIONS

The field of redox biochemistry in general and the investigation of metal ion interactions with the Trx and Grx system in particular have advanced considerably during the last few years. The identification of iron-sulfur glutaredoxins and the accumulating evidence for their role in the biosynthesis of these clusters as well as in regulation of cellular iron homeostasis represent milestones in our understanding of the physiological role of these proteins in various organisms.

Undoubtedly, the future will bring more and more interactions between redox and inorganic biochemistry to light. A key role in our understanding of the interplay between these two worlds will be the identification of protein-protein interaction networks and detailed investigations of how metal interactions affect and regulate these interactions. What are the factors and signals that mediate between metal and redox homeostasis and what do they respond to? In the future, this knowledge may provide new strategies to combat diseases in which the regulatory circuits controlling redox and metal homeostasis fail.

ACKNOWLEDGMENTS

The authors wish to thank Karin Beimborn, Gisela Lesch, and Lena Ringdén for excellent administrative assistance and Eva-Maria Hanschmann for reading the manuscript. The authors' own work was funded by the Deutsche Forschungsgemeinschaft, Karolinska Institutet, the Kempkes Foundation, the Swedish Children's Cancer Foundation (Barncancerfonden), and the Swedish Society for Medical Research.

ABBREVIATIONS

Aft	activator of ferrous transport
APS	adenylylsulfate
ASK1	apoptosis signal-regulating kinase 1
CD	circular dichroism
CIA	cytosolic iron sulfur cluster assembly
COG	clusters of orthologous groups
Fdx	ferredoxin
Fur	ferric uptake regulator
GR	glutathione reductase

Grx	glutaredoxin
GSH	glutathione
GSSG	glutathione disulfide
IRE	iron responsive element
IRP	iron regulating protein
ISC	iron-sulfur cluster
MT	metallothionein
NADPH	nicotinamide adenine dinucleotide phosphate (reduced)
NF-κB	nuclear factor κB
PAPS	phospho-adenylylsulfate
PICOT	protein kinase C interacting cousin of thioredoxin
Prx	peroxiredoxin
RNR	ribonucleotide reductase
SpTrx	spermatocyte/spermatid-specific thioredoxin
Trx	thioredoxin
TrxR	thioredoxin reductase
Txl	thioredoxin-like

REFERENCES

1. A. Holmgren, B. O. Söderberg, H. Eklund and C. I. Bränden, *Proc. Natl. Acad. Sci. USA*, 1975, **72**, 2305–2309.
2. A. Holmgren, *Structure*, 1995, **15**, 15.
3. J. L. Martin, *Structure*, 1995, **3**, 245–250.
4. H. Eklund, C. Cambillau, B. M. Sjöberg, A. Holmgren, H. Jörnvall, J. O. Höög and C. I. Bränden, *EMBO J.*, 1984, **3**, 1443–1449.
5. P. Sodano, T. H Xia, J. H. Bushweller, O. Björnberg, A. Holmgren, M. Billeter and K. Wüthrich, *J. Mol. Biol.*, 1991, **221**, 1311–1324.
6. J. L. Martin, J. C. A. Bardwell and J. Kuriyan, *Nature*, 1993, **365**, 464–468.
7. J. Kemmink, N. J. Darby, K. Dijkstra, M. Nilges and T. E. Creighton, *Biochemistry*, 1996, **35**, 7684–7691.
8. A. A. McCarthy, P. W. Haebel, A. Törrönen, V. Rybin, E. N. Baker and P. Metcalf, *Nature Struc. Biol.*, 2000, **7**, 196–199.
9. H. J. Choi, S. W. Kang, C. H. Yang, S. G. Rhee and S. E. Ryu, *Nat. Struc. Biol.*, 1998, **5**, 400–406.
10. P. Reinemer, H. W. Dirr, R. Ladenstein, J. Schäffer, O. Gallay and R. Huber, *EMBO J.*, 1991, **10**, 1997–2005.
11. O. Epp, R. Ladenstein and A. Wendel, *Eur. J. Biochem.*, 1983, **133**, 51–69.
12. E. Balatri, L. Banci, I. Bertini, F. Cantini and S. Ciofi-Baffoni, *Structure*, 2003, **11**, 1431–1443.
13. R. H. Ashley, *Mol. Membr. Biol.*, 2003, **20**, 1–11.
14. Y. Qi and N. V. Grishin, *Proteins*, 2005, **58**, 376–388.
15. A. Holmgren, *Annu. Rev. Biochem.*, 1985, **54**, 237–271.
16. D. E. Fomenko and V. N. Gladyshev, *Biochemistry*, 2003, **42**, 11214–11225.

17. C. Berndt, C. Hudemann, E. M. Hanschmann, R. Axelsson, A. Holmgren and C. H. Lillig, *Antioxid. Redox Signal.*, 2007, **9**, 151–157.
18. G. Roos, A. Garcia-Pino, K. Van Belle, E. Brosens, K. Wahni, G. Vandenbussche, L. Wyns, R. Loris and J. Messens, *J. Mol. Biol.*, 2007, **368**, 800–811.
19. M. Wunderlich and R. Glockshuber, *Protein Sci.*, 1993, **2**, 717–726.
20. G. Krause, J. Lundström, J. L. Barea, C. Pueyo de la Costa and A. Holmgren, *J. Biol. Chem.*, 1991, **266**, 9494–9500.
21. E. Mössner, M. Huber-Wunderlich and R. Glockshuber, *Protein Sci.*, 1998, **7**, 1233–1244.
22. M. Huber-Wunderlich and R. Glockshuber, *Fold. Des.*, 1998, **3**, 161–171.
23. J. Lundstrom, G. Krause and A. Holmgren, *J. Biol. Chem.*, 1992, **267**, 9047–9052.
24. J. B. Charbonnier, P. Belin, M. Moutiez, E. A. Stoura and E. Quemeneur, *Protein Sci.*, 1999, **8**, 96–105.
25. C. Nathaniel, L. A. Wallace, J. Burke and H. W. Dirr, *Biochem. J.*, 2003, **372**, 241–246.
26. H. Kadokura, L. Nichols and J. Beckwith, *J. Bacteriol.*, 2005, **187**, 1519–1522.
27. S. D. Copley, W. R. Novak and P. C. Babbitt, *Biochemistry*, 2004, **43**, 13981–13995.
28. T. Nittis, G. N. George and D. R. Winge, *J. Biol. Chem.*, 2001, **276**, 42520–42526.
29. A. H. Teh, M. Kimura, M. Yamamoto, N. Tanaka, I. Yamaguchi and T. Kumasaka, *Biochemistry*, 2006, **45**, 7825–7833.
30. A. Vlamis-Gardikas and A. Holmgren, *Methods Enzymol.*, 2002, **347**, 286–296.
31. H. Masutani, K. Hirota, T. Sasada, Y. Ueda-Taniguchi, Y. Taniguchi, H. Sono and J. Yodoi, *Immunol. Lett.*, 1996, **54**, 67–71.
32. A. Rubartelli, A. Bajetto, G. Allavena, E. Wollman and R. Sitia, *J. Biol. Chem.*, 1992, **267**, 24161–24164.
33. G. Spyrou, E. Enmark, A. Miranda-Vizuete and J. Gustafsson, *J. Biol. Chem.*, 1997, **272**, 2936–2941.
34. B. B. Buchanan and Y. Balmer, *Annu. Rev. Plant Biol.*, 2005, **56**, 187–200.
35. Y. Meyer, J. P. Reichheld and F. Vignols, *Photosynth. Res.*, 2005, **86**, 419–433.
36. S. D. Lemaire, L. Michelet, M. Zaffagnini, V. Massot and E. Issakidis-Bourguet, *Curr. Genet.*, 2007, **51**, 343–365.
37. Y. Meyer, F. Vignols and J. P. Reicheld, *Methods Enzymol.*, 2002, **347**, 394–402.
38. T. C. Laurent, E. C. Moore and P. Reichard, *J. Biol. Chem.*, 1964, **239**, 3436–3444.
39. A. Holmgren, *Eur. J. Biochem.*, 1968, **6**, 475–484.
40. M. F. Jeng, A. P. Campbell, T. Begley, A. Holmgren, D. A. Case, P. E. Wright and H. J. Dyson, *Structure*, 1994, **2**, 853–868.
41. J. Qin, G. M. Clore and A. M. Gronenborn, *Structure*, 1994, **2**, 503–522.
42. A. Smeets, C. Evrard, M. Landtmeters, C. Marchand, B. Knoops and J. P. Declercq, *Protein Sci.*, 2005, **14**, 2610–2621.
43. L. Stryer, A. Holmgren and P. Reichard, *Biochemistry*, 1967, **6**, 1016–1020.
44. C. Berndt, C. H. Lillig and A. Holmgren, *Am. J. Physiol. Heart Circ. Physiol.*, 2007, **292**, 1227–1236.

45. A. Holmgren, *J. Biol. Chem.*, 1977, **252**, 4600–4606.
46. J. Haendeler, J. Hoffmann, V. Tischler, B. C. Berk, A. M. Zeiher and S. Dimmler, *Nat. Cell Biol.*, 2002, **4**, 743–749.
47. D. A. Mitchell and M. A. Marletta, *Nat. Chem. Biol.*, 2005, **1**, 154–158.
48. S. Casagrande, V. Bonetto, M. Fratelli, E. Gianazza, I. Eberini, T. Massignan, M. Salmona, G. Chang, A. Holmgren and P. Ghezzi, *Proc. Natl. Acad. Sci. USA*, 2002, **99**, 9745–9749.
49. K. Pekkari, R. Gurunath, E. S. Arner and A. Holmgren, *J. Biol. Chem.*, 2000, **275**, 37474–37480.
50. K. Pekkari, M. T. Goodarzi, A. Scheynius, A. Holmgren and J. Avila-Carino, *Blood*, 2005, **105**, 1598–1605.
51. A. Holmgren, *Proc. Natl. Acad. Sci. USA*, 1976, **73**, 2275–2279.
52. J. O. Höög, H. Jörnvall, A. Holmgren, M. Carlquist and M. Persson, *Eur. J. Biochem.*, 1983, **136**, 223–232.
53. A. P. Fernandes and A. Holmgren, *Antioxid. Redox Signal.*, 2004, **6**, 63–74.
54. A. Holmgren, *J. Biol. Chem.*, 1979, **254**, 3664–3671.
55. F. Aslund, B. Ehn, A. Miranda-Vizuete, C. Pueyo and A. Holmgren, *Proc. Natl. Acad. Sci. USA*, 1994, **91**, 9813–9817.
56. A. P. Fernandes, M. Fladvad, C. Berndt, C. Andresen, C. H Lillig, P. Neubauer, M. Sunnerhagen, A. Holmgren and A. Vlamis-Gardikas, *J. Biol. Chem.*, 2005, **280**, 24544–24552.
57. S. Luikenhuis, G. Perrone, I. W. Dawes and C. M. Grant, *Mol. Biol. Cell.*, 1998, **9**, 1081–1091.
58. M. Luthman and A. Holmgren, *J. Biol. Chem.*, 1982, **257**, 6686–6689.
59. V. N. Gladyshev, A. Liu, S. V. Novoselov, K. Krysan, Q. A. Sun, V. M. Kryukov, G. V. Kryukov and M. F. Lou, *J. Biol. Chem.*, 2001, **276**, 30374–30380.
60. M. Lundberg, C. Johansson, J. Chandra, M. Enoksson, G. Jacobsson, J. Ljung, M. Johansson and A. Holmgren, *J. Biol. Chem.*, 2001, **276**, 26269–26275.
61. E. Herreo and M. A. de la Torre-Ruiz, *Crit. Rev. Biochem. Mol. Biol.*, 2007, **64**, 1518–1530.
62. M. T. Rodriguez-Manzaneque, J. Ros, E. Cabiscol, A. Sorribas and E. Herrero, *Mol. Cell. Biol.*, 1999, **19**, 8180–8190.
63. N. Mesecke, S. Mittler, E. Eckers, J. M. Herrmann, M. Deponte, *Biochemistry*, 2008, in press.
64. R. A. Wingert, J. L. Galloway, B. Barut, H. Foott, P. Fraenkel, J. L. Axe, G. J. Weber, K. Dooley, A. J. Davidson, B. Schmid, B. H. Paw, G. C. Shaw, P. Kingsley, J. Palis, H. Schubert, O. Chen, J. Kaplan, Tübingen Screen Consortium, and L. I. Zon, *Nature*, 2005, 436, 1035–1039.
65. N. Rouhier, E. Gelhaye and J. P. Jacquot, *Cell. Mol. Life Sci.*, 2004, **61**, 1266–1277.
66. C. Johansson, C. H. Lillig and A. Holmgren, *J. Biol. Chem.*, 2004, **279**, 7537–7543.
67. M. D. Shelton, P. B. Chock and J. J. Mieyal, *Antioxid. Redox Signal.*, 2005, **7**, 348–366.
68. T. H. Xia, J. H. Bushweller, P. Sodano, M. Billeter, O. Björnberg, A. Holmgren and K. Wüthrich, *Protein Sci.*, 1992, **1**, 310–321.

69. J. J. Kelley, T. M. Caputo, S. F. Eaton, T. M. Laue and J. H. Bushweller, *Biochemistry*, 1997, **36**, 5029–5044.
70. A. M. Benham, *Antioxid. Redox Signal.*, 2005, **7**, 835–858.
71. C. H. Lillig and A. Holmgren, *Antioxid. Redox Signal.*, 2007, **9**, 25–47.
72. A. Miranda-Vizuete, C. M. Sadek, A. Jimenez, W. J. Krause, P. Sutovsky and R. Oko, *Antioxid. Redox Signal.*, 2004, **6**, 25–40.
73. K.-K. Lee, M. Murakawa, S. Takahashi, S. Tsubuki, S. Kawashima, K. Sakamaki and S. Yonehara, *J. Biol. Chem.*, 1998, **273**, 19160–191664.
74. H. Kurooka, K. Kato, S. Minoguchi, Y. Takahashi, J. Ikeda, S. Habu, N. Osawa, A. M. Buchberg, K. Moriwaki, H. Shisa and T. Honjo, *Genomics*, 1997, **39**, 331–339.
75. Q. A. Sun, L. Kirnarskydagger, S. Shermandagger and V. N. Gladyshev, *Proc. Natl. Acad. Sci. USA*, 2001, **98**, 3673–3678.
76. D. Su and V. N. Gladyshev, *Biochemistry*, 2004, **43**, 12177–12188.
77. S. Witte, M. Villalba, K. Bi, Y. Liu, N. Isakov and A. Altman, *J. Biol. Chem.*, 2000, **275**, 1902–1909.
78. S. Black, E. M. Harte, B. Hudson and L. Wartofsky, *J. Biol. Chem.*, 1960, **235**, 2910–2916.
79. L. G. Wilson, T. Asahi and R. S. Bandurski, *J. Biol. Chem.*, 1961, **236**, 1822–1829.
80. T. Asahi, R. S. Bandurski and L. G. Wilson, *J. Biol. Chem.*, 1961, **236**, 1830–1835.
81. E. C. Moore, P. Reichard and L. Thelander, *J. Biol. Chem.*, 1964, **239**, 3445–3452.
82. L. Thelander and P. Reichard, *Annu. Rev. Biochem.*, 1979, **48**, 133–158.
83. A. Jordan and P. Reichard, *Annu. Rev. Biochem.*, 1998, **67**, 71–98.
84. A. Holmgren, *J. Biol. Chem.*, 1979, **327**, 3672–3678.
85. F. Aslund, K. D. Berndt and A. Holmgren, *J. Biol. Chem.*, 1997, **272**, 30780–30786.
86. A. Holmgren, *J. Biol. Chem.*, 1979, **327**, 3672–3678.
87. A. Potamitou, A. Holmgren and A. Vlamis-Gardikas, *J. Biol. Chem.*, 2002, **277**, 18561–18567.
88. E. G. Muller, *J. Biol. Chem.*, 1991, **266**, 9194–9202.
89. A. Koc, C. K. Mathews, L. J. Wheeler, M. K. Gross and G. F. Merrill, *J. Biol. Chem.*, 2006, **281**, 15058–15063.
90. B. Rozell, H. A. Hansson, M. Luthman and A. Holmgren, *Eur. J. Cell Biol.*, 1985, **38**, 79–86.
91. H. A. Hansson, B. Rozell, S. Stemme, Y. Engström, L. Thelander and A. Holmgren, *Exp. Cell Res.*, 1986, **163**, 363–369.
92. B. Rozell, J. A. Barcena, E. Martinez-Galisteo, C. A. Padilla and A. Holmgren, *Eur. J. Cell Biol.*, 1993, **62**, 314–323.
93. J. L. Wray, E. I. Campbell, M. A. Roberts and J. F. Gutierrez-Marcos, *Chem. Biol. Interact.*, 1998, **109**, 153–167.
94. T. Rausch and A. Wachter, *Trends Plant Sci.*, 2005, **10**, 503–509.
95. M. L. Tsang, *J. Bacteriol.*, 1981, **146**, 1059–1066.
96. M. Russel, P. Model and A. Holmgren, *J. Bacteriol.*, 1990, **172**, 1923–1929.

97. C. H. Lillig, A. Prior, J. D. Schwenn, F. Aslund, D. Ritz, A. Vlamis-Gardikas and A. Holmgren, *J. Biol. Chem.*, 1999, **274**, 7695–7698.
98. S. G. Rhee, S. W. Kang, L. E. Netto, M. S. Seo and E. R. Stadtman, *Biofactors*, 1999, **10**, 207–209.
99. B. Hofmann, H. J. Hecht and L. Flohe, *Biol. Chem.*, 2002, **383**, 347–364.
100. S. G. Rhee, H. Y. Chae and K. Kim, *Free. Radic. Biol. Med.*, 2005, **38**, 1543–1552.
101. N. Brot and H. Weissbach, *Arch. Biochem. Biophys.*, 1983, **223**, 271–281.
102. R. L. Levine, J. Moskovitz and E. R. Stadtman, *IUBMB Life*, 2000, **50**, 301–307.
103. E. R. Stadtman, H. Van Remmen, A. Richardson, N. B. Wehr and R. L. Levine, *Biochim. Biophys. Acta.*, 2005, **1703**, 135–140.
104. Y. Kabe, K. Ando, S. Hirao, M. Yoshida and H. Handa, *Antioxid. Redox Signal.*, 2005, **7**, 395–403.
105. J. R. Matthews, N. Wakasugi, J. L. Virelizier, J. Yodoi and R. T. Hay, *Nucleic Acids Res.*, 1992, **20**, 3821–3830.
106. T. Hayashi, Y. Uedo and T. Okamato, *J. Biol. Chem.*, 1993, **268**, 11380–11388.
107. E. Pineda-Molina, P. Klatt, J. Vazquez, A. Martina, M. G. de Lacoba, D. Perez-Sala and S. Lamas, *Biochemistry*, 2001, **40**, 14134–14142.
108. K. Hirot M. Matsui, M. Murata, Y. Takashima, F. S. Cheng, T. Itoh, K. Fukuda and Y. Yodoi, *Biochem. Biophys. Res. Commun.*, 2000, **274**, 177–182.
109. H. Ichijo, E. Nishida, K. Irie, P. ten Dijke, M. Saitoh, T. Moriguchi, M. Takagi, K. Matsumoto, K. Miyazono and Y. Gotoh, *Science*, 1997, **275**, 90–94.
110. M. Saitoh, H. Nishitoh, M. Fujii, K. Takeda, K. Tobiume, Y. Sawada, M. Kawabata, K. Miyazono and H. Ichijo, *EMBO J.*, 1998, **17**, 2596–2606.
111. J. J. Song, J. G. Rhee, M. Suntharalingam, S. A. Walsh, D. R. Spitz and Y. J. Lee, *J. Biol. Chem.*, 2002, **277**, 46566–46575.
112. J. J. Song and Y. J. Lee, *Biochem. J.*, 2003, **373**, 854–853.
113. M. Casadei, T. Persichini, F. Polticelli, G. Musci and M. Colsanti, *Exp. Geront.*, 2008, in press.
114. C. M. Whitacre, *Anal. Biochem.*, 1996, **234**, 177–183.
115. T. W. Wilhemsen, P. A. Olsvik, B. H. Hansen and R. A. Andersen, *J. Chromatogr. A.*, 2002, **979**, 249–254.
116. S. Gromer, S. Urig and K. Becker, *Med. Res. Rev.*, 2004, **24**, 40–89.
117. A. Patenuade, M. R. V. Murthy and M. E. Mirault, *Cell. Mol. Life Sci.*, 2005, **62**, 1063–1080.
118. H. Masutani, J. Bai, Y. C. Kim and J. Yodoi, *Mol. Neurobiol.*, 2004, **29**, 229–242.
119. H. Tomimoto, I. Akiguchi, H. Wakita, J. Kimura, K. Hori and J. Yodoi, *Brain Res.*, 1993, **625**, 1–8.
120. Y. Takagi, T. Tokime, K. Nozaki, Y. Gon, H. Kikuchi and J. Yodoi, *J. Cereb. Blood Flow Metab.*, 1998, **8**, 206–214.
121. Y. Takagi, T. Nakamura, A. Nishiyama, K. Nozaki, T. Tanaka, N. Hashimoto and J. Yodoi, *Biochem. Biophys. Res. Commun.*, 1999, **258**, 390–394.
122. C. J. World, H. Yamawaki and B. C. Berk, *J. Mol. Med.*, 2006, **84**, 997–1003.

123. M. Kobayashi-Miura, K. Shioji, Y. Hoshino, H. Masutani, H. Nakamura and J. Yodoi, *Am. J. Physiol. Heart Circ. Physiol.*, 2007, **292**, H2040–2050.
124. M. F. Lou, *J. Ocul. Pharmacol. Ther.*, 2000, **16**, 137–148.
125. M. F. Lou, *Prog. Retin. Eye Res.*, 2003, **22**, 657–682.
126. H. Masutani, S. Ueda and J. Yodoi, *Cell Death Differ.*, 2005, **12**, 991–998.
127. I. Rahman, S. K. Biswas and A. Kode, *Eur. J. Pharmacol.*, 2006, **533**, 222–239.
128. T. Nakamura, H. Nakamura, T. Hoshino, S. Ueda, H. Wada and J. Yodoi, *Antioxid. Redox Signal.*, 2005, **7**, 60–71.
129. T. Hoshino, M. Okamoto, S. Takei, Y. Sakazaki, T. Iwanaga and H. Aizawa, *Antioxid. Redox Signal.*, 2008, in press.
130. E. S. Arner and A. Holmgren, *Semin. Cancer Biol.*, 2006, **16**, 420–426.
131. M. E. Lonn, C. Hudemann, C. Berndt, V. Cherkasov, F. Capani, A. Holmgren and C. H. Lillig, *Antioxid. Redox Signal.*, 2008, **10**, 547–558.
132. T. Tanaka, H. Nakamura, A. Nishiyama, F. Hosoi, H. Masutani, H. Wada, J. Yodoi, 2000, **33**.
133. T. Yoshida, S. Oka, H. Masutani, H. Nakamura and J. Yodoi, *Antioxid. Redox Signal.*, 2003, **5**, 563–570.
134. T. Yoshida, N. Kondo, S. Oka, M. K. Ahsan, T. Hara, H. Masutani, H. Nakamura and J. Yodoi, *Biofactors*, 2006, **27**, 47–51.
135. A. Mitsui, J. Hamuro, H. Nakamura, N. Kondo, Y. Hirabayashi, S. Ishizaki-Koizumi, T. Hirakawa, T. Inoue and J. Yodoi, *Antioxid. Redox Signal.*, 2002, **4**, 693–696.
136. C. Camaschella, A. Campanella, L. De Falco, L. Boschetto, R. Merlini, L. Silvestri, S. Levi and A. Iolascon, *Blood*, 2007, **110**, 1353–1358.
137. N. Romero-Isart and M. Vašák, *J. Inorg. Chem.*, 2002, **88**, 388–396.
138. S. S. Krishna, I. Majumdar and N. V. Grishin, *Nucleic Acids Res.*, 2003, **31**, 532–550.
139. P. C. F. Graf and U. Jacob, *Cell. Mol. Life Sci.*, 2002, **59**, 1624–1631.
140. D. Su, C. Berndt, D. E. Fomenko, A. Holmgren and V. N. Gladyshev, *Biochemistry*, 2007, **46**, 6903–6910.
141. L. Masip, J. L. Pan, S. Halder, J. E. Penner-Hahn, M. P. DeLisa, G. Georgiou, J. C. A. Bardwell and J. F. Collet, *Science*, 2004, **303**, 1185–1189.
142. N. Rouhier, H. Unno, S. Bandyopadhyay, L. Masip, S. K. Kim, M. Hirasawa, J. M. Gualberto, V. Lattard, M. Kusunoki, D. B. Knaff, G. Georgiou, T. Hase, M. K. Johnson and J. P. Jacquot, *Proc. Natl. Acad. Sci. USA*, 2007, **104**, 7379–7378.
143. A. P. Yeh, C. Chatelet, S. M. Soltis, P. Kuhn, J. Meyer and D. C. Rees, *J. Mol. Biol.*, 2000, **300**, 587–595.
144. J.-F. Collet, J. C. D'Souza, U. Jakob and J. C. A. Bardwell, *J. Biol.Chem.*, 2003, **278**, 45325–45332.
145. J.-F. Collet, D. Peisach, J. C. A. Bardwell and Z. Xu, *Protein Sci.*, 2005, **14**, 1863–1869.
146. J.-F. Collet and J. C. A. Bardwell, *Mol. Microbiol.*, 2002, **44**, 1–8.
147. A. L. Pinto, H. W. Hellinga and J. P. Caradonna, *Proc. Natl. Acad. Sci. USA*, 1997, **94**, 5562–5567.

148. D. E. Benson, M. S. Wisz, W. Liu and H. W. Hellinga, *Biochemistry*, 1998, **37**, 7070–7076.
149. C. D. Coldren, H. W. Hellinga and J. P. Caradonna, *Proc. Natl. Acad. Sci. USA*, 1997, **94**, 6635–6640.
150. D. E. Benson, M. S. Wisz and H. W. Hellinga, *Proc. Natl. Acad. Sci. USA*, 2000, **97**, 6292–6297.
151. C. H. Lillig, C. Berndt, O. Vergnolle, M. E. Lönn, C. Hudemann, E. Bill and A. Holmgren, *Proc. Natl. Acad. Sci. USA*, 2005, **102**, 8168–8173.
152. S. I. Hashemy, C. Johansson, C. Berndt, C. H. Lillig and A. Holmgren, *J. Biol. Chem.*, 2007, **282**, 14428–14436.
153. C. Johansson, K. L. Kavanagh, O. Gileadi and U. Oppermann, *J. Biol. Chem.*, 2007, **282**, 3077–3082.
154. Y. Feng, N. Zhong, N. Rouhier, T. Hase, M. Kusunoki, J. P. Jacquot, C. Jin and B. Xia, *Biochemistry*, 2006, **45**, 7998–8008.
155. A. Picciocchi, C. Saguez, A. Boussac, C. Cassier-Chauvat and F. Chauvat, *Biochemistry*, 2007, **46**, 15018–15026.
156. A. P. Yeh, X. L. Ambroggio, S. L. A. Andrade, O. Einsle, C. Chatelet, J. Meye and D. C. Rees, *J. Biol. Chem.*, 2002, **277**, 34499–34507.
157. M. S. Wisz, C. Z. Garrett and H. W. Hellinga, *Biochemistry*, 1998, **37**, 8269–8277.
158. H. W. Hellinga, J. P. Caradonna and F. M. Richards, *J. Mol. Biol.*, 1991, **222**, 787–803.
159. H. W. Hellinga, *J. Am. Chem. Soc.*, 1998, **120**, 10055–10066.
160. Z. Lu, E. A. DiBlasio-Smith, K. L. Grant, N. W. Warne, E. R. La Vallie, L. A. Collins-Racie, M. T. Follettie, M. J. Williamson and J. M. McCoy, *J. Biol. Chem.*, 1996, **271**, 5059–5065.
161. C. Berndt, C. H. Lillig and A. Holmgren, *Biochim. Biophys. Acta.*, 2008, in press.
162. E. K. Dickinson, D. L. Adams, E. A. Schon and D. M. Glerum, *J. Biol. Chem.*, 2000, **275**, 26780–26785.
163. H. S. Carr and D. R. Winge, *Acc. Chem. Res.*, 2003, **36**, 309–316.
164. M. Brunori, A. Giuffre and P. Sarti, *J. Inorg. Biochem.*, 2005, **99**, 324–336.
165. S. C. Leary, P. A. Cobin, B. A. Kaufman, G. H. Guercin, A. Mattman, J. Palaty, G. Lockitch, D. R. Winge, P. Rustin, R. Horvath and E. A. Shoubridge, *Cell Metab.*, 2007, **5**, 9–20.
166. M. Jaksch, C. Paret, R. Stucka, N. Horn, J. Müller-Höcker, R. Horvath, N. Trepesch, G. Stecker, P. Freisinger, C. Thirion, J. Müller, R. Lunkwitz, G. Rödel, E. A. Shoubridge and H. Lochmüller, *Hum. Mol. Genet.*, 2001, **10**, 3025–3035.
167. Y. C. Horng, S. C. Leary, P. A. Cobine, F. B. Young, G. N. George, E. A. Shoubridge and D. R. Winge, *J. Biol. Chem.*, 2005, **280**, 34113–34122.
168. Y. V. Chinenov, *J. Mol. Med.*, 2000, **78**, 239–242.
169. J. C. Williams, C. Sue, G. S. Banting, H. Yang, D. M. Glerum, W. A. Hendrickson and E. A. Schon, *J. Biol. Chem.*, 2005, **280**, 15202–15211.
170. L. Banci, I. Bertini, V. Calderone, S. Ciofi-Baffoni, S. Mangani, M. Martinelli, P. Palumaa and S. Wang, *Proc. Natl. Acad. Sci. USA*, 2006, **103**, 8595–8600.

171. C. Abajian and A. C. Rosenzweig, *J. Biol. Inorg. Chem.*, 2006, **11**, 459–466.
172. L. Banci, I. Bertini, S. Ciofi-Baffoni, I. P. Gerothanassis, I. Leontari, M. Martinelli and S. Wang, *Structure*, 2007, **15**, 1132–1140.
173. I. Imriskova-Sosova, D. Andrews, K. Yam, D. Davidson, B. Yachnin and B. C. Hill, *Biochemistry*, 2005, **44**, 16949–16959.
174. L. Andruzzi, M. Nakano, M. J. Nilges and N. J. Blackburn, *J. Am. Chem. Soc.*, 2005, **127**, 16548–16558.
175. C. Yang, D. Carlow, R. Wolfenden and S. A. Short, *Biochemistry*, 1992, **31**, 4168–4174.
176. S. Vincenzetti, A. Cambi, J. Neuhard, K. Schnorr, M. Grelloni and A. Vita, *Protein Expr. Purif.*, 1999, **15**, 8–15.
177. A. A. Smith, D. C. Carlow, R. Wolfenden and S. A. Short, *Biochemistry*, 1994, **33**, 6468–6474.
178. S. Vincenzetti, A. Cambi, J. Neuhard, E. Garattiniand and A. Vita, *Protein Expr. Purif.*, 1996, **8**, 247–253.
179. R. L. Tatusov, D. A. Natale, I. V. Garkautsev, T. A. Tatusova, U. T. Shankavaram, B. S. Rao, B. Kiryotin, M. Y. Galperin, N. D. Federova and E. V. Koonin, *Nucleic Acids Res.*, 2001, **29**, 22–28.
180. R. Overbeek, M. Fonstein, M. D'Souza, G. D. Pusch and N. Maltsev, *Proc. Natl. Acad. Sci. USA*, 1999, **96**, 2896–2901.
181. L. N. Kinch, D. Baker and N. V. Grishin, *Proteins*, 2003, **52**, 323–331.
182. J. Lowe, H. Li, K. H. Downing and E. Nogales, *J. Mol. Biol.*, 2001, **313**, 1045–1057.
183. H. Beinert, R. H. Holm and E. Münck, *Science*, 1997, **277**, 653–659.
184. H. Beinert, *J. Biol. Inorg. Chem.*, 2000, **5**, 2–15.
185. R. Lill and G. Kispal, *Trends Biochem. Sci.*, 2000, **25**, 352–356.
186. U. Mühlenhoff and R. Lill, *Biochim. Biophys. Acta.*, 2000, **1459**, 370–382.
187. M. T. Rodriguez-Manzaneque, J. Tamarit, G. Belli, J. Ros and E. Herrero, *Mol. Biol. Cell*, 2002, **13**, 1109–1121.
188. R. Alves, E. Herrero and A. Sorribas, *Proteins*, 2004, **57**, 481–492.
189. F. Vilella, R. Alves, R. T. Rodríguez-Manzaneque, G. Bellí, S. Swaminathan, P. Sunnerhagen and E. Herrero, *Comp. Funct. Genomics*, 2004, **5**, 328–341.
190. U. Mühlenhoff, J. Gerber, N. Richhardt and R. Lill, *EMBO J.*, 2003, **22**, 4815–4825.
191. N. H. Cheng, J. Z. Liu, A. Brock, R. S. Nelson and K. D. Hirschi, *J. Biol. Chem.*, 2006, **281**, 26280–26288.
192. M. M. Molina-Navarro, C. Casas, L. Piedrafita, G. Belli and E. Herrero, *FEBS Lett.*, 2006, **580**, 2273–2280.
193. C. Krebs, J. N. Agar, A. D. Smith, J. Frazzon, D. R. Dean, B. H. Huynh and M. K. Johnson, *Biochemistry*, 2001, **40**, 14069–14080.
194. M. Wollenberg, C. Berndt, E. Bill, J. D. Schwenn and A. Seidler, *Eur. J. Biochem.*, 2003, **270**, 1662–1671.
195. S. Ollagnier-de Choudens, L. Nachin, Y. Sanakis, L. Loiseau, F. Barras and M. Fontecave, *J. Biol. Chem.*, 2003, **278**, 17993–18001.
196. H. Ding, R. J. Clark and B. Ding, *J. Biol. Chem.*, 2004, **279**, 37499–37504.
197. J. Yang, J. P. Bitoun and H. Ding, *J. Biol. Chem.*, 2006, **281**, 27956–27963.

198. J. Lu, J. Yang, G. Tan and H. Ding, *Biochem. J.*, 2008, **409**, 535–543.
199. H. Ding, J. Yang, L. C. Coleman and S. Yeung, *J. Biol. Chem.*, 2007, **282**, 7997–8004.
200. B. Ding, E. S. Smith and H. Ding, *Biochem. J.*, 2005, **389**, 797–802.
201. H. Ding, K. Harrison and J. Lu, *J. Biol. Chem.*, 2005, **280**, 30432–30437.
202. V. Braun, *Front. Biosci.*, 2003, **8**, 1409–1421.
203. J. C. Rutherford and A. J. Bird, *Eukaryot. Cell.*, 2004, **3**, 1–13.
204. J. Kaplan, D. McVey Ward, R. J. Crisp and C. C. Philpott, *Biochim. Biophys. Acta.*, 2006, **1763**, 646–651.
205. E. Chen and S. C. Ekker, *Curr. Pharm. Curr. Pharm. Biotechnol.*, 2004, **5**, 409–413.
206. L. Ojeda, G. Keller, U. Muhlenhoff, J. C. Rutherford, R. Lill and D. R. Winge, *J. Biol. Chem.*, 2006, **281**, 17661–17669.
207. N. Pujol-Carrion, G. Belli, E. Herrero, A. Nogues and M. A. de la Torre-Ruiz, *J. Cell. Sci.*, 2006, **119**, 4554–4564.
208. M. W. Hentze, M. U. Muckenthaler and N. C. Andrews, *Cell*, 2004, **17**, 285–297.
209. K. Pantopoulos, *Ann. N. Y. Acad. Sci.*, 2004, **1012**, 1–13.
210. T. A. Rouault, *Nat. Chem. Biol.*, 2006, **2**, 406–414.
211. C. Bouton, H. Hirling and J. C. Drapier, *J. Biol. Chem.*, 1997, **272**, 19969–19975.
212. C. Bouton, M. Raveau and J. C. Drapier, *J. Biol. Chem.*, 1996, **271**, 2300–2306.
213. L. Oliveira, C. Bouton and J. C. Drapier, *J. Biol. Chem.*, 1999, **274**, 516–521.
214. R. Lill, R. Dutkiewicz, H. P. Elsässer, A. Hausmann, D. J. Netz, A. J. Pierik, O. Stehling, E. Urzica and U. Mühlenhoff, *Biochim. Biophys. Acta.*, 2006, **1763**, 652–667.

Met. Ions Life Sci. **2009**, *5*, 441–481

15

Metal Ion-Binding Properties of Phytochelatins and Related Ligands

Aurélie Devez, Eric Achterberg, and Martha Gledhill

School of Ocean and Earth Sciences, National Oceanography Centre, University of Southampton, Southampton SO14 3ZH, UK

<aude@noc.soton.ac.uk>

<eric@noc.soton.ac.uk>

<mzg@noc.soton.ac.uk>

Metal Ions in Life Sciences, Volume 5 Edited by Astrid Sigel, Helmut Sigel, and Roland K. O. Sigel

Published by the Royal Society of Chemistry, www.rsc.org DOI: 10.1039/9781847558992-00441

ABSTRACT: The development of human activities and industrialization has led to an increased release of metals to the aquatic environment. Several metals (such as copper, zinc, and iron) are essential for many physiological processes but can be toxic at enhanced concentrations; others (such as cadmium, lead, and mercury) are not physiologically essential and are toxic at very low concentrations in both plant and algal cells. To cope with the deleterious effects of metals, eukaryotic cells produce strong metal-binding proteins and peptides (including the thiol phytochelatins) involved in metal tolerance and detoxification mechanisms. Plants and algae are also able to maintain the homeostasis of essential metal ions in different cellular compartments by interactions between metal transport, chelation, trafficking and sequestration activities, which regulate the uptake and distribution of these metal ions.

In this chapter, we present an overview of the metal ion-binding properties of phytochelatins and related ligands and their involvement in metal ion homeostasis, metal tolerance and detoxification mechanisms. Enzymatic processes implicated in thiol biosynthesis and regulation, and in metal ion sequestration activities, are also described. In addition, this chapter assesses our state of knowledge on the induction of thiols by phytoplankton in laboratory and field experiments.

KEYWORDS: metal detoxification and tolerance · metal ion homeostasis · phytochelatin · phytochelatin synthase · thiols · vacuolar sequestration and compartmentalization

1. INTRODUCTION

Metal ions are highly reactive with S, O, and N atoms on amino acid side chains and consequently can alter cellular functions. Metals are classified in three classes, depending on their reactivity with the functional groups of biomolecules. Class A metals (Al(III), Ca(II), Sr(II), Ba(II), La(III)) show more reactivity with oxygen (O > N > S), class B metals (Cu(I), Hg(II), Ag(I)) prefer sulfur (S > N > O), and class C metals (Fe(III), Ni(II), Zn(II), Cd(II), Cu(II)) have an intermediate affinity [1,2]. Some metals, in particular Cu(II), Co(II), Fe(III), Mn(II), and Zn(II) are essential micronutrients and act either as enzyme cofactors, mediate in redox reactions and/or interact with nucleic acids and proteins. Others, such as Cd(II), Pb(II), Hg(II), and Ag(I) are not essential. The toxic effects of metals depend on the time period and concentration to which organisms are exposed. Most of the effects of metals are related to the free metal ion properties, their ability to directly or indirectly generate free radicals and hence induce oxidative stress, and to their interactions with carboxyl and thiol groups of proteins [2].

Profound effects of metals on organisms at enhanced (toxic) concentrations include growth inhibition, lipid peroxidation, protein denaturation, DNA mutation, reduced fecundity, and death [3]. Consequently, organisms have evolved a suite of elaborate mechanisms (e.g., complex regulated network of metal transport, chelation, trafficking, and sequestration activities) that control and respond to the uptake, distribution, regulation, accumulation, and detoxification of both essential and non-essential metal ions [4]. Strategies for organisms to maintain intracellular free metal ion concentrations at levels that do not exceed cellular requirements include changes in ion permeability of the cell membrane, active extrusion, biotransformation, extra- and intracellular chelation, compartmentalization, and sequestration [5]. More specifically, phytoplankton species can respond to metal toxicity through the production of antioxidant compounds [6], exudation of organic ligands [7], and production of intracellular metal-binding thiol peptides [8,9].

The majority of proteins and peptides that function either in the uptake, transport, distribution or storage of metal ions, or in the detoxification mechanisms, possess strong high-affinity metal-binding sites. Given the high reactivity of metal ions with thiol, carboxyl, hydroxyl, and amino groups, it is not surprising that molecules that carry these functional metal-binding sites, have been described as intra- and extracellular metal chelators in plants and algae [4,10–13] and as metallochaperones [14,15]. In many cases, the known high-affinity metal-binding molecules contain cysteine, glutamate or histidine residues [16].

The first section of this chapter will provide a non-exhaustive description of strong high-affinity metal binding proteins and peptides produced in response to metal stress, and hence being involved in processes controlling the regulation of metal ions in plants and algae.

1.1. Sulfhydryl Compounds

Compounds which contain the functional –SH sulfhydryl group are classified as thiols. They are known to be strong complexing ligands for class C

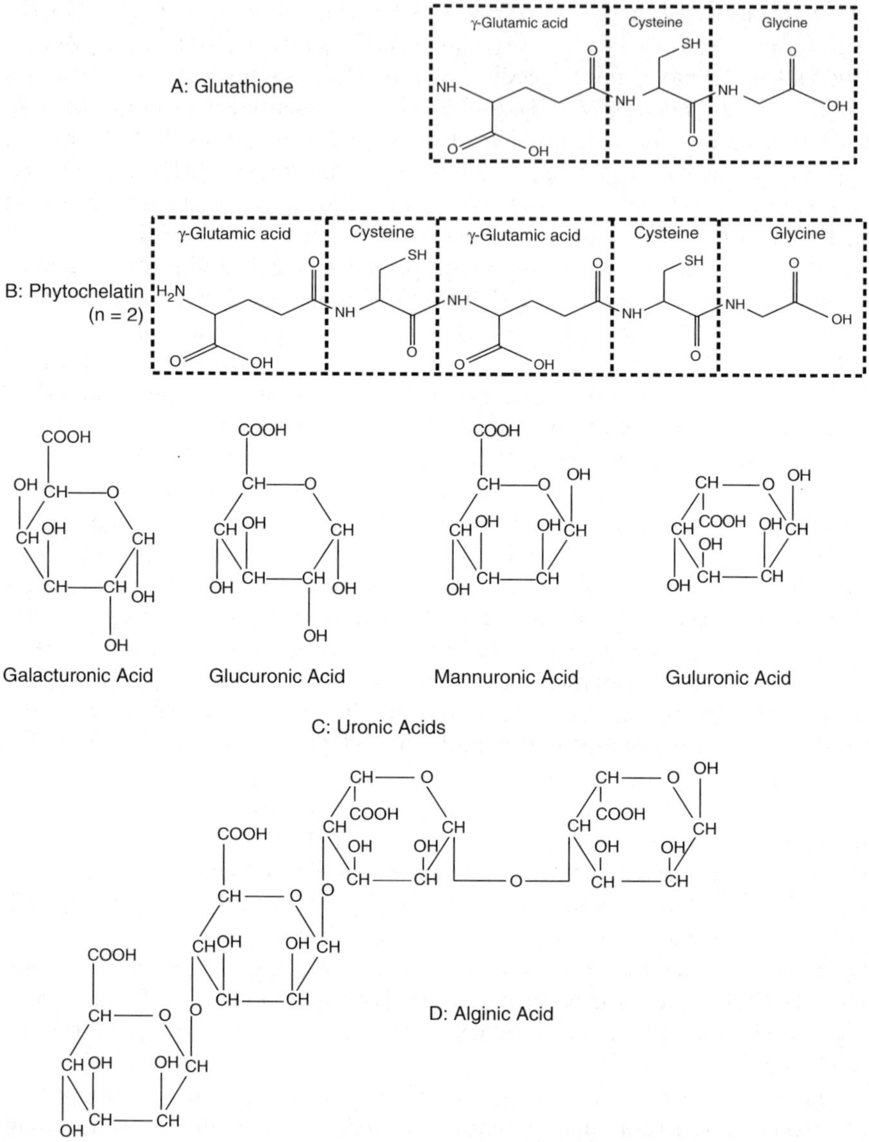

Figure 1. Structure of some metal-binding ligands considered in this chapter.

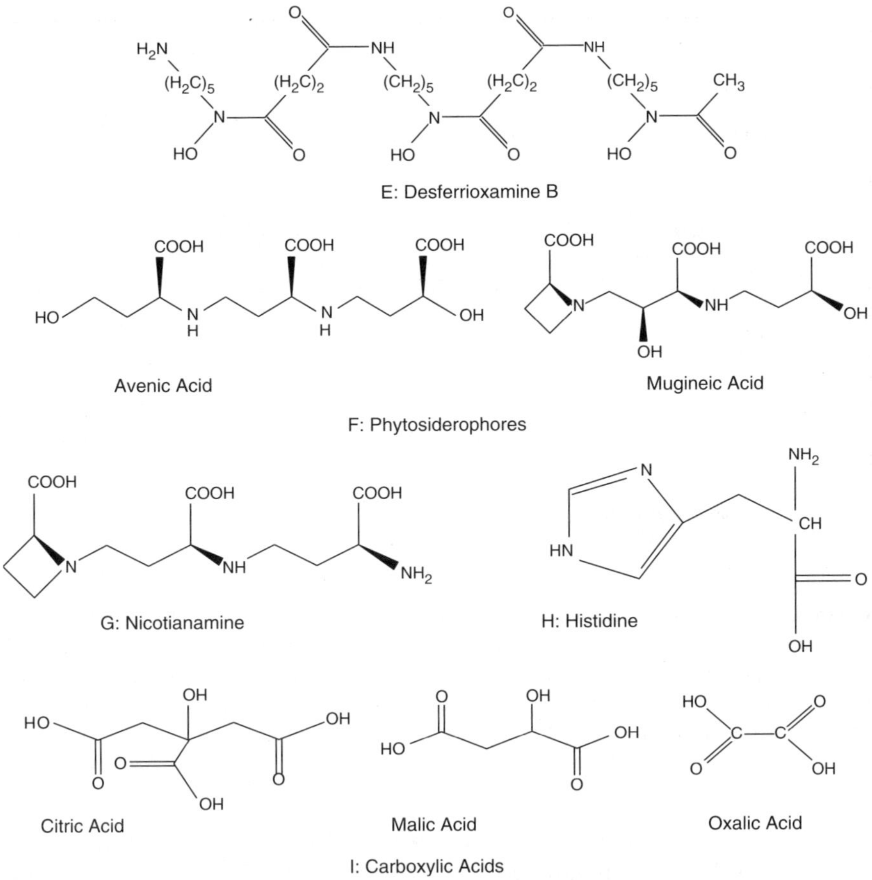

Figure 1. (Continued).

transition metals and for soft B class metals [17] because of the high affinity of the –SH groups for these metal ions.

1.1.1. *Glutathione*

Glutathione (GSH, γ-glutamylcysteinylglycine, Figure 1A) is a thiol which is typically present in all eukaryotes, and forms a major reservoir of non-protein reduced sulfur produced by animals, plants, algae and bacteria [18,19]. The synthesis of this tripeptide (Figure 2), formed by glutamic acid (Glu), cysteine (Cys), and glycine (Gly), requires the sulfur assimilation pathway (SAP) and the cysteine biosynthesis pathway [20]. Glutathione is

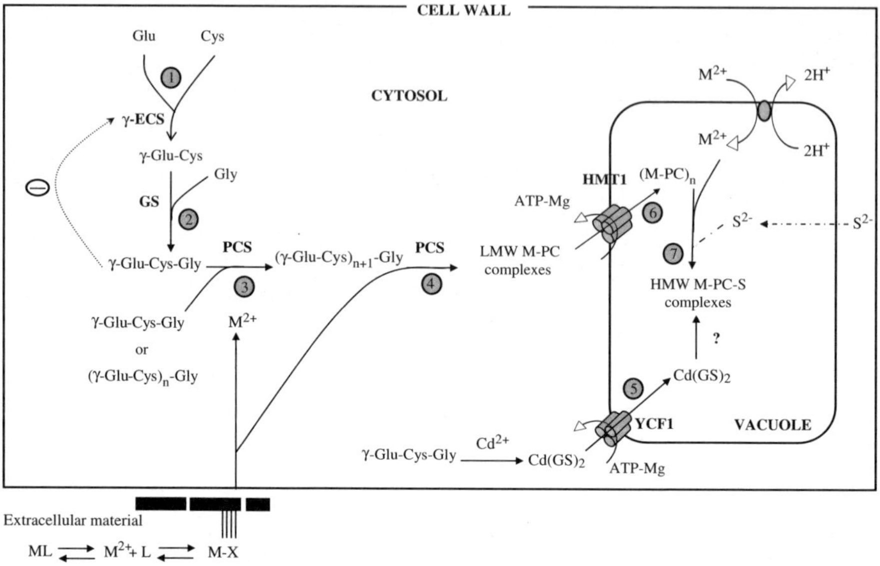

Figure 2. A schematic model for the biosynthesis of glutathione, phytochelatin and resulting complexes and for metal resistance and detoxification mechanisms mediated by PCs in eukaryotic cells. Numbered reactions are described in the text. Enzymes abbreviations: γ-ECS, γ-glutamylcysteine synthetase; GS, glutathione synthetase; PCS, phytochelatin synthase. M^{2+}, free metal ion; ML, metal complex in solution; L, ligand; X, biotic extracellular ligand.

synthesized from cysteine by two consecutive ATP-dependent reactions. In the first step, γ-glutamylcysteine (γ-EC or γ-Glu-Cys) is formed from L-glutamate and L-cysteine by γ-glutamylcysteine synthetase (γ-ECS) (Figure 2, reaction 1). The second step is catalyzed by glutathione synthetase (GS) which adds glycine to the C-terminal of γ-EC forming glutathione (γ-Glu-Cys-Gly) (Figure 2, reaction 2). GSH is synthesized in the cytosol and chloroplast, where its constituent amino acids and required enzymes have been detected, and can subsequently be transported to other cell compartments for biochemical functions [19]. The γ-linkage within GSH provides a high resistance to intracellular peptidases and therefore relatively high GSH concentrations can be maintained in cells [18].

GSH is involved in multiple metabolic processes. One of its important roles involves the maintenance of reducing conditions by intracellular redox state regulation of cell division. As an antioxidant, GSH also acts as a line of defence against reactive oxygen derivatives and radiation damage because the reduced form of GSH exists interchangeably with the oxidized form, GSSG. The ratio GSH/GSSG is often used as a sensitive index of oxidative stress *in vivo* and as an indicator of cell damage and some diseases. The maintenance of GSH levels is due to reduction of GSSG by NADPH, a reaction catalyzed by glutathione reductase (GR). GSH may be related to transport of proteins and GSH-conjugated amino acids and to sulfur or cysteine storage [19]. GSH also plays a putative role as a carrier in the sulfate assimilation pathway [18]. Further roles are the protection against elevated concentrations of metals and xenobiotic organic compounds and the production of PCs [19]. It plays an essential role in important biological processes including the synthesis of proteins and DNA, enzyme activity, and metabolism. GSH is also the substrate of an enzyme family called glutathione-S-transferases [21].

The complex biochemical and genetic regulations of cysteine and glutathione biosynthesis pathways and sulfur assimilation are affected by different stress situations such as metal ion exposure, oxidative stress, and sulfur or nitrogen deficiency (see the review [20]). Moreover, all γ-ECS enzymes described in bacteria, yeast, plants, and animals are physiologically inhibited through a feedback mechanism involving GSH (Figure 2) [22].

1.1.2. Metallothioneins and Phytochelatins

Metallothioneins (MTs) are ubiquitous low molecular weight (LMW) cysteine-rich proteins, which bind metal ions in metal-thiolate clusters by mercaptide bonds. The thiolato sulfur atoms of the cysteinyl side chains act as ligands for metal ions in these clusters [23]. MT proteins are classified based on the arrangement of Cys residues as it has been hypothesized that the differences in this arrangement could account for differences in metal affinities [24]. According to the MT classification of Robinson et al. [24], four types of MTs exist in plants (see also Chapter 5).

In some MT classification systems, phytochelatins (PCs) are, somewhat confusingly, described as class III MTs, however, PCs are not proteins like MTs. PCs are produced by plants, fungi, and algae (see Section 2.1) and have similar functions as MTs in animals and cyanobacteria, notably complexation of metals [25]. Although PCs are structurally similar to MTs, they are synthesized enzymatically, while MTs are encoded by a family of genes [12].

Phytochelatins are small cysteine-rich peptides which normally contain only three amino acids: glutamic acid, cysteine, and glycine. PCs form a family of compounds with increasing repetitions of the Glu-Cys dipeptide linked through a γ-carboxylamide bond and include a terminal Gly. The general formula is $(\gamma\text{-Glu-Cys})_n$-Gly where n has been reported as being as high as 11, but generally to be in the range of 2 to 5 (Figure 1B) [26]. Phytochelatin polypeptide chains with $n = 2$, 3, and 4 are predominant in phytoplankton [9]. These peptides resemble the ubiquitous tripeptide glutathione, γ-Glu-Cys-Gly, indicating the involvement of glutathione in phytochelatin biosynthesis [27] (see also Section 2.2.1 for details).

In addition, a number of structural variants called *iso*-PCs have been identified [27,28], such as the homo-PC $(\gamma\text{-Glu-Cys})_n$-β-Ala ($n = 2$–7) isolated from Phaseoleae plants and related to the homo-GSH (γ-Glu-Cys-β-Ala), the hydroxymethyl-PC $(\gamma\text{-Glu-Cys})_n$-Ser detected in the Poaceae plant family and related to the hydroxymethyl-GSH (γ-Glu-Cys-Ser) [29], and the *iso*-PC $(\gamma\text{-Glu-Cys})_n$-Glu isolated from maize ($n = 2$ to 4) [30]. In addition to PCs and *iso*-PCs, a series of peptides that lack the C-terminal amino acid have been discovered in maize (*Zea mays*) and characterized by Meuwly and coworkers [30] using tandem mass spectrometry. These compounds are represented by the formula (γ-Glu-Cys), and are called desglycine phytochelatins (desGly-PCs) [28].

1.2. Other Types of Organic Ligands

It is well known that large fractions of trace metals in natural waters are complexed by natural organic ligands influencing the distribution of these metals (see details in the section *Complexation with organic ligands* in [31] and references therein). The chemical composition and structure of the organic ligands in natural waters are largely unknown and it has been difficult to directly characterize the ligands. There is compelling evidence that the dissolved organic ligands are of recent biological origin [32]. Besides, there have been recent advances in studies into the structure and function of marine siderophores, a group of strong Fe(III) complexing ligands [33–35]. Furthermore, some organic ligands, such as PCs [8,9,36], other exudated thiols [37–39], including cysteine [40,41] and GSH [42,43] and extracellular polysaccharides [44], have been indirectly observed in metal stress experiments and fieldwork.

1.2.1. Acid Polysaccharides

Recent evidence suggests that extracellular carbohydrate polymers, mainly acid polysaccharides, secreted by algae and bacteria in response to low nutrient or high metal concentrations, play a significant role in heavy metal detoxification in aquatic environments [44–46].

A group of acid polysaccharides, namely uronic acids, are widely distributed in vascular plants and marine organisms and are the primary structural units in commercially important alginic and pectic acids [47]. They are composed of galacturonic, glucuronic, mannuronic, and guluronic acids (Figure 1C). Glucuronic and galacturonic acids are the most abundant and widely distributed uronic acids [48]. They differ from aldoses in the replacement of the methanolic carbon opposite the carbonyl carbon with a carboxylic acid group. Alginic acid (1,4- linked β-D-mannuronic and α-L-guluronic acid residues, Figure 1D) is the common name given to a family of linear polysaccharide-like polymers [49]. They contain three different functional groups: $-COO^-$ (carboxylate), –C–O–C– (ether) and –OH– (alcohol). As carboxylate groups have been suggested to be strong binding sites, polysaccharides possess several characteristics that make them important binding agents in marine aggregates [44]. Moreover, polysaccharides have been recognized to play important roles in the production of biofilms and the formation of mucilaginous aggregates and marine snow flocs [44,49–52], the destabilization of inorganic colloids through flocculation [53] and the complexation with trace metals [45]. These compounds are also involved in other environmental processes including binding extracellular enzymes in their active forms, scavenging trace metals from the water, immobilizing toxic substances, altering the surface characteristics of suspended particles, and modifying the solubility of associated molecules [44,46,49,54].

1.2.2. Siderophores and Phytosiderophores

Siderophores are low molecular weight chelators, presenting functional groups that bind to Fe(III) with high affinity and specificity. The non-conditional stability constants, or formation constants, of most siderophore-Fe(III) complexes are very high (log K_f ranges from 22.9 to 52) and, consequently, they are some of the best ligands for ferric ions [55]. Other metals may also complex with siderophores, some with strong affinities [55]. For example, the strong cobalt-binding ligands observed by Saito and Moffett [56] and Ellwood and van den Berg [57] are a type of "cobalophore" that are proposed to play a role similar to siderophores with iron. Several microorganisms synthesize siderophores, such as enteric, pathogenic and nitrogen fixing bacteria, fungi, Gram-positive and -negative species, phytoplankton, as well as certain species of plants (phytosiderophores) and some kinds of yeasts [58]. Siderophores are produced with the function of facilitating iron acquisition, sequestering and transporting Fe(III), which is essential for cell growth and metabolism. Most siderophores are hexadentate (see desferrioxamine B, Figure 1E) and can be distinguished by the type of ligands that constitute the binding subunits, being classified as hydroxamate, hydroxy carboxylate, and catecholate types [33]. Plants produce specific multidentate ligands known as phytosiderophores such as mugineic and avenic acids (Figure 1F) which are synthesized in response to iron and possibly zinc deficiencies [59].

Nicotianamine (Figure 1G), a non-proteinaceous amino acid precursor of phytosiderophores, has the ability to form complexes with various metal ions including Cu, Ni, Co, Zn, Fe, and Mn and is reportedly involved in distributing them in young plant tissues [59]. This molecule is formed by the condensation of three S-adenosyl-methionine molecules. This synthesis is catalyzed by the enzyme nicotianamine synthase. Subsequently, nicotianamine aminotransferase catalyzes the conversion of nicotianamine to mugineic acid derivatives. The products are converted to a range of potential ligands featuring alternating amino and carboxylato groups. There are six alternating carboxylate and amino groups whose relative positions favor the formation of six-coordinate metal complexes. Phytosiderophores appear to be linked to Fe homeostasis, as in cases where Fe(III) reaches toxic levels, nicotianamine may act as an Fe(III) scavenger to protect cells from oxidative stress and may play a role in hyperaccumulation as a long-distance transporter [13].

1.2.3. Carboxylic and Amino Acids

Many studies have implied that carboxylic and amino acids play a role in metal ion sequestration through N- and O-based donor ligands (see reviews [4,11,13]). Among them, histidine (His, Figure 1H) is considered to be the most important free amino acid involved in metal hyperaccumulation (the so-called histidine response is involved in Ni-tolerance mechanisms in plants) [60]. It can act as a tridentate ligand via its carboxylato, amine, and imadazole functions [61]. Histidine is known to have high affinity for transition metal ions such as Zn(II), Co(II), Ni(II), and Cu(II) [62].

The carboxylic acids, which include citric, isocitric, oxalic, tartaric, malic, malonic, and aconitic acids are present in high concentrations in the vacuoles of photosynthetic tissues and can be secreted from roots in response to aluminum stress [63]. Citrate, malate, and oxalate (Figure 1I) have been implicated in a range of processes, including differential metal tolerance, metal transport through the xylem and vacuolar metal sequestration [11] but they may only play a part in sequestration within isolated compartments and are unlikely to act as long-distance transporters [13].

2. PHYTOCHELATINS AND RELATED LIGANDS

2.1. Structure and Occurrence

Phytochelatins are low molecular weight (ranging from 2 to 10 kDa) metal-binding peptides, classified as thiols because of their –SH functional group. First identified in the fission yeast, *Schizosaccharomyces pombe*, and termed cadystins [64], PCs are structurally related to GSH (γ-Glu-Cys-Gly, see also Sections 1.1.1 and 1.1.2). Numerous physiological, biochemical, and genetic studies have confirmed that GSH [65], or in some cases the related compounds homo-glutathione [28,30], hydroxymethyl-glutathione [29] or γ-glutamylcysteine [66], play the role of substrate for PC biosynthesis (see the review [12]).

The capacity to synthesize PCs is thought to be present in all higher plants (angio- and gymnosperms) [26], including bryophytes [67], lichens [68], and marine macrophytes [69–76], but not in mosses [77]. The majority of marine [8] and freshwater algae [78–83], including macroalgae [84–86], produce PCs (see Section 4 and Table 1 therein). Phytochelatins have also been detected in several filamentous fungi, including *Candida glabrata* [87], *Mucor racemosus*, and *Artciulospora tetracladia* [88], in some aquatic hyphomycetes [89,90], and more recently in the macromycete *Boletus edulis* [91].

2.2. Biosynthesis and Regulation

2.2.1. Biosynthesis

While MTs are primary gene products, phytochelatins are synthesized post-translationally by the constitutive enzyme phytochelatin synthase (PCS also called γ-glutamylcysteine dipeptidyl transpeptidase) [65] which catalyzes the transpeptidation of the γ-Glu-Cys moiety of GSH either onto a second GSH molecule to form $PC_{(n=2)}$ or onto a previous synthesized PC molecule to produce a $PC_{(n+1)}$ oligomer [26] (Figure 2, reaction 3). The enzyme is a 95 kDa tetramer with a K_m of 6.7 mM for GSH [65]. A variant of PCS, called homophytochelatin synthase (hPCS), with β-Ala instead of Gly as the terminal amino acid has also been identified in *Glycine max*. This enzyme is able to use GSH or homoglutathione (hGSH; γ-Glu-Cys-β-Ala) as substrate [92].

It has been almost 20 years since investigations by Grill and coworkers [65] yielded partially purified preparations of an enzyme capable of catalyzing phytochelatin synthesis in cultured cells of a higher plant *Silene cucubalis*. However, it is only recently that its molecular identity was determined by the independent cloning and characterization of genes encoding PCS. Three research groups simultaneously isolated genes encoding for phytochelatin synthase activity in *Schizosaccharomyces pombe* [93], *Arabidopsis thaliana* [94], and *Triticum aestivum* [95]. In addition, the nematode, *Caenorhabditis elegans*, appeared to possess a PCS gene [96,97] suggesting that functional PCS genes may be present in certain animal species (see the review [98]). Furthermore, a group of proteins distantly related to PCs is present in bacteria [99,100]. These proteins, which are about 23–35 kDa in size, show similarity to the PCS *Pfam* domain 05023 and are annotated as "PC synthase-like" [99,100]. The main activity of this bacterial PCS-like enzyme is the cleavage of glycine from GSH [98].

PC synthase is constitutive, present in the cells even when there is no metal stress [10], however, a physiologically crucial and biochemically intriguing property of this enzyme is that activation requires a metal ion. The variable C-terminal region of PCS contains multiple cysteine residues, which are often present in pairs that are adjacent or in close proximity. The C-terminal region is presumed to be involved in the activation of the enzyme by metal ions, and hence is termed the "metal-sensing domain" [93], whereas the highly conserved N-terminal domain is proposed to be the PC synthase catalytic activity domain [93,101]. It seems that the less-conserved Cys residues together with some Glu residues present in the C terminal domain play a role as binding sensor for heavy metals [16]. Although many heavy metals can act as enzyme activators, Cd is the most effective followed by Ag, Bi, Pb, Zn, Cu, Hg, and Au cations [12,20].

The kinetics of PC synthesis are also consistent with a mechanism in which metal glutathione thiolates (e.g., $Cd(GS)_2$ or $Zn(GS)_2$) and free GSH act as γ-Glu-Cys acceptor and donor [101]. Firstly, in the presence of both physiological concentrations of GSH and Cd (micromolar), essentially all the cadmium would be present in a glutathione thiolate form [$Cd(GS)_2 =$ bis(glutathionato)-cadmium]. Secondly, S-alkylglutathiones can participate in PC biosynthesis in the absence of metals. These two observations are consistent with a model in which metal thiolates or alkylated thiolates can also act as substrates for PC biosynthesis [20]. Thus, the role of metal ions in enzyme activation is as an integral part of the substrate, rather than interacting directly with the enzyme itself [101]. In this way, any metal ions that form thiolate bonds with GSH may have the capacity to activate PC biosynthesis, subject to possible steric constraints in binding at the active site of the enzyme [101].

Early work suggested that PC biosynthesis *in vitro* was ultimately terminated by the PC products chelating the activating metal ions or could be prematurely terminated by the addition of a metal chelator such as EDTA [102], which provides a mechanism to auto-regulate the biosynthesis of PCs. Viewed from a perspective where the metal ion forms part of the substrate, termination of the reaction results simply from exhaustion of substrate [101].

Organisms must be able to maintain a relatively constant intracellular environment. When perturbations in the external medium occur, the organisms adjust their internal functions. Thus, both metabolic and genetic control mechanisms regulate the effects of the external changes for the intracellular environment, allowing the organisms to adapt. The cellular control machinery pursues the modulation (activation or inhibition) of critical enzyme activities through (a) short-term (biochemical) mechanism(s) consisting of non-covalent interactions with some metabolites and covalent enzyme modification and (b) long-term (genetic) mechanisms consisting in the change of the rates of synthesis and degradation of enzymes.

2.2.2. *Metabolic and Genetic Regulations*

There are a number of mechanisms by which the PC biosynthetic pathway may be regulated. The first of these is likely to be regulation of GSH biosynthesis. For GSH synthesis, it has been assumed that feedback inhibition of γ-glutamylcysteine synthetase activity by GSH is the prime regulation mechanism of the pathway [19,22,103]. Different studies (see review [20]) have demonstrated that GSH levels and sensitivity to metals change by modifying γ-ECS and glutathione synthetase activities. Moreover, changes in the GSH/GSSG ratio do not affect the transcription levels of γ-ECS and GS. Glutathione S-transferases and PCS catalyze reactions that consume GSH and as a consequence their activation (by increasing availability of their substrates or by over-expression) can overcome the GSH feedback inhibition of γ-ECS resulting in an enhanced pathway flux. Cysteine and glycine availability may also contribute to modulate GSH synthesis (see the detailed review [20] concerning the regulation mechanisms of cysteine biosynthesis pathways and ref. [19]).

Regulation of PCS activity is expected to be the primary point at which PC synthesis is regulated [104] since it is the slowest enzyme in the pathway [20]. In some species, PC synthase activity may be regulated at both transcriptional and post-translational levels [12]. In general, all the genes encoding enzymes of GSH and PC biosynthesis pathways are transcriptionally up-regulated by sulfate starvation and, in the cases where Cd(II) response has been analyzed, most of them also respond by increasing transcriptional activity [20]. Expression of ATP sulfurylase genes is also stimulated by sulfate starvation ref. [105].

2.3. Vacuolar Sequestration and Compartmentalization

Compartmentalization in the vacuole appears to be the most important mechanism for Cd(II) resistance in yeasts, fungi, plants, and algae. Cadmium can be transported into the vacuole as a free ion or associated with thiol compounds (GSH conjugates or PCs). In *S. pombe, C. glabrata*, and in a number of different algal and plant species, sulfide incorporated in the vacuole, free $Cd^{2+}$and LMW thiol complexes will form high molecular weight cadmium-sulfide bridged complexes (Figure 2). These complexes are an ultimate and stable storage reservoir of Cd^{2+} inside the cell (reviewed in [20]).

This process was observed in *Dunaliella bioculata* [106], but has been most clearly demonstrated through studies of the sensitive mutant fission yeast, *hmt1*$^-$, which is unable to form HMW complexes on exposure to Cd^{2+} [107]. This *HMT1* (heavy metal tolerance) gene encodes a member of the family of ATP-binding cassette membrane transport proteins (ABC-type transporter), and the HMT1 transporter is located in the vacuolar membrane (Figure 2) [107,108]. This gene is specific for PC transport and *hmt1*$^-$ cells are Cd(II) sensitive, which indicates that the resistance mechanism in *S. pombe* depends on the proper storage of PCs in vacuoles. *HMT1* gene expression is not Cd(II) inducible, although its over-expression enhances Cd(II) accumulation and resistance [107]. In *S. cerevisiae*, Cd(II) is also stored into the vacuole as a complex, but in contrast to *S. pompe* and plants, it is transported as bis(glutathionato)-cadmium ($Cd(GS)_2$, Figure 2, reaction 5) by YCF1 (yeast cadmium factor) also a member of the ABC family of transporters. The *YCF1* gene possesses a similar amino acid sequence as the *HMT1* one [109,110]. YCF1 may transport $Cd(GS)_2$ and some GSH-conjugates such as dinitrophenyl-GSH but no PCs [110]. In contrast, HMT1 may transport PCs and Cd-PC complexes, but apparently no other GSH conjugates (Figure 2, reaction 6) [108].

There is also increasing evidence that vacuolar localization of metal ions plays an important role in plants [104]. An ATP-dependent, proton gradient-independent activity has been identified, similar to that of HMT1 transporters and is capable of transporting both PCs and PC-Cd complexes into tonoplast vesicles derived from oat roots [111]. Nonetheless, no *HMT1* or *YCF1* homolog genes or another plant gene encoding this function has yet been identified [112].

Five protein families have also been implicated in the transport of the free Cd^{2+} ion through cell membranes: (1) cation/H^+ antiporter family; (2) CPx-type heavy metal ATPases which share the common feature of a conserved intra-membranous cysteine-proline-cysteine, cysteine-proline-histidine or cysteine-proline-serine (CPx) motif; (3) natural resistance-associated macrophage proteins (Nramp), which are a novel family of related proteins that have been implicated in the transport of divalent metal ions; (4) cation diffusion facilitator family (CDF), which is also termed the cation-efflux family, and are proteins implicated in the transport of Zn, Co, and Cd; (5) ZIP gene family and ZRT-IRT-like proteins able to transport Fe, Zn, Mn, and Cd. Some of the proteins, members of these families, have been related to uptake and storage of free metal ions (like Cd^{2+} but also Cu^{2+}, Zn^{2+}, and Co^{2+}) into the vacuole, and tolerance and detoxification mechanisms (for a detailed description of these families and functions see the reviews [4,20,113]). In algae, accumulation of metals into the vacuole is associated with the presence of sulfur too [106,114,115].

2.4. Role of Sulfide Ions

In plants, yeasts, and algae, sulfide ions are an essential component of HMW Cd-PC complexes that play an important role in the efficiency of Cd detoxification by PCs [116–119]. The presence of labile sulfide (S^{2-}) in Cd(II)-PCs was first noted by Murasugi and coworkers [120]. The incorporation of acid-labile sulfide into the HMW Cd-PC complexes increases the amount of Cd per molecule, provides a higher Cd^{2+}-binding capacity and enhances the stability of the complex [117]. The HMW complexes with a comparatively high ratio of S^{2-} to Cd [121] consist of aggregates of 10 to 20 Å diameter particles which themselves consist of a CdS crystalline core coated with PCs [117–119,121,122]. The biosynthesis of the sulfide complexes of PCs (PC-CdS) is preceded by the formation of GSH-CdS complexes, which are less stable than those formed by PCs [117,121,123]. First, Cd is chelated by GSH or PC and then labile sulfide is incorporated in the metal-thiolate clusters (Figure 2, reaction 7) [108].

Genetic evidence for the importance of sulfide in the functioning of PCs has been obtained from the analysis of Cd-sensitive mutants deficient in Cd-PC complexes [104]. The inability of cells to produce PC-CdS complexes might be either due to defects in transport of PCs across the vacuolar membrane [107] or in the biosynthetic processes involved in sulfide metabolism [124,125]. Impairment in any of these processes, i.e., sequestration, transport, and HMW complex formation, will result in a Cd-hypersensitive phenotype [107,124,125]. These include some mutants affected in steps in the adenine biosynthetic pathway [124].

HMW complexes can be formed *in vitro* by mixing sulfide, Cd, and PCs. However, in *S. pombe*, two purine biosynthetic enzymes, adenylosuccinate synthetase and succinoaminoimidazole carboximide synthetase, are required for HMW complex formation [124,125]. Biochemical characterization of the enzymes involved in this pathway indicates that in addition to catalyzing the conversion of aspartate to intermediates in adenine biosynthesis, they could also utilize cysteine sulfinate, a sulfur-containing equivalent of aspartate, to form other sulfur-containing compounds. They are believed to be intermediates or carriers in the pathway of sulfide incorporation into HMW complexes [125].

More recently, investigators have identified additional functions important in sulfide metabolism by using Cd-sensitive mutants isolated in *S. pombe* and *Candida glabrata*. In *C. glabrata*, the mutant *hem2*⁻ is deficient in porphobilinogen synthase, an enzyme involved in siroheme biosynthesis [126]. Siroheme is a cofactor for sulfite reductase required for sulfide biosynthesis. This deficiency may contribute to the Cd-sensitive phenotype. In *S. pombe*, the mutant *hmt2*⁻ hyperaccumulates sulfide in both the presence and absence of Cd [127]. The *HMT2* gene encodes a mitochondrial sulfide/quinone oxidoreductase, which was suggested to function in the detoxification of endogenous sulfide. The role of the *HMT2* gene in Cd tolerance is uncertain, but one possibility is that it detoxifies excess sulfide generated during the formation of HMW Cd-PC complexes after Cd exposure. Further additional studies are required to establish the precise influence of sulfide metabolism pathway on PC function, particularly in plants in which Cd-sensitive sulfide metabolism mutants have not been identified [104].

Together, these observations confirm the importance of sulfide in the mechanism of PC detoxification for Cd. Whether or not sulfide is involved in the detoxification of other metal ions by PCs is unknown [12]. Scarano and Morelli [119] concluded that Pb was not capable of inducing sulfide-containing complexes in a way that Cd does, even if Pb-PC complexes were accumulated in *Phaeodactylum tricornutum* cells.

2.5. Sequestration of Metal Thiol Complexes in the Chloroplast and Mitochondria

The photosynthetic protist, *Euglena gracilis* possesses a high Cd tolerance and a high Cd accumulation capacity, probably due to the absence of a specialized organelle such as a plant-like vacuole [128–130]. In this protist, PCs and GSH are found in the cytosol, chloroplasts, and even mitochondria following Cd^{2+} exposure [128–130]. These findings may be explained by either of the following mechanisms: (1) PCs are synthesized in the cytosol where they sequester Cd; the Cd-PC complexes are subsequently transported into the chloroplast and mitochondria. (2) PCs are synthesized inside the organelles where they bind Cd^{2+}, which is transported as a free ion, and form HMW complexes [20]. Observations made by Mendoza-Cozatl et al. [20,129] support the existence of the second mechanism in *Euglena* chloroplasts, but they do not exclude the existence of the first mechanism. Both processes can co-exist and PCs can be synthesized in the three cellular compartments [20]. The mechanism of how PCs are synthesized and stored in *E. gracilis* is still under investigation in Mendoza-Cozatl's laboratory [20].

In addition to *E. gracilis*, it has been found that 60% of accumulated Cd resides inside the chloroplast in a cell wall deficient strain of *Chlamydomonas reinhardtii* [131]. HMW complexes were also found in this organelle, but the origin of plastid PCs and the mechanism by which Cd is transported into the chloroplast of *Chlamydomonas* is still unknown [131].

Even if PCs have been observed in several groups of algae [8,67], there is no information about their potential implication for Cd transport to other intracellular organelles than vacuoles. But Soldo et al. [132] found that *Oocystis nephrocytioides* exposed to Cu accumulated high concentrations of the metal in the thylakoids and pyrenoids. They concluded that this localization suggests interaction of Cu with ligands localized in the chloroplast.

3. IMPORTANCE OF PHYTOCHELATINS AND RELATED LIGANDS IN METAL TOLERANCE

The major processes involved in metal homeostasis, tolerance, and detoxification include a regulated network of metal transport, trafficking, chelation, accumulation and storage activities, biotransformation, and cellular repairs mechanisms [4,59,133]. Phytochelatins are part of known metal-binding peptides in plants and algae and form key molecules which take part in these activities. The role of GSH is much more complex likely due to its multi-functionality. This present section will focus on the relative importance of phytochelatins in metal ion homeostasis and their involvement in both metal tolerance and detoxification through chelation and sequestration processes.

3.1. Metal Ion Homeostasis

Organisms possess mechanisms to maintain the optimal concentrations of essential metal ions in different cellular compartments and to minimize the damage from exposure to non-essential metal ions. Metal ion homeostasis requires intracellular complexation of metals when there is a cellular surplus with subsequent release to metal requiring apoproteins and final storage sites within cells [27]. The presence of PCs at low metal concentrations has been put forward as evidence that they may be involved in metal ion homeostasis [10,27,28,104,134]. This notion has been reinforced by field data showing that both terrestrial and aquatic plants have basal levels of phytochelatins [10]. But evidence of a role of PCs in metal ion homeostasis is not abundant.

The capability for PCs to serve as a metal ion "reservoir" for proteins has been demonstrated in *in vitro* experiments. Indeed, Cu and Zn present in phytochelatin complexes were able to reactivate enzymes in which metals had been removed from the active site [135]. Some support for the involvement of PCs in this process was obtained from various cultured plant cells transferred to a variety of fresh media containing different low concentrations of Cu and Zn [136]. The participation of γ-Glu-Cys peptides in the protection of metal-sensitive enzymes is another aspect of metal homeostasis [27]. Metal transporters (like ATPases of the CPx-type, see Section 2.3) are also involved in the overall metal ion homeostasis processes [113]. Further work is therefore necessary in this area before the function of PCs in maintaining intracellular metal homeostasis and the balance between essential and non-essential metals is fully understood [137].

3.2. Metal Detoxification

As PC concentrations increase in direct response to metal additions and accumulation to culture medium and in the cell, respectively, their importance in metal tolerance and thus detoxification is high [28]. Nevertheless, Schat et al. [134] do not provide evidence in favor of a role for PCs in the detoxification of the essential metal micronutrients Zn, Ni, and Co in plants. In fact, these elements are considered to be relatively weak activators of phytochelatin synthase, both *in vivo* and *in vitro* [9,136]. The stability of the Zn-PC complex is comparatively low [138,139]. The stabilities of Ni-PC and Co-PC complexes are unknown at present, but may be even lower, as suggested by the relatively low affinities of Ni and Co to other cysteine-based ligands [140]. Also, it is highly unlikely that PCs are essential in the detoxification of Fe, Mo, and Mn [65,141], suggesting that in general intracellular PCs might not be involved in the detoxification of excessively accumulated micronutrients in plants and algae. Although Cu, when present at toxic concentrations, induces a considerable PC accumulation (see Section 4) and apparently forms stable complexes with PCs, Schat et al. [134] suggest that PCs may not effectively contribute to Cu detoxification in most algae and higher plants, although they appear to do so in fission yeast [95]. The reason for this may lie in the presence or absence of more effective efflux- or MT-based alternative detoxification systems [134].

On the other hand, in plants and algae, intracellular PCs are required for the detoxification of certain non-essential metals and metalloids with relatively high affinities to sulfur, such as Cd, Hg [71,142], and in particular As. This metalloid has been shown to induce high levels of PC accumulation in a variety of plant species [138,143–146] and algae [147–149]. As-PC complexes have been isolated from *Silene vulgaris* [150] and from *Holcus lanatus* and *Pteris cretica* [151]. PC synthesis is supposed to be essential for As detoxification in plants [145,152,153]. It is also likely that PCs serve an additional role in metal detoxification in algae by acting as a shuttle to the vacuole or to the cytoplasmic membrane [10]. The clearest evidence for the role of PCs in detoxification (particularly Cd and As) comes from characterization of PC and phytochelatin synthase-deficient mutants of *Arabidopsis* and *S. pombe* (for reviews see [12,28]). Zenk [28] concluded that PCs and all the known *iso*-PCs as well as the desGly-PCs take part in the detoxification of metals.

In marine phytoplankton, thiol excretion (as a thiol-metal complex) appeared to occur as part of the metal detoxification mechanism [36,154]. Furthermore, the possible dependence of thiol production and exudation on metal-induced cell damage was illustrated by Tang et al. [155].

3.3. Metal Tolerance and Hyperaccumulation

A basic level of metal tolerance is ubiquitous, in addition some organisms possess naturally selected levels of tolerance called "hypertolerance" which is typically specific to certain metals [4]. Although molecular mechanisms accounting for hypertolerance are not well understood, this capacity is believed to be controlled by a small number of genes [156]. Some organisms, in particular plants, not only tolerate higher levels of metals but also hyperaccumulate them. These organisms are described as "hyperaccumulators" [157].

Compartmentalization of metals in the vacuole, noticed in both plants and algae (Section 2.3), is assumed to play a role in the tolerance mechanism of some metal hyperaccumulators [113,158]. Enhanced vacuolar metal uptake and increased uptake transporter expression are also determinants in plant metal hypertolerance and hyperaccumulation (for reviews see [4,113]). In Ni, Cd, and Zn hyperaccumulating plant species, cellular wall binding plays an important role in detoxification and enhancing metal tolerance [113]. Nicotianamine (Section 1.2.2.) may play a role in hyperaccumulation as a long-distance transporter [13]. As described in Section 1.2.3, Ni hypertolerance is also correlated with histidine response [60]. Other metal-binding molecules that are involved in metal complexation in the vacuole are organic acids [159]. For example, aluminum tolerance is linked to efflux of organic acids like malate and citrate from roots (Section 1.2.3), but they may only play a part in sequestration and are unlikely to act as long-distance transporters [13]. However, a clear correlation between the concentration of organic and amino acids produced and the degree of exposure to a metal ion has not been observed. Strong and unequivocal evidence has not been provided to support their widespread function in metal tolerance and hyperaccumulation [4,13,133].

On the other hand, the formation of intracellular PCs, involved in metal detoxification, contributes to metal tolerance. Moreover, the fact that PC synthase activity has an important role in Cd tolerance was shown in *Vigna angularis* [160]. Recently, modification or over-expression of the enzymes (e.g., γ-ECS, GS, and PCS), that are involved in the synthesis of GSH and PCs, have been undertaken with success in Indian mustard and yeast [94,95,161,162]. These studies give further evidence of the role of PCs and related peptides in metal tolerance (for review see [113]) but there is no clear evidence that these molecules play a role in Zn and Ni hyperaccumulation

[13]. In addition, increased glutathione production appears to reduce the damaging oxidative effect of high metal concentration [163]. Moreover over-expression of metal transporters involved in the PC-based metal sequestration has been shown to increase metal tolerance [108,113]. The roles of transporters, briefly described in Section 2.3, should not be neglected in both detoxification and tolerance mechanisms [4,20,113]. Transporters represent good candidates for intracellular regulation of the metal ion level by acting as efflux pump and can be responsible for compartmentalization through sequestration [113].

Although a high PC-producing capacity has not been considered necessarily a feature of tolerant organisms [164], early rapid formation and overproduction of PCs and PC-metal complexes, and expression of PC synthase genes can confer a dramatic enhanced metal tolerance for a number of plants and transgenic species [12,95]. In algae, PC production capacity is likely a necessary but not a sufficient mechanism for metal tolerance [10]. Therefore, simultaneous examination of PC content and their turnover rate is required for a better assessment of their contribution to metal detoxification [165].

In addition, roles of PCs in Fe or sulfur metabolism have also been proposed [28,166] as well as protection against oxidative stress in algae [167].

4. PHYTOCHELATIN INDUCTION IN PHYTOPLANKTON IN RESPONSE TO METAL STRESS

In this last section, we will assess the current state of knowledge on the occurrence of thiol compounds in algal species under laboratory and field conditions. An overview of the principal techniques of identification, quantification and characterization of thiols, metal-thiol complexes and their metal binding capacities, will be presented.

4.1. Identification and Detection of Metal-Thiol Complexes and Their Metal-Binding Capacities

A variety of techniques has been used for the identification and quantification of the different metal-thiol complexes and their metal-binding capacities (see the review [168]). The classical approach is reversed phase liquid chromatography (RPLC) with either post-column derivatization of the sulfhydryl groups with 5,5′- dithiobis(2-nitrobenzoic acid) (DTNB, Ellman's reagent) and UV-Vis spectrophotometric detection at 410 nm [169] or precolumn derivatization with monobromobinane (mBrB) or SBD-F (ammonium 7-fluorobenzo-2-oxa-1,3-diazole-4-sulfonate) and fluorescence detection [170–173]. Good results were also found with electrophoresis coupled with UV-Vis detection [174,175] and laser induced fluorescence [176].

Due to the presence of the electroactive –SH and/or –SS– groups, thiol compounds have also been directly analyzed by electrochemical methods, without the need of a derivatization step (see the review [177]). Current electrochemical approaches for PC determination include cathodic stripping voltammetry (CSV) [178], CSV with adsorptive accumulation of complex onto Hg [179] or using copper and silver solid amalgam electrodes [180], stripping chronopotentiometry (SCP) [181] with adsorptive accumulation [182], and differential pulse voltammetry (DPV) [183] assisted by multivariable curve resolution with alternating least squares (MCR-ALS) [184,185]. The electrochemical techniques have been coupled with electrospray ionization-mass spectrometry (ESI-MS) [186] or circular dichroism spectroscopy [187] to allow the determination of the structures of the metal-thiol complexes formed [188]. Using the modified Brdika procedure based on the polarographic catalytic hydrogen evolution in mixtures of cobalt salts and cysteine-containing proteins, Dorcak and Sestakova [189] studied recently the electrochemical behavior of PCs and related peptides at a mercury drop electrode in the presence of Co(II) ions.

In addition to these detection methods, metal-PC clusters have been analyzed directly by HP and capillary LC coupled with ESI-MS or tandem MS [190–192] or via capillary zone electrophoresis (CZE) coupled with ESI-tandem MS [193] which allow structural characterization of PCs.

Recently, RPLC and size exclusion chromatography (SEC) coupled with a metal detection such as inductively coupled plasma (ICP)-MS [139,194–196], ICP-AES/ESI-MS [138], ICP-MS/ESI-MS [151] or positive-ion fast-atom bombardment/tandem MS [29] have offered a substantial improvement in terms of sensitivity, resolution, speciation, and convenience for the analysis of PCs and their chain lengths.

Recent methods involving HPLC coupled with CoulArray electrochemical detection (HPLC-ED) [197–199] with an electrode modified with functionalized carbon nanotubes as electrochemical detector [200] offer excellent detection limits (fmol to pmol/L) with reduced analysis times in comparison with methods based on fluorescence detection. This simple, rapid and sensitive method allows the simultaneous detection of reduced and oxidized forms of glutathione [197,200], cysteine, and phytochelatins [198], cystine, N-acetylcysteine, homocysteine, and desglycine-phytochelatin [199].

The characteristics of the metal-thiol complexes have also been investigated by using optical approaches [201] and spectroscopic techniques. X-ray absorption [202], ^{1}H NMR [203], ICP-AES coupled with extended

X-ray absorption fine structure (EXAFS) [204], UV-Vis, NMR spectroscopy, and circular dichroism [205], are helpful techniques to determine the nature and properties of the chemical binding of heavy metals to PCs and allow a structural characterization of the metal-PC complexes and PC analogues. For example, Pickering et al. and Cheng et al. [202,205] revealed a predominantly tetrahedral coordination of metal by sulfur in the phytochelatin analogues and complexes. These studies showed that the free carboxylate groups were responsible for the hydrophilic poly-anionic character, with consequent extreme water solubility of the PC-metal complexes, but these groups did not participate in the complexation of Cd [202,205].

4.2. Laboratory Experiments

The production of PCs and GSH by a range of marine phytoplankton species and macrophytes has been extensively investigated in metal exposure experiments conducted under controlled laboratory culturing conditions (see the references compiled in Table 1).

4.2.1. Is Phytochelatin Induction Metal-Dependent?

Before phytochelatins had been characterized, there were several observations of intracellular metallothionein-like sulfur-rich metal chelators in microalgae. Once the structure was known, many studies confirmed the ubiquity of phytochelatin synthesis in response to Cd additions for a wide range of microalgae (see Table 1 in Section 4.2.2 and for reviews [10,137]), with the exception of a few Cd-resistant marine strains reported by Wikfords, Neeman, and Jackson [206].

Metals other than Cd also induce phytochelatin production in algae (see Table 1). Clear relationships have been observed between metal exposure and PC induction in metal exposure experiments conducted under controlled laboratory conditions, with Cd, Cu, and Zn [8,9,79,171,207–211] and to a lesser extent with As [147–149], Cr [82,212], Hg [207,213,214], Co [8,9], Ag, Ni [8,9,207,215], and Pb [8,9,208,216,217]. Phytochelatin concentrations generally increase in phytoplankton with increasing metal exposure. The PC synthesis is a function of the free or inorganic metal concentration in the medium [10]. The incubations of phytoplankton monocultures with metal ions have been conducted, in general, in well defined artificial media which allow a thorough calculation of metal speciation [8,171,218]. A direct relation between free aqueous Cd^{2+} and Cu^{2+} species and PC production has been observed in laboratory experiments indicating that the determination of metal speciation forms a key part of studies into metal-organism interactions [78,219]. The highest intracellular PC production for a number of representative coastal phytoplankton species (*Thalassiosira weissflogii, Tetraselmis maculata, Emiliania huxleyi*) has been obtained with Cd and to a lesser extent Cu, Zn, and Pb exposures at sub-lethal concentrations (see a detailed review on exposure concentrations [137]). Based on laboratory observations, the metals most likely to induce PC production in coastal waters are Cu, Cd, and Zn, since other toxic metals such as Pb, Ag, and Hg typically occur at levels below the concentrations which stimulate PC production [220].

As metals must be transported into the cell to stimulate phytochelatin synthesis, the combination of metals, notably Cu, Zn, Cd, Co, and Pb at certain concentration thresholds, can cause antagonistic and suppressing effects on PC production, probably as a result of competition for cellular binding sites and metal transporters [79,209,211,220,221]. Thus it follows that phytochelatin production will be lower if two or more metals are competing for uptake sites, especially if one metal induces phytochelatins and the other not. For example, studies have demonstrated that both Cd and Zn compete with Mn (and with each other) for phytoplankton uptake by the Mn transport system, which can decrease individual metal toxicity [222,223] and could then reduce PC production [209,211] (see also the review [10]). On the other hand, synergistic effects on PC production have been observed for combinations of Cd and moderate concentrations of Cu, but the reasons for such effects have not been fully elucidated [209,211].

While metal interactions are important at very high concentrations, different metal interactions occur at very low metal concentrations. Ahner and Morel [10] observed that decreasing Zn and Co led to increasing phytochelatin concentrations in marine diatoms exposed to low Cd concentrations. The shift-up in the Zn or Co transporters may have led to an increase in the Cd uptake rate, requiring the organisms to synthesize more phytochelatins [10].

An indirect effect of metals on GSH and PC production via generation of reactive oxygen species (ROS) was recently suggested for the green alga *Dunaliella tertiolecta* [224]. The authors hypothesized that the activation of the GSH formation enzymes stimulated by Zn could consequently lead to a higher flux rate of GSH synthesis and thus contribute to a higher PC production.

4.2.2. *Is Phytochelatin Induction Dependent on the Algal Species?*

It has been observed in marine systems that PCs are produced by chlorophytes, rodophytes, haptophytes, diatoms, prymnesiophytes, dinoflagellates,

Table 1. Studies which have observed thiol production by micro and macro phytoplankton species in laboratory metal exposure experiments.

Species (occurrence)	Metal exposure	References
Bacillariophyte (diatoms)		
Nitzschia closterium (marine)	Zn	[235]
Ditylum brightwellii (coastal)	Cd, Cu	[171]
D. brightwellii	Cd, Cu	[236]
Skeletonema costatum (coastal)	Cu	[171]
S. costatum	Cd, Cu, Zn	[220]
S. costatum	Cu, Cd, Zn (alone and in mixture)	[209]
Phaeodactylum tricornutum (estuarine)	Cd, Cu	[171]
P. tricornutum	Cd	[237]
P. tricornutum	Cu	[238]
P. tricornutum	Cd, Pb, Zn	[208]
P. tricornutum	Cu, Zn	[225]
P. tricornutum	Cd, Cu, Zn	[211]
P. tricornutum	Cd	[195]
P. tricornutum	As	[148]
Thalassiosira pseudonana (coastal)	Cd, Cu	[171]
T. pseudonana	Cu, Cd, Zn (alone and in mixture)	[209]
Thalassiosira weissflogii (coastal)	Cd, Cu, Zn, Ag, Pb, Co, Ni, Hg	[9]
T. weissflogii	Cu	[173]
T. weissflogii	Cd	[225]
T. weissflogii	Cu	[225]
T. weissflogii	Cu	[165]
Thalassiosira oceanica (oceanic)	Cd	[8]
Chlorophyte (green algae)		
Dunaliella tertiolecta (estuarine)	Cd	[8,9]
D. tertiolecta	Cu	[239]
D. tertiolecta	Cd, Zn, Ni	[215]
D. tertiolecta	Cu, Cd, Zn (alone and in mixture)	[209]
D. tertiolecta	Cd, Zn	[210]
Tetraselmis maculata (estuarine)	Cd, Pb, Cu, Zn	[8]
Tetraselmis tetrathele (marine)	Cd, Hg	[213]
T. tetrathele	Cd, Hg	[214]
Tetraselmis suecica (marine)	Cd	[240]
Chlamydomonas reinhardtii (freshwater)	Cd, As	[241]
C. reinhardtii	Cd	[242]
Chlamydomonas acidophila (freshwater)	Cd	[243]

Table 1. (*Continued*).

Species (occurrence)	Metal exposure	References
Scenedesmus acutiformis (freshwater)	Cu, Zn, Pb, Ag, Hg	[207]
Scenedesmus acuminatus (freshwater)	Cu	[244]
Scenedesmus subspicatus (freshwater)	Cu	[78]
Scenedesmus acutus (freshwater)	Cd	[80]
S. acutus	Cr	[82,212]
Scenedesmus vacuolatus (freshwater)	Cd	[81]
S. vacuolatus	As	[149]
Scenedesmus armutus (freshwater)	Cd	[83]
Stigeoclonium tenue (freshwater)	Mix: Cd, Pb, Zn	[79]
Stigeoclonium sp. (freshwater)	Mix: Cd, Pb, Zn	[79]
Stichococcus bacillaris (freshwater)	Pb	[216,217]
S. bacillaris	As	[147]
Rodophyte (red algae)		
Porphyridium purpureum (marine)	Hg	[214]
Dinoflagellate		
Heterocapsa pygmaea (estuarine)	Cd	[8]
Prorocentrum micans (estuarine)	Cu	[226]
P. micans	Cd	[227]
Prymnesiophyte (coccolithophores)		
Pleurochrysis carterae (coastal)	Cd	[8]
Emiliania huxleyi (coastal/oceanic)	Cd, Cu, Zn, Pb	[9]
E. huxleyi	Cd, Cu	[225]
E. huxleyi	Cu	[41]
Haptophyte		
Pavlova lutheri (estuarine)	Cd	[8]
Pavlova sp. (marine)	Hg	[214]
Pleurochrysis carterae (marine)	Hg	[214]
Isochrysis sp. (marine)	Hg	[214]
Macroalgae		
Sargassum muticum (marine)	Cu, Zn, Pb, Ag, Hg	[207]
Kappaphycus alvarezii (marine)	Cd	[230]
Enteromorpha sp. (marine/estuarine)	Cu	[229]
Enteromorpha linza (marine)	Cd, Zn	[84]
Enteromorpha prolifera (estuarine)	Cd, Zn	[84]
Codium fragile (marine)	Cd	[85]
Ulva spp. (marine)	Cd	[85]
Rhizoclonium tortuosum (marine)	Cd	[85]
Solieria chordalis (marine)	Cd	[85]
Fucus sp. (marine)	Cd	[85]
Gracilaria cornea (marine)	Cd	[86]
Gracilaria gracilis (marine)	Cd	[85]
Chondrophycus poiteaui (marine)	Cd	[86]

and macroalgae, with diatoms being the most important producer [8]. Only a few reports have been produced on phytochelatin production in freshwater algal species [78–83,147,207,216,217].

Table 1 presents the phytoplankton species generally used to conduct laboratory metal exposure experiments. Chlorophytes and prymnesiophytes readily produce PCs, with the prymnesiophyte *Emiliania huxleyi* showing a pronounced PC production under Cu and Cd exposure [225]. The only two studied dinoflagellate species, *Heterocapsa pygmaea* and *Prorocentrum micans*, showed a low PC production under Cd exposure, and no PC production under Cu exposure [8,9,226,227]. These observations suggest that these species have an alternative mechanism for tolerating metals [8,9,226]. *Pavlova lutheri* produce relatively few PCs under Cd exposure compared to the other species tested by Ahner, Kong, and Morel [8,9], but possess the ability to exclude or efficiently export Cd. The diatoms *Thalassiosira weissflogii* and *Phaeodactylum tricornutum* showed a high PC production in metal exposure experiments [171,225]. These species have been extensively used by many workers to assess the effects of a range of metals on PC production (see references in Table 1) because of their well-known physiology and relative ease of culturing. High concentrations of Zn induce a high level of PC synthesis in *Dunaliella tertiolecta* which is more tolerant to metals than other chlorophytes [210,215]. Moreover, in this species, PC synthase seems to be activated more strongly by Zn^{2+} than in higher plants. However, further study using isolated enzymes is required to elucidate the difference of PC synthase between *D. tertiolecta* and higher plants [210].

D. tertiolecta [210] and *Tetraselmis suecica* [175] are distinct from other algal species in their ability to synthesize longer PCs. In these studies, $n = 3$ or $n = 4$ oligomers dominate the phytochelatin pool at high Cd concentration [10,175,210]. Other species such as *E. huxleyi* and *Tetraselmis maculata* synthesize predominantly the shorter $n = 2$ oligomers [8,9]. Rijstenbil and Wijnholds [171] reported that oligomer chain lengths vary between species and depend on the metal to which the algae have been exposed. Péres-Rama, Vaamonde, and Alonso [175] suggested that the high tolerance to Cd could be due to the ability of some algal species to synthesize longer PCs. In contrast, phytochelatins are undetectable in *Tetraselmis tetrathele* and in two Haptophycean species, *Isochrysis galbana* and *P. lutheri* after Hg exposure [213,214].

In freshwaters, in the absence of relevant laboratory data, it is not possible to determine which metal is the most important inducer of PCs or to establish which freshwater species shows a pronounced PC production under metal stress experiments [10]. Among freshwater phytoplankton species, the chlorophyte *Scenedesmus sp.* is currently used to conduct laboratory experiments. The chromium-tolerant strain *S. acutus* has been tested by Torricelli et al. [80] and Gorbi et al. [82,212] and they suggested that the tolerance to Cr(VI) could be related to both cysteine and glutathione biosynthetic capacity. The studies initiated by Pawlik-Skrowronska and coworkers [79,147,216,217,228] were mainly focused on *Stigeoclonium* and *Stichoccoccus* strains. According to Pawlik-Skrowronska [228], the tolerance to high Zn concentrations in a Zn-tolerant ecotype of the filamentous green alga *Stigeoclonium tenue* relies on the complexation of metal with a cysteine-rich derivative of phytochelatins (Cys-PC_{1-3}), a novel phytochelatin-related compound produced in larger amounts than in the sensitive ecotype.

More recently, studies have been conducted with macroalgae [84–86,229,230]. A low PC induction has been observed in the green macroalgae *Enteromorpha sp.* (now *Ulva)* after exposure to Cu [229], Cd [86], and Zn [84]. In the *Enteromorpha* species, metals promote the exudation of cysteine and glutathione-like ligands [37], which may be the major thiol compounds involved in the detoxification of most of the intracellular metal ions, whereas PCs may only play a minor role [86]. Pawlik-Skowronska and coworkers [85] reported for the first time in natural assemblages, PC synthesis in native Phaeophyceae (*Fucus* sp.), Rhodophyceae (*Solieria chordalis*), and Chlorophyceae (*Rhizoclonium tortuosum*) but not in thalli of *Ulva* sp. and *Codium fragile* (Chlorophyceae). These results imply that other mechanisms of detoxification may play a more important role in *Ulva* sp. and *Codium fragile*, e.g., those associated with the thickness of the cell wall and polysaccharide content of the thallus, both of which bind metals and act as barriers to intracellular accumulation [85].

4.3. Field Experiments

Natural populations of phytoplankton contain phytochelatin concentrations that are similar to those measured in the laboratory at low metal levels [10]. Few reports have been written on thiol and phytochelatin production in freshwater organisms at relevant metal concentrations, and a limited number of studies have determined the production of PCs and GSH by phytoplankton in estuarine and marine waters (see Table 2 for references). The reported sampling areas comprise diverse environmental conditions, from pristine oceanic waters to highly metal polluted coastal/estuarine systems including mining areas [79,218] and harbors [219,220]. The presence and concentrations of PCs and other thiols have not yet been investigated in detail in all seawater environments (see Table 2 for details); only data for the North Atlantic [42], Equatorial Pacific [221], Subartic Pacific Ocean [231], and the Aegean Sea [84] are available. Thiols were also produced by freshwater phytoplankton in metal-contaminated lakes [78,232] and streams [233]. The concentration of PCs in the periphytic freshwater green algae

Table 2. Thiol production in field experiments.

Field Location	Comments	References
Saanich Inlet (USA)	Anoxic fjord	[245]
Boston and Massachusetts Bay (USA)	Harbor and bay	[219]
New England (USA)	Various harbors	[220]
Southern California Bight (USA)	Coastal area	[245]
Coastal England	Coastal area	[38]
Galveston Bay (USA)	Blooms of cyanobacteria	[246]
Equatorial Pacific	Metal-depleted waters	[221]
North Atlantic	Seawater	[42]
Thermaikos Gulf, North Aegean Sea (Greece)	Seawater	[84]
Subarctic Pacific Ocean (UK)	Seawater	[231]
Elizabeth River, Virginia (USA)	Estuarine waters 2000 and 2002	[36,209]
Eel Pond (USA)	Estuarine waters	[209]
Scheldt Estuary (The Netherlands)	Estuarine waters	[39,247]
Scheldt Estuary (The Netherlands, Belgium)	Estuarine waters	[84]
San Diego Bay, Cape Fear and Norfolk Estuaries (USA)	Estuarine waters	[248]
Elizabeth River (USA)	Urban estuary	[249]
Fal Estuary (UK)	Metal mine area	[218]
Tamar Estuary (UK)	Metal mine area	[218]
Southern Poland	Metal mine area	[79]
Switzerland and Italy	Various freshwater lakes	[78]
Connecticut (USA)	Lakes	[232]
Switzerland	Rainwater, stream	[233]

Stigeoclonium tenue in contaminated mining water reflected the bioavailability of the metals [79,228].

However, little information regarding phytoplankton composition that could be related to PC production is available for the sites listed in Table 2. In general, the dominant phytoplankton groups in coastal and estuarine waters are diatoms and dinoflagellates, while other smaller but important groups include cryptophytes, chlorophytes (green algae), and chrysophytes (cyanobacteria) [234]. It is therefore expected that an important contribution to PC production in these studies has come from diatoms, prymnesiophytes, and green algae, as the few dinoflagellate species investigated so far have not produced elevated PC concentrations on exposure to metals [8,9]. Moreover, shifts in species composition could also account in part for important variations in thiol and PC occurrence and detoxification mechanisms [78,137,209,211,228].

In marine waters, organic ligands dominate Cu and Zn speciation resulting in low free Cu^{2+} and Zn^{2+} ion concentrations. However, in areas subject to anthropogenic waste water inputs, and acid mine run-off, the total dissolved Cu and Zn concentrations can exceed the buffering capacity of the natural organic ligands and account for elevated free ionic metal concentrations [32]. Most field observations have indicated that in complex coastal systems, where bioavailable metals occur at enhanced concentrations, PC concentrations typically correlate positively with dissolved metal species, in particular Cd and Cu [137]. But relationships between PC and metal concentrations in the field are not always straightforward [137]. In addition, the concentration of Cu^{2+} is often enhanced in estuarine and coastal waters, but this does not preclude the possibility that other metals are responsible for PC production in marine systems and ideally, the concentrations of all PC-inducing metals should be studied when investigating PC distributions in the field.

From the above considerations, it becomes clear that the activation of PC synthase and resulting PC production in phytoplankton depends not only on the bioavailable metal species, but also on the phytoplankton species, metal combinations and potentially the fraction of metal that actually reaches the cytoplasm [137].

5. CONCLUDING REMARKS AND FUTURE DIRECTIONS

There is no single and simple mechanism employed by plants and algae that can account for tolerance to a wide range of metals. It has become evident that a complex network controls the regulation of metal ions in these organisms. A range of components and key processes involved in this network have yet to be identified and several fundamental questions remain to be answered. However, the advent of genomic research involving plants and phytoplankton and the recent work focusing on cellular/molecular mechanisms of both metal tolerance and resistance will change our perception of metal uptake, acquisition, distribution, accumulation and sequestration strategies. Moreover, to further our knowledge, the research should be extended to the investigation of the full range of genes potentially involved in trace metal homeostasis, detoxification, and hyperaccumulation. Indeed, the increased availability of gene deletion mutants and plant/algae over- or underexpression of certain key genes and enzymes [161,162] will provide valuable information concerning metal tolerance and resistance mechanisms. This understanding will allow detailed models to be constructed of the various responses by both sensitive and tolerant plants and algae that are subjected to metal stress [133]. Another important area for future research work is the involvement of metallochaperones in sequestration and intracellular trafficking of Cu and other metals and in metal homeostasis [14,15].

Some more interesting questions relating to the roles of PC synthase and PCs themselves in different organisms, possibly including animal species, remain to be answered. However, the mechanisms of PC biosynthesis and sulfide incorporation are still not fully understood. The isolation of PC synthase genes from a number of species will be important to further our understanding of the mechanisms of metal activation of PC biosynthesis and the catalytic nature of the mechanism itself [98].

In addition, while hyperaccumulators appear to offer beneficial solutions in a number of fields including bio- and phytoremediation, crop improvement and phytomining, a better understanding of their remarkable metal selectivity and metal accumulation pathways is needed to optimize their full potential.

Increasing worldwide concerns about water quality have resulted in targets to control metal contamination of aquatic ecosystems. For instance, metals such as Cd and Pb and their derivatives are currently considered priority substances under the European Union Water Framework Directive and are required to be monitored on a regular basis in natural waters. The US Environmental Protection Agency has overhauled the criteria to assess water quality by adopting and recommending models, such as the Biotic Ligand Model, to consider metal speciation and thus more accurately assess the effects of toxic metals on aquatic life. In this respect, the metal-binding thiol peptides produced by phytoplankton are relevant components to be taken into account for the refinement of such models and as feasible additional biomarkers for environmental risk assessment. However, before PCs can be widely applied as part of a suite of biomarkers of metal stress, further multidisciplinary research work is required on the influence of the environmental conditions on PC production [137].

ACKNOWLEDGMENTS

Dr. A. Devez was supported by a Marie Curie Intra-European Fellowship (FP6-023215-QWSTRESS), Dr. M. Gledhill is a NERC Advanced Research Fellow (UK), and Prof. E.P. Achterberg is a professor in Marine Biogeochemistry at the National Oceanography Centre, University of Southampton (UK). A.D. is grateful to N. Augis for preparing the figures and for his helpful advice during the preparation of this chapter.

ABBREVIATIONS

Ala	alanine
ATP	adenosine 5′-triphosphate
$Cd(GS)_2$	bis(glutathionato)-cadmium
CD	circular dichroism
CDF	cation diffusion facilitator family
CPx-type	cysteine-proline structure
CSV	cathodic stripping voltammetry
Cys	cysteine
CZE	capillary zone electrophoresis
DesGly	desglycine
DPV	differential pulse voltammetry
DTNB	5,5′-dithiobis(2-nitrobenzoic) acid or Ellman's reagent
EDTA	ethylenediamine-N,N,N′,N′-tetraacetate
ED	electrochemical detector
ES	electrospray
ESI	electrospray ionization
EXAFS	extended X-ray absorption fine structure
γ-EC	γ-glutamylcysteine or γ-Glu-Cys
γ-ECS	γ-glutamylcysteine synthetase
Glu	glutamic acid
Gly	glycine
Gln	glutamine
GR	glutathione reductase
GS	glutathione synthetase
GSH	glutathione or γ-glutamylcysteinylglycine or γ-Glu-Cys-Gly (reduced form)
GSSG	oxidized form of glutathione
His	histidine
hGSH	homoglutathione
HMT	heavy metal tolerance
HMW	high molecular weight
hPCS	homophytochelatin synthase
HPLC	high phase liquid chromatography
ICP-AES	inductively coupled plasma-atomic emission spectrometry
ICP-MS	inductively coupled plasma-mass spectrometry
IES-MS	ion electrospray-mass spectrometry
IRT	iron-regulated transporter
K_m	Michaelis constant
LC	liquid chromatography
LMW	low molecular weight

Log K_f	non-conditional stability constant or formation constant
MBrB	monobromobiname
MCR-ALS	multivariate curve resolution with alternating least squares
MS	mass spectrometry
MS/MS	tandem mass spectrometry
MT(s)	metallothionein(s)
N	azote
NADPH	nicotinamide adenine dinucleotide phosphate (reduced)
NMR	nuclear magnetic resonance
Nramp	natural resistance-associated macrophage proteins
PC(s)	phytochelatin(s) or $(\gamma\text{-Glu-Cys})_n$-Gly
PCS	phytochelatin synthase or γ-glutamylcysteine dipeptidyl transpeptidase
ROS	reactive oxygen species
RPLC	reverse phase liquid chromatography
S^{2-}	labile sulfide
SAP	sulfur assimilation pathway
SBD-F	ammonium 7-fluorobenzo-2-oxa-1,3-diazole-4-sulfonate
SCP	stripping chronopotentiometry
SEC	size exclusion chromatography
Ser	serine
UV/Vis	ultraviolet-visible
YCF	yeast cadmium factor
ZIP	zinc iron permease
$Zn(GS)_2$	bis(glutathionato)-zinc
ZRT	zinc-regulated transporter

REFERENCES

1. E. Nieboer and D. H. S. Richardson, *Environ. Pollut.*, 1980, **1**, 3–8.
2. J. J. R. Frausto da Silva and R. J. P. Williams, *The Biological Chemistry of the Elements. The Inorganic Chemistry of Life*, 2nd edn, Oxford, University Press, Oxford, UK, 2001, pp. 1–557.
3. G. W. Bryan and W. J. Langston, *Environ. Pollut.*, 1992, **76**, 89–131.
4. S. Clemens, *Planta*, 2001, **212**, 475–486.
5. R. H. Reed and G. M. Gadd, in: *Heavy Metal Tolerance in Plants: Evolutionary Aspects*, Ed. A. J. Shaw, CRC Press, Boca Raton, 1990, pp. 105–118.
6. O. K. Okamoto, E. Pinto, L. R. Latorre, E. J. H. Bechara and P. Colepicolo, *Environ. Contam. Toxicol.*, 2001, **40**, 18–24.
7. P. L. Croot, J. W. Moffett and L. E. Brand, *Limnol. Oceanogr.*, 2000, **45**, 619–627.

8. B. A. Ahner, S. Kong and F. M. M. Morel, *Limnol. Oceanogr.*, 1995, **40**, 649–657.
9. B. A. Ahner, S. Kong and F. M. M. Morel, *Limnol. Oceanogr.*, 1995, **40**, 658–665.
10. B. A. Ahner and F. M. M. Morel, *Progress in Phycological Research*, Ed. F. E. Round and D. J. Chapman, Biopress Ltd, Bristol, 1999, pp. 1–31.
11. W. E. Rauser, *Cell Biochem. Biophys.*, 1999, **31**, 19–48.
12. C. S. Cobbett and P. B. Goldsbrough, *Annu. Rev. Plant Biol.*, 2002, **53**, 159–182.
13. D. L. Callahan, A. J. M. Baker, S. D. Kolev and A. G. Wedd, *J. Biol. Inorg. Chem.*, 2006, **11**, 2–12.
14. T. V. O'Halloran and V. C. Culotta, *J. Biol. Chem.*, 2000, **275**, 25057–25060.
15. A. C. Rosenzweig, *Acc. Chem. Res.*, 2001, **34**, 119–128.
16. T. Maier, C. Yu, G. Kullertz and S. Clemens, *Planta*, 2003, **218**, 300–308.
17. A. Krezel and W. Bal, *Acta. Biochim. Pol.*, 1999, **46**, 567–580.
18. J. Giovanelli, S. H. Mudd and A. H. Datko, in: *The Biochemistry of Plants: A Comprehensive Treatise,* Ed. P. E. Stumpf and E. E. Conn, Academic Press, New York, 1980, pp. 453–505.
19. G. Noctor and C. H. Foyer, *Annu. Rev. Plant Physiol. Plant Mol. Biol.*, 1998, **49**, 249–279.
20. D. G. Mendoza-Cozatl, H. Loza-Tavera, A. Hernandez-Navarro and R. Moreno-Sanchez, *FEMS Microbiol. Rev.*, 2005, **29**, 653–671.
21. K. A. Marrs, *Annu. Rev. Plant Physiol. Plant Mol. Biol.*, 1996, **47**, 127–158.
22. O. W. Griffith and T. Mulcahy, *Adv. Enzymol. Relat. Areas Mol. Biol.*, 1999, **73**, 209–267.
23. D. H. Hamer, *Annu. Rev. Biochem.*, 1986, **55**, 913–951.
24. N. J. Robinson, A. M. Tommey, C. Kuske and P. J. Jackson, *Biochem. J.*, 1993, **295**, 1–10.
25. W. E. Rauser, *Ann. Rev. Biochem.*, 1990, **59**, 61–86.
26. E. Grill, E.-L. Winnacker and M. H. Zenk, *Science*, 1985, **230**, 674–676.
27. W. E. Rauser, *Plant Physiol.*, 1995, **109**, 1141–1149.
28. M. H. Zenk, *Gene*, 1996, **179**, 21–30.
29. S. Klapheck, W. Fliegner and I. Zimmer, *Plant Physiol.*, 1994, **104**, 1325–1332.
30. P. Meuwly, P. Thibault, A. L. Schwan and W. E. Rauser, *Plant J.*, 1995, **7**, 391–400.
31. K. W. Bruland and M. C. Lohan, in *Ocean and Marine Biogeochemistry*, Vol. 6 of *Treatise on Geochemistry,* Ed. H. Elderfied, H. D. Holland and K. K. Turekian, Elsevier, Amsterdam, 2003, pp. 23–47.
32. P. B. Kozelka and K. W. Bruland, *Mar. Chem.*, 1998, **60**, 267–282.
33. K. Barbeau, E. L. Rue, C. G. Trick, K. W. Bruland and A. Butler, *Limnol. Oceanogr.*, 2003, **48**, 1069–1078.
34. P. McCormack, P. J. Worsfold and M. Gledhill, *Anal. Chem.*, 2003, **75**, 2647–2652.
35. M. Gledhill, P. McCormack, S. Ussher, E. P. Achterberg, R. Fauzi, C. Mantoura and P. J. Worsfold, *Mar. Chem.*, 2004, **88**, 75–83.
36. L. P. Wei and B. A. Ahner, *Limnol. Oceanogr.*, 2005, **50**, 13–22.

37. M. T. S. D. Vasconcelos and F. M. C. Leal, *Mar. Chem.*, 2001, **74**, 65–85.
38. R. Al-Farawati and C. M. G. van den Berg, *Environ. Sci. Technol.*, 2001, **35**, 1902–1911.
39. L. M. Laglera and C. M. G. van den Berg, *Mar. Chem.*, 2003, **82**, 71–89.
40. M. T. S. D. Vasconcelos, M. F. C. Leal and C. M. G. van den Berg, *Mar. Chem.*, 2002, **77**, 187–210.
41. C. L. Dupont, R. K. Nelson, S. Bashir, J. W. Moffett and B. A. Ahner, *Limnol. Oceanogr.*, 2004, **49**, 1754–1762.
42. A. C. A. Le Gall and C. M. G. van den Berg, *Deep-Sea Res. I*, 1998, **45**, 1903–1918.
43. M. F. C. Leal, M. T. S. D. Vasconcelos and C. M. G. van den Berg, *Limnol. Oceanogr.*, 1999, **44**, 1750–62.
44. G. G. Leppard, *Colloids and Surfaces A: Physicochemical and Engineering Aspects*, 1997, **120**, 1–15.
45. L. K. Jang, D. Nguyen and G. G. Geesey, *Water Res.*, 1995, **29**, 307–313.
46. P. H. Santschi, E. Balnois, K. J. Wilkinson, J. Zhang, J. Buffle and L. Guo, *Limnol. Oceanogr.*, 1998, **43**, 896–908.
47. T. J. Painter, in *The Polysaccharides*, Ed. G. O. Aspinall, Academic Press, New York, 1983, pp. 195–285.
48. B. A. Bergamashi, J. S. Walters and J. I. Hedge, *Geochim. Cosmochim. Acta*, 1999, **63**, 413–425.
49. G. G. Leppard, *Sci. Total Environ.*, 1995, **165**, 103–131.
50. U. Passow, B. E. Logan and A. L. Alldredge, *Deep-Sea Res. I*, 1994, **41**, 335–357.
51. K. Mopper, J. Zhou, K.S. Ramana, U. Passow, H. G. Dam and D. T. Drapeau, *Deep-Sea Res. I*, 1995, **42**, 47–73.
52. F. Baldi, A. Minacci, A. Saliot, L. Mejanelle, P. Mozetic, V. Turk and A. Malej, *Mar. Ecol. Prog. Ser.*, 1997, **153**, 45–57.
53. K. J. Wilkinson, A. Joz-Roland and J. Buffle, *Limnol. Oceanogr.*, 1997, **42**, 1714–1724.
54. P. V. Bhaskar and N. B. Bhosle, *Curr. Sci.*, 2005, **88**, 45–53.
55. B. J. Hernlem, L. M. Vane and G. D. Sayles, *Water Res.*, 1999, **33**, 951–960.
56. M. A. Saito and J. W. Moffett, *Mar. Chem.*, 2001, **75**, 69–88.
57. M. J. Ellwood and C. M. G. van den Berg, *Mar. Chem.*, 2001, **75**, 49–68.
58. *Iron Transport and Storage in Microorganisms, Plants and Animals, Metal Ions in Biological Systems*, Vol. 35, Ed. A. Sigel and H. Sigel, M. Dekker, Inc., New York, 1998, pp. 775.
59. D. E. Salt, R. D. Smith and I. Raskin, *Annu. Rev. Plant Physiol. Plant Mol. Biol.*, 1998, **49**, 643–668.
60. U. Krämer, J. D. Cotter-Howells, J. M. Charnock, A. J. M. Baker and J. A. C. Smith, *Nature*, 1996, **379**, 635–638.
61. D. E. Salt, R. C. Prince, A. J. M. Baker, I. Raskin and I. J. Pickering, *Environ. Sci. Technol.*, 1999, **33**, 713–717.
62. M. Mejare and L. Bulow, *Trends Biotechnol.*, 2001, **19**, 67–73.
63. J. F. Ma, P. R. Ryan and E. Delhaize, *Trends Plant Sci.*, 2001, **6**, 273–278.
64. N. Kondo, K. Imai, M. Isobe, T. Goto, A. Murasugi, C. W. Nakagawa and Y. Hayashi, *Tetrahedron Lett.*, 1984, **25**, 3869–3872.

65. E. Grill, S. Loeffler, E.-L. Winnacker and M. H. Zenk, *Proc. Natl. Acad. Sci. USA*, 1989, **86**, 6838–6842.
66. Y. Hayashi, C. W. Nakagawa, N. Mutoh, M. Isobe and T. Goto, *Biochem. Cell Biol.*, 1991, **69**, 115–121.
67. W. Gekeler, E. Grill, E.-L. Winnacker and M. H. Zenk, *Z. Naturforsch.*, 1989, **44c**, 361–369.
68. M. Backor, B. Pawlik-Skowronska, J. Tomko, J. Budova and L. Sanita di Toppi, *Mycol. Res.*, 2006, **110**, 994–999.
69. M. Gupta, U. N. Rai, R. D. Tripathi and P. Chandra, *Chemosphere*, 1995, **30**, 2011–2020.
70. R. D. Tripathi, U. N. Rai, M. Gupta and P. Chandra, *Bull. Environ. Contam. Toxicol.*, 1996, **56**, 505–512.
71. M. Gupta, U. N. Rai, R. D. Tripathi and P. Chandra, *Chemosphere*, 1998, **37**, 785–800.
72. B. E. Maserti, V. Ferrillo, O. Avdis, U. Nesti, A. Di Garbo, A. Catsiki and P. L. Maestrini, *Aquat. Toxicol.*, 2005, **75**, 288–292.
73. S. Srivastava, S. Mishra, R. D. Tripathi, S. Dwivedi and D. K. Gupta, *Aquat. Toxicol.*, 2006, **80**, 405–415.
74. S. Mishra, S. Srivastava, R. D. Tripathi, R. Govindarajan, S. V. Kuriakose and M. N. V. Prasad, *Plant Physiol. Biochem.*, 2006, **44**, 25–37.
75. S. Mishra, S. Srivastava, R. D. Tripathi and P. K. Trivedi, *Aquat. Toxicol.*, 2007, in press.
76. T. Alvarez-Legorreta, D. Mendoza-Cozatl, R. Moreno-Sanchez and G. Gold-Bouchot, *Aquat. Toxicol.,* in press.
77. I. Bruns, K. Friese, B. Markert and G. J. Krauss, *Sci. Total Environ.*, 1999, **241**, 215–216.
78. K. Knauer, B. A. Ahner, H. B. Xue and L. Sigg, *Environ. Toxicol. Chem.*, 1998, **17**, 2444–2452.
79. B. Pawlik-Skowronska, *Aquat. Toxicol.*, 2001, **52**, 241–249.
80. E. Torricelli, G. Gorbi, B. Pawlik-Skowronska, L. Sanità di Toppi and M. G. Corradi, *Aquat. Toxicol.*, 2004, **68**, 315–323.
81. S. Le Faucheur, R. Behra and L. Sigg, *Environ. Toxicol. Chem.*, 2005, **24**, 1731–1737.
82. G. Gorbi, C. Zanni and M. G. Corradi, *Aquat. Toxicol.*, 2007, **84**, 457–464.
83. Z. Tukaj, A. Bascik-Remisiewicz, T. Skowronski and C. Tukaj, *Environ. Experim. Botany,* in press.
84. P. Malea, J. W. Rijstenbil and S. Haritonidis, *Mar. Environ. Res.*, 2006, **62**, 45–60.
85. B. Pawlik-Skowronska, J. Pirszel and M. T. Brown, *Aquat. Toxicol.*, 2007, **83**, 190–199.
86. V. Garcia-Rios, Y. Freile-Pelegrin, D. Robledo, D. Mendoza-Cozatl, R. Moreno-Sanchez and G. Gold-Bouchot, *Aquat. Toxicol.*, 2007, **81**, 65–72.
87. R. K. Mehra, E. B. Tabet, W. R. Gray and D. R. Winge, *Proc. Natl. Acad. Sci. USA*, 1988, **85**, 8815–8819.
88. J. Miersch, M. Tschimedbalshir, F. Barlocher, Y. Grams, B. Pierau, A. Schierhorn and G.-J. Krauss, *Mycol. Res.*, 2001, **105**, 883–889.

89. J. Miersch, F. Barlocher, I. Bruns and G.-J. Krauss, *Hydrobiologia*, 1997, **346**, 77–84.
90. L. Guimaraes-Soares, C. Pascoal and F. Cassio, *Ecotoxicol. Environ. Saf.*, 2007, **66**, 36–43.
91. C. Collin-Hansen, S. A. Pedersen, R. A. Andersen and E. Steinnes, *Mycol.*, 2007, **99**, 161–174.
92. M. Oven, J. E. Page, M. H. Zenk and T. M. Kutchan, *J. Biol. Chem.*, 2002, **277**, 4747–4754.
93. S. B. Ha, A. P. Smith, R. Howden, W. M. Dietrich, S. Bugg and M. J. O'Connell, *Plant Cell.*, 1999, **11**, 1153–1163.
94. O. K. Vatamaniuk, S. Mari, Y.-P. Lu and P. A. Rea, *Proc. Natl. Acad. Sci. USA*, 1999, **96**, 7110–7115.
95. S. Clemens, E. J. Kim, D. Neumann and J. I. Schroeder, *EMBO J.*, 1999, **18**, 3325–3333.
96. S. Clemens, I. Schroeder and T. Degenkolb, *Eur. J. Biochem*, 2001, **268**, 3640–3643.
97. O. K. Vatamaniuk, E. A. Bucher, J. Ward and P. A. Rea, *J. Biol. Chem.*, 2001, **276**, 20817–20820.
98. S. Clemens, *J. Plant Physiol.*, 2006, **163**, 319–332.
99. N. Tsuji, S. Nishikori, O. Iwabe, K. Shiraki, H. Miyasaka, M. Takagi, K. Hirata and K. Miyamoto, *Biochem. Biophys. Res. Comm.*, 2004, **315**, 751–755.
100. E. Harada, E. von Roepenack-Lahaye and S. Clemens, *Phytochem.*, 2004, **65**, 3179–3185.
101. O. K. Vatamaniuk, S. Mari, Y.-P. Lu and P. A. Rea, *J. Biol. Chem.*, 2000, **275**, 31451–31459.
102. S. Loeffler, A. Hochberger, E. Grill, E.-L. Winnacker and M. H. Zenk, *FEBS Lett.*, 1989, **258**, 42–46.
103. A. Meister, *Methods Enzymol.*, 1995, **251**, 3–13.
104. C. S. Cobbett, *Plant Physiol.*, 2000, **123**, 825–832.
105. E. Harada, Y. Yamaguchi, N. Koizumi and H. Sano, *J. Plant Physiol.*, 2002, **159**, 445–448.
106. E. Heuillet, A. Moreau, S. Halpren, N. Jeanne and S. Puiseaux-Dao, *Biol. Cell.*, 1986, **58**, 79–86.
107. D. F. Ortiz, L. Kreppel, D. M. Speiser, G. Scheel, G. McDonald and D. W. Ow, *EMBO J.*, 1992, **11**, 3491–3499.
108. D. F. Ortiz, T. Ruscitti, K. F. MacCue and D. W. Ow, *J. Biol. Chem.*, 1995, **270**, 4721–4728.
109. Z. S. Li, M. Szczypka, Y. P. Lu, D. J. Thiele and P. A. Rea, *J. Biol. Chem.*, 1996, **271**, 6509–6517.
110. Z. S. Li, Y. P. Lu, R.-G. Zhen, M. Szczypka, D. J. Thiele and P. A. Rea, *Proc. Natl. Acad. Sci. USA*, 1997, **94**, 42–47.
111. D. E. Salt and W. E. Rauser, *Plant Physiol.*, 1995, **107**, 1293–1301.
112. R. Sanchez-Fernandez, T. G. Davies, J. O. Coleman and P. A. Rea, *J. Biol. Chem.*, 2001, **276**, 30231–30244.
113. X. Yang, Y. Feng, Z. He and P. J. Stoffell, *J. Trace Elem. Med. Biol.*, 2005, **18**, 339–353.

114. C. Ballan-Dufrançais, C. Marcaillou and C. Amiard-Triquet, *Biol. Cell.*, 1991, **72**, 103–112.
115. Y. Nassari, J. L. Mansot, J. Wery, T. Ginsburger-Vogel and J. C. Amiard, *Arch. Environ. Contam. Toxicol.*, 1997, **33**, 147–155.
116. D. N. Weber, C. F. Shaw III and D. H. Petering, *J. Biol. Chem.*, 1987, **262**, 6962–6964.
117. C. T. Dameron, R. N. Reese, R. K. Mehra, A. R. Kortan, P. J. Carroll, M. L. Steigerwald, L. E. Brus and D. R. Winge, *Nature*, 1989, **338**, 596–597.
118. R. N. Reese, C. A. White and D. R. Winge, *Plant Physiol.*, 1992, **98**, 225–229.
119. G. Scarano and E. Morelli, *Plant Sci.*, 2003, **165**, 803–810.
120. A. Murasugi, C. Wada and Y. Hayashi, *J. Biochem.*, 1993, **93**, 661–664.
121. W. Bae and R. K. Mehra, *J. Inorg. Biochem.*, 1998, **69**, 33–43.
122. R. K. Mehra, P. Mulchandani and T. C. Hunter, *Biochem. Biophys. Res. Commun.*, 1994, **200**, 1193–1200.
123. J. Barbas, V. Santhanagopalan, M. Blaszczynski, W. R. Ellis Jr. and D. R. Winge, *J. Inorg. Biochem.*, 1992, **48**, 95–105.
124. D. M. Speiser, D. F. Ortiz, L. Kreppel and D. W. Ow, *Mol. Cell.Biol.*, 1992, **12**, 5301–5310.
125. R.-H. Juang, K. F. MacCue and D. W. Ow, *Arch. Biochem. Biophys.*, 1993, **304**, 392–401.
126. T. C. Hunter and R. K. Mehra, *J. Inorg. Biochem.*, 1998, **69**, 293–303.
127. J. G. Vande Weghe and D. W. Ow, *J. Biol. Chem.*, 1999, **274**, 13250–13257.
128. O. Coppelloti, *Comp. Biochem. Physiol.*, 1989, **94C**, 35–40.
129. D. Mendoza-Cozatl, S. Devars, H. Loza-Tavera and R. Moreno-Sanchez, *Physiol. Plant*, 2002, **115**, 276–283.
130. C. Aviles, H. Loza-Tavera, N. Terry and R. Moreno-Sanchez, *Arch. Microbiol.*, 2003, **180**, 1–10.
131. K. Nagel, U. Adelmeier and J. Voigt, *J. Plant Physiol.*, 1996, **149**, 86–90.
132. D. Soldo, R. Hari, L. Sigg and R. Behra, *Aquat. Toxicol.*, 2005, **71**, 307–317.
133. J. L. Hall, *J. Exp. Bot.*, 2002, **53**, 1–11.
134. H. Schat, M. Llugany, R. Vooijs, J. Hartley-Whitaker and P. M. Bleeker, *J. Exp. Bot.*, 1992, **53**, 2381–2392.
135. J. Thurmann, E. Grill, E.-L. Winnacker and M. H. Zenk, *FEBS Lett.*, 1991, **284**, 66–69.
136. E. Grill, J. Thumann, E.-L. Winnacker and M. H. Zenk, *Plant Cell*, 1988, **7**, 375–378.
137. S. K. Kawakami, M. Gledhill and E. P. Achterberg, *J. Phycol.*, 2006, **42**, 975–989.
138. T. Maitani, H. Kubota, K. Sato and T. Yamada, *Plant Physiol.*, 1996, **110**, 1145–1150.
139. I. Leopold and D. Gunther, *Fresenius J. Anal. Chem.*, 1997, **359**, 364–370.
140. D. D. Perrin, in *Stability Constants of Metal-Ion Complexes*, IUPAC Chemical Data Series no 22, Part B: Organic Ligands, Pergamon Press, Oxford, 1979.
141. A. Brune, W. Urbach and K. J. Dietz, *New Phytol.*, 1995, **129**, 403–409.
142. R. Howden and C. S. Cobbett, *Plant Physiol.*, 1992, **99**, 100–107.

143. E. Grill, E.-L. Winnacker and M. H. Zenk, *Proc. Natl. Acad. Sci. USA*, 1987, **84**, 439–443.
144. R. Nakazawa, M. Ikawa, K. Yasuda and H. Takenaga, *Soil Sci. Plant Nutrition*, 2000, **46**, 271–275.
145. M. E. V. Schmöger, M. Oven and E. Grill, *Plant Physiol.*, 2000, **122**, 793–801.
146. Y. Cai, J. Su and L. Q. Ma, *Environ. Pollut.*, 2004, **129**, 69–78.
147. B. Pawlik-Skowronska, J. Pirszel, R. Kalinowska and T. Skowronski, *Aquat. Toxicol.*, 2004, **70**, 201–212.
148. E. Morelli, M. C. Mascherpa and G. Scarano, *Biometals*, 2005, **18**, 587–593.
149. S. Le Faucheur, F. Schildknecht, R. Behra and L. Sigg, *Aquat. Toxicol.*, 2006, **80**, 355–381.
150. F. E. C. Sneller, L. M. Van Heerwaarden, F. J. L. Kraaijeveld-Smit, W. M. Ten Bookum, P. L. M. Koevoets, H. Schat and J. A. C. Verkleij, *New Phytol.*, 1999, **144**, 223–232.
151. A. Raab, J. Feldmann and A. A. Meharg, *Plant Physiol.*, 2004, **134**, 1113–1122.
152. J. Hartley-Whitaker, G. Ainsworth, R. Vooijs, W. M. Ten Bookum, H. Schat and A. A. Meharg, *Plant Physiol.*, 2001, **126**, 299–306.
153. R. D. Tripathi, S. Srivastava, S. Mishra, N. Singh, R. Tuli, D. K. Gupta and F. J. M. Maathuis, *Trends Biotechnol.*, 2007, **25**, 158–165.
154. J. G. Lee, B. A. Ahner and F. M. M. Morel, *Environ. Sci. Technol.*, 1996, **30**, 1814–1821.
155. D. Tang, M. M. Shafer, D. A. Karner and D. E. Armstrong, *Limnol. Oceanogr.*, 2005, **50**, 516–525.
156. M. R. Macnair, *New Phytol.*, 1993, **124**, 541–559.
157. A. J. M. Baker and R. R. Brooks, *Biorecovery*, 1989, **1**, 81–126.
158. Y. P. Tong, R. Kneer and Y. G. Zhu, *Trends Plant Sci.*, 2001, **9**, 7–9.
159. U. Krämer, I. J. Pickering, R. C. Prince, I. Raskin and D. E. Salt, *Plant Physiol.*, 2000, **122**, 1343–1353.
160. M. Inouhe, R. Ito, S. Ito, N. Sasada, H. Tohoyama and M. Joho, *Plant Physiol.*, 2000, **23**, 1029–1036.
161. Y. L. Zhu, E. A. H. Pilon-Smits, L. Jouanin and N. Terry, *Plant Physiol.*, 1999, **119**, 73–79.
162. Y. L. Zhu, E. A. H. Pilon-Smits, L. Jouanin and N. Terry, *Plant Physiol.*, 1999, **121**, 1169–1177.
163. J. L. Freeman, M. W. Persans, K. Nieman, C. Albrecht, W. Peer, I. J. Pickering and D. E. Salt, *Plant Cell*, 2004, **16**, 2176–2191.
164. J. A. De Knecht, N. Vanbaren, W. M. Ten Bookum, H. W. W. F. Sang, P. L. M. Koevoets, H. Schat and J. A. C. Verkleij, *Plant Sci.*, 1995, **106**, 9–18.
165. A.-J. Miao and W.-X. Wang, *Environ. Sci. Technol.*, 2007, **41**, 1777–1782.
166. L. Sanità di Toppi and R. Gabbrielli, *Environ. Exp. Bot.*, 1999, **41**, 105–130.
167. E. Pinto, T. C. S. Sigaud-Kutner, M. A. S. Leitao, O. K. Okamoto, D. Morse and P. Colepicolo, *J. Phycol.*, 2003, **39**, 1008–1018.
168. E. Bramanti, D. Toncelli, E. Morelli, L. Lampugnani, R. Zamboni, K. E. Miller, J. Zemetra and A. D'Ulivo, *J. Chromatogr. A.*, 2006, **1133**, 195–203.
169. E. Grill, E.-L. Winnacker and M. Z. Zenk, *Methods Enzymol.*, 1991, **205**, 333–341.

170. R. C. Fahey and G. L. Newton, *Methods Enzymol.*, 1987, **143**, 85–96.
171. J. W. Rijstenbil and J. A. Wijnholds, *Mar. Biol.*, 1996, **127**, 45–54.
172. F. E. C. Sneller, L. M. Van Heerwaarden, P. L. Koevoets, R. Vooijs, H. Schat and J. A. C. Verkleij, *J. Agric. Food Chem.*, 2000, **48**, 4014–4019.
173. D. Tang, L. S. Wen and P. H. Santschi, *Anal. Chim. Acta.*, 2000, **408**, 299–307.
174. H. Kubota, K. Sato, T. Yamada and T. Maitani, *J. Chromatogr. A*, 1998, **803**, 315–320.
175. M. Pérez-Rama, E. T. Vaamonde and J. A. Alonso, *Electrophoresis*, 2005, **26**, 610–620.
176. J. J. Hart, R. M. Welch, W. A. Norvell and L. V. Kochian, *Electrophoresis*, 2002, **23**, 81–87.
177. P. C. White, N. S. Lawrence, J. Davis and R. G. Compton, *Electroanalysis*, 2002, **14**, 89–98.
178. G. Scarano and E. Morelli, *Anal. Chim. Acta*, 1996, **319**, 13–18.
179. I. Sestakova, H. Vodickova and P. Mader, *Electroanalysis*, 1998, **10**, 764–770.
180. B. Yosypchuk, I. Sestakova and L. Novotny, *Talanta*, 2003, **59**, 1253–1258.
181. M. Fojta, M. Fojtova, L. Havran, H. Pivonkova, V. Dorcak and I. Sestakova, *Anal. Chim. Acta*, 2006, **558**, 171–178.
182. N. Serrano, I. Sestakova, J. M. Diaz-Cruz and C. Ariño, *J. Electroanal. Chem.*, 2006, **591**, 105–117.
183. G. Scarano and E. Morelli, *Electroanalysis*, 1998, **10**, 39–43.
184. S. M. Diaz-Cruz, J. Mendieta, A. Monjonell, R. Tauler and M. Esteban, *J. Inorg. Biochem.*, 1998, **70**, 91–98.
185. B. H. Cruz, J. M. Diaz-Cruz, I. Sestakova, J. Velek, C. Ariño and M. Esteban, *J. Electroanal. Chem.*, 2002, **520**, 111–118.
186. E. Chekmeneva, J. M. Diaz-Cruz, C. Ariño and M. Esteban, *Electroanalysis*, 2007, **19**, 310–317.
187. A. Alberich, C. Ariño, J. M. Diaz-Cruz and M. Esteban, *Anal. Chim. Acta*, 2007, **584**, 403–409.
188. M. Esteban, C. Ariño, J. M. Diaz-Cruz, S. M. Diaz-Cruz and R. Tauler, *Trends Anal. Chem.*, 2000, **19**, 49–61.
189. V. Dorcak and I. Sestakova, *Bioelectrochem.*, 2006, **68**, 14–21.
190. T.-Y. Yen, J. A. Villa and J. G. DeWitt, *J. Mass Spectrom.*, 1999, **34**, 930–941.
191. V. Vacchina, H. Chassaigne, R. Lobinski, M. Oven and M. H. Zenk, *Analyst*, 1999, **124**, 1425–1430.
192. H. A. El-Zohri, R. Cabala and H. Frank, *Anal. Bioanal. Chem.*, 2005, **382**, 1871–1876.
193. S. Mounicou, V. Vacchina, J. Szpunar, M. Potin-Gautier and R. Lobinski, *Analyst*, 2001, **126**, 624–632.
194. V. Vacchina, K. Pole and J. Szpunar, *Analyst*, 1999b, **14**, 1557–1566.
195. V. Loreti, D. Toncelli, E. Morelli, G. Scarano and J. Bettmer, *Anal. Bioanal. Chem.*, 2005, **383**, 398–403.
196. R. Raml, W. Goessler and K. A. Francesconi, *J. Chromatogr. A*, 2006, **1128**, 164–170.
197. L. Manna, L. Valvo and P. Betto, *J. Chromatogr. A*, 1999, **846**, 59–64.

198. D. Potesil, J. Petrlova, V. Adam, J. Vacek, B. Klejdus, J. Zehnalek, L. Trnkova, L. Havel and R. Kizek, *J. Chromatogr. A*, 2005, **1084**, 134–144.
199. J. Petrlova, R. Mikelova, K. Stejskal, A. Kleckerova, O. Zitka, J. Petrek, L. Havel, J. Zehnalek, V. Adam, L. Trnkova and R. Kizek, *J. Separation Sci.*, 2006, **29**, 1166–1173.
200. W. Zhang, F. Wan, W. Zhu, H. Xu, X. Ye, R. Cheng and L.-T. Jin, *J. Chromatogr. B*, 2005, **818**, 227–232.
201. R. K. Mehra, J. Miclat, V. R. Kodati, R. Abdullah, T. C. Hunter and P. Mulchandani, *Biochem. J.*, 1996, **314**, 73–82.
202. I. J. Pickering, R. C. Prince, G. N. George, W. E. Rauser, W. A. Wickramasinghe, A. A. Watson, C. T. Dameron, I. G. Dance, D. P. Fairlie and D. E. Salt, *Biochim. Biophys. Acta-Protein Structure Mol. Enzymol.*, 1999, **1429**, 351–364.
203. N. J. C. Bailey, M. Oven, E. Holmes, J. K. Nicholson and M. H. Zenk, *Phytochem.*, 2003, **62**, 851–858.
204. G. De la Rosa, J. R. Peralta-Videa, M. Montes, J. G. Parsons, I. Cano-Aguilera and J. L. Gardea-Torresdey, *Chemosphere*, 2004, **55**, 1159–1168.
205. Y. Cheng, Y.-B. Yan and J. Liu, *J. Inorg. Biochem.*, 2005, **99**, 1952–1962.
206. G. H. Wikfors, A. Neeman and P. J. Jackson, *Mar. Ecol. Progr. Ser.*, 1991, **79**, 163–170.
207. W. Gekeler, E. Grill, E.-L. Winnacker and M. H. Zenk, *Arch. Microbiol.*, 1988, **150**, 197–202.
208. E. Morelli and G. Scarano, *Mar. Environ. Res.*, 2001, **52**, 383–395.
209. L. Wei, J. Donat, G. Fones and B. A. Ahner, *Environ. Sci. Technol.*, 2003, **37**, 3609–3618.
210. N. Tsuji, N. Hirayanagi, O. Iwabe, T. Namba, M. Tagawa, S. Miyamoto, H. Miyasaka, M. Takagi, K. Hirata and K. Miyamoto, *Phytochem.*, 2003, **62**, 453–459.
211. S. K. Kawakami, M. Gledhill and E. P. Achterberg, *Biometals*, 2005, **19**, 51–60.
212. G. Gorbi, E. Torricelli, B. Pawlik-Skowronska, L. Sanità di Toppi, C. Zanni and M. G. Corradi, *Aquat. Toxicol.*, 2006, **79**, 132–139.
213. M. Satoh, E. Karaki, M. Kakehashi, E. Okazaki, T. Gotah and Y. Oyama, *J. Phycol.*, 1999, **35**, 989–994.
214. M. Satoh, Y. Hirachi, A. Yoshioka, M. Kobayashi and Y. Oyama, *Phycol.*, 2002, **98**, 983–990.
215. K. Hirata, Y. Tsujimoto, T. Namba, T. Ohta, N. Hirayanagi, H. Miyasaka, M. H. Zenk and K. Miyamoto, *J. Biosci. Bioeng.*, 2001, **92**, 24–29.
216. B. Pawlik-Skowronska, *Aquat. Toxicol.*, 2000, **50**, 221–230.
217. B. Pawlik-Skowronska, *Environ. Pollut.*, 2002, **119**, 119–127.
218. S. K. Kawakami, M. Gledhill and E. P. Achterberg, *Trends Anal. Chem.*, 2005, **25**, 133–142.
219. B. A. Ahner, N. M. Price and F. M. M. Morel, *Proc. Natl. Acad. Sci. USA*, 1994, **91**, 8433–8436.
220. B. A. Ahner, F. M. M. Morel and J. W. Moffett, *Limnol. Oceanogr.*, 1997, **42**, 601–608.
221. B. A. Ahner, J. G. Lee, N. M. Price and F. M. M. Morel, *Deep Sea Res. I*, 1998, **45**, 1779–1796.

222. W. G. Sunda, *Biol. Oceanogr.*, 1988, **6**, 411–442.
223. W. G. Sunda and S. A. Huntsman, *Limnol. Oceanogr.*, 1996, **41**, 373–387.
224. N. Tsuji, N. Hirayanagi, M. Okada, H. Miyasaka, K. Hirata, M. H. Zenk and K. Miyamoto, *Biochem. Biophys. Res. Comm.*, 2002, **293**, 653–659.
225. B. A. Ahner, L. Wei, J. R. Oleson and N. Ogura, *Mar. Ecol. Progr. Ser.*, 2002, **232**, 93–103.
226. O. M. Lage, A. M. Parente, M. T. S. D Vasconcelos, C. A. R. Gomes and R. Salema, *J. Phycol.*, 1996, **32**, 416–423.
227. R. Pistocchi, M. A. Mormile, F. Guerrini, G. Isani and L. Boni, *J. Appl. Phycol.*, 2000, **12**, 469–477.
228. B. Pawlik-Skowronska, *Aquat. Botany*, 2003, **75**, 189–198.
229. J. W. Rijstenbil, F. Dehairs, R. Ehrlich and J. A. Wijnholds, *Aquat. Toxicol.*, 1998, **42**, 187–209.
230. S. Hu and M. Wu, *Mol. Mar. Biol. Biotechnol.*, 1998, **7**, 97–104.
231. C. L. Dupont, J. W. Moffett, R. R. Bidigare and B. A. Ahner, *Deep Sea Res. I*, 2006, **53**, 1961–1974.
232. H. Hu, S. E. Mylon and G. Benoit, *Limnol. Oceanogr.*, 2006, **51**, 2763–2774.
233. S. Le Faucheur, R. Behra and L. Sigg, *Environ. Sci. Technol.*, 2005, **39**, 8099–8107.
234. G. E. Fogg and B. Thake, in: *Algal Cultures and Phytoplankton Ecology*, University of Wisconsin Press, London, 1987, pp. 81–121.
235. J. L. Stauber and T. M. Florence, *Mar. Biol.*, 1990, **105**, 519–524.
236. J. W. Rijstenbil and L. J. A. Gerringa, *Aquat. Toxicol.*, 2002, **56**, 115–131.
237. E. Torres, A. Cid, P. Fidalgo, C. Herrero and J. Abalde, *Aquat. Toxicol.*, 1997, **39**, 231–246.
238. E. Morelli and E. Pratesi, *Bull. Environ. Contam. Toxicol.*, 1997, **59**, 657–664.
239. M. Gonzalez-Davila, J. M. Santana-Casiano, J. Perez-Pena and F. J. Miller, *Environ. Sci. Technol.*, 1995, **29**, 289–301.
240. M. Pérez-Rama, C. Herrero, J. Abalde and E. Torres, *Environ. Toxicol. Chem.*, 2001, **20**, 2061–2066.
241. I. Kobayashi, S. Fujiwara, H. Saegusa, M. Inouhe, H. Matsumoto and M. Tsuzuki, *Mar. Biotechnol.*, 2006, **8**, 94–101.
242. K. Nishikawa, H. Machida, Y. Yamakoshi, R. Ohtomo, K. Saito, M. Saito and N. Tominaga, *Plant Sci.*, 2006, **170**, 307–313.
243. K. Nishikawa, A. Onodera and N. Tominaga, *Chemosphere*, 2006, **63**, 1553–1559.
244. A. T. Lombardi, T. M. R. Hidalgo and A. A. H. Vieira, *Chemosphere*, 2005, **60**, 453–459.
245. P. A. Matrai and R. D. Vetter, *Limnol. Oceanogr.*, 1988, **33**, 624–631.
246. D. Tang, C. C. Hung, K. W. Warnken and P. H. Santschi, *Limnol. Oceanogr.*, 2000, **45**, 1289–1297.
247. L. M. Laglera and C. M. G. van den Berg, *Mar. Chem.*, 2006, **101**, 130–140.
248. D. Tang, M. M. Shafer, D. A. Karner, J. Overdier and D. E. Armstrong, *Environ. Sci. Technol.*, 2004, **38**, 4247–4253.
249. C. L. Dryden, A. S. Gordon and J. R. Donat, *Mar. Chem.*, 2007, **103**, 276–288.

Met. Ions Life Sci. **2009**, *5*, 483–514

Subject Index

A

Metal Ions in Life Sciences, Volume 5 Edited by Astrid Sigel, Helmut Sigel, and Roland K. O. Sigel

Published by the Royal Society of Chemistry, www.rsc.org DOI: 10.1039/9781847558992-00483

B

C

E

F

G

I

K

L

N

O

Q

R

S

T

U

V

W